Physics of Organic Semiconductors

Edited by

W. Brütting

Physics of Organic Semiconductors

Edited by
Wolfgang Brütting

WILEY-VCH
VCH

WILEY-VCH Verlag GmbH & Co. KGaA

Editor

Prof. Dr. Wolfgang Brütting
Institute of Physics
University of Augsburg, Germany

Cover Picture
Courtesy of R.W.I. de Boer et al. (see Chapter 14)

■ All books published by Wiley-VCH are carefully
produced. Nevertheless, authors, editors, and
publisher do not warrant the information contained in
these books, including this book, to be free of errors.
Readers are advised to keep in mind that statements,
data, illustrations, procedural details or other items
may inadvertently be inaccurate.

Library of Congress Card No.: Applied for
British Library Cataloguing-in-Publication Data:
A catalogue record for this book is available from the
British Library.

**Bibliographic information published by
Die Deutsche Bibliothek**
Die Deutsche Bibliothek lists this publication in the
Deutsche Nationalbibliografie; detailed bibliographic
data is available in the Internet at
<http://dnb.ddb.de>.

Printed in the Federal Republic of Germany.
Printed on acid-free paper.

Composition   Kühn & Weyh, Satz und Medien,
Freiburg
Printing   Strauss GmbH, Mörlenbach
Bookbinding   J. Schäffer GmbH i.G., Grünstadt

ISBN-13:   978-3-527-40550-3
ISBN-10:   3-527-40550-X

Contents

List of Contributors   *XV*

Introduction to the Physics of Organic Semiconductors   *1*
W. Brütting

1	History *1*	
2	Materials *2*	
3	Basic Properties of Organic Semiconductors *4*	
3.1	Optical Properties *4*	
3.2	Charge Carrier Transport *6*	
3.3	Device Structures and Properties *9*	
4	Outline of this Book *12*	

Part I   **Growth and Interfaces**   *15*

1   **Organic Molecular Beam Deposition: Growth Studies beyond the First Monolayer**   *17*
F. Schreiber

1.1	Introduction *17*
1.2	Organic molecular beam deposition *18*
1.2.1	General concepts of thin film growth *18*
1.2.2	Issues specific to organic thin film growth *20*
1.2.3	Overview of popular OMBD systems *22*
1.3	Films on oxidized silicon *24*
1.3.1	PTCDA *24*
1.3.2	DIP *24*
1.3.3	Phthalocyanines *26*
1.3.4	Pentacene *27*
1.4	Films on aluminium oxide *27*
1.4.1	PTCDA *28*
1.4.2	DIP *28*
1.4.3	Phthalocyanines *29*
1.4.4	Pentacene *29*

Physics of Organic Semiconductors. Edited by W. Brütting
Copyright © 2005 WILEY-VCH Verlag GmbH & Co. KGaA, Weinheim
ISBN 3-527-40550-X

1.5   Films on metals   *30*
1.5.1   PTCDA   *30*
1.5.1.1   Structure and epitaxy of PTCDA/Ag(111)   *30*
1.5.1.2   Comparison with other substrates   *31*
1.5.1.3   Dewetting and thermal properties   *31*
1.5.1.4   Real-time growth   *31*
1.5.2   DIP   *33*
1.5.3   Phthalocyanines   *34*
1.5.4   Pentacene   *34*
1.6   Films on other substrates   *34*
1.7   More complex heterostructures and technical interfaces   *35*
1.8   Summary and Conclusions   *37*

2   Electronic Properties of Interfaces between Model Organic Semiconductors and Metals   *41*
           M. Knupfer and H. Peisert

2.1   Introduction   *41*
2.2   Experimental methods   *42*
2.2.1   Photoemission spectroscopy   *43*
2.2.2   Polarization dependent XAS   *45*
2.3   Adsorption geometry and substrate roughness   *46*
2.4   Organic/metal interfaces: general properties and trends   *50*
2.5   Reactive surfaces: indium-tin oxide and PEDOT:PSS   *58*
2.6   Concluding remarks   *64*

3   Kelvin Probe Study of Band Bending at Organic Semiconductor/Metal Interfaces: Examination of Fermi Level Alignment   *69*
           H. Ishii, N. Hayashi, E. Ito, Y. Washizu, K. Sugi, Y. Kimura, M. Niwano, O. Ouchi, and K. Seki

3.1   Introduction   *69*
3.1.1   Energy level alignment at organic/metal interfaces   *69*
3.2   Basic aspects of band bending   *74*
3.2.1   Band bending in thermal equilibrium   *74*
3.2.2   Band bending in nonequilibrium   *75*
3.3   KP measurements for insulating surface   *76*
3.4   Experimental   *78*
3.5   Results ans Discussion   *78*
3.5.1 C_{60}/metal interface: case of band bending   *78*
3.5.2   TPD/metal interface: case for no band bending   *81*
3.5.3   Potential gradient built in Alq_3 film   *84*
3.5.4   Concluding Remarks   *90*

4 **Thermal and Structural Properties of the Organic Semiconductor Alq$_3$ and Characterization of its Excited Electronic Triplet State** *95*
M. Cölle and W. Brütting

4.1 Introduction *95*
4.2 Crystalline Phases of Alq$_3$ *97*
4.3 Thermal Properties of Alq$_3$ *100*
4.4 The Molecular Structure of ô-Alq$_3$ *106*
4.4.1 High resolution powder diffraction using synchrotron radiation *106*
4.2 Vibrational Analysis *111*
4.5 Population and properties of the electronic excited triplet state *114*
4.5.1 Population of the triplet states *114*
4.2 Phosphorescence of Alq$_3$ *118*
4.6 Summary *125*

Part II **Photophysics** *129*

5 **Ultrafast Photophysics in Conjugated Polymers** *131*
G. Lanzani, G. Cerullo, D. Polli, A. Gambetta, M. Zavelani-Rossi, C. Gadermaier

5.1 Introduction *131*
5.2 Femtosecond laser systems *132*
5.3 The pump-probe experiment *135*
5.4 Coherent vibrational spectroscopy *136*
5.5 Vibrational relaxation *137*
5.6 Electronic relaxation *139*
5.6.1 Triplet and soliton pairs *140*
5.6.2 Charge state dynamics *142*
5.6.2.1 A brief overview about charge photogeneration *142*
5.6.2.2 Field-assisted pump-probe spectroscopy *144*
5.6.2.3 The role of higher-lying states *145*
5.7 Conclusion *149*

6 **The Origin of the Green Emission Band in Polyfluorene Type Polymers** *153*
S. Gamerith, C. Gadermaier, U. Scherf, E. J. W. List

6.1 Introduction *153*
6.2 Photo-physical Properties of Polyfluorene-type Polymers *155*
6.2.1 Optical properties of pristine polyfluorenes *156*
6.2.2 Optical properties of polyfluorenes in the ô-phase *157*
6.2.3 Fluorescent on chain defects in polyfluorene-type polymers *160*
6.2.3.1 Formation of keto-type defects in PF-type polymers *160*
6.2.3.2 The formation of keto defects during the synthesis *160*
6.2.3.3 Formation of keto defects during thermal degradation *161*
6.2.3.4 Formation of keto defects due to photo-oxidative degradation *165*
6.2.4 The origin of the green emission band at 2.2-2.3 eV *166*

6.2.5   Degradation in polyfluorene-based electroluminescent devices   *176*

6.3   Conclusion   *178*

7   **Exciton Energy Relaxation and Dissociation in Pristine and Doped Conjugated Polymers**   *183*

V. I. Arkhipov and H. Bässler

7.1   Introduction   *183*

7.2   Field-induced exciton dissociation probed by luminescence quenching   *185*

7.2.1   Steady state fluorescence quenching   *185*

7.2.2   Time resolved fluorescence quenching   *188*

7.2.3   A model of field-assisted dissociation of optical excitations   *191*

7.3   Exciton dissociation in doped conjugated polymers   *195*

7.3.1   Experiments on photoluminescence quenching in doped polymers   *195*

7.3.2   Exciton energy relaxation and quenching by Förster energy transfer   *198*

7.3.3   Exciton quenching at charge transfer centers   *204*

7.4   Photoconductivity in pristine and weakly doped polymers   *208*

7.4.1   Intrinsic photogeneration   *210*

7.4.2   Dopant assisted photogeneration   *215*

7.4.3   A model of dopant-assisted charge photogeneration   *220*

7.5   Photoconductivity in polymer donor/acceptor blends   *224*

7.5.1   Enhanced exciton dissociation at high concentration of electron acceptors   *224*

7.5.2   A model of efficient exciton dissociation at a donor/acceptor interface   *226*

7.6   Conclusions   *230*

8   **Polarons in ŏ-conjugated Semiconductors: Absorption Spectroscopy and Spin-dependent Recombination**   *235*

M. Wohlgenannt

8.1   Introduction   *235*

8.1.1   Polarons in ŏ-conjugated polymers and oligomers   *235*

8.1.2   Organic electroluminescence   *236*

8.1.2.1   Spin dependent exciton formation cross-sections   *239*

8.2   Polaron absorption spectra   *240*

8.2.1   Experimental   *240*

8.2.2   Experimental results   *241*

8.2.2.1   Polaron absorption spectra in doped oligomers; non-degenerate ground state systems   *242*

8.2.2.2   Polaron absorption spectra in doped oligomers; degenerate ground state systems   *242*

8.2.2.3   Polaron absorption spectra in photodoped polymer films   *244*

8.3   Is polaron recombination (exciton formation) spin-dependent?   *245*

8.3.0.1   Experimental   *245*

8.3.0.2   Experimental Results   *246*
8.4   Conclusions   *252*

9   **Phosphorescence as a Probe of Exciton Formation and Energy Transfer in**
    Organic Light Emitting Diodes   *257*
    M. Baldo and M. Segal

9.1   Introduction   *257*
9.2   Phosphorescence, the Spin Dependence of Exciton Formation
    and Triplet Energy Transfer   *258*
9.3   Phosphorescent Studies of Exciton Formation   *260*
9.4   Energy Transfer in Blue Phosphorescent OLEDs   *264*
9.5   Conclusions   *267*

Part III   **Transport and Devices**   *271*

10   **Electronic Traps in Organic Transport Layers**   *273*
    R. Schmechel and H. von Seggern

10.1   Introduction   *273*
10.2   Origin of trap states   *275*
10.3   Trap detection techniques   *276*
10.4   Traps in non-doped and doped small molecule semiconductors   *279*
10.4.1   Pristine traps and their electronic structure   *279*
10.4.2   Effect of doping on electronic traps   *283*
10.3.3   Doping Induced Electrical and Optical properties   *288*
10.3.4   Structure related traps and their influence on device performance   *293*
10.5   Traps in polymeric semiconductors   *297*
10.6   Conclusions   *301*

11   **Charge Carrier Density Dependence of the Hole Mobility**
    in poly(*p*-phenylene vinylene)   *305*
    C. Tanase, P. W. M. Blom, D. M. de Leeuw, E. J. Meijer

11.1   Introduction   *305*
11.2   Experimental results and Discussion   *306*
11.2.1   Charge carrier mobility in hole-only diodes   *307*
11.2.2   Local mobility versus field-effect mobility in FETs   *309*
11.2.3   Comparison of the hole mobility in LEDs and FETs   *314*
11.3   Conclusions   *317*

12   **Analysis and Modeling of Organic Devices**   *319*
    Y. Roichman, Y. Preezant, N. Rappaport, and N. Tessler

12.1   Introduction   *319*
12.1.1   Motivation   *320*
12.2   Charge transport   *321*

12.2.1	The Gaussian DOS	322
12.2.2	The Equilibrium Charge Density	323
12.2.3	The diffusion coefficient	324
12.2.4	The mobility coefficient	324
12.3	The operation regime of organic devices	326
12.3.1	Contact workfunction	329
12.4	Organic Light-Emitting Diodes	330
12.4.1	Space charge regime	330
12.4.2	Contact limited regime	331
12.4.3	I-V Analysis of LEDs	333
12.5	Field Effect Transistors	334
12.6	Photo-Cells	339

13   Fabrication and Analysis of Polymer Field-effect Transistors 343
S. Scheinert and G. Paasch

13.1	Introduction	343
13.2	Models	345
13.2.1	The Shockley current characteristics	345
13.2.2	Material models	346
13.2.3	Drift-diffusion-model and numerical simulation	350
13.3	Analytical estimates	352
13.3.1	Basic dependencies	352
13.3.2	Interrelation between material and device properties	353
13.3.3	Cut-off frequency	356
13.4	Experimental	358
13.4.1	Preparation of transistors and capacitors	358
13.4.2	Measuring techniques	359
13.5	Device characterization and analyzes	360
13.5.1	Basic procedure: P3OT devices as an example	360
13.5.1.1	Layer characterization	360
13.5.1.2	MOS capacitor	360
13.5.1.3	Thin film transistors	363
13.5.2	Devices made from MEH-PPV	365
13.5.3	Further transistor designs	369
13.5.3.1	Finger structure	369
13.5.3.2	Short channel design	370
13.5.4	Subthreshold currents and trap recharging	373
13.5.5	Hysteresis	377
13.6	Selected simulation studies	379
13.6.1	Formation of inversion layers	379
13.6.2	Trap distributions in top and bottom contact transistors with different source/drain work functions	381
13.6.3	Short-channel effects	383
13.7	Conclusions	385

A   Non-degenerate approximation for the Gaussian distribution   *387*

B   DDM with traps and trap distributions   *388*

14   Organic Single-Crystal Field-Effect Transistors   *393*
          R. W. I. de Boer, M. E. Gershenson, A. F. Morpurgo, and V. Podzorov

14.1   Introduction   *393*
14.2   Fabrication of single-crystal organic FETs   *395*
14.2.1   Single-crystal growth   *396*
14.2.2   Crystal characterization   *399*
14.2.2.1 Polarized-light microscopy   *400*
14.2.2.2 The time-of-flight experiments   *401*
14.2.2.3 Space charge limited current spectroscopy   *402*
14.2.3   Fabrication of the field-effect structures   *405*
14.2.3.1 Electrostatic bonding technique   *405*
14.2.3.2 "Direct" FET fabrication on the crystal surface   *409*
14.3   Characteristics of single-crystals OFETs   *412*
14.3.1   Unipolar operation   *413*
14.3.2   Field-effect threshold   *415*
14.3.3   Sub-threshold slope   *416*
14.3.4   Double-gated rubrene FETs   *417*
14.3.5   Mobility   *419*
14.3.6   Mobility anisotropy on the surface of organic crystals.   *422*
14.3.7   Preliminary results for the OFETs with high-*k* dielectrics.   *424*
14.4   Conclusion   *426*
14.5   Acknowledgments   *429*

**15   Charge Carrier Photogeneration and Transport in Polymer-fullerene
            Bulk-heterojunction Solar Cells**   *433*
          I. Riedel, M. Pientka, and V. Dyakonov

15.1   Introduction   *433*
15.2   Experimental   *434*
15.3   Photoinduced charge transfer in polymer: fullerene composites   *435*
15.4   Photovoltaic characteristics and charge carrier photogeneration efficiency
          of polymer-fullerene bulk heterojunction solar cells   *439*
15.5   Analysis of the electrical transport properties   *442*
15.5.1   Illumination intensity dependence   *442*
15.5.2   Temperature dependence   *443*
15.5.3   Bulk-heterojunction solar cells with increased absorber thickness   *446*
15.5.4   Dependence of IPCE on reverse bias   *447*
15.6   Conclusions   *448*

16   **Modification of PEDOT:PSS as Hole Injection Layer in Polymer LEDs**   *451*
M. M. de Kok, M. Buechel, S. I. E. Vulto, P. van de Weijer, E. A. Meulenkamp,
S. H. P. M. de Winter, A. J. G. Mank, H. J. M. Vorstenbosch, C. H. L. Weijtens,
and V. van Elsbergen

16.1   Introduction   *451*
16.2   Experimental   *454*
16.3   Results   *456*
16.3.1   Chemistry and bulk properties   *456*
16.3.2   Surface properties   *460*
16.3.3   Device performance   *463*
16.4   Discussion   *466*
16.4.1   Effect of NaOH on PEDOT:PSS properties   *466*
16.4.2   Effect of NaOH addition on PLED device performance   *470*
16.5   Conclusions   *470*

17   **Insights into OLED Functioning Through Coordinated Experimental**
Measurements and Numerical Model Simulations   *475*
D. Berner, H. Houili, W. Leo, and L. Zuppiroli

17.1   Introduction   *475*
17.2   Device Fabrication   *477*
17.3   The Model   *479*
17.3.1   Coulomb effects   *480*
17.3.2   Equations of motion and injection   *480*
17.3.3   Transport in the organic material   *482*
17.3.4   Recombination   *484*
17.3.5   Numerical procedure   *484*
17.4   Case Studies   *486*
17.4.1   Electrode effects   *486*
17.4.2   Time of flight simulations   *487*
17.4.3   Internal electric field   *489*
17.4.3.1   Experimental analysis   *490*
17.4.3.2   Discussion of Simulations   *491*
17.4.3.3   Charge and current distribution inside the device   *492*
17.4.3.4   Electric field density distribution   *494*
17.4.4   Dye doping   *494*
17.4.4.1   Experimental results: *494*
17.4.4.2   Analysis Through Simulation   *498*
17.4.4.3   Mobility variation inside the doped part   *499*
17.4.4.4   Recombination rate distribution   *502*
17.4.4.5   Energy level shift inside the doped region   *502*
17.4.4.6   General recombination picture   *505*
17.4.4.7   Transport characteristics   *505*
17.5   Summary   *507*

18 **Optimizing OLED Structures for a-Si Display Applications via Combinatorial Methods and Enhanced Outcoupling** *511*
W. Rieß, T. A. Beierlein, and H. Riel

18.1 Introduction *511*
18.2 Experimental *512*
18.3 Results and Discussion *514*
18.3.1 Combinatorial Device Optimization *514*
18.3.1.1 Layer-Thickness Variations: Charge-Carrier Balance *515*
18.3.1.2 Layer-Thickness Variation: Optical Effects *518*
18.3.2 Optical Outcoupling *519*
18.3.3 20-inch a-Si AMOLED Display *524*
18.4 Conclusions *525*

Index *529*

List of Contributors

V. I. Arkhipov
IMEC
Heverlee-Leuven, Belgium

M. Baldo
Department of Electrical Engineering
and Computer Science
Massachusetts Institute of Technology,
USA

H. Bäßler
Institute of Physical, Nuclear and
Macromolecular Chemistry
Universität Marburg, Germany

T. A. Beierlein
IBM Research GmbH
Rüschlikon, Switzerland

D. Berner
CFG S.A. Microelectronic
Morges, Switzerland

P. W. M. Blom
Materials Science Centre
University of Groningen,
The Netherlands

R. W. I. de Boer
Department of Nanoscience
Delft University of Technology,
The Netherlands

W. Brütting
Institute of Physics
University of Augsburg, Germany

M. Buechel
Philips Research Laboratories
Eindhoven, The Netherlands

G. Cerullo
National Laboratory of Ultrafast and
Ultraintense Optical Science and INFM
Politecnico di Milano, Italy

M. Cölle
Philips Research
Eindhoven, The Netherlands

V. Dyakonov
Experimental Physics
(Energy Research)
University of Würzburg, Germany

V. van Elsbergen
Philips Forschungslaboratorien
Aachen, Germany

C. Gadermaier
Nat'l Lab. of Ultrafast and Ultraintense
Optical Science – INFM, Milano, Italy

Physics of Organic Semiconductors. Edited by W. Brütting
Copyright © 2005 WILEY-VCH Verlag GmbH & Co. KGaA, Weinheim
ISBN 3-527-40550-X

A. Gambetta
National Laboratory of Ultrafast and
Ultraintense Optical Science and
INFM
Politecnico di Milano, Italy

S. Gamerith
Christian-Doppler Laboratory
Advanced Functional Materials
Technische Universität Graz, Austria

M. E. Gershenson
Department of Physics and Astronomy
Rutgers University, Piscataway, USA

N. Hayashi
Venture Business Laboratory
Nagoya University, Japan

H. Houili
Laboratory of Optoelectronics of
Molecular Materials, EPFL
Lausanne, Switzerland

H. Ishii
Research Institute of Electrical
Communication
Tohoku University, Japan

E. Ito
Local Spatio-Temporal Functions
Laboratory
Wako, Japan

Y. Kimura
Research Institute of Electrical
Communication
Tohoku University, Japan

M. M. de Kok
Philips Research Laboratories
Eindhoven, The Netherlands

M. Knupfer
Leibniz Institute for Solid State and
Materials Research
Dresden, Germany

G. Lanzani
National Laboratory of Ultrafast and
Ultraintense Optical Science and
INFM
Politecnico di Milano, Italy

D. M. de Leeuw
Philips Research Laboratories
Eindhoven, The Netherlands

W. Leo
CFG S.A. Microelectronic
Morges, Switzerland

E. J. W. List
Christian-Doppler Laboratory
Advanced Functional Materials
Technische Universität Graz, Austria

A. J. G. Mank
Philips Centre for Manufacturing
Technology
Eindhoven, The Netherlands

E. J. Meijer
Philips Research Laboratories
Eindhoven, The Netherlands

E. A. Meulenkamp
Philips Research Laboratories
Eindhoven, The Netherlands

A. F. Morpurgo
Department of Nanoscience
Delft University of Technology,
The Netherlands

M. Niwano
Research Institute of Electrical
Communication
Tohoku University, Japan

Y. Ouchi
Graduate School of Science
Nagoya University, Japan

G. Paasch
Leibniz Institute for Solid State and
Materials Research
Dresden, Germany

H. Peisert
Leibniz Institute for Solid State and
Materials Research
Dresden, Germany

M. Pientka
Energy- and Semiconductor Research
Laboratory
University of Oldenburg, Germany

V. Podzorov
Department of Physics and Astronomy
Rutgers University, Piscataway, USA

D. Polli
National Laboratory of Ultrafast and
Ultraintense Optical Science and INFM
Politecnico di Milano, Italy

Y. Preezant
Electrical Engineering Dept.
Nanoelectronic center, Technion,
Haifa, Israel

N. Rappaport
Electrical Engineering Dept.
Nanoelectronic center, Technion,
Haifa, Israel

I. Riedel
Energy- and Semiconductor Research
Laboratory
University of Oldenburg, Germany

H. Riel
IBM Research GmbH
Rüschlikon, Switzerland

W. Rieß
IBM Research GmbH
Rüschlikon, Switzerland

Y. Roichman
Electrical Engineering Dept.
Nanoelectronic center, Technion,
Haifa, Israel

S. Scheinert
Institute of Solid State Electronics
Ilmenau Technical University, Germany

U. Scherf
Makromolekulare Chemie
Bergische Universität Wuppertal,
Germany

R. Schmechel
Institut für Material- und
Geowissenschaften
Technische Universität Darmstadt,
Germany

H. von Seggern
Institut für Material- und
Geowissenschaften
Technische Universität Darmstadt,
Germany

F. Schreiber
Institut für Angewandte Physik
Universität Tübingen, Germany

M. Segal
Department of Electrical Engineering
and Computer Science
Massachusetts Institute of Technology,
USA

K. Seki
Research Center for Materials Science
Nagoya University, Japan

K. Sugi
Research Institute of Electrical
Communication
Tohoku University, Japan

C. Tanase
Materials Science Centre
University of Groningen,
The Netherlands

N. Tessler
Electrical Engineering Dept.
Nanoelectronic center, Technion,
Haifa, Israel

H. J. M. Vorstenbosch
Philips Centre for Manufacturing
Technology
Eindhoven, The Netherlands

S. I. E. Vulto
Philips Research Laboratories
Eindhoven, The Netherlands

Y. Washizu
Venture Business Laboratory
Nagoya University, Japan

P. van de Weijer
Philips Research Laboratories
Eindhoven, The Netherlands

C. H. L. Weijtens
Philips Forschungslaboratorien
Aachen, Germany

S. H. P. M. de Winter
Philips Research Laboratories
Eindhoven, The Netherlands

M. Wohlgenannt
Department of Physics and Astronomy
University of Iowa, USA

M. Zavelani-Rossi
National Laboratory of Ultrafast and
Ultraintense Optical Science and
INFM
Politecnico di Milano, Italy

L. Zuppiroli
Laboratory of Optoelectronics of
Molecular Materials, EPFL
Lausanne, Switzerland

Introduction to the Physics of Organic Semiconductors

W. Brütting

1
History

With the invention of the transistor around the middle of the last century, inorganic semiconductors like Si or Ge began to take over the role as dominant material in electronics from the before prevailing metals. At the same time, the replacement of vacuum tube based electronics by solid state devices initiated a development which by the end of the 20th century has lead to the omnipresence of semiconductor microelectronics in our everyday life. Now at the beginning of the 21st century we are facing a new electronics revolution that has become possible due to the development and understanding of a new class of materials, commonly known as *Organic Semiconductors*. The enormous progress in this field has been driven by the expectation to realize new applications, such as large area, flexible light sources and displays, low-cost printed integrated circuits or plastic solar cells from these materials.

Strictly speaking organic semiconductors are not new. The first studies of the dark and photoconductivity of anthracene crystals (a prototype organic semiconductor, see e.g. [1]) date back to the early 20th century [2,3]. Later on, triggered by the discovery of electroluminescence in the 1960s [4,5], molecular crystals were intensely investigated by many researchers. These investigations could establish the basic processes involved in optical excitation and charge carrier transport (for a review see e.g. [6,7]). Nevertheless, in spite of the principal demonstration of an organic electroluminescent diode incorporating even an encapsulation similar to the ones used in nowadays commercial display applications [8], there were several draw-backs preventing practical use of these early devices. For example, neither high enough current densities and light output nor sufficient stability could be achieved. The main obstacles were the high operating voltage as a consequence of the crystal thickness in the micrometre to millimetre range together with the difficulties in scaling up crystal growth as well as preparing stable and sufficiently well-injecting contacts to them.

Since the 1970s the successful synthesis and controlled doping of conjugated polymers [9] established the second important class of organic semiconductors which was honoured with the Nobel Prize in Chemistry in the year 2000. Together with organic photoconductors (molecularly doped polymers) these conducting poly-

Physics of Organic Semiconductors. Edited by W. Brütting
Copyright © 2005 WILEY-VCH Verlag GmbH & Co. KGaA, Weinheim
ISBN 3-527-40550-X

mers have initiated the first applications of organic materials as conductive coatings [10] or photoreceptors in electrophotography [11].

The interest in undoped organic semiconductors revived in the 1980s due to the demonstration of an efficient photovoltaic cell incorporating an organic hetero-junction of p- and n-conducting materials [12] as well as the first successful fabrication of thin film transistors from conjugated polymers and oligomers [13-15]. The main impetus, however, came from the demonstration of high-performance electroluminescent diodes from vacuum-evaporated molecular films [16,17] and from conjugated polymers [18,19]. Owing to the large efforts of both academic and industrial research laboratories during the last 15 years, organic light-emitting devices (OLEDs) have progressed rapidly and meanwhile lead to first commercial products incorporating OLED displays [20]. Other applications of organic semiconductors e.g. as logic circuits with organic field-effect transistors (OFETs) or organic photovoltaic cells (OPVCs) are expected to follow in the near future (for an overview see e.g. [21]).

2
Materials

As already mentioned above, there are two major classes of organic semiconductors: low molecular weight materials and polymers. Both have in common a conjugated π-electron system being formed by the p_z-orbitals of sp^2-hybridized C-atoms in the molecules (see Fig. 1). As compared to the σ-bonds forming the backbone of the molecules, the π-bonding is significantly weaker. Therefore, the lowest electronic excitations of conjugated molecules are the π-π^*-transitions with an energy gap typically between 1.5 and 3 eV leading to light absorption or emission in the visible spectral range. As shown in Table 1 for the family of the polyacenes the energy gap can be controlled by the degree of conjugation in a molecule. Thus chemistry offers a wide range of possibilities to tune the optoelectronic properties of organic semiconducting materials. Some prototype materials which are also discussed in this book are given in Fig. 2.

An important difference between the two classes of materials lies in the way how they are processed to form thin films. Whereas small molecules are usually deposited from the gas phase by sublimation or evaporation, conjugated polymers can only be processed from solution e.g. by spin-coating or printing techniques. Additionally, a number of low-molecular materials can be grown as single crystals allowing intrinsic electronic properties to be studied on such model systems (see e.g. [23] or [24] for an overview of the different classes of materials). The controlled growth of highly ordered thin films either by vacuum deposition or solution processing is still subject of ongoing research, but will be crucial for many applications (see e.g. [25]).

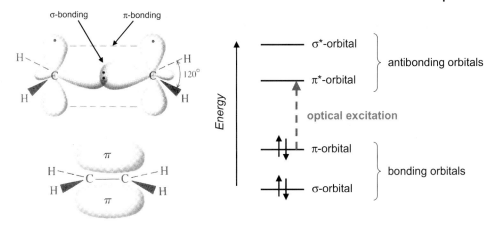

Figure 1   Left: σ- and π-bonds in ethene, as an example for the simplest conjugated π-electron system. The right viewgraph shows the energy levels of a π-conjugated molecule. The lowest electronic excitation is between the bonding π-orbital and the antibonding π*-orbital (adopted from [22]).

Table 1   Molecular structure of the first five polyacenes, together with the wavelength of the main absorption peak (taken from [7]).

Molecule	Structure	Absorption Maximum
Benzene		255 nm
Naphthalene		315 nm
Anthracene		380 nm
Tetracene		480 nm
Pentacene		580nm

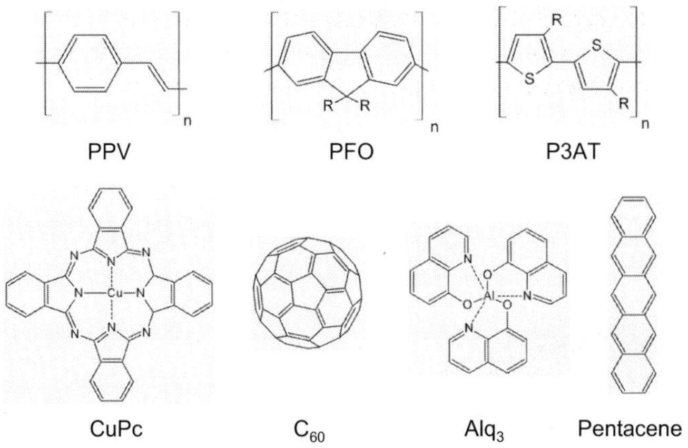

PPV PFO P3AT

CuPc C$_{60}$ Alq$_3$ Pentacene

Figure 2 Molecular structure of some prototype organic semiconductors: PPV: poly(p-phenylenevinylene), PFO: polyfluorene, P3AT: poly(3-alkylthiophene), Alq$_3$: tris(8-hydroxyquinoline)aluminium, fullerene C$_{60}$, CuPc: Cu-phthalocyanine, pentacene.

3
Basic Properties of Organic Semiconductors

The nature of bonding in organic semiconductors is fundamentally different from their inorganic counterparts. Organic molecular crystals are van der Waals bonded solids implying a considerably weaker intermolecular bonding as compared to covalently bonded semiconductors like Si or GaAs. The consequences are seen in mechanical and thermodynamic properties like reduced hardness or lower melting point, but even more importantly in a much weaker delocalization of electronic wavefunctions among neighbouring molecules, which has direct implications for optical properties and charge carrier transport. The situation in polymers is somewhat different since the morphology of polymer chains can lead to improved mechanical properties. Nevertheless, the electronic interaction between adjacent chains is usually also quite weak in this class of materials.

3.1
Optical Properties

Owing to the weak electronic delocalization, to first order the optical absorption and luminescence spectra of organic molecular solids are very similar to the spectra in the gas phase or in solution (apart from the trivial solvent shift). In particular, intramolecular vibrations play an important role in solid state spectra and often these vibronic modes can be resolved even at room temperature. Thus the term "oriented gas" is sometimes used for molecular crystals. Nevertheless, solid state spectra can differ in detail with respect to selection rules, oscillator strength and energetic position; moreover, due to the crystal structure or the packing of polymer chains a pronounced anisotropy can be found. Additionally disordered organic solids usually show a considerable spectral broadening. This is schematically shown in Fig. 3.

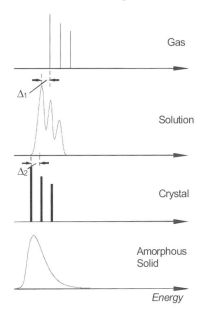

Figure 3 Schematical representation of optical spectra of organic molecules in different surroundings. Δ_1 and Δ_2 denote the respective solvent shift in solution and solid state.

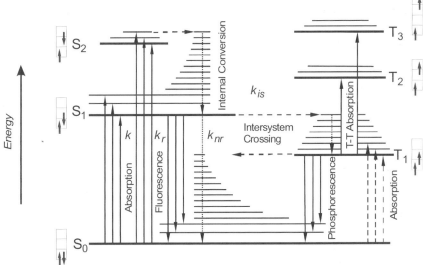

Figure 4 Energy level scheme of an organic molecule (left: singlet manifold, right: triplet manifold). Arrows with solid lines indicate radiative transitions, those with broken lines nonradiative transitions (taken from [7]). Typical lifetimes of the S_1 state are in the range 1...10 ns. Triplet lifetimes are usually in the millisecond range for pure aromatic hydrocarbons, but can be considerably shorter in molecules incorporating heavy atoms, like e.g. Pt or Ir.

Frenkel Exciton   Electron-Hole Recombination/
                                    Exciton Dissociation

$$S_1 \quad \text{—————————} \quad \begin{array}{c} \text{Exciton} \\ \text{Binding Energy} \end{array} \updownarrow$$

$$S_0 \quad \text{————}$$

Figure 5 The energetic difference between an excited state sitting on one molecule (sometimes called a Frenkel exciton) and a pair of uncorrelated negative and positive carriers sitting on different molecules far apart defines the exciton binding energy. A simple estimation as the Coulomb energy of an electron-hole pair localized at a distance of about 10 Å in a medium with a dielectric constant of 3 yields a value of about 0.5 eV for the exciton binding energy.

As a consequence of this weak electronic delocalization, organic semiconductors have two important peculiarities as compared to their inorganic counterparts. One is the existence of well-defined spin states (singlet and triplet) like in isolated molecules which has important consequences for the photophysics of these materials (see Fig. 4). However, since intersystem crossing is a weak process, this also sets an upper limit for the electroluminescence quantum efficiency in OLEDs. A second important difference originates from the fact that optical excitations ("excitons") are usually localized on one molecule and therefore have a considerable binding energy of typically 0.5 to 1 eV. Thus in a photovoltaic cells this binding energy has to be overcome before a pair of independent positive and negative charge carriers is generated (see Fig. 5).

3.2
Charge Carrier Transport

When transport of electrons or holes in an organic molecular solid is considered, one has to bear in mind that this involves ionic molecular states. E.g. in order to create a hole, an electron has to be removed to form a radical cation M^+ out of a neutral molecule M. This defect electron can then move from one molecule to the next. In the same way, electron transport involves negatively charged radical ions M$^-$. (Qualitatively, the same arguments hold for polymers, however, in this case charged states are usually termed positive or negative polarons.) As compared to isolated molecules in the gas phase, these ionic states are stabilized in the solid by polarization energies leading to an energy level scheme as shown in Fig. 6. From this picture one can clearly see that due to the already mentioned exciton binding energy the optical gap between the ground state and the first excited singlet state is considerably less than the single particle gap to create an uncorrelated electron-hole pair. In going from molecular crystals to disordered organic solids one also has to consider locally varying polarization energies due to different molecular environments which lead to a Gaussian density of states for the distribution of transport sites as shown in Fig. 7.

Thus, depending on the degree of order the charge carrier transport mechanism in organic semiconductors can fall between two extreme cases: band or hopping transport. Band transport is typically observed in highly purified molecular crystals at not too high temperatures. However, since electronic delocalization is weak the bandwidth is only small as compared to inorganic semiconductors (typically a few kT at room temperature only). Therefore room temperature mobilities in molecular crystals reach only values in the range 1 to 10 cm²/Vs [27]. As a characteristic feature of band transport the temperature dependence follows a power law behaviour

$$\mu \propto T^{-n} \text{ with } n = 1 \dots 3 \tag{1}$$

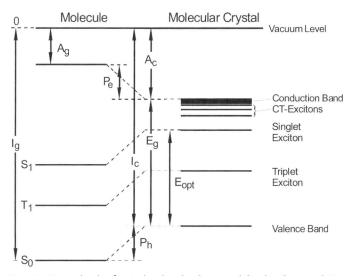

Figure 6 Energy levels of an isolated molecule (left) and a molecular crystal (right). I_g and A_g denote the ionization potential and electron affinity in the gas phase, I_c and A_c the respective quantites in the crystal. Due to the polarization energies P_h and P_e charged states are stabilized in the crystal. E_g is the single particle gap being relevant for charge carrier generation, whereas E_{opt} denotes the optical gap measured in absorption and luminescence. Their difference is the so-called exciton binding energy. (Figure adopted from [7]).

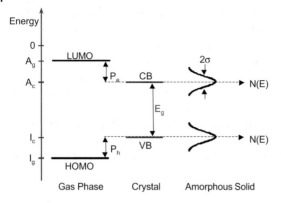

Figure 7 Energy levels of an isolated molecule (left), a molecular crystal (middle) and an amorphous solid (right). The width of the Gaussian density of states in an amorphous solid is typically in the range of σ = 80...120 meV, whereas the band width in molecular crystals is less than 100 meV. (Figure adopted from [26]).

upon going to lower temperature. However, in the presence of traps significant deviations from such a behaviour are observed [28].

In the other extreme case of an amorphous organic solid hopping transport prevails which leads to much lower mobility values (at best around 10^{-3} cm^2/Vs, in many cases however much less). Instead of a power law the temperature dependence then shows an activated behaviour and the mobility also depends on the applied electric field:

$$\mu(F, T) \propto \exp(-\Delta E/kT) \cdot \exp(\beta\sqrt{F}/kT) \qquad (2)$$

Depending on the model slightly different temperature dependencies for the mobility have been suggested (for a review see [11]). Furthermore, space-charge and trapping effects as well as details of the charge carrier injection mechanism have to be considered for describing electrical transport in organic solids [29-33].

On a macroscopic level, the current through a material is given by the charge carrier density n and the carrier drift velocity v, where the latter can be expressed by the mobility μ and the electric field F:

$$j = env = en\mu F \qquad (3)$$

One has to bear in mind that in contrast to metals this is usually not a linear relation between j and F since both the carrier density and mobility can depend on the applied field. According to this equation, apart from the field, the two parameters n and μ determine the magnitude of the current. Thus it is instructive to compare their typical values with inorganic semiconductors and discuss different ways to control them.

As already discussed above, the mobility strongly depends on the degree of order and purity in organic semiconductors and therefore to a great deal on the preparation and growth conditions. It can reach values of 1-10 cm^2/Vs in molecular crystals (see the contribution of de Boer et al. in this book), but values of only 10^{-5} cm^2/Vs in amorphous materials are also not unusual. The highest mobility values achievable in thin films are nowadays comparable to amorphous silicon which is of course orders of magnitude less than crystalline Si [34-36].

The second parameter is the charge carrier density n. The intrinsic carrier density in a semiconductor with an energy gap E_g and an effective density of states N_0 (which is strictly speaking the product of valence and conduction band densities) is given by:

$$n_i = N_0 \cdot \exp(-E_g/2kT) \tag{4}$$

Taking typical values for an organic semiconductor with E_g= 2.5 eV and N_0= 10^{21} cm^{-3} leads to a hypothetical carrier density of n_i= 1 cm^{-3} at room temperature, which is of course never reachable since impurities will lead to much higher densities in real materials. Nevertheless, the corresponding value for Si (E_g= 1.12 eV and N_0= 10^{19} cm^{-3}) is with n_i= 10^{10} cm^{-3} many orders of magnitude higher, which demonstrates that organic semiconductors should have extremely low conductivity if they are pure enough.

In order to overcome the limitations posed by the low intrinsic carrier density, different means to increase the carrier density in organic semiconductors can be applied:

1. (electro-)chemical doping,
2. carrier injection from contacts,
3. photo-generation of carriers, and
4. field-effect doping.

In the following section these methods will be briefly discussed together with their realization in different device structures.

3.3
Device Structures and Properties

Controlled doping has been one of the keys for the success of semiconductor microelectronics. There have been efforts to use tools like ion implantation doping also for organic semiconductors [37], however, due to the concomitant ion beam damages and the need for sophisticated equipment this method is probably not compatible with organic devices. Other techniques, like chemical doping by adding strong electron donors or acceptors as well as by electrochemical means have been successfully applied [38,39]. At this point one should also mention that often unintentional doping of organic materials already occurs during the synthesis or handling of the materials since in many cases ambient oxygen causes p-type doping of organic materials. Thus at present, controlled doping in organic semiconductors is still in its infancy

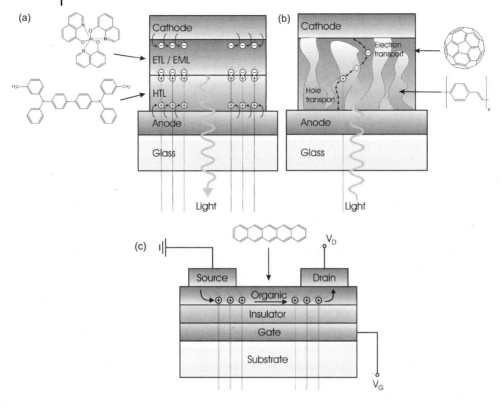

Figure 8 Different types of organic semiconductor devices are shown. (a) Organic light-emitting diode (OLED): Typically, a heterolayer structure is used, where HTL stands for hole transport layer and ETL for electron transport layer, EML denotes the emission layer. Instead of the displayed combination of a triphenyl-amine derivative and Alq_3, polymeric OLEDs usually employ a conductive polymer (PEDOT:PSS) together with luminescent polymers like PPV or PFO derivatives. (b) Organic photovoltaic cell (OPVC): The so-called bulk-heterojunction devices usually consist of a mixture of soluble PPV (or P3AT) and fullerene derivatives. Alternatively, mixed layers of evaporated small molecules like CuPc and C_{60} can be used. (c) Organic field-effect transistor (OFET): Prototypical materials in p-channel OFETs are pentacene as a low molecular weight material and P3AT as a conjugated polymer, respectively. Among others, e.g. C_{60} can be employed in n-channel transistors.

and needs further investigations to employ it as a powerful tool for organic electronics.

Injection of charge carriers from contacts is essentially the process that governs device operation in organic light-emitting devices (OLEDs) (see Fig. 8a). This requires low energetic barriers at the metal-organic interfaces for both contacts to inject equally high amounts of electrons and holes, which is required for a balanced charge carrier flow. Thus the interface energetic structure plays a very crucial role for achieving efficient OLEDs. Another process that comes into play is space-charge limitation of the current. Due to relatively high electric fields being applied to

OLEDs (typically 5 to 10 V across a layer thickness of 100 nm yield $F =$ 0.5...1 MV/cm) materials with low mobility such as Alq$_3$ (having an electron mobility of 10^{-5} cm^2/Vs) still yield high enough current densities for display applications. This is a consequence of the space-charge limited current scaling with the 3rd power of the reciprocal thickness [40]:

$$j_{SCLC} = \frac{9}{8} \, \varepsilon \varepsilon_0 \, \mu \, \frac{V^2}{d^3} \tag{5}$$

Apart from charge carrier transport, the efficiency of OLEDs is also strongly influenced by photophysical processes. First of all, materials with a high fluorescence quantum yield are required. However, since a large fraction of the excited states formed by charge carrier recombination are triplets, the most efficient OLEDs nowadays make use of energy transfer to so-called triplet emitters, where the presence of heavy metals renders the transition from the triplet state to the ground state via phosphorescence an allowed process [41].

The second important device application of organic semiconductors is in organic photovoltaic cells (OPVCs) (see Fig. 8b). In spite of their high absorption coefficient, which exceeds 10^5 cm^{-1} in most materials, the application of organic semiconductors in OPVCs faces the problem of the large exciton binding energy which prohibits efficient exciton dissociation. This can be overcome by making use of a photoinduced charge transfer between an electron donor like PPV and the fullerene C$_{60}$ as an acceptor [42]. Due to the short exciton diffusion length of typically 10 nm only, efficient OPVCs use the so-called bulk-heterojunction concept of mixing donor and acceptor in one single layer. In spite of the huge progress recently achieved, there are still challenges to achieve sufficient lifetime of OPVCs under ambient conditions or the availability of low-band gap materials to make better use of the solar spectrum [43,44].

Organic field-effect transistors (OFETs) (see Fig. 8c) are 3-terminal devices in which the charge carrier density in the channel between source and drain contacts can be controlled by the applied gate voltage across a thin dielectric [45]. The current is then given by

$$I_d = \frac{W}{L} \, C_i \, \mu \, (V_g - V_t) V_d \tag{3}$$

in the linear region, and by

$$I_d = \frac{W}{2L} \, C_i \, \mu \, (V_g - V_t)^2 \tag{3}$$

in the saturation regime. Thus the performance of OFETs can be tuned to some degree by using suitable geometries with short channel length L or thin insulating layers (leading to higher values for C_i), but it is clear that also the mobility needs to be high (in the range of amorphous Si) to realize switching at frequencies higher than about 100 kHz which will be needed for more demanding applications in the future [46,47]. This requires materials and methods to grow highly ordered organic semiconductor films. A further challenge will be to realize CMOS-like organic integrated circuits by using materials with ambipolar charge transport properties.

4

Outline of this Book

The following chapters are intended to give an up-to-date overview of the different facets of the subject of this book. They are based on articles published previously in a special issue of *physica status solidi* (a) (Vol. 201 (6), May 2004) devoted to this topic. The contents of the book is divided into three parts: (I) Growth and Interfaces, (II) Photophysics, (III) Transport and Devices.

In the first chapter of part I, F. Schreiber describes organic molecular beam deposition with emphasis on growth studies beyond the first monolayer, i.e. the evolution of the film structure and morphology. The following two contributions by M. Knupfer and H. Peisert and by H. Ishii et al. deal with the electronic structure at interfaces between organic semiconductors and metals. In particular, issues like energy level alignment, formation of interface dipoles, band bending as well as interactions and molecular orientation at interfaces are studied. In the next article, M. Cölle and W. Brütting present thermal and structural investigations on the isomerism of the Alq$_3$ molecule together with the first direct observation of the triplet state in this prototype organic electroluminescent material.

The latter subject directly leads to part II, which is devoted to the *Photophysics* of organic semiconductors. In the first article, G. Lanzani et al. give an overview on the contributions of ultrafast spectroscopy to the understanding of photophysics in conjugated polymers. S. Gamerith et al. investigate the emission properties of polyfluorene type polymers and their implications for polymer light-emitting diodes. V.I. Arkhipov and H. Bässler present experimental and theoretical aspects on the problem of exciton dissociation and charge photogeneration in conjugated polymers. The final two articles of this block deal with the problem of the singlet-triplet ratio in OLEDs: M. Wohlgenannt presents investigations on polarons in π-conjugated polymers and oligomers providing strong evidence for spin-dependent recombination, while M. Baldo and M. Segal study exciton formation and energy transfer processes using phosphorescent molecular materials which are involved in highly efficient phosphorescent OLEDs.

Part III of this book is dealing with *Transport and Devices*. The topic is introduced by an article by R. Schmechel and H. von Seggern on electronic traps in organic semiconductors, which is a key issue for understanding and optimizing charge transport in organic devices. In the following paper, C. Tanase et al. elucidate the fundamental differences in charge carrier transport between OLEDs and OFETs by studying the dependence of the charge carrier mobility on carrier density. The next two articles deal with modelling of organic devices: First Y. Roichman et al. present a self consistent picture of charge injection and transport in low mobility disordered organic semiconductors. Then S. Scheinert and G. Paasch nicely demonstrate the strength of a combination of device design and fabrication, experimental data analysis and numerical simulation for understanding and improving OFETs. High mobility organic single-crystal field-effect transistors are studied by R.W.I. de Boer et al. Their investigations show that single-crystal OFETs are ideal model systems for the study of intrinsic electronic transport properties of organic molecular semiconductors. The following paper by I. Riedel and V. Dyakonov gives an overview of the state-of-the-art in bulk heterojunction photovoltaic devices based on polymer-fullerene blend systems. Device applications of organic semiconductors in OLEDs are the subject of the remaining three articles: First, M.M. de Kok et al. report on the importance of the hole injection layer for efficiency and lifetime of polymer LEDs. Then D. Berner et al. provide a combined experimental study together with modelling to understand organic multi-layer OLEDs. Finally, W. Rieß et al. give an impressive demonstration of how combinatorial device optimization has lead to the development of a prototype 20-inch active-matrix full-colour OLED display driven by amorphous silicon thin-film transistors.

Acknowledgements

I would like to thank all contributing authors of this book for their efforts to present their work on the various aspects of this topic. Furthermore, I thank the editorial office of Wiley-VCH, in particular Ron Schulz, for their cooperation to publish this book.

References

[1] N. Karl, *Organic Semiconductors*, in O. Madelung, M. Schulz, and H. Weiss (Eds.), Landolt-Boernstein (New Series), Group III, Vol. 17 Semicondcuctors, Sub-volume 17i, page 106. Springer, Berlin, 1985.

[2] J. Koenigsberger, K. Schilling, Ann. Physik 32, 179 (1910).

[3] M. Volmer, Ann. Physik 40, 775 (1913).

[4] M. Pope, H. Kallmann, and P. Magnante, J. Chem. Phys. 38, 2042 (1963).

[5] W. Helfrich and W.G. Schneider, Phys. Rev. Lett. 140, 229 (1965).

[6] E.A. Silinsh. *Organic Molecular Crystals*. Springer, Berlin 1980.

[7] M. Pope and C.E. Swenberg. *Electronic Processes in Organic Crystals*. Clarendon Press, Oxford 1982.

[8] D.F. Williams and M. Schadt, Proc. IEEE (Lett.) 58, 476 (1970).

[9] C. K. Chiang, C. R. Fincher, Jr., Y. W. Park, A. J. Heeger, H. Shirakawa, E. J. Louis, S. C. Gau, and Alan G. MacDiarmid, Phys. Rev. Lett. 39, 1098 (1977).

[10] T.A. Skotheim (Ed.). *Handbook of Conducting Polymers*. M. Dekker, New York 1986.

[11] P. M. Borsenberger and D.S. Weiss. *Organic Photoreceptors for Imaging Systems*. M. Dekker, New York 1993.

[12] C.W. Tang, Appl. Phys. Lett. 48, 183 (1986).

[13] H. Koezuka, A. Tsumara, and T. Ando, Synth. Met. 18, 699 (1987).

[14] J.H. Burroughes, C.A. Jones, and R.H. Friend, Nature 335, 137 (1988).

[15] G. Horowitz, D. Fichou, X.Z. Peng, Z. Xu, and F. Garnier, Solid State Communications 72, 381 (1989).

[16] C.W. Tang and S.A. VanSlyke, Appl. Phys. Lett. 51, 913 (1987).

[17] C.W. Tang and S.A. VanSlyke, J. Appl. Phys. 65, 3610 (1989).

[18] J.H. Burroughes, D.D.C. Bradley, A.R. Brown, R.N. Marks, K. Mackay, R.H. Friend, P.L. Burn, and A.B. Holmes, Nature 347, 539 (1990).

[19] D. Braun and A.J. Heeger, Appl. Phys. Lett. 58, 1982 (1991).

[20] Pioneer Co. (Japan). In November 1997 Pioneer Co. in Japan commercialized a monochrome 256x64 dot matrix OLED display for automotive applications.

[21] Materials Today, September 2004: C. Reese et al. p. 20-27; I.D.W. Samuel et al. p. 28–35; N.S. Sariciftci p. 36-40; J.K. Borchardt p. 42–46 (www.materialstoday.com).

[22] K.P.C. Vollhardt, *Organische Chemie*, VCH-Verlag, Weinheim 1990

[23] M. Pope and C.E. Swenberg. *Electronic Processes in Organic Crystals and Polymers*. Oxford University Press, Oxford 1999.

[24] R. Farchioni, G. Grosso (Eds.), *Organic Electronic Materials*, Springer 2001.

[25] F. Faupel, C. Dimitrakopoulos, A. Kahn, C. Wöll (Eds.), *Organic Electronics*, Special Issue of J. Mater. Res. 19 (7), 2004.

[26] H. Bässler, phys. stat. sol. (b) 107, 9 (1981).

[27] N. Karl, J. Marktanner, Mol. Cryst. Liq. Cryst. 355, 149 (2001).

[28] N. Karl: *Charge Carrier Mobility in Organic Crystals*, in R. Farchioni, G. Grosso (Eds.), *Organic Electronic Materials*, Springer 2001.

[29] W. Brütting, S. Berleb, A. Mückl, Organic Electronics 2, 1 (2001).

[30] J.C. Scott, J. Vac. Sci. Technol. A 21, 521 (2003).

[31] U. Wolf, V.I. Arkhipov, H. Bässler, Phys. Rev. B 59, 7507 (1999); ibid. 59, 7514 (2001).

[32] P.W.M. Blom, M.C.J.M. Vissenberg, Mater. Sci. Eng. 27, 53 (2000).

[33] M.A. Baldo, S.R. Forrest, Phys. Rev. B 64, 085201 (2001).

[34] C.D. Dimitrakopoulos, P.R.L. Malenfant. Adv. Mater. 14, 99 (2002).

[35] H. Klauk, M. Halik, U. Zschieschang, G. Schmid, C. Dehm, Appl. Phys. Lett. 84, 2673 (2004).

[36] T.W. Kelley, D.V. Muyers, P.F. Baude, T.P. Smith, T.D. Jones, Mat. Res. Soc. Symp. Proc. 771, 169 (2003).

[37] A. Moliton. *Ion implantation doping of electroactive polymers and device fabrication*. In J.R. Reynolds T.A. Skotheim, and R.L. Elsenbaumer (Eds.), *Handbook of Conducting Polymers*. Marcel Dekker, New York 1998.

[38] M. Pfeiffer, K. Leo, X. Zhou, J.S. Huang, M. Hofmann, A. Werner, J. Blochwitz-Nimoth, Organic Electronics 4, 89 (2003).

[39] M. Gross, D.C. Müller, H.-G. Nothofer, U. Scherf, D. Neher, C. Bräuchle, K. Meerholz, Nature 405, 661 (2000).

[40] M.A. Lampert and P. Mark, *Current Injection in Solids*, Academic Press, New York 1970.

[41] M.A. Baldo, D.F. O'brien, Y. You, A. Shoustikov, S. Sibley, M.E. Thompson, S.R. Forrest, Nature 395, 151 (1998); S.R. Forrest, Organic Electronics 4, 45 (2003).

[42] N.S. Sariciftci, L. Smilowitz, A.J. Heeger, F. Wudl, Science 258, 1474 (1992).

[43] C.J. Brabec, N.S. Sariciftci, J.C. Hummelen, Adv. Funct. Mater. 11, 15 (2001).

[44] P. Peumans, S. Uchida, S.R. Forrest, Nature 425, 158 (2003).

[45] G. Horowitz, Adv. Mater. 10, 365 (1998).

[46] G.H. Gelinck, et al. Nature Mater. 3, 106 (2004).

[47] A. Knobloch, A. Manuelli, A. Bernds, W. Clemens, J. Appl. Phys. 96, 2286 (2004).

Part I
Growth and Interfaces

Physics of Organic Semiconductors. Edited by W. Brütting
Copyright © 2005 WILEY-VCH Verlag GmbH & Co. KGaA, Weinheim
ISBN 3-527-40550-X

1

Organic Molecular Beam Deposition: Growth Studies beyond the First Monolayer

F. Schreiber

1.1
Introduction

Organic semiconductors exhibit a range of interesting properties, and their application potential is rather broad, as seen in many other contributions in this book. For the crystalline 'small molecule' systems, grown by organic molecular beam deposition (OMBD), subject of the present contribution, it is generally agreed that the structural definition is important for the functional properties. The following list should serve to illustrate the various aspects:

1. The definition of interfaces (degree of interdiffusion and roughness)
   (a) Organic-organic (e.g., in organic diodes)
   (b) Organic-metal (e.g., for electrical contacts)
   (c) Organic-insulator (e.g., in transistors (insulating layer between gate and semiconductor))
2. The crystal structure
   (a) Which structure is present? (Note that polymorphism is very common in organics).
   (b) Are different structures coexisting?
   (c) Orientation of the structure (epitaxy)?
   (d) Is the structure strained (epitaxy)?
3. Crystalline quality/defect structure
   (a) Mosaicity (Note that in a thin film one has to distinguish between quality in the xy plane and in z direction (surface normal).)
   (b) Homogeneity within a given film (density of domain boundaries etc.)
   (c) Density of defects (and their nature), which also impacts the electronic properties.

Since the structure has a strong impact on the functional properties, understanding the structure formation, i.e., the growth process, and finding ways to optimise the structural definition is a prerequisite for technological progress. Moreover, understanding the physics of the growth process provides several fundamental challenges.

Physics of Organic Semiconductors. Edited by W. Brütting
Copyright © 2005 WILEY-VCH Verlag GmbH & Co. KGaA, Weinheim
ISBN 3-527-40550-X

We will mostly focus on 'thicker' films, their growth modes and the evolution of the morphology for thickness ranges which are typically employed in organic semiconductor applications. We will discuss only to a limited extent the work on the first monolayer, although as the 'seed layer' for the following layers this is obviously important. Thus, some of the classical surface science issues, such as binding site, epitaxial relation etc., are not in the focus of this review. For these issues and also for information on the history of the field, we refer to Refs. [1–10]. Also, we will not discuss issues related to chirality, although they are undoubtedly intriguing [9–13].

In terms of growth technology, the equipment is essentially the same as for inorganic molecular beam epitaxy. Evaporation cells on a vacuum chamber are used to provide a flux of molecules at the substrate surface (typically some range around 1 Å/sec to 1 Å/min), and ideally the growth can be monitored *in situ*. Virtually all surface and interface techniques have been used for OMBD-grown films, and we refer to standard textbooks for details of the experimental methodology.

This review is organised as follows. We first present some of the general issues in thin film growth and then what is specific and potentially different for organics (Sec. 1.2). In Sec. 1.2.3, we give an overview of the most popular systems. Sec. 1.3–1.6 contains a number of case studies, trying to highlight the issues that we feel are particularly relevant and typical for OMBD. The case studies are based on four different compounds. They are organised according to the (inorganic) substrates, covering, insulators, metals, and semiconductors. In Sec. 1.7, we briefly indicate the issues for organics-based heterostructures. Some conclusions are given in Sec. 1.8.

In a review with limited space such as the present one, it cannot be our goal to give a complete and exhaustive overview. Instead, the examples are centered mostly around our own work, which we try to discuss in the context of the general field. This selection is obviously unbalanced, and we apologise for omissions of important other work.

1.2
Organic molecular beam deposition

1.2.1
General concepts of thin film growth

Crystal and thin film growth are enormously rich subjects with many different facets and theoretical approaches. For a thorough treatment of the underlying concepts, we refer to Refs. [14–17]. Here we shall only briefly touch on selected aspects which we feel are important in the present context and help to appreciate the issues related to thin film growth (see also Fig. 1).

One approach to describe the various relevant interactions uses the concept of surface and interface energies, γ, similar to what is done for wetting phenomena. Typically, the surface energies (i.e., the relative contributions of the free substrate surface, γ_s, the film surface, γ_f, and the film-substrate interface, γ_i) are then related to the different growth modes, i.e. Frank-van-der-Merwe (layer-by-layer), Stranski-

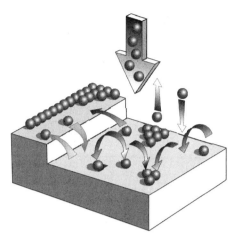

Figure 1 Schematic of processes relevant in thin film growth, such as adsorption (as a result of a certain impingement rate), (re-)desorption, intra-layer diffusion (on a terrace), inter-layer diffusion (across steps), nucleation and growth of islands.

Krastanov (layer plus islands after a certain critical thickness), and Vollmer-Weber (islands starting at the first monolayer).

Other issues are related to epitaxy, which however we will not discuss in great detail. (For clarity, we should emphasize that under epitaxial relation we understand the crystallographic relation between film and substrate, which does not necessarily imply smooth film growth.) Nevertheless, we should point out that, generally, the surface energies depend on the strain field induced by the lattice mismatch at the film-substrate interface, and thus also on the number of layers of the film. Therefore, the epitaxial relation of film and substrate is important not only in a crystallographic sense but also for the growth behaviour.

It should be emphasized that growth is actually a non-equilibrium phenomenon, and equilibrium or near-equilibrium energy considerations alone cannot properly account for all growth scenarios. Thus, a dynamic description is needed. This description has to take into account the flux of adsorbates towards the surface (corresponding to a certain supersaturation), the adsorption and re-desorption probabilities, and the diffusion processes on the surface (interlayer and intralayer) and their respective barriers. In the last two decades a theoretical framework has been established, which relates growth mechanisms to a set of scaling exponents describing the dependence of the surface roughness on film thickness and lateral length scale. Much effort has been spent to theoretically predict scaling exponents for certain growth models, as well as to determine them experimentally [14–19].

The scaling theory of growth-induced surface roughness is based on the behaviour of the height difference correlation function (HDCF), the mean square height difference $g(R) = \langle [h(x, y) - h(x', y')]^2 \rangle$ of pairs of points laterally separated by

$R = \sqrt{(x-x')^2 + (y-y')^2}$. The HDCF displays distinct behaviours for $R \ll \xi$ and $R \gg \xi$, where ξ denotes the correlation length. For $R \ll \xi$ one expects a power law increase as $g(R) \approx a^2 R^{2\alpha}$, where α is the static roughness exponent and the prefactor a is a measure of the typical surface slope. For $R \gg \xi$ the heights at distance R become uncorrelated. Hence $g(R)$ saturates at the value $g(R \gg \xi) = 2\sigma^2$, where $\sigma = \langle (h - \langle h \rangle)^2 \rangle^{1/2}$ is the standard deviation of the film height (or 'rms roughness'). The three parameters σ, ξ and a evolve with film thickness according to the power laws $\sigma \sim D^\beta$, $\xi \sim D^{1/z}$ and $a \sim D^\lambda$, defining the growth exponent β, the dynamic exponent z and the steepening exponent λ. Assuming that the regimes $R \ll \xi$ and $R \gg \xi$ are connected through a scaling form $g(R) = 2\sigma^2 \tilde{g}(R/\xi)$, it follows that the scaling exponents are related by $\beta = \alpha/z + \lambda$. For $\lambda = 0$ (no steepening) one has $\beta = \alpha/z$. Scaling with $\lambda > 0$ is referred to as anomalous [16]. The HDCF can be determined experimentally by real space methods (such as atomic force microscopy) or diffuse scattering, each having their advantages.

1.2.2
Issues specific to organic thin film growth

While the general considerations presented above apply to inorganic as well as organic thin film systems, there are a few issues specific to organics (Fig. 2), which can lead to quantitatively and qualitatively different growth behaviour.

1. Organic molecules are 'extended objects' and thus have *internal degrees of freedom*. This is probably the most fundamental difference between growth of atomic and growth of organic systems.
   (a) The *orientational degrees of freedom* which are not included in conventional growth models can give rise to qualitatively new phenomena, such as the change of the molecular orientation during film growth (Fig. 2). Also, even without considering a *transition* during the growth, the distinction of 'lying-down' and 'standing-up' films is important and obviously only possible for molecular systems.
   (b) The *vibrational degrees of freedom* can have an impact on the interaction with the surface as well as the thermalisation upon adsorption and the diffusion behaviour.
2. The *interaction potential* (molecule-molecule and molecule-substrate) is generally different from the case of atomic adsorbates, and van-der-Waals interactions are more important.
   (a) The response to strain is generally different. Potentially, more strain can be accomodated and in those systems, where the build-up of strain leads to a 'critical thickness' (before the growth mode changes), this thickness can be greater for 'softer' materials.
   The different ('softer') interactions with the substrate and the corrugation of the potential have also been discussed in terms of 'van-der-Waals epitaxy' and 'quasi-epitaxy' [3].

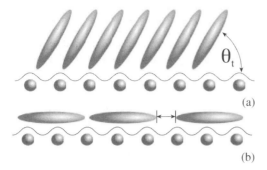

Figure 2 Issues specific for organics in the context of thin film growth. (a) Orientational degrees of freedom, potentially leading to orientational domains (additional source of disorder). They can also give rise to orientational transitions during growth. (b) Molecules larger than the unit cells of (inorganic) substrates, thus leading to translational domains. Generally, this can also lead to a smearing-out of the corrugation of the substrate potential experienced by the adsorbate.

    (b) The importance of van-der-Waals interactions implies that the relevant temperature scales (for evaporation from a crucible and also for diffusion on the substrate) are usually lower. It should be emphasized, however, that the *total* interaction energy of a molecule (integrated over its 'contact area' with a surface) can be substantial and comparable to that of strongly interacting (chemisorbing) atomic adsorbates. Nevertheless, in terms of interaction energies *per atom* the organic molecules considered here are usually more weakly bound.

    (c) Since we are concerned with closed-shell molecules and van-der-Waals-type crystals, there are no dangling bonds at the organic surface, and thus the surface energies are usually weaker than for inorganic substrates.

    (d) Importantly, however, if the surface of the *substrate* is 'strongly interacting', this results in limited diffusion and thus the evolution of well-ordered films is hampered. In the extreme case of a 'very reactive' surface (e.g., with dangling bonds available), the molecules may even dissociate upon adsorption.

3. The *size of the molecules and the associated unit cells* are greater than that of typical (inorganic) substrates.

    (a) The effective lateral variation of the potential is smeared out (i.e., averaged over the size of the molecule), making the *effective corrugation* of the substrate as experienced by the molecule generally *weaker* than for atomic adsorbates.

    (b) The size difference of the unit cells of adsorbate and substrate implies that there are more translational domains (see Fig. 2).

    (c) Moreover, organics frequently crystallize in low-symmetry structures, which again can lead to multiple domains (not only translational, but also orientational domains). Importantly, both are a source of disorder, *in addition* to those known from inorganic systems (e.g., vacancies).

Generally, most of the above points directly or indirectly impact the interactions and thus also the barriers experienced during diffusion. Thus, not only the static structure, but also the growth dynamics exhibit differences compared to inorganic systems.

1.2.3
Overview of popular OMBD systems

Organic chemistry provides obviously a vast number of dyes and semiconductors, which are potentially interesting for thin film studies, and there is the additional possibility of specifically modifying certain functionalities. A fairly large number of compounds has indeed been employed for thin film work, but not for all of these have detailed growth studies been performed. We will limit ourselves to only selected systems, largely based on examples from our own work (see Fig. 3).

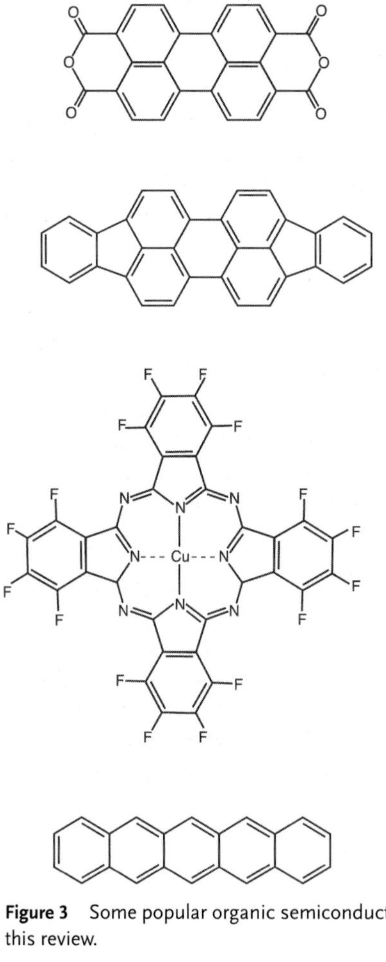

PTCDA

DIP

$F_{16}CuPc$

Pentacene

Figure 3 Some popular organic semiconductors discussed in this review.

1. PTCDA

   The perylene-derivative PTCDA (3,4,9,10-perylene-tetracarboxylic dianhy-dride, $C_{24}H_8O_6$, a red dye) has long been regarded as a model system for OMBD [2, 3, 20–27]. Its bulk structure (actually α and β phase) exhibits layered molecular planes, and it was expected that the regular stacking of these planes (along the [102] direction in α phase notation) is favourable for well-behaved film growth, which turned out to be not necessarily correct. The optical properties [28–32] as well as the vibrational properties [32–35] have been thoroughly studied.

2. DIP

   Diindeno(1,2,3,-cd,1′,2′,3′-lm)perylene ($C_{32}H_{16}$, DIP, a red dye) has the same perylene core as PTCDA, but it has been studied much less. It has recently been shown to exhibit excellent out-of-plane ordering behaviour [19, 36–38] and, associated with this, very good charge carrier transport properties [39], suggesting that DIP will be studied and exploited more in the future.

3. Phthalocyanines

   Phthalocyanines (Pc's) are rather popular [40–45], and some of the early work on OMBD has employed Pc's [40]. They exhibit a certain degree of "specific tunability", due to the possible central metal ion, which can be changed with-in a broad range, and due to the choice of the sidegroup(s) [41, 42]. $F_{16}CuPc$ is particularly attractive, since it is considered a good candidate as an n-type conducting organic material [46]. As a blue dye [44] it is also interesting for optoelectronic applications [45, 41].

4. Oligoacenes (anthracene, tetracene, pentacene)

   The oligoacenes and in particular pentacene have recently attracted consider-able attention, since their charge transport properties were reported to be excellent [7, 39]. An important feature of pentacene seems to be that it can be grown in well-ordered thin films, although the 'bulk structure' and a 'thin film structure' appear to be competing. It does not appear to be entirely clear whether or not there is indeed an identifyable special feature of pentacene that makes it superior to other compounds in terms of transport properties.

There are, of course, many other popular systems, which, however, we cannot dis-cuss due to the limited space. These include, e.g., oligothiophenes, oligophenyls and also 'sheets of graphite'. Besides the crystalline systems, there are also amorphous small-molecule organic semiconductors prepared by OMBD, such as Alq$_3$ and TPD. In terms of the growth physics, amorphous systems exhibit obviously some differ-ences (no strain due to epitaxy; different diffusion barriers; no crystallographic domains; etc.). They are worth studying in their own right, but we cannot discuss them here. For examples from various other systems we refer to Refs. [1–10].

1.3
Films on oxidized silicon

Silicon wafers are among the most common substrates for thin film growth. They are stable in air with their oxidised surface layer, the thickness of which can be 'tuned' by thermal oxidation (from some 15 Å (native oxide) to several 1000 Å). Also, they are very flat and relatively easy to clean.

In the context of organic electronics, of course, they are very popular as a substrates for thin-film transistors (TFTs), since the oxide can serve as the insulating layer between the silicon as the bottom contact (gate) and the active organic semiconductor on top.

We should also mention that oxidised silicon surfaces are suitable for surface modification using self-assembled monolayers (SAMs) [47, 48], which has been exploited, e.g., for the growth of pentacene [49].

1.3.1
PTCDA

It was expected that the regular stacking of PTCDA molecules in the [102] direction (in α phase notation) of the bulk structure would give rise to well-behaved film growth. This regular stacking is indeed observed on siliconoxide and many other substrates, unless the growth temperature is too low and no well-defined structure evolves or a too strong interaction with a very 'reactive' substrate leads to other orientations of the first PTCDA monolayer. However, it is important to realise that a regular stacking and well-defined orientation of the molecules within the films does not necessarily imply smooth surfaces.

In an early study, it was already found that PTCDA on oxidised silicon exhibits smooth surfaces only for growth at low temperatures ($T < 50\,°C$ for deposition rates around 1 Å/s), where the crystallinity was not very good [50]. For growth at higher temperatures, the films exhibited good crystallinity, but showed a tendency to island growth ('dewetting').

These results demonstrate a not uncommon feature of growth on substrates with low surface energies. If the films tend to dewet from the substrate near equilibrium, then the above pattern (relatively flat, but low-crystallinity films for low T, and dewetting (i.e. rough) morphologies with good crystallinity for high T) is quite frequently found.

1.3.2
DIP

DIP has the same perylene core as PTCDA, but the indeno endgroups instead of the anhydride endgroups give rise to a completely different structural behaviour compared to PTCDA. DIP has recently been studied in detail [19, 36–38, 51–53], and it was found to exhibit excellent out-of-plane order on siliconoxide surfaces.

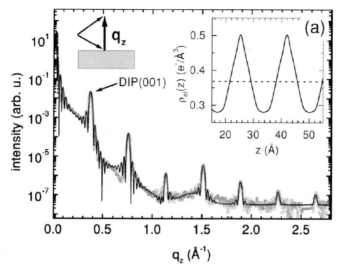

Figure 4 Specular X-ray scan of a 206 Å thick DIP film. Many higher-order Bragg reflections are observed, which can be used for the reconstruction of the electron density profile using the various Fourier components (close-up shown in the inset). From Ref. [36] with permission.

Films with various thicknesses ($69\,\text{Å} \leq D \leq 9000\,\text{Å}$) were prepared on oxidized (4000 Å) Si(100) substrates at a substrate temperature of $145 \pm 5°\text{C}$ and at a deposition rate of 12 ± 3 Å/min. The out-of-plane X-ray spectra exhibit well-defined Bragg reflections corresponding to a lattice spacing of $d_{DIP} \approx 16.55$ Å (suggesting essentially upright-standing molecules) and associated Laue oscillations, the analysis of which shows that the films are coherently ordered across the entire thickness [36]. The rocking width, which is a measure of the distribution of the out-of-plane lattice planes, is $0.01°$ and lower [36, 38]. The lattice spacing is consistent with a model of molecules standing essentially upright with a tilt angle θ_{tilt} presumably around $15°$-$20°$. The large number of higher-order Bragg reflections could be used to deconvolute the out-of-plane electron density distribution in a Fourier series (Fig. 4)

$$\rho_{\text{el}}(z) = \rho_0 + \sum_n A_n \cos\left(n\frac{2\pi}{d_{\text{DIP}}}z + \phi_n\right) \tag{1}$$

where the Fourier amplitude, A_n, is associated with the intensity of the nth Bragg reflection [36]. We can speculate that the shape of DIP with its slightly narrow head and tail may be favourable for an ordering mechanism with some degree of interdigitation of molecules from neighbouring (i.e., top and bottom) lattice planes.

On siliconoxide, the in-plane structure is, of course, a 2D powder. The packing appears to follow a herringbone motiv. The structure will be discussed also in the context of growth on Au (Ref. [37] and Sec. 1.5).

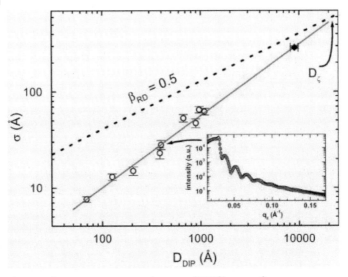

Figure 5 Root-mean-square roughness σ of DIP films as a function of thickness Δ_{DIP}. The inset shows a typical X-ray reflectivity dataset and a fit to the data. The solid line in the main plot is a linear fit to the data and the growth exponent is obtained as β = 0.748 0.05. The dotted line denotes the random deposition limit β_{RD} = 0.5. From Ref. [19] with permission.

The growth including the evolution of the HDCF and the associated growth exponents, α, β, and $1/z$, were studied using AFM and X-ray scattering (specular and diffuse) [19]. Whereas the static roughness exponent α (average of AFM and X-rays 0.684 ± 0.06) is similar to that observed in many other growth experiments [15], the values for $1/z$ (0.92 ± 0.20) and β (0.748 ± 0.05) were found to be rather large (Fig. 5). Specifically, the DIP films belong to the class of systems which display the phenomenon of *rapid roughening*, for which $\beta > 1/2$, i.e., the roughness increases faster with thickness D than the random deposition limit $\beta_{RD} = 0.5$ [16]. This effect appears hard to rationalise in the absence of a thermodynamic driving force (e.g., dewetting). A model which is consistent with the scaling exponents involves random spatial inhomogeneities in the local growth rate, which are fixed during the growth process [54, 19]. It is plausible that when certain regions of the surface persistently grow faster than others, the surface will roughen very rapidly. It was suggested that the spatial inhomogeneities may be related to the different tilt domains of the film and the inevitable grain boundaries in between these [19].

1.3.3
Phthalocyanines

Phthalocyanines also tend to grow in a standing-up configuration in thicker films on 'inert' substrates. Films of F16CuPc between 120 and 450 Å were recently found to exhibit very good crystalline out-of-plane order with rocking widths around 0.01°

and well-defined Kiessig interferences and Laue satellites around the out-of-plane Bragg reflection [55].

The in-plane structure is, of course, azimuthally disordered, since the substrate is isotropic. One of the complications for phthalocyanines is a strong anisotropy of the crystal structure and the associated growth properties, which can lead to needle-like features, both for F16CuPc [56] as well as for H16CuPc [57].

1.3.4
Pentacene

Pentacene on siliconoxide has been studied intensely due to its relevance for OFETs [7]. Ruiz et al. studied the initial stages of the growth [58]. Their analysis of the island distribution in (sub)monolayer films by dynamic scaling showed that the smallest stable island consists of four molecules. Meyer zu Heringdorf et al. showed that under appropriate growth conditions the single-crystal grain sizes can approach 0.1 mm [59].

For thicker films, pentacene thin films exhibit some complication in the sense that there is a 'thin film structure' and a 'bulk structure', which can coexist, depending on the growth conditions.

An interesting idea is that of surface modification involving self-assembled monolayers (SAMs) [48]. Shtein et al. studied the effects of film morphology and gate dielectric surface preparation on the electrical characteristics of organic-vapour-phase-deposited pentacene thin-film transistors including surface modification using SAMs [49]. Meyer zu Heringdorf et al. employed cyclohexene-saturation of Si(001) to modify the growth dynamics [59]. Voigt et al. studied the growth of tetracene on oil-covered surfaces [60]. While they actually used ITO as solid substrates, the concept might equally well be applicable to siliconoxide surfaces.

1.4
Films on aluminium oxide

Interfaces of organics with insulators are of obvious relevance for organic electronics, and aluminiumoxide is one of the most commonly used insulators. Unfortunately, sputtered aluminiumoxide layers frequently exhibit a rather high roughness and not well-defined starting conditions for growth studies. Sapphire is aluminiumoxide (Al_2O_3) in its purest and best ordered form. It is also a popular substrate for epitaxy of metals and inorganic semiconductors, and it can be obtained in very high crystalline quality. We will focus here on sapphire, since it is very suitable for model studies of the growth of organics on insulator surfaces (see Sec. 1.6 for other substrates).

Surfaces of ionic substrates, which are not charge balanced, tend to be unstable and/or exhibit strong relaxations/reconstructions. In the case of sapphire, the ($11\bar{2}0$) surface ('A plane') is charge balanced and rather inert, and it has been used for growth studies. An important feature to realise for surfaces of crystals is that they

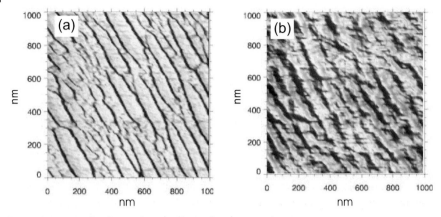

Figure 6 Topography of an A-plane sapphire substrate (a) and an F16CuPc film (120 Å) film (b) on this substrate determined by non-contact AFM. The step pattern of the substrate serves to azimuthally align the film (see text). From Ref. [64] with permission.

commonly exhibit a miscut, i.e. a difference between the physical surface and the (low-index) crystallographic plane. This gives rise to a step pattern, which in the case of essentially perfect crystals like sapphire, is the dominating feature of the surface morphology (Fig. 6). Issues related to the surface preparation have been discussed in Ref. [61].

1.4.1
PTCDA

PTCDA on sapphire has, to our knowledge, not been studied in detail. Test results, however, indicate that the overall behaviour is similar to that for PTCDA on oxidized silicon, i.e., that for growth at high temperatures the films tend to (partially) dewet [62].

The overall growth scenario is most likely not changed significantly by the presence of steps, but the in-plane order of PTCDA may be affected. However, even with alignment at the step edges PTCDA would most likely still exhibit multiple domains (see also the discussion of PTCDA on metals).

1.4.2
DIP

Based on the results for DIP on siliconoxide it is expected that DIP would also exhibit good out-of-plane ordering on the similarly 'inert' sapphire. Preliminary data indicate that this is, in fact, the case [63]. In addition, the stepped sapphire substrate can induce in-plane ordering, as first demonstrated for the growth of phthalocyanines [64] (see below), which was indeed also found for DIP [63].

1.4.3
Phthalocyanines

As described above, the regular surface steps associated with miscut sapphire can serve as templates for film growth with azimuthal alignment. While the concept of stepped substrates has been used frequently for monolayer adsorbates, its use for comparatively thick films (5 to 50 ML) of relatively large molecules was first demonstrated by Osso et al. for $F_{16}CuPc$ on A-plane sapphire [64]. The resulting azimuthal ordering has been shown by four methods sensitive to different aspects [64]. AFM was used to image the surface morphology of the bare substrate. After film growth, the characteristic step pattern of the bare substrate was shown to be basically replicated, suggesting an azimuthal coupling of the film structure to the substrate morphology (Fig. 6). In-plane X-ray diffraction (GIXD) showed that the crystal structure of the film was indeed not a 2D powder, but was aligned. The distribution width ('mosaicity' of the in-plane lattice) was several degrees broad, which suggests a rather weak driving force for the ordering. The in-plane order was also visible in the azimuthal intensity distribution of the vibrational modes detected by Raman scattering. Finally, the resulting anisotropy of the dielectric function was studied by spectroscopic ellipsometry, offering the chance to study the *'intrinsic'* behaviour of these systems without a strongly reduced disorder-induced broadening of the optical transitions. We should note that the strong optical anisotropies of these systems are an interesting field of study in their own right, and give rise to non-trivial effects in the propagation of light [44].

A discussion of the mechanisms responsible for the azimuthal alignment and in particular the possible competition between the effects of the step edges themselves (morphology) and the underlying lattice (epitaxial driving forces) is found in Ref. [56].

The out-of-plane ordering was similarly good as for F16CuPc or DIP on siliconoxide, i.e. a well-defined Bragg reflection with Laue oscillations and mosaicities around $0.01°$, although the tendency of phthalocyanines to grow in needles can cause some complications. We note that the tilt angle of the molecules as well as the out-of-plane lattice parameter was found to depend on the growth temperature (and are different from the bulk structure parameters), indicating that the structure may not be in full equilibrium.

1.4.4
Pentacene

Similar concepts and mechanisms as observed for DIP and $F_{16}CuPc$ in terms of azimuthal alignment should be applicable to pentacene on sapphire, but to our knowledge there are no published results yet.

1.5
Films on metals

Interfaces with metals are of obvious relevance for contacting organic semiconductors. The choice of the metal is frequently determined more by the desired work function and thus electron or hole injection properties than by growth considerations. Nevertheless, there is a wide variety of metals in terms of behaviour as substrates for organic thin film growth, and it is important to realise that this can have a profound impact on the growth and the resulting structural and functional properties. Besides issues related to the surface morphology, crystalline quality, potentially crystalline orientation and size of the unit cell (epitaxy), it is very important how 'reactive' or 'inert' the metal is, since this determines the mobility of the molecules on the surface and thus the growth.

For strongly reactive substrates, the molecules tend to behave almost in a 'hit-and-stick' fashion, i.e. without significant mobility and thus no long-range order. Less reactive metals such as noble metals, to which we will limit ourselves here, turned out to be rather popular and suitable.

We will concentrate on metal single crystals. From a practical point of view, for growth studies it is important that their surfaces can be 'recycled' by sputtering and annealing, i.e. several growth experiments can be performed on the same substrate and on (essentially) the same surface. Less reactive metals are also easier to keep clean before growth. Obviously, with metal substrates the application of electron-based surface science methods is possible, since the signal does not suffer from charging effects. This has been used excessively by the surface science community in particular for molecular monolayers on surfaces of metal single crystals.

We should also mention that metal surfaces are suitable for surface modification using self-assembled monolayers (SAMs) [47, 48], which has been employed in particular for Au(111). Examples include the growth of PTCDA on alkanethiol SAMs [65–68].

1.5.1
PTCDA

PTCDA on metal surfaces has been thoroughly studied, with the noble metals being particularly popular.

1.5.1.1 Structure and epitaxy of PTCDA/Ag(111)

On Ag(111), very well-defined epitaxial growth of PTCDA(102) has been observed [2, 20, 24, 26]. The 2D structure is characterised by a herringbone arrangement of the flat-lying molecules, which corresponds to a layer of the (102) plane of the bulk structure, with a small degree of distortion (strain). Possible mechanisms leading to the well-behaved 2D structure of PTCDA on Ag(111) were recently discussed in Ref. [69]. The vertical PTCDA-Ag(111) spacing was found to be 2.8 ... 2.9 Å, depending on T, based on X-ray diffraction [25], but it may differ for low-temperature deposition if the adsorption state differs.

For growth extending beyond the monolayer, a more complex azimuthal distribution arises, and, depending on the growth temperature, also domains non-collinear with principal axes of the substrate can form to relief strain [24]. Interestingly, the epitaxial relations could be rationalised similar to the Nishiyama-Wassermann vs. Kurdjumov-Sachs relations for fcc(111)/bcc(110) growth, although, of course, the PTCDA structure is not bcc [24].

1.5.1.2 Comparison with other substrates

The comparison with PTCDA/Au(111) yields a qualitatively similar picture [22, 27, 30], although details of the epitaxy appear to differ, which is not too surprising given that the structure is a result of a rather delicate balance of different factors and given that the corrugation of the substrate potential experienced by PTCDA is different. Of course, Au(111) exhibits a reconstruction, and also the substrate-molecule interaction may differ slightly if the overlap of the outermost orbitals is different.

On the more open Ag(110) surface, an entirely different structure was found already in the monolayer, characterised by a 'brick-stone' arrangement, [20]. Phase transitions of PTCDA/Ag(110) were studied in Ref. [71].

Growth on Cu(110) was studied by Möller's group [72–74]. The monolayer was found to differ from those known from other substrates. For thicker films, Stranski-Krastanov growth was found, similar to the case on Ag(111) (see below).

1.5.1.3 Dewetting and thermal properties

While the structure and epitaxy in the monolayer regime are well-defined, the later stages of the growth (potentially after a certain threshold thickness) can, of course, exhibit islanding and a very rough resulting morphology. It was recently found that PTCDA on Ag(111), a well-behaved system in the monolayer regime, exhibits indeed Stranski-Krastanov growth. At growth temperatures $T \lesssim 350$ K, relatively smooth epitaxial films have been found, whereas at $T \gtrsim 350$ K, well-separated crystallites with bulk crystalline structure on top of a 2 ML thick wetting layer have been observed [24, 25, 75, 76]. These results are qualitatively the same as those for PTCDA on Au(111) [22].

The thermally-induced post-growth dewetting of 'low-temperature' grown films was also studied, confirming that the films tend to dewet if given sufficient thermal energy [25]. In these experiments, also the thermal expansion of PTCDA was determined ($1.06 \pm 0.06 \times 10^{-4}$ K^{-1} out-of-plane) [25]. For a comparison with other systems (Alq$_3$ and TPD), see Ref. [77]. While islanding of the films is usually not desirable, it should be pointed out there might also be ways to exploit islanding or dewetting and the formation of small crystallites for 'self-organised nanostructures' (similar to Si-Ge quantum dots).

1.5.1.4 Real-time growth

In order to shed light on the dynamics and the temperature dependence of the 2D-3D transition (layer-by-layer to islanding), a real-time X-ray diffraction study of the growth of PTCDA on Ag(111) was performed [76]. The idea is as follows (Fig. 7). In

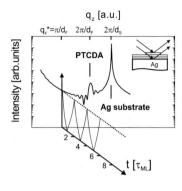

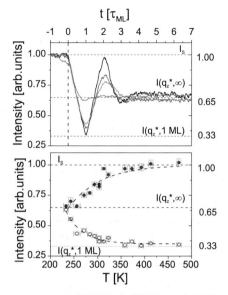

Figure 7 Simulation of the specular rod scattering of a thin PTCDA film on Ag(111) as a function of the out-of-plane momentum transfer q_z (top left). The time axis in this figure (for a fixed q_z) indicates the intensity oscillations at the anti-Bragg point during growth (see text for explanation). The top right figure shows the time dependence (in units of monolayer deposition times) of the normalised scattered intensity at the anti-Bragg point for various temperatures (233 K (red); 283 K (green); 303 K (blue); 358 K (black)). The bottom figure shows the temperature dependence of the deviation from layer-by-layer growth expressed in terms of the intensity of the minimum (open symbols) and of the maximum (filled symbols) of the scattered intensity at 1 ML and 2 ML, respectively (see text for details). From Ref. [76] with permssion.

kinematic theory the specular x-ray scattering intensity is the sum of the scattering contributions from the film and the substrate,

$$I(q_z, t) = |F(q_z, t)|^2 = \left| f_F \sum_{n=1}^{\infty} e^{iq_z d_F (n-1)} \theta_n(t) + f_S \frac{1}{1 - e^{-iq_z d_0}} e^{-iq_z d_0} \right|^2 . \qquad (2)$$

f_F and f_S are the form factors of the film and the substrate, d_F and d_S are the corresponding lattice spacings, and $d_0 = 2.8$ Å is the distance between the substrate and the first layer of the film [76]. $\theta_n(t)$ is the time-dependent fractional coverage of the nth layer within the organic film. At the anti-Bragg point of the PTCDA film ($q_z^* = \pi/d_F$) the first term in Eq. 2 equals $f_F \sum (-1)^{(n-1)} \theta_n(t)$. Therefore, the coverage difference

$$\Delta\theta(t) = \sum_m \theta_{2m+1}(t) - \sum_m \theta_{2m}(t) = \theta_{\text{odd}}(t) - \theta_{\text{even}}(t) \qquad (3)$$

can be deduced from the measured intensity $I(q_z^*, t)$. Specifically, it is possible to distinguish the coverage of the first and the second layer in the initial stage of the growth. In the case of layer-by layer growth, characteristic intensity oscillations are observed.

Fig. 7 shows typical time-dependent intensity data during growth in a dedicated chamber [78], measured at various substrate temperatures between 233 and 258 K [76]. $t = 0$ is defined as the starting time of the deposition. The signal is normalized to the substrate scattering intensity, $I_S = I(q_z^*, t < 0)$, and the time is normalized to the deposition time, τ_{ML}, of one monolayer, which corresponds to the intensity minimum. A typical growth measurement exhibits distinct intensity oscillations for $t \lesssim 3\,\tau_{ML}$, followed by a constant intensity during further deposition, similar to the observations for PTCDA/Au(111) [79]. The intensity oscillations correspond to layer-by-layer growth. The transition to a constant intensity indicates the breakdown of layer-by-layer growth and the onset of islanding characteristic of SK growth. As can be seen from the transition to a time-independent scattering signal (associated with an equal probability for a given molecule to be accomodated in even and odd layers), the islanding starts rapidly after completion of a 2 ML "wetting" layer.

Comparing the growth data for different temperatures (Fig. 7), we find that for $T \geq 358$ K the oscillations are not visibly damped for $t < 2\tau_{ML}$. They are followed by a sharp transition to a time-independent intensity (islanding). For lower temperatures, the oscillations are progressively damped, and the 2D-3D transition is smeared out as the temperature is lowered.

The experimental data, i.e., in particular the 2D-3D transition, could be modelled by kinetic Monte-Carlo simulations using a relatively simple model for the energies/barriers, the most important feature of which is the dependence of the interlayer transport barrier, E_{inter}, on the layer number n, namely $E_{inter}(n \leq 3) = 0$ and $E_{inter}(n > 3) > 0$ [76].

Moreover, for elevated temperatures strong post-growth diffusion was observed [80].

1.5.2
DIP

In the monolayer regime, DIP as many other organic semiconductors, was studied by STM. The molecules were found to be lying down flat on the substrate [81]. The interaction of DIP with Au was found to be physisorptive [37]. In the regime of thicker films, DIP was recently studied in detail *on* Au contacts [37] (and as substrate for Au contacts evaporated *on* DIP [37, 51–53]). Importantly, in contrast to growth on siliconoxide, due to the stronger interaction with the Au substrate, the lying-down configuration tends to prevail not only for monolayers, but also for multilayers. Since the standing-up configuration (which again followed a herringbone-like motiv) appears to have the more favourable surface energy (as seen on siliconoxide), there is obviously a competition between the two configurations (standing-up vs. lying-down), and they are found to coexist [37]. From the point of view of growth kinetics this competition is very interesting, but it is certainly a further complication and an additional source of disorder which is usually undesirable.

1.5.3
Phthalocyanines

Phthalocyanines were among the first 'large' molecules that were studied by STM with (sub)molecular resolution [40]. In the monolayer regime, the molecules are lying down flat on the surface, and the 2D structures have been thoroughly studied. Recently, the (vertical) bonding distance to the metal substrate was determined using XSW [82]. For thicker films, there is a competition between the lying-down configuration of the first layer and the tendency to stand up. The impact of roughness on the ordering behaviour was studied in Ref. [83].

1.5.4
Pentacene

Acenes on metal substrates was studied by several groups. Early work on the orientation of various aromatic hydrocarbons including tetracene on metal surfaces using NEXAFS was done by Koch and collaborators [84].

More recent work focussed on pentacene. Pentacene structures on Au(111) as a function of coverage (up to the equivalent of around 2 ML) were studied by Parkinson's group [85]. In the monolayer regime of pentacene on Cu(110), Lukas et al. reported a novel mechanism giving rise to long-range order on Cu(110), based on the modulation of the adsorption energy due to charge-density waves related to a surface state [86].

While it is not too surprising that the molecules in the monolayer regime tend to be lying more or less flat on the surface, importantly, for the growth of thicker films on Cu(110) an orientational transition from a lying-down configuration to an essentially standing-up configuration was observed [87].

An interesting study of the 'hyperthermal' growth of pentacene (exhibiting hyperthermal energies in a seeded supersonic molecular beam) on Ag(111) was presented by Casalis et al. [88]. They found that at low substrate temperatures (200 K) highly ordered films can be grown by hyperthermal deposition when thermal deposition leads only to disordered films. The results were interpreted as a result of 'local annealing' due to the impact of the hyperthermal molecules. This technique appears to have the potential to tailor the growth of molecular systems in addition to what is possible by changing the impingement rate and the substrate temperature, and it may be further tested in the future.

1.6
Films on other substrates

Many other substrates than the above have been employed, which we cannot all review. We shall only mention some of the most important other substrates.

Quite popular for growth studies is graphite, since it is easy to prepare. In our general classification of substrates, graphite would be 'weakly interacting', and, e.g.,

PTCDA would probably also exhibit (partial) dewetting. Other examples from this group of layered substrates are MoS_2, GeS, and Sb_2S_3 [4].

Also rather weakly interacting would be MgO, which falls essentially in the same category as sapphire and siliconoxide. Mica, which can be easily prepared by cleavage, may also be seen in the category of rather inert substrates.

Alkalihalogenides, such as NaCl and KCl, are quite popular as simple substrates for growth studies, since they are easy to prepare. For some studies, they have the additional benefit that they can be easily dissolved and the film can be studied by TEM.

Metals, as indicated above, span a broad range from the noble metals to very reactive substrates.

A very important class of substrates are certainly (inorganic) semiconductors such as Si and GaAs, since they may be used in the integration of organic-inorganic hybrid devices. Moreover, they are very well-defined in terms of their surface and overall structural quality, which is favourable for growth studies. If the surface is clean, however, they can exhibit 'dangling bonds' and be rather reactive. In these cases, the organic adsorbates then tend to 'hit and stick', i.e. they usually do not diffuse over significant distances, hence they do not form long-range ordered structures. A strategy to avoid these problems, but still benefit from the above advantages, is the use of surface-passivated semiconductors, such as H-Si or Se-GaAs.

Polymeric substrates and possible routes for oriented growth of pentacene have been studied in Ref. [89].

1.7
More complex heterostructures and technical interfaces

Organic semiconductor devices frequently do not only consist of a film on a substrate, but involve additional layers such as metal contacts or insulating layers or also different organic components in a multilayer structure.

Metal contacts are one obvious requirement for many applications of organic semiconductors. It turns out that the controlled deposition of metals *on* organics ('top electrode') is non-trivial. In order to reduce problems related to interdiffusion (and ultimately short-circuiting) and traps related to surface states, different strategies can be pursued.

1. Deposition at low temperatures to 'freeze in' the interdiffusion;
2. Deposition at (moderately) high rates with the idea that the metals are quickly forming larger aggregates which are then less mobile and diffuse less deep into the organic film;
3. Use of 'suitably reactive' metals and/or organics, so that a strong interaction at the top layer(s) of the organic material prevents interdiffusion;
4. 'Soft deposition' by 'thermalising' or at least reducing the energy of the impinging metal atoms by 'baffling' these using a noble gas or other means [90];

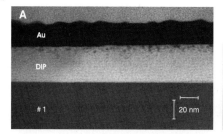
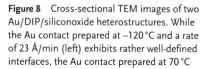
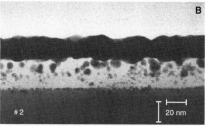

Figure 8 Cross-sectional TEM images of two Au/DIP/siliconoxide heterostructures. While the Au contact prepared at −120 °C and a rate of 23 Å/min (left) exhibits rather well-defined interfaces, the Au contact prepared at 70 °C and a rate of 0.35 Å/min (right) shows strong interdiffusion. Note that individual lattice planes of the DIP film can be resolved. From Ref. [51] with permission.

5. Miscellaneous other non-thermal deposition strategies including, e.g., electrochemical deposition may be attempted.

Recently, we performed studies of the deposition of gold, which is widely used as a hole injection material, onto well-defined DIP thin film surfaces to study the interdiffusion (Fig. 8). The study followed the 'classical' approach without specific precaution against interdiffusion except for variation of the temperature and the rate [37, 38, 51–53]. The important result was that under rather typical typical deposition conditions near room temperature the metal interdiffusion was already significant, and the layers would exhibit electrical shorts (Fig. 8). If the substrate, however, is cooled, fairly well-defined interfaces could be obtained. We note that Faupel's group studied similar issues in detail for metal deposition on polymers [91].

Recently, Sellner et al. have studied aluminium-oxide/DIP interfaces, which apparently exhibit a very different interdiffusion behaviour and which can very effectively encapsulate the organic film and enhance its thermal stability [92].

We should note that for device structures one also has to take into account the effect of the morphology of technical interfaces and surfaces on the growth behaviour of organics [83].

Another important interface, which has not been excessively studied with regard to growth and structure, is the organic-organic interface as found, e.g., in OLEDs. Some early work on superlattices and bilayers can be found, e.g., in Refs. [93–96]. PTCDA on self-assembled monolayers (SAMs) as well-defined organic model surfaces has been studied in Refs. [65–68]. PTCDA on hexa-peri-hexabenzocoronene (HBC) was investigated in Ref. [97]. A number of different polynuclear aromatic hydrocarbons including DIP and perylene were studied by Kobayashi's group [98]. Other studies were concerned with the post-growth stability of the organic-organic interface and the interdiffusion behaviour [77, 99].

1.8
Summary and Conclusions

This review does not claim to be complete in any way. We have rather presented a few case studies, which we hope serve to highlight a few of the issues specific to the growth of organic thin films. We shall summarise some of these.

1. Epitaxial relations can be complicated, and the films can exhibit a large number of symmetry-equivalent domains. Moreover, the coexistence of different phases can give rise to complications.
2. Islanding (after some critical thickness) is not uncommon, and is, of course, not prevented by well-defined structural relation between film (or the first monolayer) and substrate.
3. Even for systems that tend to 'wet' the substrate, overproportional roughening may occur, and the growth exponents may be very different from those expected based on conventional theories.
4. The controlled preparation of organics-based heterostructures can be particularly difficult, given the tendency for interdiffusion of, e.g., metal contacts.

Despite these in some regard 'additional complications' of organics well-ordered thin films can be grown by OMBD. We hope the improvement of the understanding of the growth mechanisms will further promote the applications of organics.

Moreover, organics with their specific features promise to give rise to fundamentally new growth phenomena such as orientational transitions and new universality classes (scaling exponents) which is an exciting subject in its own right.

Acknowledgements

We wish to thank the many students, collaborators, and colleagues who contributed in various ways to the work reviewed here.

We are particularly grateful to H. Dosch. Valuable comments on the manuscript were made by A. Gerlach, S. Kowarik, and J. Pflaum. We wish to acknowledge financial support from the Max-Planck-Gesellschaft (Germany), the Deutsche Forschungsgemeinschaft (Focus Programme 'Organic Field-Effect Transistors', Germany), the EPSRC (UK), and the University of Oxford (UK).

References

[1] A. Koma, Prog. Crystal Growth Charact. **30**, 129 (1995).

[2] E. Umbach, S. Sokolowski, and R. Fink, Appl. Phys. A **63**, 565 (1996).

[3] S. R. Forrest, Chem. Rev. **97**, 1793 (1997).

[4] N. Karl and C. Günther, Cryst. Res. Technol. **34**, 243 (1999).

[5] D. E. Hooks, T. Fritz, and M. D. Ward, Adv. Mater. **13**, 227 (2001).

[6] J. Fraxedas, Adv. Mater. **14**, 1603 (2002).

[7] C. D. Dimitrakopoulos and P. R. L. Malenfant, Adv. Mater. **14**, 99 (2002).

[8] F. Rosei, M. Schunack, Y. Naitoh, P. Jiang, A. Gourdon, E. Laegsgaard, I. Stensgaard, C. Joachim, and F. Besenbacher, Prog. Surf. Sci. **71**, 95 (2003).

[9] S. M. Barlow and R. Raval, Surf. Sci. Rep. **50**, 201 (2003).

[10] G. Witte and C. Wöll, J. Mater. Res. **19** (2004), 1889.

[11] M. O. Lorenzo, C. J. Baddeley, C. Muryn, and R. Raval, Nature **404**, 376 (2000).

[12] Q. Shen, D. Frankel, and N. V. Richardson, Surf. Sci. **497**, 37 (2002).

[13] R. L. Toomes, J. H. Kang, D. P. Woodruff, M. Polcik, M. Kittel, and J. T. Hoeft, Surf. Sci. Lett. **522**, L9 (2003).

[14] A.-L. Barabäsi and H. E. Stanley, Fractal Concepts in Surface Growth (Cambridge University Press, Cambridge, 1995).

[15] J. Krim and G. Palasantzas, Int. J. Mod. Phys. B **9**, 599 (1995).

[16] J. Krug, Adv. Phys. **46**, 139 (1997).

[17] A. Pimpinelli and J. Villain, Physics of crystal growth (Cambridge University Press, Cambridge, 1999).

[18] F. Biscarini, P. Samori, O. Greco, and R. Zamboni, Phys. Rev. Lett. **78**, 2389 (1997).

[19] A. C. Dürr, F. Schreiber, K. A. Ritley, V. Kruppa, J. Krug, H. Dosch, and B. Struth, Phys. Rev. Lett. **90**, 016104 (2003).

[20] K. Glöckler, C. Seidel, A. Soukopp, M. Sokolowski, E. Umbach, M. Böhringer, R. Berndt, and W.-D. Schneider, Surf. Sci. **405**, 1 (1998).

[21] M. Möbus, N. Karl, and T. Kobayashi, J. Cryst. Growth **116**, 495 (1992).

[22] P. Fenter, F. Schreiber, L. Zhou, P. Eisenberger, and S. R. Forrest, Phys. Rev. B **56**, 3046 (1997).

[23] C. Seidel, C. Awater, X. D. Liu, R. Ellerbrake, and H. Fuchs, Surf. Sci. **371**, 123 (1997).

[24] B. Krause, A. C. Dürr, K. Ritley, F. Schreiber, H. Dosch, and D. Smilgies, Phys. Rev. B **66**, 235404 (2002).

[25] B. Krause, A. C. Dürr, F. Schreiber, H. Dosch, and O. H. Seeck, J. Chem. Phys. **119**, 3429 (2003).

[26] B. Krause, A. C. Dürr, K. A. Ritley, F. Schreiber, H. Dosch, and D. Smilgies, Appl. Surf. Sci. **175**, 332 (2001).

[27] T. Schmitz-Hübsch, T. Fritz, F. Sellam, R. Staub, and K. Leo, Phys. Rev. B **55**, 7972 (1997).

[28] M. Hoffmann, K. Schmidt, T. Fritz, T. Hasche, V. M. Agranovich, and K. Leo, Chem. Phys. **258**, 73 (2000).

[29] M. Leonhardt, O. Mager, and H. Port, Chem. Phys. Lett. **313**, 24 (1999).

[30] I. Vragović, R. Scholz, and M. Schreiber, Europhys. Lett. **57**, 288 (2002).

[31] M. I. Alonso, M. Garriga, N. Karl, J. O. Ossó, and F. Schreiber, Org. Electron. **3**, 23 (2002).

[32] R. Scholz, Habilitation Thesis, Chemnitz (2003).

[33] R. Scholz, A. Y. Kobitski, T. U. Kampen, M. Schreiber, D. R. T. Zahn, G. Jungnickel, M. Elstner, M. Sternberg, and T. Frauenheim, Phys. Rev. B **61**, 13659 (2000).

[34] F. S. Tautz, S. Sloboshanin, J. A. Schaefer, R. Scholz, V. Shklover, M. Sokolowski, and E. Umbach, Phys. Rev. B **61**, 16933 (2000).

[35] V. Wagner, T. Muck, J. Geurts, M. Schneider, and E. Umbach, Appl. Surf. Sci. **212**, 520 (2003).

[36] A. C. Dürr, F. Schreiber, M. Münch, N. Karl, B. Krause, V. Kruppa, and H. Dosch, Appl. Phys. Lett. **81**, 2276 (2006).

[37] A. C. Dürr, N. Koch, M. Kelsch, A. Rühm, J. Ghijsen, R. L. Johnson, J. J. Pireaux, J. Schwartz, F. Schreiber, H. Dosch, and A. Kahn, Phys. Rev. B **68**, 115428 (2003).

[38] A. C. Dürr, F. Schreiber, M. Kelsch, and H. Dosch, Ultramicroscopy **98**, 51 (2003).

[39] N. Karl, in: Organic Electronic Materials, edited by R. Farchioni and G. Grosso (Springer, Berlin, 2001), Vol. 2.

[40] P. H. Lippel, R. J. Wilson, M. D. Miller, C. Wöll, and S. Chiang, Phys. Rev. Lett. **62**, 171 (1989).

[41] C. C. Leznoff and A. B. P. Lever, Phthalo-cyanines: Properties and Applications (VCH Publishers Inc., New York, 1989).

[42] D. Schlettwein, in: Supramolecular Photosensitive and Electroactive Materials, edited by H. S. Nalwa (Academic Press, San Diego, 2001).

[43] S. Yim and T. S. Jones, Surf. Sci. **521**, 151 (2002).

[44] M. I. Alonso, M. Garriga, J. O. Ossó, F. Schreiber, E. Barrena, and H. Dosch, J. Chem. Phys. **119**, 6335 (2003).

[45] J. A. Rogers, Z. Bao, A. Dodabalapur, and A. Makhija, IEEE Electron Device Lett. **21**, 100 (2000).

[46] Z. Bao, A. J. Lovinger, and J. Brown, J. Am. Chem. Soc. **120**, 207 (1998).

[47] F. Schreiber, J. Phys.: Condens. Matter, **16** (2004), R 881.

[48] F. Schreiber, Prog. Surf. Sci. **65**, 151 (2000).

[49] M. Shtein, J. Mapel, J. B. Benzinger, and S. R. Forrest, Appl. Phys. Lett. **81**, 268 (2002).

[50] M. Möbus and N. Karl, Thin Solid Films **215**, 213 (1992).

[51] A. C. Dürr, F. Schreiber, M. Kelsch, H. D. Carstanjen, and H. Dosch, Adv. Mater. **14**, 961 (2002).

[52] A. C. Dürr, F. Schreiber, M. Kelsch, H. D. Carstanjen, H. Dosch, and O. H. Seeck, J. Appl. Phys. **93**, 5201 (2003).

[53] N. Koch, A. C. Dürr, J. Ghijsen, R. L. Johnson, J. J. Pireaux, J. Schwartz, F. Schreiber, H. Dosch, and A. Kahn, Thin Solid Films **441**, 145 (2003).

[54] J. Krug, Phys. Rev. Lett. **75**, 1795 (1995).

[55] J. O. Ossó, F. Schreiber, M. Garriga, M. I. Alonso, E. Barrena, and H. Dosch, Org. Electron. **5** (2004), 135.

[56] E. Barrena, J. O. Ossó, F. Schreiber, M. I. Alonso, M. Garriga, and H. Dosch, J. Mater. Res., **19** (2004), 2061.

[57] S. Kowarik et al., in preparation.

[58] R. Ruiz, B. Nickel, N. Koch, L. C. Feldmann, R. Haglund, Jr., A. Kahn, F. Family, and G. Scoles, Phys. Rev. Lett. **91**, 136102 (2003).

[59] F.-J. Meyer zu Heringdorf, M. C. Reuter, and R. M. Tromp, Nature **412**, 517 (2001).

[60] M. Voigt, S. Dorfsfeld, A. Volz, and M. Sokolowski, Phys. Rev. Lett. **91**, 026103 (2003).

[61] T. Becker, A. Birkner, G. Witte, and C. Wöll, Phys. Rev. B **65**, 115401 (2002).

[62] B. Krause et al., unpublished.

[63] J. O. Ossó et al., in preparation.

[64] J. O. Ossó, F. Schreiber, V. Kruppa, H. Dosch, M. Garriga, M. I. Alonso, and F. Cerdeira, Adv. Funct. Mater. **12**, 455 (2002).

[65] F. Schreiber, M. C. Gerstenberg, H. Dosch, and G. Scoles, Langmuir **19**, 10004 (2003).

[66] F. Schreiber, M. C. Gerstenberg, B. Edinger, B. Toperverg, S. R. Forrest, G. Scoles, and H. Dosch, Phys. Condens. Matter **283**, 75 (2000).

[67] M. C. Gerstenberg, F. Schreiber, T. Y. B. Leung, G. Bracco, S. R. Forrest, and G. Scoles, Phys. Rev. B **61**, 7678 (2000).

[68] R. Staub, M. Toerker, T. Fritz, T. Schmitz-Hübsch, F. Sellam, and K. Leo, Surf. Sci. **445**, 368 (2000).

[69] M. Eremtchenko, J. Schaefer, and F. S. Tautz, Nature **425**, 602 (2003).

[70] I. Chizhov, A. Kahn, and G. Scoles, J. Cryst. Growth **208**, 449 (2000).

[71] C. Seidel, J. Poppensieker, and H. Fuchs, Surf. Sci. **408**, 223 (1998).

[72] M. Stöhr, M. Gabriel, and R. Möller, Europhys. Lett. **59**, 423 (2002).

[73] M. Stöhr, M. Gabriel, and R. Möller, Surf. Sci. **507–510**, 330 (2002).

[74] M. Stöhr, M. Gabriel, and R. Möller, Appl. Phys. A **74**, 303 (2002).

[75] L. Chkoda, M. Schneider, V. Shklover, L. Kilian, M. Sokolowski, C. Heske, and E. Umbach, Chem. Phys. Lett. **371**, 548 (2003).

[76] B. Krause, F. Schreiber, H. Dosch, A. Pimpinelli, and O. H. Seeck, Europhys. Lett. **65**(3), 372 (2004).

[77] P. Fenter, F. Schreiber, V. Bulovic, and S. R. Forrest, Chem. Phys. Lett. **277**, 521 (1997).

[78] K. A. Ritley, B. Krause, F. Schreiber, and H. Dosch, Rev. of Sci. Instrum. **72**, 1453 (2001).

[79] P. Fenter, P. Eisenberger, P. Burrows, S. R. Forrest, and K. S. Liang, Physica B **221**, 145 (1996).

[80] B. Krause, A. C. Dürr, F. Schreiber, H. Dosch, and O. H. Seeck, Surf. Sci. **572** (2004) 385.

[81] R. Strohmaier, PhD thesis, Stuttgart (1997).

[82] A. Gerlach, F. Schreiber, S. Sellner et al., Phys. Rev. B, in press (2005).

[83] H. Peisert, T. Schwieger, J. M. Auerhammer, M. Knupfer, M. S. Golden, J. Fink, P. R. Bressler, and M. Mast, J. Appl. Phys. **90**, 466 (2001).

[84] P. Yannoulis, R. Dudde, K. H. Frank, and E. E. Koch, Surf. Sci. **189/190**, 519 (1987).

[85] C. B. France, P. G. Schroeder, J. C. Forsythe, and B. A. Parkinson, Langmuir **19**, 1274 (2003).

[86] S. Lukas, G. Witte, and C. Wöll, Phys. Rev. Lett. **88**, 028301 (2002).

[87] S. Lukas, S. Söhnchen, G. Witte, and C. Wöll, Chem. Phys. Chem. (2004), **5** (2004), 266.

[88] L. Casalis, M. F. Danisman, B. Nickel, G. Bracco, T. Toccoli, S. Iannotta, and G. Scoles, Phys. Rev. Lett. **90**, 206101 (2003).

[89] M. Brinkmann, S. Graff, C. Straupe, J.-C. Wittmann, C. Chaumont, F. Nuesch, A. Aziz, M. Schaer, and L. Zuppiroli, J. Phys. Chem. B **107**, 10531 (2003).

[90] N. Okazaki and J. R. Sambles, Extended Abstr. Int. Symp. Org. Mol. Electron. (Nagoya, Japan), 18–19 May 2000, pp. 66–67.

[91] F. Faupel, R. Willecke, and A. Thran, Mat. Sci. Eng. R **22**, 1 (1998).

[92] S. Sellner, A. Gerlach, F. Schreiber, M. Kelsch, N. Kasper, H. Dosch, S. Meyer, J. Pflaum, M. Fischer, and B. Gompf, Adv. Mater. **16** (2004), 1750.

[93] F. F. So, S. R. Forrest, Y. Q. Shi, and W. H. Steier, Appl, Phys. Lett. **56**, 674 (1990).

[94] M. L. Anderson, V. S. Williams, T. J. Schuerlein, G. E. Collins, C. D. England, L.-K. Chau, P. A. Lee, K. W. Nebesny, and N. R. Armstrong, Surf. Sci. **307–309**, 551 (1994).

[95] H. Akimichi, T. Inoshita, S. Hotta, H. Noge, and H. Sakaki, Appl. Phys. Lett. **63**, 3158 (1993).

[96] T. Nonaka, Y. Mori, N. Nagai, Y. Nakagawa, M. Saeda, T. Takahagi, and A. Ishitani, Thin Solid Films **239**, 214 (1994).

[97] F. Sellmann, T. Schmitz-Hübsch, M. Toerker, S. Mannsfeld, H. Proehl, T. Fritz, K. Leo, C. Simpson, and K. Müllen, Surf. Sci. **478**, 113 (2001).

[98] A. Hoshino, S. Isoda, and T. Kobayashi, J. Cryst. Growth **115**, 826 (1991).

[99] S. Heutz, G. Salvan, T. S. Jones, and D. R. T. Zahn, Adv. Mater. **15**, 1109 (2003).

[100] Y. Li, G. Wang, J. A. Theobald, and P. H. Beton, Surf. Sci. **537**, 241 (2003).

2
Electronic Properties of Interfaces between Model Organic Semiconductors and Metals

M. Knupfer and H. Peisert

2.1
Introduction

The fundamental properties and the application of organic materials in electronic and optoelectronic devices have been an issue of intensive research and development since many years (e.g. Refs. [1–5]). The potential devices comprise organic light-emitting diodes, organic photovoltaic cells and organic transistors as well as combinations of these as, for instance, active organic matrix diplays that are driven by organic transistors (e.g. Refs. [2,4,6–12]). These devices are anticipated to enable flexible, low cost electronics. While many aspects of the performance of organic devices are quite well understood, there are properties which are still subject of research worldwide. One of the open issues are the structural and electronic properties of interfaces between the various organic and inorganic components of the devices [13–15].

The importance of interfaces for the device performance cannot be overestimated as they determine charge injection and charge flow in the devices [13,15]. Numerous studies of thin films of organic molecular semiconductors on metals have been performed in order to investigate the electronic properties of various organic/electrode interfaces. The work of several research groups all over the world [e.g. 13,14,16–29] demonstrated the breakdown of the assumption of a common vacuum level at these interfaces. Although the alignment of the chemical potentials at organic/metal interfaces is now clear for many cases, the interface electronic structures are often discussed ignoring the possible formation of an interfacial electric dipole layer. Moreover, not only the knowledge of barrier heights between organic solids and metals is of enormous importance for an understanding and improvement of organic semiconductor devices, the type of the interaction at the interface – physisorption or chemisorption – can also significantly affect the device performance. Indeed, a charge transfer or chemical reaction was found for organic molecules in contact with several potential electrode materials [13,15,22,30–38], a situation which is of tremendous importance for the electronic characterization and/or simulation of the related devices, since resulting localized electronic states may represent charge carrier traps at these interfaces [39,40].

Physics of Organic Semiconductors. Edited by W. Brütting
Copyright © 2005 WILEY-VCH Verlag GmbH & Co. KGaA, Weinheim
ISBN 3-527-40550-X

In addition, it has been shown that the ordering and orientation of the organic molecules can also be crucial for the device efficiency and it has been demonstrated that the device characteristics are dependent upon the anisotropy of the transport properties in the organic thin film [41]. A model system for the entire class of low molecular weight, organic molecules is the family of the phthalocyanines (Pc's) and their fluorinated relatives (see Fig. 1). They represent one of the most promising candidates for ordered organic thin films, as these systems possess advantageous attributes such as chemical stability and excellent film growth which can result in optimized electronic properties [42,43]. Whereas thin phthalocyanine films are known as p-type materials [44] in air, it has been shown recently that perfluorinated copper phthalocyanine may be favorable as n-channel material for organic electronics [45]. Especially for the realization of complementary logic circuits both p- and n-channel materials are required.

On the other hand, known since the earliest work on organic light emitting diodes [6], devices are often based on amorphous or polycrystalline organic thin films and until now they have reached a remarkable performance. In this context, the organic molecules TPD (N,N′-diphenyl-N,N′-bis (3-methylphenyl) – (1,1′) –biphenyl-4,4′diamine) and Alq$_3$ (8-tris-hydroxyquinoline aluminium) can be regarded as model substances for hole transport layers (HTL) and electron transport layers (ETL) or emission layers (EML) in organic light emitting diodes, respectively (for their structure see Fig. 1), and thus the electronic structure of the related electrode/organic and organic/organic interfaces has been investigated in detail [e.g. 13,14,22,46].

In this review we summarize recent experimental studies of the orientation of copper phthalocyanines that were deposited on a number of different electrode or gate materials, and of the energy level alignment at the resulting interfaces. The main experimental techniques used were x-ray absorption spectroscopy and photoemission spectroscopy.

2.2
Experimental methods

For the investigation of the chemistry and electronic properties of interfaces, X-ray photoemission spectroscopy (XPS) and ultraviolet photoemission spectroscopy (UPS) are valuable tools. In particular, photoemission spectroscopy (PES) is the most important technique for the determination of band offsets at semiconductor heterojunctions and at metal/semiconductor interfaces [47]. Furthermore, polarization dependent x-ray absorption spectroscopy (XAS, often also referred to as near-edge absorption fine structure – NEXAFS) represents an ideal tool to investigate the geometry of molecular adsorbates and especially of thin and ultra-thin films on single and polycrystalline substrates [48–50]. We will briefly introduce these methods below as far as it is necessary for the understanding of the following chapters. In all cases (XPS, UPS and XAS) the sample is placed in an ultra-high vacuum system (the base pressure is typically about 10^{-10} mbar) and irradiated by monochromated light. Whereas in photoemission (or photoelectron) spectroscopy the energy distribution of the emitted photoelectrons at a fixed excitation energy is analyzed, in XAS the absorption of the X-rays as a function of their energy is monitored.

2.2.1
Photoemission spectroscopy

In photoemission spectroscopy, electrons from occupied states are excited above the vacuum level and can thus escape from the sample. In a first approximation, the measured kinetic energy (E_{kin}) allows the determination of the binding energy (E_B) of the photoelectron via a simple equation:

$$E_B = h\nu - E_{kin} - \phi_{SP} \tag{1}$$

where hν is the photon energy and ϕ_{SP} is a spectrometer specific constant, the work function of the spectrometer. The very small escape depth of the photoelectrons is responsible for the surface sensitivity of the method. Assuming that due to the removal of the considered electron in the orbital i the rest of the electron system is not affected (frozen orbital approximation), E_B corresponds to the orbital energies $-\varepsilon(i)$ theoretically accessible using the Hartree-Fock approximation. However, the remaining electrons of the environment can screen the photohole and thus, comparing the total energies of the N and N-1 electron system, an additional relaxation contribution R_D has to be taken into account.

$$E_B(i) = E(N-1,i) - E(N) = -\varepsilon(i) + R_D \tag{2}$$

Since PES does not exactly probe E_B of the initial state, related effects are usually called final state (FS) screening effects. Although the occurrence of FS screening effects is well-known in photoemission spectroscopy, measured E_B are very often discussed neglecting these effects and indeed the interpretation of energy shifts succeed in most cases. However, FS effects impact both the measured absolute E_B in solids [51] and especially the measured E_B at interfaces [52]. The probably most important result in this context was the successfully assertion of energetic shifts observed in Xe multilayers to FS effects [52].

Only photoelectrons whose kinetic energy is higher than the work function Φ of a sample can escape from the surface, consequently Φ can be determined in photoemission by the difference between the photon energy and the width of the spectrum. To guarantee that electrons with the lowest kinetic energies can be analyzed an additional bias voltage of 3 to 10 V is usually applied to the sample. The width of the photoemission spectrum is given by the energetic separation of the high binding energy (secondary electron) cutoff and the Fermi energy. In this context, possible shifts of the cutoff and thus of the vacuum level suggest the formation of an interfacial dipole layer Δ. The determination of important energetic values at interfaces using photoemission spectroscopy is illustrated in Fig. 2. Similar to the case of inorganic semiconductors, a linear extrapolation of the lower binding energy edge of the

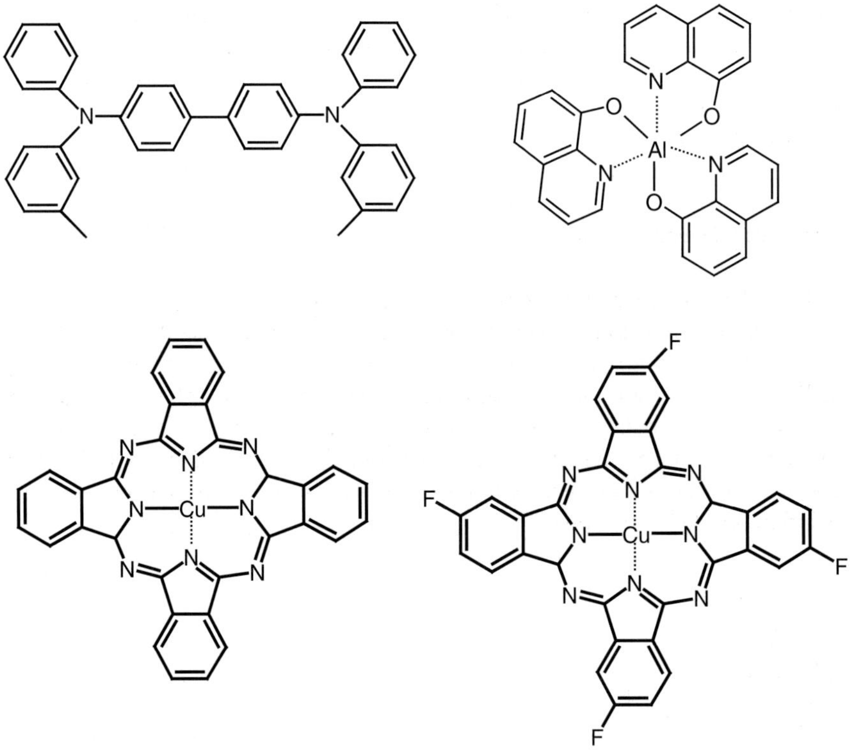

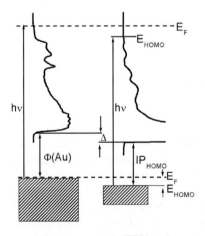

Figure 1 Molecular structure of model organic semiconductors: TPD, Alq$_3$, CuPc and CuPcF$_4$.

Figure 2 Schematic determination of the energy level alignment at interfaces using photoemission spectroscopy. Example: TPD on Au.

highest occupied molecular orbital (HOMO), the highest occupied valence state, is used to define the energy of the top of the HOMO (E_{HOMO}). Finally, in the case of homogeneous films core level photoemission (XPS) additionally offers the ability to determine the film thickness of the organic by monitoring the attenuation of the intensity of the substrate peaks due to the organic overlayer [22].

2.2.2
Polarization dependent XAS

In XAS, the absorption of the x-rays due to excitations from occupied into unoccupied levels as a function of energy is probed. Since a tunable, polarized excitation source is necessary for this method, synchrotron radiation is almost the only practical light source. In our case, the measurements were performed at the third generation synchrotron radiation source BESSY II (Berlin). Generally, two modes of the detection are possible, the fluorescence yield and the electron yield method, where due to its higher surface sensitivity the latter is of advantage for the study of thin and ultra-thin (organic) films.

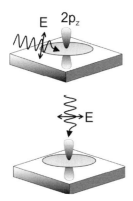

Figure 3 Polarisation dependent XAS: the chosen angle of incidence favors either excitations into the π* (top) or the σ* (bottom) orbitals.

Possible core excitations into vacant states are governed by selection rules, such as the change of the quantum number of the angular momentum ($\Delta l = \pm 1$) or the direction of the electric vector **E** of the linearly polarized synchrotron radiation. Thus, the excitation from e.g. a C 1s to a π* orbital in a planar π conjugated carbon system is allowed for **E** vertical to the molecular plane (parallel to $2p_z$ orbitals), whereas the transition into σ* is allowed for **E** parallel to the molecular plane and to the chemical bond as is shown schematically in Fig. 3. Consequently, in addition to the information about the unoccupied electronic structure of the material in question the molecular orientation can be probed by monitoring the anisotropic absorption: the relative intensities of excitations from occupied levels into either the π* or σ* molecular orbital are investigated as a function of the angle of incidence of the linearly polarized synchrotron radiation by variation of the measurement geometry, i.e. by rotation of the sample. In the last years polarization dependent XAS became almost a standard tool for the investigation of the geometry of molecular adsorbates [48–50]. Finally, we note that this method can also be applied to investigate other materials in which the bonding and/or orbital arrangement are anisotropic (e.g. [53,54]).

2.3
Adsorption geometry and substrate roughness

From a number of studies it is known that crystal perfection has considerable impact on the electronic and transport properties of organic systems, with extremely high mobilities in highly ordered materials [55]. For several well ordered organic molecules, in addition a strong anisotropy of the transport properties can be expected and has been demonstrated e.g. for perylenetetracarboxylic dianhydride (PTCDA) [41]. For this system, it was found that electron transport occurs predominantly along the molecular in-plane direction, whereas hole transport was observed mainly in the direction normal to the molecular planes [41]. Thereby the adsorption geometry of PTCDA on oxidized silicon was monitored by x-ray and electron diffraction. A similar anisotropy of the charge transport properties has been theoretically predicted for other organic thin film systems [56].

In this chapter the molecular orientation of the organic semiconductor CuPc grown as thin film on various substrates is summarized. The substrates include technically relevant materials such as indium-tin-oxide (ITO), oxidized Si, polycrystalline gold and the conducting polymer PEDOT:PSS (poly-3,4-ethylenedioxy-thiophene:polystyrenesulfonate), as well as well defined single crystal surfaces for comparison [57]. In order to study the molecular orientation, polarization-dependent x-ray absorption spectroscopy was used. There have been numerous studies indicating a planar Pc adsorption geometry on many single crystalline substrates [42,43,58–63]. However, as CuPc is used in thin film form in organic transistors, e.g. on oxidized Si and with polycrystalline gold or PEDOT:PSS contacts, and in multilayer light emitting devices on ITO or PEDOT:PSS, it is important to know the molecular orientation in thin films that were deposited on these technically relevant substrates.

The molecular orientation of CuPc has been studied measuring the relative intensity of core level excitations from the N1s level into either the π* or σ* molecular orbital manifolds of CuPc as a function of the angle of incidence, θ, of the linearly polarized synchrotron radiation (normal incidence gives $\theta = 90°$ and grazing incidence, $\theta = 10°$). This enables an analysis of the molecular orientation in the thin film. For a planar CuPc adsorption geometry, the dipolar transitions observed at grazing x-ray incidence are those from the N1s level into groups of individual π* states, as has been described in detail for the closely related system NiPc [58].

In Fig. 4 the N 1s excitation spectra of 20 - 50 nm thick CuPc films on two selected substrates [Au(110) in Fig. 4a and ITO in Fig. 4b] are displayed as a function of the

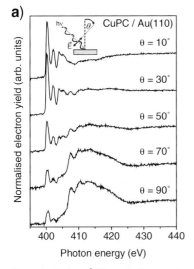
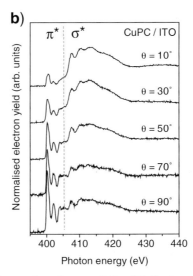

Figure 4 Series of N1s excitation spectra of CuPc films grown on (a) Au(110) and (b) ITO as a function of the angle θ between the surface normal and the electric field vector of the soft x-ray radiation [57]. The lower energy features (398–405 eV) represent the π^* resonances, whereas those features above 405 eV are related to the σ^* resonances.

incidence angle of the synchrotron radiation [57]. The absorption was monitored indirectly by measuring the total electron yield in the form of the drain current. For both substrates a very clear angular dependence of the π^* resonances (E = 402–408 eV) is observed: the maximum intensity of the π^* resonances for CuPc on Au(110) is observed at grazing incidence, whereas for the ITO substrate the highest intensity was found at normal incidence. Furthermore, the angle-dependence of the N $1s \rightarrow \sigma^*$ intensities (E = 408–430 eV) behave in an analogous, but opposite manner for both substrates. These observations demonstrate that the organic molecules are well ordered on *both* substrates, but that the adsorbate geometry in each case is radically different: for Au(110) the CuPc molecular plane is parallel to the substrate surface and for ITO it is perpendicular. This means that even on a relatively rough, polycrystalline substrate such as ITO, CuPc forms well-ordered films in which the molecules 'stand' on the substrate surface.

Importantly, we observe analogous behavior for the other non-ordered substrates studied: oxidized silicon, polycrystalline gold and PEDOT:PSS. In Fig. 5 we present a quantitative analysis of the angle-dependence of the N1s-π^* resonance intensity for CuPc on polycrystalline as well as on single crystal substrates. Assuming 100% linear polarization of the synchrotron radiation, the intensity should be described by a simple $\cos^2\theta$ function for flat lying molecules and $\cos^2(\theta+90°)$ for a perfect standing orientation. As can be seen from Fig. 5, for CuPc on Au(110) the data follow the theoretical curve expected for lying molecules quite well. Also, for CuPc/GeS(001) a preferential orientation parallel to the surface is clearly visible, even if there are clear deviations from the ideal curve. The clear angle-dependence of the intensity of the

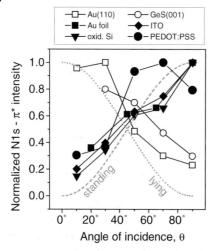

Figure 5 Symbols: angle-dependence of the intensity of the π*
resonances of CuPc on different substrates. The expected intensity profiles for standing and lying CuPc molecules are indicated
by dotted and dashed lines, respectively.

first pair of π* resonances points to a high degree of ordering for all substrates.
Surprisingly, the 20–50 nm thick CuPc films on the technologically relevant substrates are almost as highly ordered as on the single crystals.

The different growth modes observed on various substrates can be understood in
terms of different molecule-substrate interactions, since the molecule-molecule interaction strength within the CuPc films remains unchanged. If the molecule-substrate interaction is stronger than the molecule-molecule interaction, the adsorption
of the molecule occurs in a lying geometry (i.e. the molecular plane is parallel to the
substrate surface). This has been observed for numerous metal, semiconductor or
alkali halide substrates [43,57–61]. If the molecule-substrate interaction is weak, as
in the case of van-der-Waals substrates like highly oriented pyrolitic graphite
(HOPG) or MoS_2, then increased tilt-angles of 10–25° have been observed [62,63].
For thicker films of NiPc on HOPG, even a standing molecular orientation has been
proposed [63].

On polycrystalline substrates, there are no well defined adsorption sites for large
organic molecules due to the roughness of the surface on a molecular scale and a
standing geometry then is favored for a significant part of the first monolayer, thus
locking-in the molecular orientation for the remainder of the film growth process.
The opposite adsorption geometry on the polycrystalline Au foil and on the Au(110)
single crystal shows impressively that properties of the substrate surface such as
roughness or symmetry are crucial for the strength of the molecule-substrate interaction, and consequently they affect the growth direction significantly. Indeed, a partially standing orientation has also been observed for CuPc on the structurally and
morphologically ill defined quartz glass substrate [64] and for an organic transistor
device consisting of a 300 nm CuPc layer on oxidized silicon [44].

Note, that a standing molecular orientation of CuPc was observed on very rough surfaces. Therefore, the scale of the roughness seems to be important for the determination of the adsorption geometry. On atomically flat substrates there exist adsorption sites, which enable in most cases strong substrate-molecule interactions. Consequently, a lying adsorption geometry is favored and the orientation of the molecules is defined by the substrate surface structure. If the number of surface steps is increased, firstly the orientation of the lying molecules relative to the substrate may be changed as shown recently for CuPc on Au(111) [65] whereby in addition step bunching or macroscopic facetting might occur. A roughness larger than the molecule size causes a weakened substrate-molecule interaction, and thus a "standing" adsorption geometry becomes favorable. This was reported for CuPc on rough Si(111) surfaces with a root-mean-square roughness (RMS) of 0.2 nm [66] and is also the case for the technical relevant substrates such as ITO or oxidized silicon. For instance, Fig. 6 shows the surface morphology of the sputter-cleaned ITO-substrate used in our studies. The maximal roughness is about 8 nm (RMS ~ 1.5 nm) and there are no terraces in the magnitude of the molecule size which would enable a strong substrate-molecule interaction for a molecular orientation parallel to the substrate.

8 nm

0 nm

Figure 6 STM image (370 x 370 nm) of sputter-cleaned ITO [57]: The relatively rough surface (RMS ~ 1.5 nm) does not enable strong interaction sites for subsequent absorption of CuPc molecules.

To summarize, thin CuPc films on technical substrates such as ITO, oxidized Si polycrystalline gold and PEDOT:PSS are almost as highly ordered as on the single crystals but the molecular orientation in the two cases is radically different: the CuPc molecules stand on the technical substrates and lie on the single crystalline substrates, as schematically depicted in Fig. 7. (Note, that thicker films of evaporated CuPc usually adopt the so-called α modification with a herringbone structure [67].) This shows that the extrapolation of single crystal adsorption data (in which the CuPc molecules lie) to 'real' device substrates could be highly misleading. Upon the reasonable assumption of an anisotropy of the transport properties of CuPc, the different adsorption geometries would be crucial for the performance of devices based on highly ordered organic films.

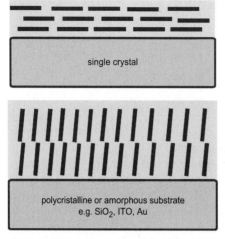

Figure 7 Schematic representation of the adsorption geometry of CuPc molecules on technical substrates (standing) and on single crystalline substrates (lying).

2.4
Organic/metal interfaces: general properties and trends

Bringing two materials into contact, the interface properties are determined by several factors. The contact allows the achievement of a thermodynamic equilibrium via charge flow across the interface to equalize the chemical potentials on both sides. This usually results in the formation of interface dipoles or band bending near the interface [68] as illustrated in Fig. 8. Provided the absence of chemical interaction and surface reconstruction on either side of the interface, the electronic energy level alignment is determined by fundamental properties of the two materials: their work function and charge carrier densities.

During the last years, there has been an intensive discussion how interfaces between *organic* semiconductors and metals can be rationalized. The vast majority of experimental studies using photoemission spectroscopy reported drastic variations of the energy levels of the organic semiconductor under investigation within very few nanometers from the interface and small variations at larger distances [e.g. 13,14,16–29,69]. Therefore, these observations were often interpreted as the formation of a local interface dipole. The mechanism of the dipole formation however remained unraveled. A number of possible factors that drive the dipole formation were proposed, among them chemical reactions, ion formation, mirror forces or surface electronic rearrangement as well as the presence of permanent dipoles at the interface in the case of some organic materials [14].

In this chapter we discuss studies of the interfaces between gold and differently fluorinated copper phthalocyanines (CuPc, $CuPcF_4$, $CuPcF_{16}$). It has been shown that these are free from chemical interactions [70,71] and thus provide us with an

without contact   with contact

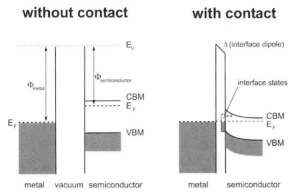

Figure 8 Formation of a metal-semiconductor interface.

ideal model system to investigate general and fundamental properties of organic/ metal interfaces. Please note, that in the studies discussed below abrupt interfaces are formed in contrast to cases where metals are deposited onto organic films. We start the discussion with core level photoemission spectra of CuPc as a function of CuPc film thicknesses on Au [70]. As one example for the data sets, Fig. 9 depicts the C1s core level spectra as a function of CuPc deposition on Au. All C1s spectra

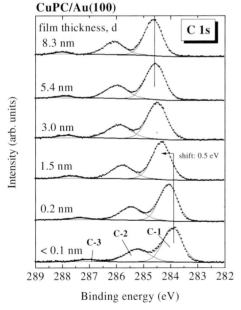

Figure 9 C 1s XPS spectra (hv=1486.6eV) of CuPc on Au(100) [70]. The spectra for low coverages are strongly shifted to lower binding energies. The absence of additional features at the earliest stages of deposition and the only very small broadening rule out chemical interactions at the interface.

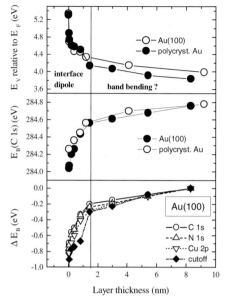

Figure 10 Comparison of the energy level shifts of CuPc on gold as obtained from XPS and UPS measurements [70].

consist of three different components (labelled C-1, C-2, and C-3). They can be attributed to the aromatic carbon of the benzene rings (C-1), pyrrole carbon linked to nitrogen (C-2) and a π-π* satellite of the latter [63,72]. Note, that for an exact explanation of the C 1s features in phthalocyanines the consideration of an additional satellite structure arising from the aromatic carbon is necessary [73]. Since changes of the peak shape or the energetic position of the main core level component allow the analysis of interactions and the energy level alignment at interfaces, we do not discuss the more complicated satellite structure here.

The absence of special features at the earliest stages of deposition of CuPc on Au(100) in both the C 1s (Fig. 9) and the corresponding N 1s and Cu 2p spectra (not shown) signals that no interaction occurs at these interfaces independent on the orientation of the CuPc molecules. The very small symmetrical broadening (5 %) of all core level spectra can be explained by inequivalent adsorption sites and by the fact, that the spectroscopic information in XPS contains contributions from several organic layers since the mean free path for the organic is in the range of 1.5–2.5 nm depending on the kinetic energy of the photoelectrons. However, we do observe strong energetic shifts of all core level features towards higher binding energies with increasing layer thickness d especially for coverages up to 2 nm. Equivalent results were obtained for a polycrystalline gold substrate and strong energetic shifts as a function of deposition were also observed for other organic/metal interfaces [e.g. 13,14,16–29,69].

The binding energy of all core levels as well as the high binding energy cutoff (i.e. the work function) as a function of the CuPc layer thickness is summarized in Fig. 10.

The upper two panels compare the changes resulting from CuPc deposition on poly-crystalline and single crystalline Au substrates. From Fig. 10 it is clear that the behavior on both substrates is very similar. The lower panel compares the shifts of all core levels and the high binding energy cutoff for CuPc on Au(100). The distinct kink in all curves at a nominal thickness of about 1.5 nm clearly points to two different underlying reasons for the observed shifts and had let to the assignment of the shifts at low coverages to the formation of interface dipoles [70]. Equivalent results were observed for the fluorinated relatives $CuPcF_4$ and $CuPcF_{16}$ on gold [73,74]. In Fig. 11 a summary of the core level binding energy shifts as a function of $CuPcF_{16}$ layer thickness on polycrystalline Au is shown. Clearly, rapid changes occur within the first 1–2 nm whereas the variations at thicknesses larger than 2 nm are small.

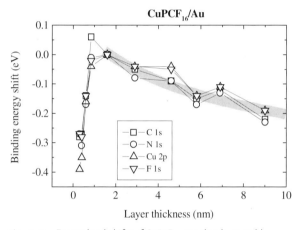

Figure 11   Energy level shifts of $CuPcF_{16}$ core levels on gold.

The data presented so far harbor a wealth of information on fundamental interface properties at organic/metal interfaces. We firstly focus on the rapid changes as e.g. observed in Figs. 10 and 11 and other studies at low film thicknesses. It can be shown that these changes to a significant portion are caused by screening effects whereby the photohole in the photoemission final state is more efficiently screened near a metal as is the case at the discussed interfaces. The contribution of these final state (FS) screening effects to the observed shifts can be estimated via the extra-atomic relaxation energy, which is roughly proportional to both the ability for screening by the environment and in the case of our interface systems the distance from the metal surface, r. The former is determined in macroscopic dielectric models by the polarization charge $(1-1/\varepsilon)e$, where ε is the dielectric constant of the environment (i.e. for ultra-thin films the substrate dielectric constant is predominantly responsible for screening). In interface systems, the screening ability of the substrate is indirectly proportional to the distance from the substrate surface via $e^2/4r$. Thus, both the distance from the substrate *and* the type of the substrate is important. For metal substrates, ε becomes infinite (also known as image charge screening), and the FS screening is proportional to $e^2/4r$, whereas for semiconducting substrates the dielectric constant is similar to the organic and therefore no layer dependent changes in the FS effects are expected. This can be nicely demonstrated using GeS(001) instead of Au(100) as a substrate and carrying out the equivalent experiment. Then, the strong binding energy variations at low film thicknesses are drastically reduced in accordance with the lower screening ability of GeS as compared to Au [70].

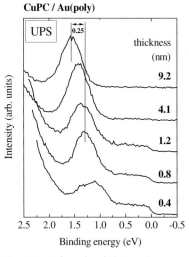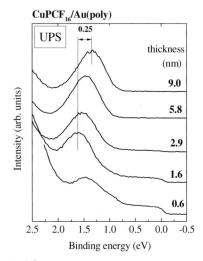

Figure 12   Valence band photoemission spectra of CuPc (left panel) and CuPcF$_{16}$ (right panel) deposited on gold as a function of film thickness.

It is now important to realize that this image charge screening is affecting all photoemission final states, i.e. also those created in valence states of the organic semiconductor. This is demonstrated in Fig. 12 where the valence band photoemission spectra of the systems CuPc/Au and CuPcF$_{16}$/Au are shown as a function of film thickness. Obviously, the valence band features related to emission from the highest occupied molecular orbitals of the phthalocyanines substantially shift to higher binding energy for small film thicknesses similar to the core level data presented above. Moreover, the valence band photoemission final state, one hole in the valence band, exactly is what is relevant for electrical transport (at least in the low charge carrier density limit). Consequently, the screening as described above concerns the energy of electronic transport levels in organic semiconductors in the vicinity of an organic/metal interface. Furthermore, screening changes its sign for holes and electrons, respectively, i.e. the valence and conduction band states approach each other near a metal. In other words, the transport energy gap is reduced near the metal as schematically depicted in Fig. 13. This transport gap reduction can be up to several tenth of an eV. Thus, there is *unconventional band bending* near interfaces between organic semiconductors and metals due to the screening ability of the metal as has also been proposed previously [15,75–78].

Finally, the image charge screening is substantially larger in organic semiconductors as compared to inorganic (traditional) semiconducting materials. This is firstly a direct consequence of the considerably smaller static dielectric constant (ε) of organic materials which is in the range of 3–4, while for materials like Si, Ge or GaAs ε is larger than 10, and secondly due to the more confined wave function of the molecular electronic states of organic semiconductors in comparison to the band-like states in e.g. Si.

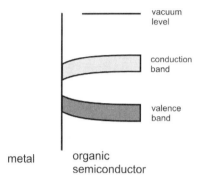

Figure 13 Variation of the valence and conduction levels near a metal-semiconductor interface as a result of the image charge potential.

While screening explains the binding energy shift near the interfaces it is negligible at distances larger than 2 nm from the interface [70,75]. Therefore, the smaller changes there have to arise from a different mechanism. They can be attributed to band bending in the conventional sense, i.e. changes of the electrostatic potential as a consequence of the equilibration of the chemical potentials of the two materials in contact. The fact that these changes are very small, of the order of 0.1–0.2 eV, is well consistent with the very small number of charge carriers in the undoped and thus intrinsic organic semiconductors discussed here. It is worth noting that an additional contribution to the observed shifts at larger layer thicknesses could result from the finite film thickness which is necessarily probed using photoemission. This leads to a change in the electrostatic potential at the outer organic surface as compared to that expected at the same distance from the interface for an infinitely thick organic semiconductor [79].

Consequently, the sign of the variations (curvature) at larger film thicknesses indicates the relative values of the work functions of the organic semiconductor and the metal. In intrinsic organic semiconductors the chemical potential can be expected in the middle of the (transport) energy gap, which is actually confirmed by a number of experimental studies. The size of the transport gap, however, is in many cases not well known because it can be much larger than what is measured using optical techniques due to relatively high exciton binding energies of the lowest optical excitations [80–82]. In the case of the CuPcF$_x$ systems discussed here, the transport gap, E_g, has only been determined for the unfluorinated CuPc using a combination of photoemission and inverse photoemission spectroscopy [80]. It is about 2.3 eV, 0.6 eV larger than the onset of the optical absorption. For the fluorinated CuPc's we assume a similar transport gap based on very similar optical absorption data [29], similar molecular and thin film structure and similar occupied valence electronic structure. Then, the work function, Φ_{sc}, of the CuPc, CuPcF$_4$ and CuPcF$_{16}$ can be calculated from the measured ionization potentials (IP): Φ_{sc}=IP-½E$_g$, which gives 3.9 eV, 4.55 eV and 4.95 eV, respectively.

Coming back now to the binding energy changes farer from the interfaces as seen in Figs. 10 and 11, the direction (curvature) of the changes for CuPc/Au (and CuPcF$_4$/Au [73]) are consistent with the sign of the differences of the corresponding work functions, Φ_m-Φ_{sc} which are positive. For CuPcF$_{16}$, however, this is not the case which implies that the value of Φ_m and/or Φ_{sc} is not correct. We attribute this inconsistency to the fact that the metal work function Φ_m at the interface to the organic semiconductor is smaller than its value into vacuum [29]. This conclusion has also been drawn recently by several research groups [83–86] and follows the long known reduction of metal work functions when covered with a monolayer of e.g. Xenon [14,87]. Thus, considering interfaces between organic semiconductors and metals, the metal work function has to be reduced as compared to its size into vacuum, the magnitude of the reduction is about 0.2–0.4 eV. Importantly, such a reduced Φ_m should also be used in transport simulations across organic/metal interfaces.

The data presented in Figs. 10 and 11 clearly reveal a kink-like change in the binding energy variations at about 1.5–2 nm. Following the discussion and arguments above, it is reasonable to analyze the vacuum level position relative to that of the chemical potential at 2 nm from the interface in order to obtain information on the existence of (short range or local) interface dipoles. The result of such an analysis is shown in Fig. 14 whereby the effect of the reduction of the metal work function is also included. From Fig. 14 it becomes clear that the significant change in the ionization potential or work function of the CuPc's upon fluorination has almost no effect for the hole injection barrier, i.e. the relative alignment of the energy levels at the interface. Instead, the change in Φ_m-Φ_{sc} (often denoted as built-in potential V$_{bi}$) is to a large extent compensated by the formation of an interface dipole Δ. Moreover, the interface dipole follows linearly the work function difference Φ_m-Φ_{sc} also for other undoped organic semiconductors as demonstrated in Fig. 15. This indicates that their linear dependence is universal for undoped organic semiconductors on metals provided the absence of interactions at the interface. In such cases the linear dependence then allows to calculate/predict the interface dipole for yet experimentally unstudied organic/metal contacts. While this is an important result, the microscopic origin of the dipole formation is still not well understood. Since chemical reactions, ion formation or permanent molecular dipoles, as being discussed, do not apply for the interfaces discussed here and image charge screening as well as surface rearrangement only contribute partly to the interface dipoles as shown above, further mechanisms need to be considered in order to rationalize the behavior as depicted in Fig. 15.

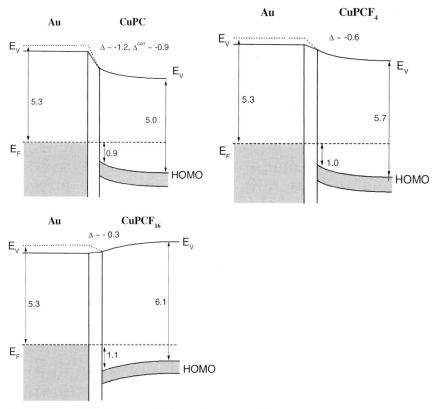

Figure 14 Energy level diagram of CuPc/Au, CuPcF₄/Au and CuPcF₁₆/Au interfaces.

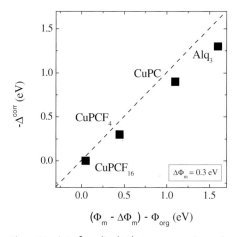

Figure 15 Interface dipoles between organic semiconductors and gold as a function of the work function difference. The reduced gold workfunction is considered (see text and Fig. 14).

Recently, we have proposed a scenario which might significantly contribute to the interface dipole formation [79]. Whereas band bending is negligible as long as the electronic states of the intrinsic organic semiconductor are well defined in energy, it was demonstrated that an exponential energy distribution of the transport energy levels can change this situation significantly. From transport studies it is well established that the electronic levels in many organic semiconductors are exponentially or Gaussian-like distributed in energy. Allowing for such a distribution with a reasonable width of the order of 0.1–0.3 eV in simulations of the energy level alignment, it is found that a band bending is caused within the first 2 nm of the organic film which can explain the observed dipoles to a considerable extent. We note that the term band bending in this context should not be taken literally, but is used to describe the energy variations of the exponentially or Gaussian-like distributed transport levels on average. Finally, the attribution of the interface dipoles to band bending in these cases is in nice agreement with the observation that the behavior as shown in Fig. 15 nearly corresponds to the Schottky-Mott rule [68] for the energy level alignment at semiconductor interfaces. Interface dipoles roughly following the Schottky-Mott rule scenario have also been reported from studies in which the metal work function has been varied using different metals in contact with an organic semiconductor, however there are also several examples where the interface properties cannot be explained by this simple approach most likely due to interactions at the respective interfaces [14,16,21,25,28,34,88–92].

In summary, the above discussion demonstrates that there has been considerable progress in recent years in the understanding of electronic properties of interfaces between organic semiconductors and metals. In detail, three contributions that determine the energy position of the electronic states as a function of the distance from the interface have been identified: (i) a strong bending of the electronic states as a consequence of image charge screening by a nearby metal for distances smaller that 1.5 nm; (ii) a reduction of the metal work function at the interface by 0.2–0.4 eV as compared to its value into vacuum and (iii) a strong "short range" interface dipole at 2 nm from the interface which might be related to the fact that the energy levels in organic molecular semiconductors are ill defined and which can be described in terms of "band bending" of the distributed transport levels.

2.5
Reactive surfaces: indium-tin oxide and PEDOT:PSS

Strong chemical interactions observed at different organic/inorganic interfaces depend upon the substrate and the organic molecule under investigation. Besides the well-known complete charge transfer from highly reactive metals to organic molecules (for instance K/Alq_3, K/CuPc, K/ZnPc or K/C_{60} [93–96]), interactions and/or charge transfer were also observed for organic materials on less reactive metals such as Ag [13,30,36,37]. In this chapter we will demonstrate that a strong interaction also occurs on sputter-cleaned indium-tin oxide (ITO) [22,32,38] and PEDOT:PSS [97], which both are of technological interest in devices such as light emitting diodes and transistors.

In devices, the hole injection efficiency is a critical parameter and depends strongly on the work function of the electrode, i.e. a high value of Φ can lower this energy barrier. The measurement of Φ for Ar ion or O ion sputter-cleaned ITO using UPS yields values from 4.2 to 4.4 eV, [22,98,99] whereas it has been shown that the work function of ITO can be varied over a wide range (3.9–5.1 eV) by chemical treatments with acids and bases [100]. In the last years, numerous studies have been performed in order to investigate and to tune the work function of ITO and thus to match the organic energy levels [see, e.g. 14,22,98–101]. In the following we discuss now the chemistry of organic/ITO interfaces.

Fig. 16 shows C 1s and N 1s XPS spectra (normalized to the same peak height) recorded after different exposure times of sputter-cleaned ITO to TPD [22]. The upper spectra are typical of thicker TPD films. In case of the C1s core level data, the most intense peak at 285.2 eV can be attributed to carbon atoms of the phenylene rings and the peak at 286.2 eV corresponds to carbon atoms co-ordinated with nitrogen. The N 1s spectrum basically consists of a single feature at 400.3 eV [22]. We observe a small shift of ~ 0.2 eV of the core level peaks towards higher binding energies with growing film thickness which may be caused by a band-bending in the first layers of the organic solid and/or by final state effects (see discussion above).

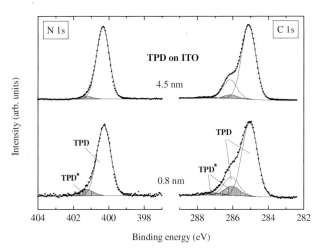

Figure 16 TPD on ITO: N 1s and C 1s XP-spectra fitted illustrating the presence of two TPD-species (TPD and TPD*) [22].

Importantly, the spectra for lower TPD film thickness are significantly different from those of the thicker films. As indicated in Fig. 16, a high energy component is present for the thin film data which can be explained by the presence of an additional TPD* species, whose spectral features are shifted by 1 eV towards higher binding energies. Furthermore, the relative intensity of TPD* obtained from the C 1s spectrum (8 %) of the 0.8 nm thick film is similar to the result of the fit of the N 1s line (9 %). The relative intensity of TPD* for a 4.5 nm thick film is reduced (~4.5 % for both N 1s and C 1s). Moreover, the intensity reduction of the TPD* related features follows smoothly the film thickness of the TPD layer [22]. The shift to higher binding energies points towards a partially oxidized TPD at the interface (TPD$^{\delta+}$) and it thus indicates a charge transfer from TPD to the ITO substrate.

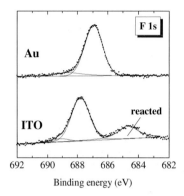

Figure 17 F 1s core level spectra for ultrathin CuPcF$_4$ films (about 0.3–0.4 nm) on Au and on ITO. The additional peak at about 684.7 eV on ITO shows the chemical reaction at the interface [38].

Chemical interaction is also observed for other organic molecules in contact with ITO. This is illustrated in Fig. 17 where we show a comparison of the F1s core level spectra for submonolayer coverages of CuPcF$_4$ on ITO and Au [38]. On ITO a clear additional F 1s feature at about 3 eV lower binding energy is observed which is completely absent for CuPcF$_4$ on Au. Note, that this observation is independent on the cleaning procedure (Ar ion or O ion etching) applied to the ITO surface [32,38]. The relative intensity of the additional feature of about 25 % shows either that 25 % of the molecules are reduced as a result of a charge transfer from ITO to the molecules or that the molecules strongly interact with ITO via one of the four fluorinated ligands. In the first case one would expect (i) a significant shift of all core levels to lower binding energies at submonolayer coverages and (ii) an additional peak in the valence band UPS spectra (not shown). As both is clearly not the case, we can rule out a reduction of the whole molecules. We are thus left with a scenario where CuPcF$_4$ and ITO interact most likely via one ligand of the standing molecule. This interpretation is consistent with the line width changes of the C1s core level spectra of CuPcF$_4$ and also of CuPc going from very small to larger film thicknesses on ITO, which in addition demonstrates that a similar interaction occurs for CuPc on ITO [32].

Such interactions now can have important consequences for the charge carrier transport across the interface as they can be connected with a charge transfer at the interface and the formation of interface states, and thus for energy levels that can trap charge carriers [39]. This is schematically drawn in Fig. 18. For instance, the threshold voltage in devices comprising such interfaces is then more determined by the potential that is needed to fill the trap levels than by the charge carrier injection barrier. Since commercially available ITO substrates are usually produced by a sput-

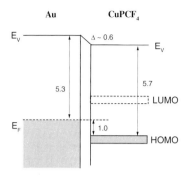 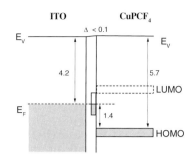

Figure 18 Energy level alignment of CuPcF$_4$/Au (left) and CuPcF$_4$/ITO (right) as obtained from UPS for 2 nm thick films. Whereas the Fermi level E$_F$ on Au is near to the mid-gap position, interface states on ITO shift E$_F$ towards the LUMO [38]. Please note that the position of the LUMO is derived from bulk values and a possible shift of the LUMO due to the interaction at the interface is not considered.

tering process and our applied cleaning procedures do not alter the surface conditions significantly, the results of our studies indicate the presence of interface states in many devices which contain standard, not specially treated ITO electrodes.

Among the conducting polymers that are discussed as potential organic electrodes for all-organic devices, PEDOT:PSS is one of the most promising candidates because PEDOT:PSS is easily processable from solution, thin films of this conducting polymer show a high morphological as well as a high redox stability and their surfaces are very smooth [102,103]. Moreover, it has been demonstrated recently that the work function of spin-cast PEDOT:PSS films can be controlledly adjusted within a fairly large energy window of about 0.6 eV by subsequent electrochemical treatment [104] whereas the work function of electropolymerized PEDOT:PSS films can even be varied by more than 1 eV [105]. This represents a possibility to tune device properties which is unknown for traditional noble metal electrodes such as Au.

In contrast to the remarkable progress that has been achieved concerning interfaces between organic materials and inorganic (metal) electrodes (see above), this is not yet the case for the understanding of organic-organic interfaces. Although, similar to organic/inorganic interfaces, electronic dipole layers (Δ) were also found at organic/organic interfaces [13,17,19,20,26,46], the size of the dipoles and thus the energy level alignment can be different in both cases. For instance, it was shown that the hole injection barrier at interfaces between some organic semiconductors deposited on PEDOT:PSS is drastically smaller than on gold, although the two metallic materials have a similar work function [86,106]. In Fig. 19 we demonstrate that the binding energy of the valence electronic levels of the organic semiconductors CuPc, CuPcF$_4$ and CuPcF$_{16}$ deposited on PEDOT:PSS significantly differ from those of these materials on Au [101,107]. Whereas the spectra for CuPc and CuPcF$_4$ on PEDOT:PSS are shifted to lower binding energies relative to those on Au, the valence band spectrum of CuPcF$_{16}$ on PEDOT:PSS is shifted in the opposite direction. In other words, the relative position of E$_F$ in the gap of the organic semiconductor near the interface to PEDOT:PSS is shifted towards the HOMO for CuPc and CuPcF$_4$ and towards the lowest unoccupied molecular orbital (LUMO) for CuPcF$_{16}$. This is illustrated in the insets of Figs. 19 a) – c) for 2 nm thick CuPcF$_x$ films: whereas E$_F$ lies near the mid-gap position on the gold substrate (dotted lines) distinct deviations in different directions are observed for the Pc's on PEDOT:PSS (solid lines). In a first approximation, a mid-gap position of E$_F$ can be expected for an intrinsic semiconductor and in the case of organic semiconductor/gold interfaces this equilibrium is virtually reached at a very small film thickness of about 2 nm (see chapter 4). However, a non-midgap position of E$_F$, as found for the phthalocyanines on PEDOT:PSS at 2 nm from the interface indicates additional charges probably resulting from a charge transfer near the interface, the direction of which corresponds to a p-type doping for CuPc and CuPcF$_4$ (electron transfer from the organic semiconductor to the substrate), and an n-type doping for CuPcF$_{16}$ (electron transfer from the substrate to the organic semiconductor). The question then arises how such a charge transfer and its direction for the different Pc molecules can be rationalized?

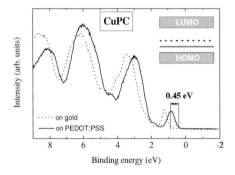
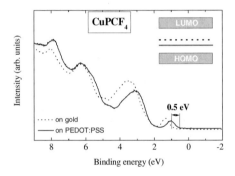

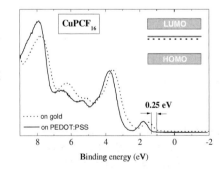

Figure 19   UPS spectra of CuPc (a), CuPcF$_4$ (b) and CuPcF$_{16}$ (c) on PEDOT:PSS (solid line) and on Au (dotted line). The film thickness of the Pc's is about 2 nm. The shift of the spectra to lower [(a), (b)] or higher (c) binding energies on PEDOT:PSS compared to Au can be explained by a different doping near the interface.

A measure for the ability to accept electrons is the electron affinity (E_A). A high value for E_A causes the material being an electron acceptor, whereas materials with a low E_A rather act as electron donors. Since the gap is nearly the same for all Pc's, the increase of IP due to the fluorination results in an equivalent increase of E_A and this causes a high ability of CuPcF$_{16}$ to accept electrons. On the other hand, the relatively low E_A of CuPc favors an electron transfer in the opposite direction. Furthermore, mobile charge carriers are already present in the PEDOT:PSS film and thus they can be easily transferred to the Pc's, if this process is energetically favorable. Such a charge transfer is possible when the corresponding molecules and polymer chains are next to each other. The extent and direction of the electron transfer between the two organic layers can also be expressed in terms of the difference of the electrochemical potentials μ. A comparison of the electrochemically measured μ of PEDOT:PSS to the calculated μ of the phthalocyanines (from the HOMO and LUMO energies) also predicts an oxidation for CuPc and CuPcF$_4$ (p-doping) and a reduction for CuPcF$_{16}$ (n-doping). Moreover, the above found tendency to accept preferably holes or electrons for the Pc is also reflected in other studies: CuPc is known as p-type material in air [44], whereas its perfluorinated relative CuPcF$_{16}$ is one of the few organic semiconductors which have demonstrated high performance and stability in air for n-channel operation [45].

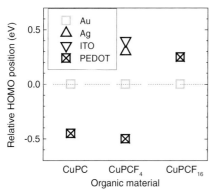

Figure 20 Position of the HOMO of differently fluorinated phthalocyanine thin films (thickness 2 nm) on reactive substrates relative to the Pc/gold interfaces.

Finally, comparing the characteristics of interfaces where interactions determine the valence energy level alignment, a variation of the metal work function (e.g. going from PEDOT:PSS to Ag or ITO) can affect the direction of the charge transfer involved. This is the case for CuPcF$_4$ on PEDOT:PSS and on Ag [97,107,108]. Since Φ(PEDOT:PSS) $> \Phi$ (CuPcF$_4$), and Φ (Ag) or Φ (ITO) $< \Phi$ (CuPcF$_4$) an opposite charge transfer can be expected: an electron transfer *from* the molecule for CuPcF$_4$/PEDOT:PSS and *to* the molecule for CuPcF$_4$/Ag. This causes a different position of E_F in the gap of the organic semiconductor which corresponds to a p-doping for CuPcF$_4$/PEDOT:PSS and a n-doping for CuPcF$_4$/Ag as indicated by the corresponding energy positions of the HOMO depicted in Fig. 20. Analogously, a change of the work function of the organic semiconductor affects the charge transfer direction and thus the position of the chemical potential in the energy gap as seen by comparing the PEDOT:PSS/CuPcF$_4$ and PEDOT:PSS/CuPcF$_{16}$ interfaces in Fig. 20.

2.6
Concluding remarks

During the last decade there has been considerable progress as regards the understanding of interfaces of organic semiconductors and various metallic materials. Some of the individual aspects are discussed in this review article. The successful experimental and theoretical work of a number of research groups worldwide notwithstanding, there remain unresolved issues, in particular concerning interfaces in all-organic devices which are of great importance for a successful commercialization of organic electronic devices.

Acknowledgements

We thank G. Paasch, J. Fink, T. Schwieger, D. Olligs, J.M. Auerhammer, M.S. Golden, A. Kahn, X. Liu, L. Dunsch, A. Petr, F. Zhang, Th. Schmidt, P. Bressler and M. Mast for fruitful discussions and their contribution to some of the experiments. Technical support from R. Hübel, D. Müller, S. Leger and K. Müller is gratefully acknowledged. Part of the work has been financially supported by the Bundesministerium für Bildung und Forschung under grant number 01BI 163.

References

[1] E.A. Silinsh, Organic Molecular Crystal, Springer Series in Solid State Science **16** (Springer, Berlin, 1978) and references therein.

[2] K. Müllen and G. Wegner (eds.), Electronic Materials: The Oligomer Approach (Wiley-VCH, Weinheim, 1998) and references therein.

[3] M. Pope and C.E. Swenberg, Electronic Processes in Organic Crystals and Polymers (Oxford University Press, Oxford, 1999) and references therein.

[4] G. Hadziioannou and P.F. van Hutten (eds.), Semiconducting Polymers (Wiley-VCH, Weinheim, 2000) and references therein.

[5] R. Farchioni and G. Grosso (eds.), Organic Electronic Materials, Springer Series in Material Science **41** (Springer, Berlin, 2001) and references therein.

[6] C.W. Tang and S.A. van Slyke, Appl. Phys. Lett. **51**, 913 (1987).

[7] R.H. Friend, R.W. Gymer, A.B. Holmes, J.H. Burroughes, R.N. Marks, C. Taliani, D.D.C. Bradley, D.A. Dos Santos, J.L. Bredas, M. Löglund, and W.R. Salaneck, Nature **397**, 121 (1999).

[8] G. Horowitz, Adv. Mater. **10**, 365 (1998).

[9] D. de Leeuw, Physics World (March issue), 31 (1999).

[10] N.S. Sariciftci, L. Smilowitz, A.J. Heeger, and F. Wudl, Science **258**, 1474 (1992).

[11] M. Granström, K. Petritsch, A.C. Arias, A. Lux, M.R. Andersson, and R.H. Friend, Nature **395**, 257 (1998).

[12] H.E.A. Huitema, G.H. Gelinck, J.B.P.H. van der Putten, K.E. Kuijk, C.M. Hart, E. Cantatore, P.T. Herwig, A.J.J.M. van Breemen, and D.M. de Leeuw, Nature **414**, 599 (2001).

[13] W.R. Salaneck, K. Seki, A. Kahn, and J.-J. Pireaux (eds.), Conjugated Polymer and Molecular Interfaces (Marcel Dekker, New York, 2002) and references therein.

[14] H. Ishii, K. Sugiyama, E. Ito, and E. Seki, K., Adv. Mater. **11**, 605 (1999).

[15] J.C. Scott, J. Vac. Sci. Technol. A **21**, 521 (2003).

[16] S. Narioka, H. Ishii, D. Yoshimura, M. Sei, Y. Ouchi, K. Seki, S. Hasegawa, T. Miyazaki, Y. Harima, and K. Yamashita, Appl. Phys. Lett. **67**, 1899 (1995).

[17] A. Rajagopal, C.I. Wu, and A. Kahn, J. Appl. Phys. **83**, 2649 (1996).

[18] Y. Park, V. Choong, E. Ettedgui, Y. Gao, B.R. Hsieh, T. Wehrmeister, and K. Müllen, Appl. Phys. Lett. **69**, 1080 (1996).

[19] R. Schlaf, B.A. Parkinson, P.A. Lee, K.W. Nebesny, and N.R. Armstrong, Appl. Phys. Lett. **73**, 1026 (1998).

[20] S.T. Lee, X.Y. Hou, M.G. Mason, and C.W. Tang, Appl. Phys. Lett. **74**, 670 (1998).

[21] Th. Kugler, W.R. Salaneck, H. Rost, and A.B. Holmes, Chem. Phys. Lett. **310**, 391 (1999).

[22] H. Peisert, T. Schwieger, M. Knupfer, M.S. Golden, and J. Fink, J. Appl. Phys. **88**, 1535 (2000).

[23] S.C. Veenstra, U. Stalmach, V.V. Krasnikov, G. Hadziioannou, H.T. Jonkman, A. Heeres, and G.A. Sawatzky, Appl. Phys. Lett. **76**, 2253 (2000).

[24] P.G. Schroeder, C.B. France, J.B. Park, and B.A. Parkinson, J. Phys. Chem. B **107**, 2253 (2003).

[25] G. Koller, R.I.R. Blyth, S.A. Sardar, F.P. Netzer, and M.G. Ramsey, Appl. Phys. Lett. **76**, 927 (2000).

[26] L. Chkoda, C. Heske, M. Sokolowski, and E. Umbach, Appl. Phys. Lett. **77**, 1093 (2000).

[27] I.G. Hill, A.J. Mäkinen, and Z.H. Kafafi, Appl. Phys. Lett. **77**, 1825 (2000).

[28] N.J. Watkins, L. Yan, and Y. Gao, Appl. Phys. Lett. **80**, 4384 (2002).

[29] H. Peisert, M. Knupfer, and J. Fink, Appl. Phys. Lett. **81**, 2400 (2002).

[30] I.G. Hill, J. Schwartz, and A. Kahn, Organic Electronics **1**, 5 (2000).

[31] C. Shen and A. Kahn, J. Appl. Phys. **90**, 4549 (2001).

[32] H. Peisert, M. Knupfer, T. Schwieger, and J. Fink, Appl. Phys. Lett. **80**, 2916 (2002).

[33] A.C. Arias, M. Granström, D.S. Thomas, K. Petritsch, and R.H. Friend, Phys. Rev. B **60**, 1854 (1999).

[34] L. Yan and Y. Gao, Thin Solid Films **417**, 101 (2002).

[35] M. Fahlmann and W.R. Salaneck, Surf. Sci. **500**, 904 (2002).

[36] S.C. Veenstra, A. Heeres, G. Hadziioannou, G.A. Sawatzky, and H.T. Jonkman, Appl. Phys. A **75**, 661 (2002).

[37] F.S. Tautz, M. Eremtchenko, J.A. Schäfer, M. Sokolowski, V. Shklover, K. Glöckler, and E. Umbach, Surf. Sci. **502–503**, 176 (2002).

[38] H. Peisert, M. Knupfer, and J. Fink, Synth. Met. **137**, 869 (2003).

[39] W. Brütting, H. Riel, T. Beierlein, and W. Riess, J. Appl. Phys. **89**, 1704 (2001).

[40] P.H. Nguyen, S. Scheinert, S. Berleb, W. Brütting, and G. Paasch, Organic Electronics **2**, 105 (2001).

[41] J.R. Ostrick, A. Dodabalapur, L. Torsi, A.J. Lovinger, E.W. Kwock, T.M. Miller, M. Galvin, M. Berggren, and H.E. Katz, J. Appl. Phys. **81**, 6804 (1997).

[42] S.R. Forrest, Chem. Rev. **97**, 1793 (1996).

[43] A. Koma, J. Cryst. Growth **201/202**, 236 (1999).

[44] Z. Bao, A.J. Lovinger, and A. Dodabalapur, Appl. Phys. Lett. **69**, 3066 (1996).

[45] Z. Bao, A.J. Lovinger, and J. Brown, J. Am. Chem. Soc. **120**, 207 (1998).

[46] H. Peisert, T. Schwieger, M. Knupfer, M.S. Golden, and J. Fink, Synth. Metals **121**, 1435 (2001).

[47] K. Horn, Appl. Surf. Sci. **166**, 1 (2000) and references therein.

[48] J. Stöhr, NEXAFS Spectroscopy (Springer, Berlin, 1992).

[49] E. Umbach, K. Glockler, and M. Sokolowski, Surf. Sci. **404**, 20 (1998).

[50] K. Seki, H. Ishii, and Y. Ouchi, in Chemical Applications of Synchrotron Radiation (T.K. Sham ed.), Advanced Series of Physical Chemistry **12A**, 386 (World Scientific Singapore, 2002).

[51] H. Peisert, T. Chassé, P. Streubel, A. Meisel, and R. Szargan, J. Electron Spectrosc. Relat. Phenom. **68**, 321 (1994).

[52] G. Kaindl, T.-C. Chiang, D.E. Eastman, and F.J. Himpsel, Phys. Rev. Lett. **45**, 1808 (1980).

[53] P.A. Brühwiler, A.J. Maxwell, C. Puglia, A. Nilsson, S. Andersson, and N. Martensson, Phys. Rev. Lett. **74**, 614 (1995).

[54] J. Fink, N. Nücker, E. Pellegrin, H. Romberg, M. Alexander, and M. Knupfer, J. Electron Spectrosc. Relat. Phenom. **66**, 395 (1994).

[55] N. Karl, in [5], p. 283.

[56] J. Cornil, J.-P. Calbert, D. Beljonne, R. Silbey, and J.-L. Brédas, Adv. Mater. **12**, 978 (2000).

[57] H. Peisert, T. Schwieger, J.M. Auerhammer, M. Knupfer, M.S. Golden, J. Fink, P.R. Bressler, and M. Mast, J. Appl. Phys. **90**, 466 (2001).

[58] M.L.M. Rocco, K.-H. Frank, P. Yannoulis, and E.-E. Koch, J. Chem. Phys. **93**, 6859 (1990).

[59] J.J. Cox, S.M. Bayliss, and T.S. Jones, Surf. Sci. **433–435**, 152 (1999).

[60] T. Shimada, A. Suzuki, T. Sakurada, and A. Koma, A., Appl. Phys. Lett. **68**, 2502 (1996).

[61] T.J. Schuerlein and N.R. Armstrong, J. Vac. Sci. Technol. **A 12**, 1992 (1994).

[62] K.K. Okudeira, S. Hasegawa, H. Ishii, K. Seki, Y. Harada, and N. Ueno, J. Appl. Phys. **85**, 6453 (1999).

[63] L. Ottavario, S.D. Nardo, L. Lozzi, M. Passacantando, P. Picozzi, and S. Santucci, Surf. Sci. **373**, 318 (1997).

[64] K. Hayashi, T. Horiuchi, and K. Matsushige, Adv. X-Ray Analysis **39**, 653 (1997).

[65] L. Chizhov, G. Scoles, and A. Kahn, 2000. Langmuir **16**, 4358 (2000).

[66] M. Nakamura, Y. Morita, Y. Mori, A. Ishitani, and H. Tokumoto, J. Vac. Sci. Technol. **B 14**, 1109 (1996).

[67] M. Ashida, N. Uyeda, and E. Suito, Bull. Chem. Soc. Japan **39**, 2616 (1996).

[68] S.M. Sze, Physics of Semiconductor Devices (John Wiley & Sons, New York, 1981).

[69] R. Schlaf, P.G. Schroeder, W.M. Nelson, B.A. Parkinson, C.D. Merritt, L.A. Crisafulli, H. Murata, and Z.H. Kafafi, Surf. Sci. **450**, 142 (2000).

[70] H. Peisert, T. Schwieger, J.M. Auerhammer, M., Knupfer, M.S. Golden, and J. Fink, J. Appl. Phys. **91**, 4872 (2002).

[71] J.M. Auerhammer, M. Knupfer, H. Peisert, and J. Fink, Surf. Sci. **506**, 333 (2002).

[72] G. Dufour, C. Poncey, F. Rochet, H. Roulet, M. Sacci, M. De Santis, and M. De Crescenzi, Surf. Sci. **319**, 251 (1994).

[73] H. Peisert, M. Knupfer, and J. Fink, Surf. Sci. **515**, 491 (2002).

[74] H. Peisert, M. Knupfer, T. Schwieger, G.G. Fuentes, D. Olligs, J. Fink, and Th. Schmidt, J. Appl. Phys. **93**, 9683 (2003).

[75] A.J. Twarowski, J. Chem. Phys. **77**, 1458 (1982).

[76] T.R. Ohno, Y. Chen, S.E. Harvey, G.H. Kroll, J.H. Weaver, R.E. Hauffler, and R.E. Smalley, Phys. Rev. **B 44**, 13747 (1991).

[77] I.G. Hill, A.J. Mäkinen, and Z.H. Kafafi, J. Appl. Phys. **88**, 889 (2000).

[78] E.V. Tsiper, Z.G. Soos, W. Gao, and A. Kahn, Chem. Phys. Lett. **360**, 47 (2002).

[79] G. Paasch, H. Peisert, M. Knupfer, J. Fink, and S. Scheinert, J. Appl. Phys. **93**, 6084 (2003).

[80] I.G. Hill, A. Kahn, Z.G. Soos, and R.A. Pascal Jr., Chem. Phys. Lett. **327**, 181 (2000).

[81] M. Knupfer, H. Peisert, and T. Schwieger, Phys. Rev. **B 65**, 033204 (2001).

[82] M. Knupfer, Appl. Phys. **A 77**, 623 (2003).

[83] X. Crispin, V. Geskin, A. Crispin, J. Cornil, R. Lazzaroni, W.R. Salaneck, and J.-L. Bredas, J. Am. Chem. Soc. **124**, 8131 (2002).

[84] P.S. Bagus, V. Staemmler, and Ch. Wöll, Phys. Rev. Lett. **89**, 096104 (2002).

[85] L. Yan, N.J. Watkins, S. Zorba, Y. Gao, and C.W. Tang, Appl. Phys. Lett. **81**, 2752 (2002).

[86] N. Koch, A. Kahn, J. Ghijsen, J.-J. Pireaux, J. Schwartz, R.L. Johnson, and A. Elschner, Appl. Phys. Lett. **82**, 70 (2003).

[87] A. Zangwill, Physics at Surfaces (Cambridge University Press, Cambridge, 1988) and references therein.

[88] Y. Harima, H. Okazaki, Y. Kunugi, K. Yamashita, H. Ishii, and K. Seki, Appl. Phys. Lett. **69**, 1059 (1996).

[89] I.G. Hill, A. Rajagopal, A. Kahn, and Y. Hu, Appl. Phys. Lett. **73**, 662 (1998).

[90] G. Greczynski, M. Fahlman, and W.R. Salaneck, Chem. Phys. Lett. **321**, 379 (2000).

[91] G. Greczynski, Th. Kugler, and
W.R. Salaneck, J. Appl. Phys. **88**, 7187
(2000).

[92] S. Park, T.U. Kampen, D.R.T. Zahn, and
W. Braun, Appl. Phys. Lett. **79**, 4241 (2001).

[93] T. Schwieger, H. Peisert, M. Knupfer,
M.S. Golden, and J. Fink, Phys. Rev. **B 63**,
165104 (2001).

[94] T. Schwieger, H. Peisert, M.S. Golden,
M. Knupfer, and J. Fink, Phys. Rev. **B 66**,
155207 (2002).

[95] T. Schwieger, M. Knupfer, W. Gao, and
A. Kahn, Appl. Phys. Lett. **83**, 500 (2003)

[96] M. Knupfer, Surf. Sci. Rep. **42**, 1 (2001).

[97] H. Peisert, M. Knupfer, F. Zhang, A. Petr,
L. Dunsch, and J. Fink, Appl. Phys. Lett. **83**,
3930 (2003).

[98] Y. Park, V. Choong, Y. Gao, B.R. Hsieh, and
C.W. Tang, Appl. Phys. Lett. **68**, 2699 (1996).

[99] D.J. Milliron, I.G. Hill, C. Shen, A. Kahn,
and J. Schwartz, J. Appl. Phys. **87**, 572
(2000).

[100] F. Nüesch, L.J. Rothberg, E.W. Forsythe,
Q.T. Le, and Y. Gao, Appl. Phys. Lett. **74**,
880 (1999).

[101] I.G. Hill and A. Kahn, J. Appl. Phys. **86**,
2116 (1999).

[102] L. Groenendaal, F. Jonas, D. Freitag,
H. Pielartzik, and J.R. Reynolds, Adv. Mater.
12, 481 (2000).

[103] P.K.H. Ho, J.-S. Kim, J.H. Burroughes,
H. Becker, S.F.Y. Li, T.M. Brown, F. Cacialli,
and R.H. Friend, Nature **404**, 481 (2002).

[104] A. Petr, F. Zhang, H. Peisert, M. Knupfer,
and L. Dunsch, Chem. Phys. Lett. **385**, 140
(2004).

[105] H. Frohne, D.C. Müller, and K. Meerholz,
ChemPhysChem **3**, 707 (2002).

[106] A.J. Mäkinen, I.G. Hill, R. Shashidhar,
N. Nikolov, and Z.H. Kafafi, Appl. Phys.
Lett. **79**, 557 (2001).

[107] H. Peisert, M. Knupfer, F. Zhang, A. Petr,
L. Dunsch, and J. Fink, Surf. Sci. **566–568**,
554 (2004)

[108] T. Schwieger, H. Peisert, and M. Knupfer,
Chem. Phys. Lett. **384**, 197 (2004).

3

Kelvin Probe Study of Band Bending at Organic Semiconductor/ Metal Interfaces: Examination of Fermi Level Alignment

H. Ishii, N. Hayashi, E. Ito, Y. Washizu, K. Sugi, Y. Kimura, M. Niwano, O. Ouchi, and K. Seki

3.1
Introduction

3.1.1
Energy level alignment at organic/metal interfaces

During the last decade organic semiconductors have been used in various type of electronic devices [1]. Especially organic light emitting diodes (OLED) have been extensively investigated and its commercial products are now extending in markets [2, 3]. The other devices such as organic thin film transistors (OTFT) [4] and organic photovoltaic cells [5, 6] also have attracted much attention. In spite of such extensive studies of organic electronic devices, their operating mechanisms are not yet well understood. A better understanding of device physics, such as the electronic structures and carrier transport properties, are indispensable for improving the existing devices and developing new types of devices.

Energy level alignment at organic/electrode interfaces, that is, *how the energy levels of organic semiconductors and electrodes are aligned at their interfaces,* is one of the fundamental issues about these devices. As a simple model to discuss this problem, the Mott-Schottky (MS) model [7, 8], which is a well-known *textbook model* for inorganic semiconductor/metal contact, has so far been applied to organic interfaces until recently without any serious discussion. This model is illustrated in Fig. 1 for an n-type semiconductor. When a neutral semiconducting organic solid and a neutral metal are isolated, the energy diagrams of the two solids are aligned at the common vacuum level between the two solids. When the two solids make contact, the MS model assumes (i) a vacuum level alignment right at the interface region and (ii) band bending in the space charge layer (SCL) to achieve the alignment of the bulk Fermi levels between them. Thus, the barrier height for hole injection, Φ_B^p (Φ_B^n for electron), corresponds to $I_s^{th} - \Phi_m$ ($\Phi_m - \chi$), where I_s^{th} and χ are the ionization energy and electron affinity of the solid, respectively. In relation to band bending, the *Fermi level alignment* is a fundamental concept in the MS model. The thermal equilibrium state, in which Fermi level is constant everywhere in the system, is assumed to be achieved. Since the bulk Fermi level of a semiconductor coincides with that of the substrate metal, the built-in potential, V_{bi}, which corresponds to the voltage

Physics of Organic Semiconductors. Edited by W. Brütting
Copyright © 2005 WILEY-VCH Verlag GmbH & Co. KGaA, Weinheim
ISBN 3-527-40550-X

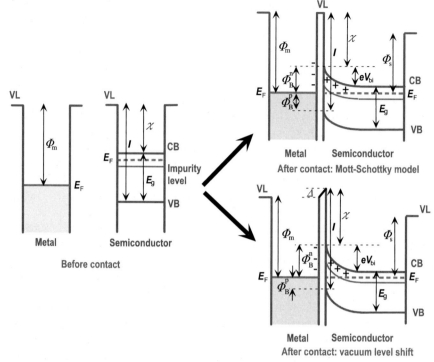

Figure 1 Models for energy level alignment at organic semiconductor/metal contact: (a) Energy diagram before contact. (b) Mott-Schottky model assuming vacuum level alignment at the interface and band bending with Fermi level alignment. (c) Realistic model including vacuum level shift at the interface.

change across the space charge layer, coincides with the difference in the work function between the electrode and the semiconducting solid. $(eV_{bi} = |\Phi_m - \Phi_s|)$

The above model is a very naive one although it has been applied for a long time in order to discuss the interfaces of organic devices. Concerning the first assumption, extensive studies using photoemission spectroscopy have recently revealed that vacuum level alignment does not occur in most organic/metal interfaces due to the formation of an *interface dipole* inducing vacuum level shift (Δ) [9–16]. The existence of Δ demonstrates that the barrier height estimated from the MS model has to be modified by the amount of Δ. V_{bi} should be also modified by the same value $(eV_{bi} = |\Phi + \Delta - \Phi_s|)$.

In contrast to the vacuum level alignment, the experimental examination of the second assumption, band bending, has been limited, and there still remains questions such as *Does band bending occur in organic semiconductors?* and *Is Fermi level alignment achieved for an organic system?*. Organic materials applied to the active layer of electronic devices are usually called organic semiconductors, but their HOMO-LUMO gap is typically large (2–3 eV). As will be discussed later, the concentration of thermally excited carriers is thus extremely small like in an insulator. The band bending effect is usually neglected at insulator/organic interfaces. Thus the experimental examination of band bending is required for organic semiconductors.

In addition, Fermi level alignment is a critical problem. This scheme requires thermal equilibrium. Organic semiconductors are basically molecular solids where molecules are bound only by van der Waals forces, and their properties such as wave function and charge density are fairly localized within each molecule. This means that the carrier exchange process between adjacent molecules is not effective in contrast to the case of inorganic semiconductors with good conductivity. Thus it is not obvious that the unbalance in the Fermi level can be compensated by the redistribution of carriers, and experimental examination of this point is also highly desired.

In the practical situation of organic devices, band bending is closely related to the potential profile across the organic layer in devices with an applied external electric field. For example, in the case of organic light emitting diodes (OLEDs), the electric field is often assumed to be constant and the potential is proportional to the position as shown in Fig. 2(a). In contrast, in the case of organic solar cells, band bending in the space charge layer has been often assumed as shown in Fig. 2(b). Basically, a semiconductor with a sufficient carrier concentration can screen the external electric field. Polarization induced by an external field can exclude the penetration of the field deep into the inside of the semiconductor. This phenomenon is closely related to the concentration of mobile carriers and space charges. In practical devices, such band bending is critical to their operation. For example, band bending can induce a large electric field in the space charge layer which enables efficient charge separation of excitons formed by photoabsorption. When the space charge layer is thin because of heavy doping, carrier injection can be enhanced due to tunnelling across the space charge layer.

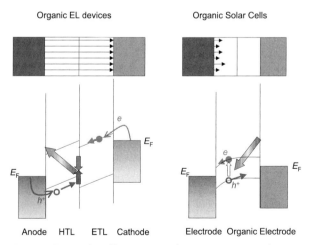

Figure 2 Potential profiles in organic devices. (a) A case with no band bending like organic light emitting diodes. (b) A case with band bending like organic solar cells. Lines of electric force are schematically shown in the upper panel.

Before going on to the main issue about band bending, it is noteworthy to consider the definition of the term *"band bending"*. Here we use the following definition. When the energy of the electronic levels of a solid with a finite energy gap shows variation as a function of the distance from the interface, we call it band bending. In a narrow meaning, the impurity-derived energy shift in the MS-model should be called band bending. However, as will be discussed later, there are other mechanisms which induce an energy shift near the organic/metal interface. The latter should be also regarded as band bending in the broad sense. It should be noted that the term *"band"* does not mean *band structure* or *energy band with electron delocalization*. Whether band bending is induced or not is governed not by the type of wave function (delocalized Bloch wave or fairly localized molecular orbital) but by the statistics of the occupation of energy levels. Although another term *"energy level bending"* may be preferred, in this paper, we use the familiar term "band bending" in this broad context.

The experimental examination of band bending in organic semiconductors has been performed using various techniques. Capacitance - voltage (*C-V*) measurement is an approach to examine the thickness of the space charge layer due to band bending under an electric field. The thickness of the SCL in organic devices was reported for several device structures [17] and the value was derived using the theory for inorganic semiconductors. There remains a question, whether or not the theory for band bending of inorganic semiconductors is applicable to organic semiconductors. In that sense, the result obtained from *C-V* measurements is dependent on the model used for the analysis. In addition, the experimental data obtained by this technique is often affected by the low carrier mobility and/or trap effect in the organic layer, making the analysis more difficult.

As another approach, photoelectron spectroscopy such as ultraviolet photoelectron spectroscopy (UPS) and X-ray photoelectron spectroscopy (XPS) have been applied to the examination of band bending [18–22]. By using this technique, so far, two kinds of interfaces, *organic on metal* and *metal on organic* have been used. In the former, the energy level shift was measured as a function of the thickness of the organic layer. In the latter case, energy level shifts have actually been reported for several kinds of interfaces. However, it is often difficult to distinguish the peak shift due to band bending from that due to sample charging during the photoemission measurements. The latter effect is induced by the incomplete neutralization of the holes left after photoelectron emission due to the rather poor conductivity of organic semiconductors.

Possible band bending can be expected to occur in a wide space charge layer in the case of organic semiconductors because of the large band gap and consequently small amount of carriers available for band bending. Thus the experimental examination for sufficiently thick organic films, which is strongly desired, is not possible by photoemission. Instead, experimental techniques which can probe the energy level shift without any emission of charged particle are necessary. Surface potential measurements using the Kelvin probe (KP) method is one of the solutions.

The KP method has been applied to various organic films under ambient atmosphere or low-to-high vacuum pressure ($\geq 10^{-5}$ Pa) [22–28], starting from the pioneering work of Kotani and Akamatu [30, 31]. However, basic studies under ultra-high vacuum (UHV) were carried out only for small molecules with up to a few molecular layers. In order to discuss the basic aspect of band bending, the examination of band bending under UHV conditions is also necessary. In previous papers, we proposed a criterion for judging the establishment of the Fermi level alignment that the energy separation between the vacuum level of the organic layer and the Fermi level should not depend on the metal substrate [28]. Only in limited number of previous studies, Fermi level alignment was experimentally confirmed by performing measurements of the surface potential of organic materials on metal substrates with different work functions [24]. If the observed surface potential is independent of the substrate metals, the alignment of the bulk Fermi level is highly probable, and we can determine the work function. However, Fermi level alignment was often assumed *a priori*, and the work functions of organic materials are often deduced from the KP measurements for only one substrate metal. KP studies of organic films have another problem. It is well established that the KP method probes the surface potential or work function in the case of metallic and semiconducting samples in which the sample surface accumulates charges and works as an electrode of the capacitor during the KP measurements. However, in the case of organic semiconducting materials, the physical meaning of the experimentally obtained values are not very evident, since Fermi level alignment between the substrate and organic layer may be not achieved or the exchange of mobile carriers between them may be limited. Thus it is highly desired to clarify *what is observed for organic/metal interfaces by the KP method.*

In the present study, we report our recent efforts to examine band bending in organic semiconductors using the KP method, bearing the above-described points in mind. After a brief summary of the framework of band bending, we will discuss the physical meaning of the KP measurements in organic films and arrive at the conclusion that the KP method still probes the position of the vacuum level of the sample. We will then discuss the band bending at C_{60}/metal and TPD (*N,N'*-bis(3-methylphenyl)-*N,N'*-diphenyl-[1,1'-biphenyl]-4,4'-diamine)/metal interfaces with part of the previously reported data [28,32]. In the former case, gradual band bending was observed, while almost a flat band was observed in the latter. On the basis of these results, we will discuss the validity of the second assumption of the Mott-Schottky model as well as the concept of Fermi level alignment. Besides the band bending in the MS model, we found another type of band bending effect at Alq_3 / metal interfaces. Previously, we reported that the permanent dipole moments of Alq_3 molecule are aligned under dark conditions, leading to a giant surface potential [29]. This phenomenon can be regarded as a new type of band bending. We found that this system is in nonequilibrium, and it is quite stable for years. By combining the results for the three interfaces, we will examine the validity of Fermi level alignment in organic systems.

3.2
Basic aspects of band bending

In this section, we will briefly survey the basic concepts related to band bending. The review article such as [33] is useful for details. In the Mott-Schottky model, the Fermi level alignment at the interface is an essential assumption, as shown in Fig. 1. The formation of a space charge layer (SPL), with space charges typically created by the ionization of dopants and/or the thermally excited carriers, leads to the Fermi level alignment between the metal and the semiconductor.

In the following, we will summarize the relation between the Fermi level and carrier density for the cases of (i) thermal equilibrium and (ii) nonequilibrium.

3.2.1
Band bending in thermal equilibrium

The carrier densities of electrons (n) and holes (p) in a semiconductor under thermal equilibrium can be expressed as

$$n = n_i \exp\left(\frac{E_F - E_i}{k_B T}\right) \qquad p = p_i \exp\left(\frac{E_i - E_F}{k_B T}\right) \tag{1}$$

Here, n_i and p_i are the intrinsic carrier densities of an electron and hole, respectively. The subscript i means the value for intrinsic semiconductors. E_F and E_i are the Fermi level energy and intrinsic Fermi level energy of the semiconductor, respectively. k_B is Boltzmann's constant and T is temperature. When the semiconductor comes in contact with a metal, the densities of the carriers and ionized dopants are changed to form the SCL, and induce the potential profile $\Phi(x) = E_F - E_i(x)$, where x is the coordinate along the axis perpendicular to the surface. The Fermi level is assumed to be constant in the whole region of the metal, interface, SCL, and bulk part of the semiconductor. Thus the $\Phi(x)$ can be principally related to the charge density at x, $\rho(x)$ as

$$\frac{d^2\Phi}{dx^2} = -\frac{\rho(x)}{\varepsilon} \tag{2}$$

$$\rho(x) = e(N_D - N_A + p - n). \tag{3}$$

where e is the elementary charge, and N_D and N_A are the densities of the donor and acceptor in the semiconductor, respectively. The actual form of $\Phi(x)$ can be derived by solving eq. (2) under a given boundary condition. In the case of a strong depletion, in which dopants in SCL are assumed to be fully ionized, Φ becomes a parabolic curve and can be estimated as

$$V_{bi} = -\frac{eN_D d^2}{\varepsilon \varepsilon_0} \tag{4}$$

In this situation, we can understand that the Fermi level is constant in the whole system, i.e. thermal equilibrium is realized. Now we discuss the n-type semiconductor as an example. There exist two kinds of currents in the SCL. One is the drift current: The electric field in SCL, E, induces the drift of carriers, and the current can be expressed as $e n \mu_n E$, where n and μ_n are the density and mobility of electrons, respectively. The other current is the diffusion current. The density of electrons in the conduction band can be changed as a function of $E_F - E_i(x)$ (eq. 1). Since $E_F - E_i(x)$ is also changed in the SCL, n is not uniform, leading to the diffusion current, $-e D_n \frac{dn}{dx}$, when D_n is the electron diffusion coefficient. In equilibrium, the currents cancel each other. The net current, J_n, then becomes zero.

$$J_n = e n \mu_n E - e D_n \frac{dn}{dx} = 0 \tag{5}$$

This is the physical meaning of thermal equilibrium at the interface.

3.2.2
Band bending in nonequilibrium

In a nonequilibrium system, i.e., the case for $J_n \neq 0, J_p \neq 0$, the Fermi level is no longer at a constant energy within the system, and cannot be defined as in the case of equilibrium. However, we can approximately discuss the system using the concept of the quasi-Fermi level if the deviation from thermal equilibrium is small. The quasi-Fermi level ($\tilde{E}_F^n$, $\tilde{E}_F^p$ for electron and hole) is an index for the number of excited carriers, and it is defined by the following relation.

$$n = n_i \exp\left(\frac{\tilde{E}_F^n - E_i}{k_B T}\right) \tag{6}$$

$$p = p_i \exp\left(\frac{E_F^p - \tilde{E}_F^p}{k_B T}\right) \tag{7}$$

Now, $\tilde{E}_F^p$ and $\tilde{E}_F^n$ are not equal in the nonequilibrium system. When the system reaches thermal equilibrium, $\tilde{E}_F^p = \tilde{E}_F^n = E_F$ holds. Again, the current J_p, J_n can be expressed as

$$J_n = e n \mu E - e D_n \frac{dn}{dx} = \frac{d\tilde{E}_F^n}{dx} \tag{8}$$

$$J_p = e p \mu E - e D_p \frac{dp}{dx} = \frac{d\tilde{E}_F^p}{dx} \tag{9}$$

Thus, the spatial gradient of $\tilde{E}_F^{p,n}$ induces the net flow of carriers, which is governed by the carrier density, local electric field, and the diffusion constant.

3.3
KP measurements for insulating surface

The principle of the Kelvin probe method was established by Lord Kelvin in 1898 [34]. Zisman first applied this principle using a vibrating capacitor in 1932 [35]. KPM was then applied to various surfaces and films to measure their work function or surface potential. The principle of the measurement is shown in Fig. 3. In this technique, the sample substrate and a metallic, vibrating reference electrode constitute a parallel plate capacitor. The vibration of the reference electrode induces an ac current. The current is nulled when the voltage applied to the reference electrode is equal to the contact potential difference between the reference and the sample.

When the sample film is a conductor, the surface part of the sample acts as a plate of the capacitor. i.e., charges are accumulated at the surface when the bias voltage is applied as shown in Fig. 3(a). On the other hand, if the sample film is an insulator or semiconductor with a poor electrical conductivity, the charges are at least not fully accumulated at the sample surface, as shown in Fig. 3(b). Charges may be accumulated at the sample surface as well as at the interface between the substrate and the

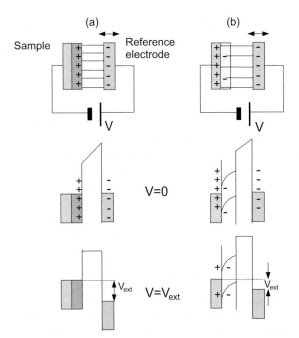

Figure 3 Principle of Kelvin probe measurements for conducting film (a) and semiconducting film with possible band bending (b). The top part shows the layout of the sample film and the reference electrode. The contact potential is formed under zero bias voltage as shown in the middle panel. Under an adjusted bias voltage, the electric field in the gap becomes null as shown in the bottom panel. When a null signal is detected in the KP measurements, the alignment of the vacuum level of the sample film and the reference electrode is guaranteed. See the text for details.

sample. Some part of the charges may move into the sample film and some polarization due to possible band bending may also exist. Thus the situation is quite different from that of a conducting sample.

Now we have to discuss the question, *what is observed by the KP method for insulating sample on a metallic substrate?* Pfeiffer and Karl proposed that the KP method gives the correct surface potential of the insulating sample if the capacitance of the sample film is larger than that of the gap between the sample and reference electrodes [23]. Very recently we have discussed the same problem in detail [36], and concluded that the vacuum level of the sample film exactly coincides with that of the reference electrode at the null-signal condition in the KP measurement, that is, *the KP method yields the correct work function at least under no external field*.

The reason why the KP method is still usable for an insulating film is briefly discussed below. Details are described in ref. [36]. When a finite charge is accumulated at the surface of the reference electrode, the counter charge must be in the sample film and/or the interface between the sample and the substrate, while the distribution of the charge in the sample and the substrate is unknown. When the null detection occurs in the KP measurement, it means at least that no charge is accumulated at the surface of the reference electrode. According to Gauss's theorem, the electric field, E in the gap region between the sample film and the reference electrode can be expressed as $\frac{\sigma_{ref}}{\varepsilon_0}$, where σ_{ref} is the charge density of the surface of the reference electrode. Thus E must be zero in the case of $\sigma_{ref} = 0$ at the null-signal condition, indicating no potential drop in the gap. Thus the vacuum level of the reference electrode exactly coincides with that of the sample layer, while possible band bending may induce charges both in the space charge layer of the semiconductor and at the surface region of the conducting substrate on which the sample film is formed. A possible charge exchange between the sample film and the substrate may occur during the KP measurement, but such an effect can be completely neglected in the case of null-detection condition, that is, the charge distribution in the sample film is not changed by the vibration. Thus we can conclude that the KP method probes the precise surface potential of the film even in the case when the sample is not a good conductor or insulator. It should be noted that the observed potential reflects the charge distribution in the sample layer and the substrate surface under the null-signal condition. For example, if there exist unexpected trap charges at the surface of the organic film, the surface potential, which includes the effect of the charges, is measured. When the field is applied to the sample and vacuum layer, the charge distribution may be changed: KP probes the band bending for the null-detection condition with zero field in the gap. For our discussion, the band bending and Fermi level alignment of organic/metal interfaces under zero or very small bias, KP measurements can mostly provide correct results for band bending. In actual devices, however, a large field is applied to the organic layer as in Fig. 2, and we have to take care of the redistribution of charges due to the field.

3.4
Experimental

The organic materials used in the studies reported in this article were C_{60}, TPD, and Alq_3. The former two were purchased from Tokyo Chemical Industry, and the latter from Aldrich, respectively. The quoted purity of C_{60} was 99.9%, and most of our experiments for C_{60} were carried out using this material as-received. Part of C_{60}, TPD and Alq_3 were purified once using vacuum sublimation under a temperature gradient. Five metals (Au, Ag, Cu, Mg, Ca) were used as the substrates, which were prepared using materials purchased from the Nilaco Corporation (purity exceeding 99.9%).

KP measurements were carried out using a UHV chamber, which consists of three parts, a measurement chamber, an evaporation chamber, and a loading chamber. The sample can be transferred among the chambers without exposure to air. The details of the system are described elsewhere [28].

In the experiments, the metal substrates were first prepared by vapor deposition onto a Si wafer ((111),1.0–3.0 $\Omega \cdot$ cm). Typical thickness of the metal layer was 40nm. One of the organic materials was then deposited using a quartz crucible heated by a W wire wound around it. The base pressure of the evaporation chamber was 2.7×10^{-8} Pa, and the working pressure during the deposition was less than 1.3×10^{-6} Pa for the metals and less than 4×10^{-7} Pa for the organic materials.

After each step of the deposition, the sample was transferred to the measurement chamber, where the surface potential was measured with a Kelvin Probe (McAllister Technical Service:KP6000). The work function of the reference electrode was determined by measuring various freshly deposited polycrystalline metals and comparing the results with the values from the literature.

3.5
Results ans Discussion

3.5.1
C_{60}/metal interface: case of band bending

The energy of the vacuum level relative to the Fermi level of the substrate, ε_{vac}^{F}, of the C_{60} film on Au, Cu, and Ag metals is shown in Fig. 4 as functions of the film thickness, d. The data in the small thickness region ($d \leq 10$ nm) are shown in the inset.

An abrupt shift in the vacuum level was observed at $d \leq 1$ nm in Fig. 4. This is due to the formation of an interfacial dipole layer. The values of ε_{vac}^{F} converge to 4.65 eV in the region $d = 2 - 10$ nm. This result is similar to that of the photoemission study by Ohno et al. [37]. Since the value of ε_{vac}^{F} is independent of the metal substrate, it seems to be plausible that the converged value is the Fermi level of the bulk C_{60}, i.e. the work function of C_{60}.

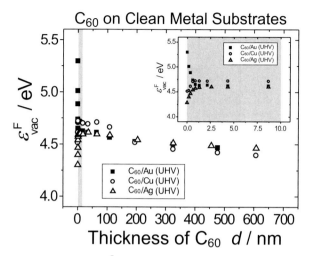

Figure 4 Variation in ε_{vac}^F for C_{60} on Au, Cu, and Ag as a function of the thickness of C_{60}, d. The data in the region for small d up to 10 nm is shown in the inset.

Additional C_{60} layers are deposited to clarify this point. The value of ε_{vac}^F shows a gradual downward shift, converging to 4.44 eV at $d = 500$ nm. Throughout this process, the vacuum level of the C_{60} film is independent of the kind of substrate, satisfying the criteria of the Fermi level alignment. At the first convergence, the Fermi level of the C_{60} *monolayer* is aligned to that of the metal substrates probably due to the efficient mixing of the wave function between the C_{60} layer and the metal surface [37]. At the second convergence, we postulate that the *bulk* Fermi level of the C_{60} multilayer corresponds to that of the metal. Thus we ascribe this gradual downward shift to the band bending leading to the bulk Fermi level alignment.

From the data in Fig. 4, we can estimate the space charge density assuming a uniform distribution of fully ionized donors in the space charge layer. According to eq. (4), the thickness of the space charge layer, W, can be expressed as

$$W = \sqrt{\frac{2\varepsilon_r \varepsilon_0 V_{bi}}{eN_D}}, \tag{10}$$

where V_{bi} is the built-in potential. ε_r is the relative dielectric constant of C_{60}, ε_0 is the dielectric permittivity of a vacuum, and N_D is the dopant density. By setting

$$W = 495\,\text{nm} \qquad V_{bi} = 0.21\,\text{V} \qquad \varepsilon_r = 4.4 \tag{11}$$

we obtain

$$N_D = \frac{2\varepsilon_r \varepsilon_0 V_{bi}}{eW^2} = 4.2 \times 10^{14}/\text{cm}^3 \tag{12}$$

Thus the concentration is on the order of ppm since C_{60} has $1.40 \times 10^{21}/\text{cm}^3$ based on the density of $1.682\,\text{g}/\text{cm}^3$ obtained from the x-ray diffraction of the fcc phase at 300 K [38]. Here it should be noted that the value is the concentration of the dopant that supplies electrons to the C_{60} molecules. Other impurities are probably also included.

Next, we discuss the positions of the Fermi level of C_{60} and the donor levels. Fig. 5(a) shows the energy diagram of the Cu/C_{60} interfaces obtained from the KP measurements. The energy of the HOMO level is assumed using the threshold ionization potential of C_{60} ($I_s^{th} = 6.17\,\text{eV}$ [39]). The HOMO-LUMO gap is estimated using the photoemission and inverse photoemission (IPES) results of C_{60} [40] ($E_g = 2.6\,\text{eV}$. Since it may include energy broadening due to the relatively poor energy resolution of IPES, the actual E_g may be slightly larger.) Since the converged values of ε_{vac}^F are common irrespective of the metal substrates as shown in Fig. 4, the Fermi level of the bulk C_{60} layer probably coincides with that of the substrate. These results demonstrate that the bulk Fermi level of C_{60} could be experimentally obtained under the confirmation of the Fermi level alignment at the interface.

If we assume that the C_{60} is an intrinsic semiconductor, the bulk Fermi level can be derived as

$$E_F = \frac{E_c + E_v}{2} + k_B T \ln\left(\frac{N_v}{N_c}\right), \tag{13}$$

where E_v denotes the energy of the top of the valence band (HOMO), and E_c depicts that of the bottom of the conduction band (LUMO). Since the HOMO and LUMO levels of C_{60} have 5-fold and 3-fold degeneracy, the effective valence- and conduction band densities of states, N_v and N_c are 7.0×10^{21} and $4.2 \times 10^{21}/\text{cm}^3$, respectively. The second term is on the order of tens meV for room temperature. Thus E_F is at the midpoint of the HOMO-LUMO gap. The observed bulk Fermi level locates in the upper half of the energy gap ($E_c - E_F = 0.82\,\text{eV}$, $E_F - E_v = 1.76\,\text{eV}$), which is characteristic of n-type semiconductors.

Next we discuss the energy position of the donor level on the basis of the energy diagram and the estimated value of N_D. The electron density of C_{60}, n, at room temperature can be estimated as

$$n = N_c \exp\left(-\frac{E_c - E_F}{k_B T}\right) \sim 2.4 \times 10^7/\text{cm}^3 \tag{14}$$

The hole concentration of bulk C_{60}, p, is negligibly small due to the large value of $E_v - E_F$. The donor is partly ionized inducing positive charges. The concentration of the ionized donor, N_D^+ can be estimated as

$$N_D^+ = N_D \left[1 - \exp\left(-\frac{E_D - E_F}{k_B T}\right)\right]. \tag{15}$$

Here, E_D is the energy of the donor level. According to the charge neutrality condition, $n \sim N_D^+$ is required. By using the value of N_D obtained above, $E_D - E_F$ can be estimated to be almost zero. (Strictly speaking, the Fermi-Dirac distribution function should be used. However, the following result is qualitatively correct.) This means that the donor level is very close to the bulk Fermi level; in other words, the bulk Fermi level is governed not by C_{60} but the impurity.

The observed position of the bulk Fermi level was found to change during sample purification. Fig. 5(b) shows the energy diagram of the C_{60}/Cu interface, where C_{60} once purified by vacuum sublimation was used. The work function of the purified C_{60} could be deduced to be 4.59 eV, if we assume bulk Fermi level alignment, although this assumption should be further confirmed by depositing the purified C_{60} on other metals. This work function value is 0.20 eV lower than that of the as-received C_{60}. The purification induces the shift in the Fermi level position towards the midpoint of the energy gap ($E_F - E_C = 1.02$ eV). The Fermi level is still above the center of the gap, indicating that this purified sample still includes a donor impurity.

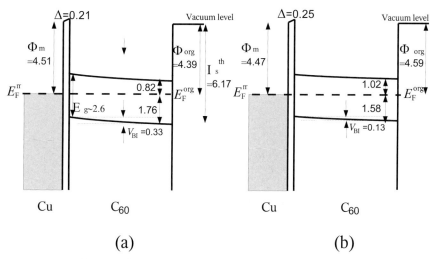

Figure 5 Energy diagrams of Cu/C_{60} interfaces obtained from the KP measurements in the case of as-received (a) and once-purified C_{60}.

3.5.2
TPD/metal interface: case for no band bending

Fig. 6 shows the change in ε_{vac}^F of TPD/metal (Au, Ag, Cu, Mg, Ca) as a function of the film thickness, d. At $d < 1$ nm, an abrupt shift in ε_{vac}^F due to the formation of an interfacial dipole layer is observed for all the substrates. After the completion of the shift, the plot becomes quite flat, indicating no thickness dependence.

The observed change in ε_{vac}^F is completed at around $d = 1$ nm, while a ten times thicker deposition of TPD is necessary for the Mg substrate. Similar slow changes have also been reported for other interfaces such as sexiphenyl (6P)/Mg [41] and 4,4'-N,N'-dicabazolyl-biphenyl (CBP)/Mg [13]. A low sticking coefficient and/or island growth of the adsorbed molecules on such substrates have been pointed out

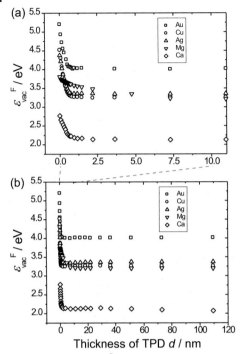

Figure 6 Variation in ε_{vac}^{F} for TPD on Au, Cu, Ag, Mg and Ca substrates as a function of TPD thickness, d. (a) The region for small d up to 10nm is expanded. (b) The entire region upto $d = 110$nm.

as the possible origin of such a slow change in ε_{vac}^{F}. In the latter case, more than one monolayer is necessary to completely cover the substrate surface. It should be noted that we have to sometimes be careful to distinguish band bending from such types of slow changes in ε_{vac}^{F}.

Fig. 6(b) shows the dependence of ε_{vac}^{F} on d over a wider thickness range up to $d = 110$ nm. The values of ε_{vac}^{F} did not show any change in the thicker region, indicating the flat band behavior. Furthermore, the values of ε_{vac}^{F} in the thick layers are significantly dependent on the metal, although those for the Cu, Ag, and Mg substrates are similar. The dependence of ε_{vac}^{F} clearly demonstrates that the alignment of the bulk Fermi level is not achieved up to $d = 110$ nm under UHV conditions. The observed flat band behavior is probably due to the high purity of the sample. Because of the very low concentration of the impurity, the space charge layer must be much wider, leading to the flat band behavior within 110 nm.

Energy diagrams of TPD/Metal interfaces constructed from the KP measurements are shown in Fig. 7. In the diagram, the values reported for the ionization threshold energy (I_s^{th} = 5.1 eV [42]) and band gap from the absorption edge (E_g = 3.3 eV [43]) are used to estimate the locations of the HOMO and the LUMO levels. The actual band gap of TPD is expected to be slightly larger: The difference between the optical gap and single particle gap is large, at least 0.5 eV, and perhaps as much as 1 eV for these materials. Here we will discuss the possibility of Fermi level alignment at these interfaces. First, we assume that the Fermi level is constant in the entire region of these interfaces. A common Fermi level, E_F is shown in the figure as the dash-dotted line. The intrinsic Fermi level of TPD layer is also shown by the dashed line.

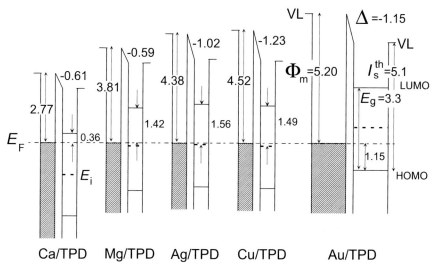

Figure 7 Energy diagram of TPD/metal (Au, Cu, Ag, Mg and Ca) interfaces constructed from the results of KP measurements. The values reported for the ionization threshold energy ($I_s^{th} = 5.1\,$eV [42]) and band gap from the absorption edge ($E_g = 3.3\,$eV [43]) are used to estimate the locations of the HOMO and LUMO levels.

For these interfaces, the V_{bi} calculated from the density of the thermal carriers is negligibly small within 100nm. Actually the energy separations between the LUMO and E_F for the Mg, Ag, and Cu substrates are greater than 1 eV, and the separation between the HOMO and E_F for the Au substrate is also greater than 1 eV. The density of the thermal carrier excited to 1eV above E_F is roughly $10^4/cm^3$, and the possible band bending in 100 nm becomes 3×10^{-13} V. Thus we cannot distinguish whether the Fermi level alignment is established over the entire region of the interfaces.

Alternatively, we can also discuss the possibility that the Fermi level alignment is locally established. If the system is not in equilibrium, a quasi-Fermi level is available as mentioned in the introduction. Because ε_{vac}^F is observed to be constant irrespective of the film thickness, the position of the HOMO and the LUMO levels of the TPD film is constant within the TPD layer. If we assume that the concentration of the impurity is negligible, it is plausible that the TPD layer is electrically neutral, suggesting that the quasi-Fermi level coincides with the intrinsic Fermi level of

TPD, i.e. midpoint of the gap. For the Ca/TPD and Au/TPD interfaces, apparently the quasi-Fermi level of the TPD layer does not coincide with the Fermi level of the metal side. Because of the large barrier for carrier injection from the metal to the TPD layer, the possible current flow across the interface to establish the balance of the Fermi level may be limited, and the system does not yet reach equilibrium. In this scheme, the relative location of the quasi-Fermi level in the HOMO-LUMO gap is independent of the metal substrates, although it is hard to experimentally determine the position of the quasi-Fermi level.

3.5.3
Potential gradient built in Alq₃ film

Alq$_3$ is one of the most representative materials for organic light emitting diodes. This molecule is known to have a permanent dipole moment [45]. In order to examine the effect of the molecular dipole on band bending, Kelvin probe measurements were performed for the Alq$_3$/metal(Au, Al) interfaces [29]. Figure 8 shows the KP results for the interfaces under light irradiation; the Alq$_3$ film was irradiated by white light from a flashlight during its deposition. The general trends in the dependence of ε_{vac}^{F} on the Alq$_3$ thickness for each metal are similar to those for TPD; an abrupt downward shift of the vacuum level was observed at $d < 1$ nm and ε_{vac}^{F} became constant in the thicker region up to $d = 600$ nm. The values of ε_{vac}^{F} are dependent on the substrates. This means no bulk-Fermi-level alignment occurs at the Alq$_3$/Au, Al interfaces within a 600 nm thickness as in the case of the TPD/metal interfaces. Again this is probably due to the very low concentration of the impurities included in the Alq$_3$.

Light irradiation is a key for this film. When the Alq$_3$ film is deposited under dark conditions, ε_{vac}^{F} of the film shows a much different behavior from that under light irradiation. Figure 9 plots ε_{vac}^{F} of the Alq$_3$ film deposited on Au and Al substrates under dark conditions. During the initial stage of deposition up to $d = 1$ nm,

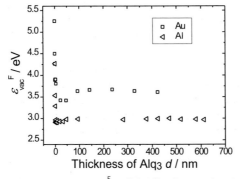

Figure 8 Variation in ε_{vac}^{F} of Alq$_3$ film deposited on Au and Al as a function of the film thickness. The deposition of Alq$_3$ was performed under light irradiation.

ε_{vac}^F shows an abrupt downward shift due to the formation of an interfacial dipole layer as in the case of "under light irradiation". In the thicker region, ε_{vac}^F shows a linear decrease over a wide range of thicknesses ($d = 5 \sim 550$ nm) with a mean slope of about -0.05 V/nm. Even at $d < 500$ nm, no saturation of the change in ε_{vac}^F was observed.

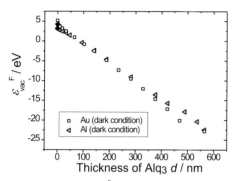

Figure 9 Variation in ε_{vac}^F of Alq$_3$ film deposited on Au and Al as a function of the film thickness. The deposition of Alq$_3$ was performed under dark conditions.

The linearity of ε_{vac}^F versus the thickness d suggests no net charge inside the film according to Poisson's equation; the second derivative of the potential becomes zero. This situation is similar to that of a capacitor. The linearity indicates uniform growth of the Alq$_3$ layer with a constant polarization, p. There remains a net charge only at both ends of the Alq$_3$ film and the charge density is equal to p. From the following equation, p can be estimated.

$$E = \frac{\phi}{d} = \frac{p}{\varepsilon_0}. \tag{18}$$

where E and ε_0 are the macroscopic electric field in the Alq$_3$ layer and the dielectric constant of the vacuum, respectively. ϕ is the potential across the Alq$_3$ film, which is experimentally determined as the total change of ε_{vac}^F from a 5 to 560 nm thickness. Using eq. (18) at the Alq$_3$/Au interface, $p = 0.44$ mC/m^2 is obtained from $\phi = 28$ V at $d = 560$ nm.

An Alq$_3$ molecule has a permanent dipole moment μ, which is estimated to be 4.1D and 7.1D for meridional and facial isomers using DFT calculation [45]. The meridional form is reported for usual thin films deposited by vacuum vapor deposition [46], while the crystal of the facial isomer has been found very recently [47]. If a perfect orientation of Alq$_3$ molecules is assumed, the polarization p can be estimated as

$$p = \frac{\mu}{\nu}, \tag{2}$$

where v is the volume per molecule ($0.5\,\text{nm}^3$ according to X-ray crystallographic analysis [46]). Thus the polarization in a perfect orientation can be estimated as $27\,\text{mC/m}^2$ and $47\,\text{mC/m}^2$ for the meridional and facial isomers, respectively. These are several tens larger than the observed value. Thus the molecular orientation is mostly random, but a finite polarization still remains on average as shown in Fig. 10. The partial formation of the domain of the ordered Alq$_3$ molecules may also explain the observed polarization. A similar domain formation was reported for the case of highly polar molecules mixed into less polar molecules prepared by the codeposition of two components including Alq$_3$ [48].

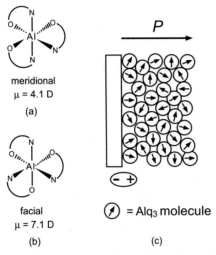

meridional
$\mu = 4.1\,\text{D}$
(a)

facial
$\mu = 7.1\,\text{D}$
(b)

P

= Alq$_3$ molecule

(c)

Figure 10 Two isomeric forms of Alq$_3$: (a) meridional and (b) facial. Model of the formation of giant surface potential of Alq$_3$ film is shown in (c).

Figure 11 shows the energy diagram constructed from the observed change in ε_{vac}^{F}. The energy levels have a constant gradient as a function of the film thickness. In the sense of a broad definition of band bending, the energy locations of the energy levels change as a function of the distance from the interface. The observed linear behavior is in contrast to the case of band bending leading to bulk Fermi level alignment, where the energy position quadratically depends on the distance from the interface under the assumption of a fully-ionized impurity of uniform concentration.

In the case of band bending leading to bulk Fermi level alignment as in the case of the C$_{60}$/metal interface, the Fermi level locates at a constant position over the entire region of the interface: The diffusion current due to the unbalance of carrier density is precisely cancelled by the drift current due to carriers in the space charge layer, leading to no net current at every place in the system. In contrast, it is clear that the Fermi level alignment is not achieved at the Alq$_3$/metal interfaces formed

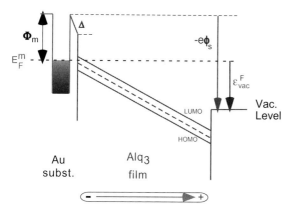

Figure 11 Energy diagram of Alq₃/Au interface constructed from the observed result. The energy levels of Alq₃ shows a linear decrease. At the surface, even the LUMO level is below the Fermi level of the substrate, indicating a nonequilibrium situation.

under dark conditions. First, if a Fermi level alignment is assumed, E_F should locate far above the vacuum level of the surface part of the Alq₃ film. Since the work function corresponds to the energy separation between E_F and the vacuum level of the solid, now the work function of the Alq₃ film is negative; this does not make physical sense. Secondly, this result means that almost all levels above the HOMO, which are usually unoccupied in the neutral Alq₃ solid, must be fully occupied by electrons, since the Fermi level locates above these levels. This situation leads to negative charging. Such charging will cancel the polarization due to the orientational polarization and finally achieve the alignment of the bulk Fermi level of Alq₃. However, the observed result is not the case.

Instead of the Fermi level E_F, a quasi-Fermi level is available to this system since thermal equilibrium is not established. If Alq₃ molecules in the films are charged, ε_{vac}^F should deviate from the linear function of d. Thus the observed linearity suggests that Alq₃ molecules are almost neutral and it is plausible that $\tilde{E}_F^p$ and $\tilde{E}_F^n$ are close to the midpoint of the HOMO-LUMO gap. ($\tilde{E}_i = \tilde{E}_F^p = \tilde{E}_F^n$). According to eq. (8), hole and electron currents are proportional to the gradient of the quasi-Fermi level and are finite. Such a current induces the accumulation of holes and electrons at both ends of the Alq₃ layer to perfectly cancel the polarization due to molecular orientation. An equilibrium state will then be finally established. However, the rate of moving to the equilibrium state is quite slow, as shown in the following experiment.

Figure 12 shows the change in the surface potential of the Alq₃ film in a vacuum without light irradiation after the deposition under dark conditions. The initial surface potential is about 7 V for the 100 nm thick film. A decrease of 0.2V was observed after 200 hours. Although the precise function form of this decay curve is not known, a 10% loss in 10 years is roughly expected. Thus it will take many years to lead to a thermal equilibrium state. Such a slow decay is probably due to the extremely small number of thermally excited carriers of Alq$_3$. When the sample is exposed to the atmosphere, the decay rate is increased. This is probably because ionized species in the atmosphere are adsorbed on the Alq$_3$ film to compensate the polarization.

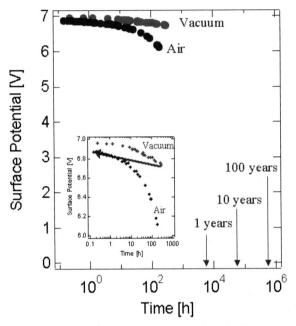

Figure 12   The change in the surface potential of Alq$_3$ film in vacuum after the deposition under dark conditions [49]. After keeping a vacuum condition for 200 hours, the chamber is purged and the sample was kept in air.

The large surface potential built up by depositing Alq$_3$ in the dark was also reduced by light irradiation with a light source of a 30-W flashlight placed near the viewing port at a distance of ca.18cm from the sample. The decay of ε_{vac}^{F} of the Alq$_3$ film under this condition is shown in Fig. 13, where the abscissa is the irradiation time t. The originally large ε_{vac}^{F} was reduced to about 3V within 1hour. It should be noted that the light irradiation and KP measurements were simultaneously performed: The sample surface was covered by the reference electrode of the KP probe and the sample film was irradiated through the narrow gap between the sample and the electrode at an extremely glancing angle, resulting in a very low rate of light irradiation. If the sample film is directly irradiated by the flashlight, the surface potential rapidly vanished.

The mechanism of the elimination of the giant surface potential by light irradiation is not clear at present, and two possible explanations are suggested. One is *the photocurrent model*, in which an incident photon is absorbed by the Alq$_3$ molecule and a photocurrent is generated under the local field due to the giant potential. Elec-

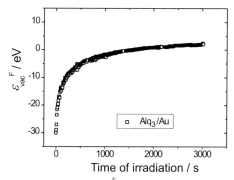

Figure 13 Variation in ε_{vac}^{F} against irradiation time t with white light using 30 W flashlight at a distance of ca. 18 cm through a viewing port.

trons and holes then travel to both ends of the Alq$_3$ film to cancel the polarization. The other is *the molecular reorientation model*, where the depolarization process is triggered by the photon absorption of Alq$_3$. Possible processes are the rotation of the molecule in the excited state and the formation of a paired structure of two Alq$_3$ molecules where each dipole moment has the opposite direction.

The photocurrent model seems to be initially plausible, but it can not explain the following result. Fig. 14(a) shows the change in the surface potential (SP), which corresponds to $\varepsilon_{vac}^{F} - \Phi_{m}$, of Alq$_3$ deposited under dark conditions as a function of the film thickness. The SP linearly increased with the increasing thickness. It reaches 27 V for the 575 nm thick film. The surface charge density, p, is estimated to be 0.44 mC/m^2. The Alq$_3$ film was irradiated by light, and the SP is reduced to almost zero. If *the photocurrent model* holds, the density of the counter charges accumulated at the surface and the substrate interface of Alq$_3$ film equal $-p$ and $+p$, respectively. Thus the polarization is completely compensated by the counter charges. Next, an additional Alq$_3$ layer was again deposited on the film under dark conditions. Here we expect that the counter charges accumulated at the surface of the depolarized film should be able to migrate to the surface of the deposited film to compensate the polarization of the additionally deposited film. Thus we can expect no variation in the SP during the additional deposition if the additional layer has the same molecular orientation as the original Alq$_3$ film. However, the increase in SP is observed by the additional deposition, and SP is proportional to the thickness of the additional layer with the same slope. This result makes this model dubious.

This experiment demonstrates the nonequilibrium behavior of the Alq$_3$ film. The energy diagram is shown in Fig. 14(b). The regions with and without polarization coexist. This means that the driving force towards thermal equilibrium is still negligible although there is no barrier for carrier exchange between the unpolarized and polarized Alq$_3$ layers. From the viewpoint of tailoring the band bending, this result indicates that a potential gradient can be controlled during the film growth by switching light irradiation, leading to possible application to control device performance.

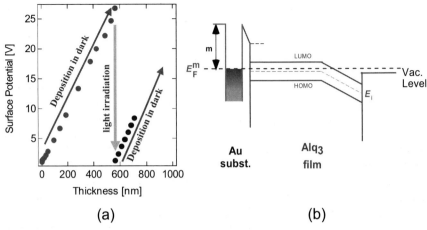

Figure 14   The change in the surface potential of Alq$_3$ film is plotted as a function of the film thickness (a) and the energy diagram constructed from the result (b). The large SP formed by deposition in the dark was removed by light irradiation. Additional deposition on the film induced an increase in the SP again.

Concerning the second model, *the molecular reorientation model*, we reported a theoretical simulation to analyze the decay rate of SP as a function of the irradiation time. The results suggest that this model is dominant, although more experimental studies should be carried out for examining the lifetime, dipole moments, and the possibility of molecular reorientation in the excited state. The details are described elsewhere [49].

3.5.4
Concluding Remarks

We have examined the band bending behavior of interfaces between organic semiconductors (C$_{60}$, TPD, Alq$_3$) and various metals (Au, Cu, Al, Mg, Ca) by measuring the change in the energy of the vacuum level under ultrahigh vacuum conditions by the Kelvin probe method. If we accept a broad definition of band bending, we observed several kinds of band bendings, in which the energies of the electronic levels of organic semiconductors change as a function of the distance from the interface with metal electrodes.

First, in the case of the C$_{60}$/metal interfaces, *band bending leading to bulk Fermi level alignment* (Fig. 15(a)) was observed. Actually, the energy of the vacuum level of a C$_{60}$ film showed a gradual shift due to band bending in the low thickness region, and then it became constant irrespective of the metal electrodes, satisfying the criteria for the examination of the bulk Fermi level alignment. This band bending is induced by the residual impurities in the film. In this case, the Fermi level of the system is constant at every place on the interface not only in the bulk but also in the interface region. This means that the work function of the C_{60} film on metal electrodes, which is defined as an energy separation between the Fermi level and the vacuum level of the film, depends on the film thickness in the space charge layer, but becomes independent in a fairly thick film. It should be noted that the work function of the bulk C_{60} was observed in the thickness on the order of 100nm. We emphasize that *band bending leading to Fermi level alignment* can be achieved in the presence of an impurity, but the thickness of the space charge layer often becomes thicker or compatible to the typical thickness of the organic devices.

Secondly, in the case of well purified organic semiconductors, a flat band feature occurs (Fig. 15(b)). No significant band bending was induced in the 100nm thick TPD film. No alignment of the bulk Fermi level is clearly established. For this situation, there are two schemes to discribe the Fermi level alignment. One is that the system is in thermal equilibrium and the Fermi level is common at the system. In this interpretation, carriers are efficiently exchanged between the electrode and organic layer, and the possible band bending is negligibly small due to the extremely small number of thermally excited carriers in the organic layer. Thus, the work function of the organic layer is not a characteristic value for the material, but strongly depends on the substrate metal. The other interpretation is that Fermi level alignment is achieved only within the organic layer, and the carrier exchange between the metal substrate and the organic layer is quite poor, and the system is not in equilibrium. In the case of TPD, it was impossible to distinguish the two models due to low concentration of the thermal carriers.

The above two models are within the scheme of band bending in the MS model: the energy levels of the organic layer can be bent by an impurity. If the organic layer is intentionally or unintentionally doped by an impurity, the energy level is bent by the space charge layer. However, the width of the space charge layer can often be larger than the typical dimension of the organic devices. In the case of intentional doping, a narrower space charge layer was reported by co-deposition with molecular dopant [21].

Third, we found a new type of band bending in a broad sense for Alq_3 as shown in Fig. 15(c). The energy level shift proportional to the distance from the interface was observed in contrast to the usual band bending due to impurity doping, in which the energy levels quadratically change. This phenomenon can be understood by orientation of the molecular dipole, and is characteristic of molecular systems. It is quite interesting that this band bending is erased by light irradiation. We demonstrated the control of band bending by light irradiation during the film deposition. As is clear in the energy diagram of Fig. 15(c), the system is definitely not in thermal equilibrium. The observed very slow decay rate of the surface potential indicates that the extremely small concentration of thermally excited carriers is not enough to supply the carriers to compensate for the band bending. Thus this result demonstrates that the time necessary to achieve thermal equilibrium, i.e., Fermi level alignment, can be very long depending on the system. This example again demonstrates that it is dangerous to naively assume Fermi level alignment for organic devices.

Finally, there are other types of band bendings in the broad sense. The screening effect to carriers near a metal electrode due to image charge is one example

(a) Band bending to leading
bulk E_F alignment

(b) Flat band

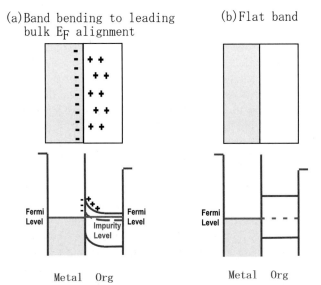

Metal   Org   Metal   Org

(c) Orientation molecular dipole   (d) Screening effect

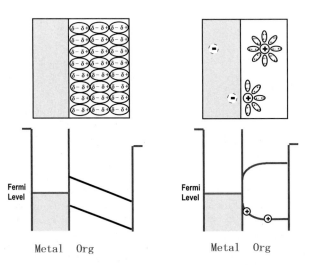

Metal   Org   Metal   Org

Figure 15 Various type of band bendings at organic semiconductor/metal interfaces. (a) Impurity-derived band bending leading to the bulk Fermi level alignment. (b) Flat band case due to high purity of organic semiconductor. (c) Linear potential profile observed for Alq$_3$ film evaporated under dark conditions. (d) Band bending effect in broad sense due to screening effect of carriers near metal electrode.

(Fig. 15(d)) [50, 51]. The effective potential of electrons and holes near the metal electrode changes as a function of the distance from the interface. The other type of band bending was reported for the metal/organic interfaces with metal clusters buried in organic layers. The clusters work as a carrier trap, forming a space charge for band bending [52].

The studies of band bending of organic semiconductors are necessary to control the internal electric field in actual organic devices during operation. It is also an important factor in organic photovoltaic cells. Spontaneous orientation of molecular dipoles and its depolarization by light irradiation of the Alq$_3$ film are interesting phenomenon, and have possibility for developing new types of devices.

Acknowledgement

The authors wish to thank Prof. N. Ueno of Chiba University and Prof. N. Karl of the University of Stuttgart for these fruitful discussions and suggestions about the Fermi level alignment problem. Part of this work was supported by Grants-in-Aid for Scientific Research (Nos.15350075, 14205007 and 13640576) and Creative Scientific Research (No.14GS0213) from the Ministry of Education, Culture, Sports, Science and Technology of Japan.

References

[1] M. Pope and C. E. Swenberg, Electronic Processes in Organic Crystals and Polymers, 2nd ed. (Oxford University, Oxford, New York, 1999)

[2] C. W. Tang, S. A. Van Slyke, Appl. Phys. Lett. **51**, 913 (1987)

[3] R. H. Friend, R. W. Gymer, A. B. Holmes, J. H. Burroughes, R. N. Marks, C. Taliani, D. D. C. Bradley, D. A. Dos Santos, J. L. Brédas, M. Lögdlund, and W. R. Salaneck, Nature **397**, 121 (1999).

[4] D. Dimitrakopoulos and P. R. L. Malenfant, Adv. Mat. **14**, 99 (2002).

[5] D. Wöhrle and D. Meissner, Adv. Mater. **3**, 129 (1991).

[6] M. Grätzel, Nature **414**, 338 (2001).

[7] N. F. Mott, Proc. Cambridge Philos. Soc. **34**, 568 (1938).

[8] W. Schottky, Phys. Z. **41**, 570 (1940).

[9] H. Ishii, K. Sugiyama, E. Ito, and K. Seki, Adv. Mater. **11**, 605 (1999).

[10] H. Ishii and K. Seki, in Conjugated Polymer and Molecular Interfaces, edited by W. R. Salaneck, K. Seki, A. Kahn, and J.-J. Pireaux, Marcel Dekker, New York (2002).

[11] R. Schlaf, B. A. Parkinson, P. A. Lee, K. W. Nebesny, G. Jabbour, B. Kippelen, N. Peyghambarian, and N. R. Armstrong, J. Appl. Phys. **84**, 6729 (1998).

[12] Y. Gao, Acc. Chem. Res. **32**, 247 (1999).

[13] I. G. Hill, A. Rajagopal, and A. Kahn, J. Appl. Phys. **84**, 3236 (1998).

[14] S. T. Lee, Y. M. Wang, X. Y. Hou, and C. W. Tang, Appl. Phys. Lett. **74**, 670 (1999).

[15] Q. T. Le, L. Yan, Y. Gao, M. G. Mason, D. J. Giesen, and C. W. Tang, J. Appl. Phys. **87**, 375 (2000).

[16] T. Kugler, M. Lögdlund, and W. R. Salaneck, Acc. Chem. Res. **32**, 225 (1999).

[17] Y. Harima, K. Takeda, and K. Yamashita, J. Phys. Chem. Sol. **56**, 1223 (1995).

[18] N. Sato and M. Yoshikawa, J. Electron Spectrosc. Relat. Phenom. **78**, 387 (1996).

[19] Y. Park, V. Choong, E. Ettedgui, Y. Gao, B. R. Hsieh, T. Wehrmeister, and K. Müllen, Appl. Phys. Lett. **69**, 1080 (1996).

[20] R. Schlaf, P. G. Schroeder, M. W. Nelson, B. A. Parkinson, C. D. Merritt, L. A. Crisafulli, H. Murata, and Z. H. Kafafi, Surf. Sci. **450**, 142 (2000).

[21] J. Blochwitz, T. Fritz, M. Pfeiffer, K. Leo, D. M. Alloway, P. A. Lee and N. R. Armstrong, Organic Electronics, **2**, 97 (2001).

[22] W. Gao and A. Kahn, J. Phys. Condens. Matter **15** S2757 (2003).

[23] M. Pfeiffer, K. Leo, and N. Karl, J. Appl. Phys. **80**, 6880 (1996).

[24] M. Iwamoto, A. Fukuda, and E. Itoh, J. Appl. Phys. **75**, 1607 (1994).

[25] Y. Harima, K. Yamashita, H. Ishii, and K. Seki, Thin Solid Films **366**, 237 (2000).

[26] M. Hiramoto, K. Ihara, and M. Yokoyama, Jpn. J. Appl. Phys., Part 1 **34**, 3803 (1995).

[27] E. Moons, A. Goossens, and T. Savenije, J. Phys. Chem. B **101**, 8492 (1997).

[28] N. Hayashi, H. Ishii, Y. Ouchi, and K. Seki, J. Appl. Phys. **92**, 3784 (2002).

[29] E. Ito, Y. Washizu, N. Hayashi, H. Ishii, N. Matsuie, K. Tsuboi, Y. Ouchi, Y. Harima, K. Yamashita, and K. Seki, J. Appl. Phys. **92**, 7306 (2002).

[30] M. Kotani and H. Akamatu, Bull. Chem. Soc. Jpn. **43**, 30 (1970).

[31] M. Kotani and H. Akamatu, Discuss. Faraday Soc. **51**, 94 (1971).

[32] N. Hayashi, E. Ito, H. Ishii, Y. Ouchi, and K. Seki, Synth. Met. **121**, 1717 (2001).

[33] L. Kronik and Y. Shapira, Surf. Sci. Rep. **37**, 1 (1999).

[34] Lord Kelvin, Philos. Mag. **46**, 82 (1898).

[35] W. A. Zisman, Rev. Sci. Instrum. **3**, 367 (1932).

[36] K. seki, N. Hayashi, E. Ito, Y. Ouchi, and H. Ishii, submitted to Organic Electronics.

[37] T. R. Ohno, Y. Chen, S. E. Harvey, G. H. Kroll, J. H. Weaver, R. E. Haufler, and R. E. Smalley, Phys. Rev. B **44**, 13747 (1991).

[38] P. A. Heiney, J. E. Fischer, A. R. McGhie, W. J. Romanow, A. M. Denenstein, J. P. McCauley, Jr., A. B. Smith, III, and D. E. Cox, Phys. Rev. Lett. **66**, 2911 (1991).

[39] N. Sato, Y. Saito, and H. Shinohara, Chem. Phys. **162**, 433 (1992).

[40] P. J. Benning, D. M. Poirier, T. R. Ohno, Y. Chen, M. B. Jost, F. Stepniak, G. H. Kroll, J. H. weaver, J. Fure, and R. E. Smalley, Phys. Rev. B **45**, 6899 (1992).

[41] H. Oji, E. Ito, M. Furuta, K. Kajikawa, H. Ishii, Y. Ouchi, and K. Seki, J. Electron. Spectrosc. Relat. Phenom. **101–103**, 517 (1999).

[42] K. Sugiyama, D. Yoshimura, T. Miyamae, T. Miyazaki, H. Ishii, Y. Ouchi, and K. Seki, J. Appl. Phys. **83**, 4928 (1998).

[43] S. Egusa, private communication.

[44] C. H. Marée, R. A. Weller, L. C. Feldman, K. Pakbaz, and H. W. H. Lee, J. Appl. Phys. **84**, 4013 (1998).

[45] A. Curioni, M. Boero, and W. Andreoni, Chem. Phys. Lett. **294**, 263 (1998).

[46] M. Brinkmann, G. Gadret, M. Muccini, C. Taliani, N. Masciocchi, and A. Sironi, J. Am. Chem. Soc. **122**, 5147 (2000).

[47] M. Colle, R. E. Dinnebier, W. Brütting, Chem. Commun. **23**, 2908 (2002).

[48] M. A. Baldo, Z. G. Soos, and S. R. Forrest, Chem. Phys. Lett. **347**, 297 (2001).

[49] K. Sugi, H. Ishii, Y. Kimura, M. Niwano, N. Hayashi, Y. Ouchi, E. Ito, and K. Seki , MRS Proceedings Volume **771**, L7.9.1 (2003).

[50] I. G. Hill, A. J. Makinen, and Z. H. Kafafi, Appl. Phys. Lett. **77**, 1825(2000).

[51] E. V. Tsiper, Z. G. Soos, W. Gao and A. Kahn, Chem. Phys. Lett. **360**, 47 (2002).

[52] H. Oji, E. Ito, M. Furuta, H. Ishii, Y. Ouchi, and K. Seki, Synth. Met., **121**, 1721 (2001).

4

Thermal and Structural Properties of the Organic Semiconductor Alq₃ and Characterization of its Excited Electronic Triplet State

M. Cölle and W. Brütting

4.1
Introduction

8-hydroxyquinoline metal chelate complexes were used for many years in analytical chemistry for a gravimetric determination of various metal cations in solution [1]. The development of more convenient spectroscopic techniques has meanwhile replaced this method and concomitantly decreased the interest of researching chemists in this reagent. Increasing interest in tris(8-hydroxyquinoline)aluminum(III) (Alq₃) shown in Figure 1 for technical applications started with a report on efficient electroluminescent devices using Alq₃ as the active medium [2]. These so-called organic light emitting diodes (OLEDs) opened the way for a new generation of flat panel displays. After nearly two decades of intensive research and development of OLEDs, Alq₃ still continues to be the workhorse in low-molecular weight materials for these devices. It is used as electron-transporting layer, as emission layer where green light emission is generated by electron-hole recombination in Alq₃, and it also serves as host material for various dyes to tune the emission color from green to red [3]. Many studies in this field have focused on the optimization of device performance with respect to efficiency and long-term stability or on the understanding of charge transport properties of amorphous thin films [4-13]. These investigations revealed that electrical transport in Alq₃ is characterized by a hopping-type charge carrier mobility displaying a Poole-Frenkel-like dependence on the electric field and on temperature. It was further found that trapping in distributed trap states is involved in charge transport, in particular at low fields. Different suggestions as to the origin of these traps were made, including a polaronic self–trapping effect, extrinsic traps due to impurities and the presence of a mixture of isomers of the Alq₃ molecule having different energy levels. However, no clear proof for one or the other possibility explaining the microscopic nature of these traps was given.

Another surprising circumstance was that in spite of the widespread usage of Alq₃ as amorphous films in OLEDs, comparatively few investigations were devoted to the material's structural, electronic and optical properties in the crystalline state, as well as to the dependence of these properties on the preparation conditions until recently [14-17]. On the other hand, it was mentioned in one of the very first publications on OLEDs based on thin films that the so-called "amorphous" film of Alq₃

Physics of Organic Semiconductors. Edited by W. Brütting
Copyright © 2005 WILEY-VCH Verlag GmbH & Co. KGaA, Weinheim
ISBN 3-527-40550-X

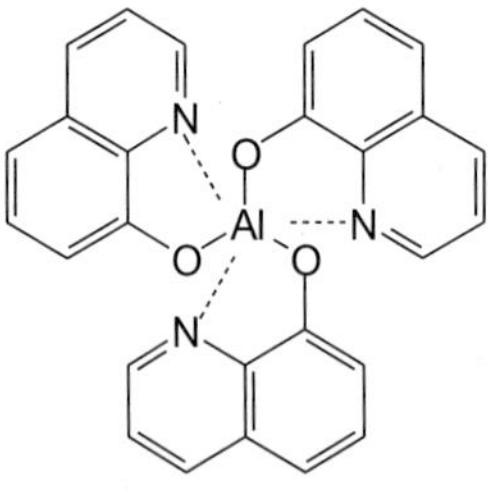

Figure 1   Chemical structure of Tris(8-hydroxyquinoline) aluminum(III) (Alq$_3$)

might have nanocrystalline domains [2], which raises questions concerning the morphology and properties of Alq$_3$. For example, what kind of crystalline phases can be formed by Alq$_3$ and what are their electronic and optical properties? What is the packing of the molecules? Packing and intermolecular interactions are important for optical properties as well as for their electrical characteristics and the transport mechanism of charge carriers.

Another unresolved issue concerns the isomerism of the Alq$_3$ molecule. It is well-known that octahedral complexes of the type MN$_3$O$_3$, where M is a trivalent metal and N and O stand for the nitrogen and oxygen atoms in the quinoline ligands, can occur in two different geometric isomers: meridional and facial, as shown in Figure 2 [18]. Nevertheless, until recently only the meridional isomer had been clearly identified and no direct experimental evidence for the facial isomer had been found. Therefore it was generally believed that the meridional isomer is predominant, both in amorphous films and crystals of Alq$_3$. The existence and the properties

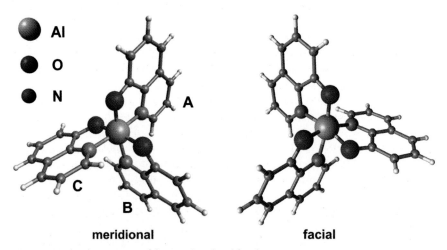

Figure 2   Molecular structure of the meridional and facial isomer of Alq$_3$. For the facial isomer the ligands are equivalent, but in the meridional molecule the ligands can be clearly distinguished and therefore the labeling of the ligands by A, B and C is given.

of the facial isomer are discussed in detail in the literature and a key issue is its possible presence in sublimed Alq$_3$ films [14,19-27]. Many suggestions have been made about its influence on trap density, charge carrier transport and thus on the characteristics and performance of OLEDs. For example, the higher dipole moment of the facial isomer is expected to influence the morphology of the film as well as the injection of charge carriers at the interface. In addition the different HOMO and LUMO levels predicted for the two isomers are expected to influence the injection barrier and could act as traps for charge carriers [24,28-32]. Therefore the question is whether the facial isomer is present in one or the other modification of Alq$_3$, and if so, if it is possible to isolate it. The isolation of the facial isomer is of great interest, as it will allow its properties to be examined separately and thus its role in OLEDs to be clarified.

4.2
Crystalline Phases of Alq$_3$

This section describes the preparation and identification of different crystalline phases of Alq$_3$ obtained by sublimation. In order to induce growth of different phases, the temperature gradient in a sublimation tube was used. Phases that grow at different temperatures were obtained and their crystal structures were investigated.

Temperature gradient sublimation is a common method for purification of organic materials. After this purification procedure polycrystalline powders of different appearance were found in the sublimation tube and thus we distinguished between three different zones in the glass tube. The materials in these zones, hereafter called fractions, differ in their shape of crystals, their color, their solubility and their fluorescence.

A typical example of these glass tubes after sublimation is shown in Figure 3 with indicated areas for the three different fractions. In the hottest zone of the growth area there is an approximately 1.5 cm wide region with very small needle-like crystals with white or slightly yellow appearance (fraction1). This zone is followed by the main fraction (about 8.5 cm) with yellow cubic crystals and dimensions up to 500x500x500µm^3, showing yellowish-green fluorescence (fraction2). In the subsequent colder zone of the sublimation tube another fraction is obtained with dark yellow-green needle-like crystals with a size of 50x50x500µm^3 (fraction 3).

These fractions have different solubility in organic solvents. While fraction 3 and (apart from a small residue) also fraction 2 are readily dissolved in chloroform at a relatively high concentration of more than 1% by weight, the solubility of fraction 1 is extremely poor. It takes several hours to dissolve a sizeable amount in chloroform, but then the color of the solution becomes similar to that of the other fractions.

Further differences between the three fractions are found in their photoluminescence (PL) spectra. Figure 4 shows the spectra measured with an excitation wavelength of 350nm at room temperature. All fractions show one broad PL band with no additional structures and a tail at the side of longer wavelengths. Their main diff-

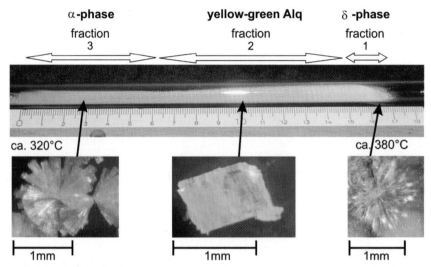

α-phase
fraction 3

yellow-green Alq
fraction 2

δ-phase
fraction 1

ca. 320°C

ca. 380°C

1mm   1mm   1mm

Figure 3 Picture of a sublimation tube. Due to the temperature gradient in the sublimation tube, the material obtained is separated into three zones, which are labeled by fraction 1, fraction 2 and fraction 3. Crystals of these fractions in the tube are also shown.

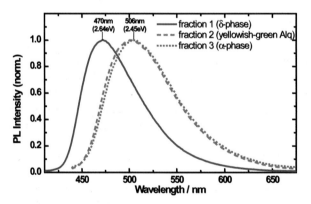

Figure 4 PL spectra of the three fractions obtained from the sublimation tube, excited at 350nm and measured at room temperature.

erence is the large blue shift of the PL maximum of about 0.19eV (36nm) from fraction 3 to fraction 1 with a PL maximum at about 506nm (2.45eV) and 470nm (2.64eV), respectively.

In order to investigate the origin of these differences, the crystallographic data of the three fractions were determined by using X-ray powder diffraction as shown in Figure 5. As a result two different phases were found. Fraction 1 and fraction 3 show the main differences. These differences are best seen for small angles below

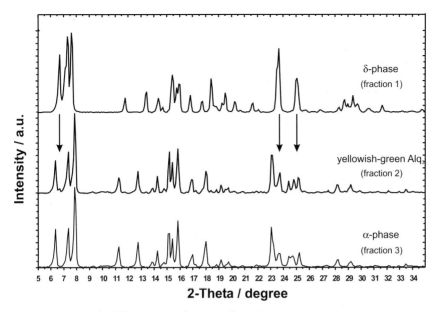

Figure 5 X-ray powder diffractograms of polycrystalline Alq$_3$
fractions 1, 2, and 3 obtained from the sublimation tube in the
2Θ-range from 5 to 35 degrees (step width $\Delta 2\Theta = 0.083°$) Arrows
mark areas with the most significant differences.

9 degrees and in the region between 22 and 26 degrees. From these two spectra the unit cells for fraction 1 and fraction 3 were determined. Indexing of the peaks observed is given in Reference [15] and [17] and the cell parameters determined for the different phases of Alq$_3$ are summarized in Table 1 together with two other phases (β- and γ-) found by Brinkmann et al. The spectrum of fraction 2 seems to be a mixture of two phases. Basically the spectrum is similar to that of fraction 3 apart from some small peaks or shoulders at positions where fraction 1 and fraction 3 are different, for example at 23.5 degrees and especially at 6.69 degrees. This suggests that fraction 2 mainly consists of the same phase as fraction 3, but has some small admixtures of material from fraction 1. The result that fraction 2 is a mixture of two different phases is relevant for applications, as it is mainly this fraction that is used for fabrication of OLEDs. From these X-ray data it becomes clear that the main difference is between fraction 1 and fraction 3, which have different unit cells given in Table 1.

It is possible to compare these crystal data obtained above with results of other researchers. Brinkmann et al. reported on three different crystalline structures called α-, β- and γ-phase [14]. The published data for the α-phase are identical to those of fraction 3. β-Alq$_3$ is grown from solution and its properties are in principle similar to the α-phase, only with a small red shift in the PL due to slightly different intermolecular interaction in the crystal. The published data of γ-Alq$_3$ are listed in Table 1 for completeness. All phases and evaporated films were identified as consisting of the meridional isomer, and therefore only the meridional molecule was found at that time.

Table 1: Crystallographic data of the polycrystalline phases of Alq$_3$.

	α-phase (fraction 3) [14, 15]	β-phase [14]	γ-phase [14]	δ-phase (fraction 1) [16, 17]
crystal system	triclinic	triclinic	trigonal	triclinic
space group	P-1	P-1	P-31c	P-1
Z	2	2	2	2
a [Å]	12.91	10.25	14.41	13.24
b [Å]	14.74	13.17	14.41	14.43
c [Å]	6.26	8.44	6.22	6.18
α [°]	89.7	97.1	90	88.55
β [°]	97.7	89.7	90	95.9
γ [°]	109.7	108.6	120	113.9
V [Å^3]	1111	1072	1118	1072.5

The denotation of the phases in our work is in accordance with these published data. Fraction 3 and the main part of fraction 2 consist of the α-phase. The structure of fraction 1 is new and no corresponding phase has been published so far. Accordingly fraction 1 is hereafter called the δ-phase of Alq$_3$.

δ-Alq$_3$ exhibits major differences to all other phases obtained from the sublimation tube. It is a whitish powder, has a different crystal structure and, importantly, a strongly blue-shifted PL. On the other hand the α- and β-phase are very similar, as reported by Brinkmann et al. Consequently it seems to be most interesting to investigate the differences and similarities of the α- and δ-phase of Alq$_3$, as will be done in the following sections of this article.

4.3
Thermal Properties of Alq$_3$

The phases discussed above were grown in different areas of the sublimation tube in regions of different temperature. Thus temperature obviously has a strong influence on the formation of these phases and it is important to learn more about the thermal properties of Alq$_3$. Therefore the formation conditions of the different phases of Alq$_3$ were investigated using differential scanning calorimetry (DSC) measurements in combination with structural and optical characterization.

Figure 6 shows the DSC measurement of polycrystalline Alq$_3$ powder (α-phase) taken at a heating rate of 20 °C/min. Coupled endothermic and exothermic peaks are observed at about 395 °C prior to the large melting transition at 419 °C. This additional phase transition has also been reported in the literature and has been attributed to polymorphism of the crystalline material [14, 34]. It is very pronounced at fast heating rates (above 15 °C/min). For slow heating rates the endothermic and exothermic transitions become broader and the peak height decreases as compared to the strong melting peak. The peaks start to intermingle and are shifted to a

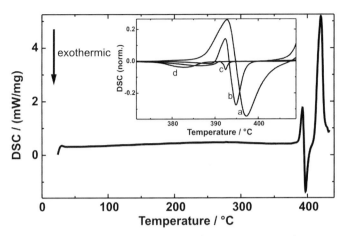

Figure 6 DSC trace of Alq₃ with pronounced thermal transitions at 393°, 396° and 419 °C measured at a heating rate of 20 °C/min. Inset: Broadening and intermingling of the endothermic and exothermic peaks around 395 °C in the DSC signal related to the sweep speed (a: 20 °C/min, b: 10 °C/min, c: 5 °C/min, d: 2 °C/min; normalized on the melting peak intensity). At low measuring speed only the more pronounced exothermic transition is visible.

slightly lower temperature, as shown in the inset of Figure 6 for heating rates of 20°, 10°, 5° and 2 °C per minute. This behavior is similar to known irreversible monotropic solid-solid transitions [35-37]. Typically, the monotropic transition is slow and is mostly observed a few degrees below the melting point. Thus it is advisable to measure the monotropic transition isothermally at very slow heating rates.

It should be noted that increasing the temperature above 430 °C results in decomposition of the material and that a small broad transition at 320 °C reported by Sapochak et al. [34] was not observed in our samples. For the following measurements a slow heating rate of 2 °C/min was used, where the shift of the peak temperatures is fairly small (see Figure 6) and where it is possible to stop the process at a defined temperature. Using this procedure the conditions for the preparation of different Alq₃ phases by a controlled thermal annealing process were investigated.

For these slow DSC measurements three different regions are distinguished in Figure 7: In the first region (A) below the exothermic phase transition Alq₃ is the usual yellowish green powder, in the second region (B) between this phase transition and the melting peak Alq₃ is a whitish powder, and finally in region C Alq₃ is a liquid melt. The glassy state of Alq₃ was obtained by quenching this melt in liquid nitrogen. Its highly amorphous character was verified by using X-ray powder diffraction measurements with an image plate detection system. Cooling down the liquid melt slowly resulted in yellowish-green powder (A) again, as was previously reported [38]. All of these materials are stable at room temperature.

Figure 8 shows the PL spectra measured at room temperature of annealed polycrystalline Alq₃ powder from regions A and B as well as of the quenched amorphous melt (C). For annealing temperatures up to 365 °C Alq₃ is a yellowish-green powder with a PL maximum at 506 nm (curve A). After the exothermic transition at about

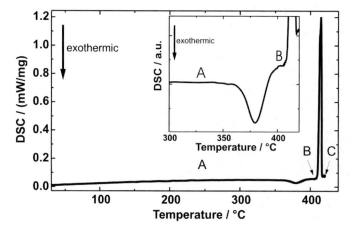

Figure 7   DSC trace of Alq₃ measured at a heating rate of 2 °C/min. The clearly pronounced exothermic phase transition at 380 °C prior to the melting point is enlarged in the inset, as it becomes broad and less intense compared to the melting peak for this slow heating rate. A, B and C mark the regions of yellowish-green Alq₃, blue Alq₃ and melt, respectively.

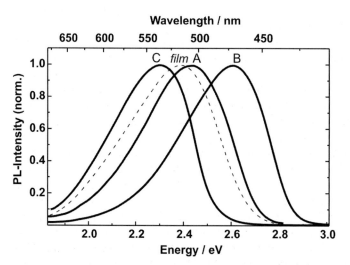

Figure 8   PL spectra of Alq₃ samples taken from regions A, B and C of Figure 7, respectively, excited at 350nm. The PL of an evaporated Alq₃ film (dashed line) is shown for purposes of comparison. All spectra were measured at room temperature.

380 °C, there is a big blue shift of 0.18eV (37nm), associated with a slight change in the shape of the PL spectrum (curve B), which is less symmetric for blue Alq₃. The quenched melt (curve C) is clearly red-shifted (0.14eV) compared to the yellowish-green Alq₃-powder (curve A). The strong difference in the emission color can be seen in Figure 9, where samples of the quenched melt, yellowish-green and blue

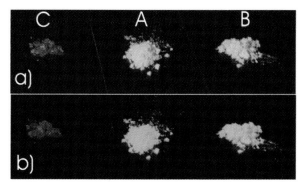

Figure 9 Photographs of Alq₃ samples taken from regions A, B and C in Figure 8: a) in usual daylight and b) under UV-irradiation (excitation wavelength: 366nm), clearly showing the strong blue shift of the luminescence of the annealed material.(CIE color coordinates for A: x=0.27, y=0.50; for B: x=0.16 , y=0.26).

Alq₃ are shown in daylight (a) and under UV-irradiation (b), respectively. The emission color is shifted from green (CIE coordinates: x=0.27, y=0.5) to blue (x=0.16, y=0.26). From Figure 9 one can also see the relatively low PL intensity of the quenched melt compared to the very intense PL emission of blue Alq₃. For PL quantum efficiency the values obtained for blue Alq₃, yellowish-green powder, evaporated film and quenched amorphous melt were 51%, 40%, 19% and 3%, respectively.

For comparison the dashed line in Figure 8 is the PL spectrum of an evaporated Alq₃-film as used in OLEDs. Although these films are commonly called "amorphous", one can clearly see that the PL maximum is located between the quenched melt and crystalline Alq₃. This is an indication of the nanocrystalline character of these films, as noted already by Tang et al. [2].

Yellowish-green Alq₃, blue Alq₃ and amorphous melt can be converted into each other. As described above, yellowish-green Alq₃ annealed above the phase transition at 380 °C results in blue Alq₃. Annealing blue Alq₃ above the melting point and cooling it down slowly, as shown in Figure 10, yields yellowish-green powder again and a pronounced recrystallization peak is observed. With the same procedure of annealing the quenched melt above the melting point and cooling it down slowly, yellowish-green powder is obtained again, and the quenched melt is converted into blue Alq₃ by annealing it between 380 °C and 410 °C. The successful conversion from one phase into the other was confirmed by measurements of the PL spectra, FT-IR spectra, Raman spectra, and X-ray diffraction.

Obviously, blue Alq₃ is formed during the phase transition at about 380 °C. This phase transition appeared when starting the measurement with yellowish-green Alq₃, as shown in Figure 10. On the other hand, when starting the annealing procedure with blue Alq₃ material no such phase transition was observed, as shown in Figure 10 i) and ii) trace b. However, measurements taken after the sample in Figure 10 b had cooled down showed the exothermic peak again, as can be seen in trace c of Figure 10 i).

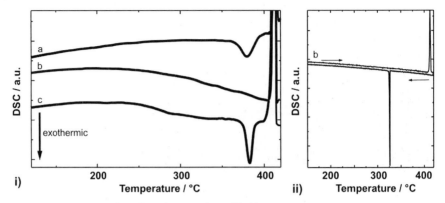

Figure 10 DSC traces of a: yellowish-green Alq$_3$ and b: blue Alq$_3$. Trace c shows a second heating cycle after cooling down the melt (b) again. By annealing blue Alq$_3$ no phase transition at 380 °C is observed (trace b in i) and ii)). Cooling down the melt gives a strong recrystallization peak.

As blue Alq$_3$ is formed in the region between the crystallographic phase transition and the melting point, the influence of temperature and preparation conditions in the region between 385 °C and 410 °C was investigated. Figure 11 shows X-ray powder diffraction spectra of blue Alq$_3$ prepared under three different conditions. For spectrum (I) yellowish-green Alq$_3$ powder (α-Alq$_3$) was annealed at 400 °C for 2h. This spectrum is similar to the one obtained for fraction 1 in the sublimation tube shown in Figure 5. The shoulder at 2θ=7.05° for different samples of blue Alq$_3$ was variably pronounced. From this one may assume another high-temperature phase to be present in these samples. To test this, Alq$_3$ was annealed for several minutes at a higher temperature of 410 °C (very close to the melting point) and a dark yellow substance was obtained, which exhibited only poor photoluminescence together with blue luminescent material. Its X-ray spectrum (Figure 11 (II)) has a number of new peaks, which become very obvious for example at 2Θ=7.05° (the position of the shoulder in spectrum (I)) and 25.85°. On the other hand, spectrum (III) shows Alq$_3$-powder annealed at 390 °C for 6 hours. The additional lines observed in spectrum (II) are no longer present in this spectrum.

Based on these investigations, blue luminescent Alq$_3$ obtained by annealing yellowish-green Alq$_3$ (α-phase) above the phase transition at about 380 °C was identified as the δ-phase of Alq$_3$ with the unit cell given in Table 1. As seen in curves (I) and (II) of Figure 11, annealing Alq$_3$ at temperatures higher than 380°, close to the melting point, results in the appearance of new peaks in the X-ray spectra, which can be attributed to an additional high temperature phase. Brinkmann et al. have reported on such a high temperature phase, namely γ-Alq$_3$ [14]. Using the given unit cell parameters from their work, the positions of all possible X-ray peaks for this phase were calculated, as indicated by the vertical bars in curve (II) of Figure 11. These calculated peaks are located at the positions where spectrum (II) and (III) are different. Therefore it suggests that in sample (II) there is a high concentration of

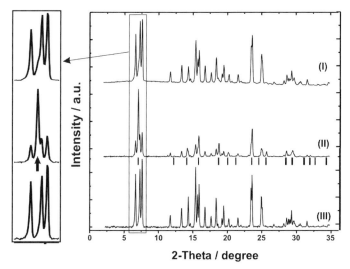

Figure 11 X ray powder diffractograms of polycrystalline blue Alq₃ prepared under different conditions. For spectrum (I) yellowish-green Alq₃-powder (α-Alq₃) was annealed at 400 °C for 2 hours. In spectrum (II) the powder was annealed at 410 °C (close to melting point). For spectrum (III) Alq₃ was annealed at 390 °C for 6 hours. The additional lines and shoulders observed in spectrum (II) are not present in spectrum (III). Bars in spectrum (II) mark calculated positions for all possible X-ray peaks of γ-Alq₃.

γ-Alq₃, whereas sample (III) is practically pure δ-Alq₃, as will be confirmed in the next section. From this it can be concluded that there are two high temperature phases of Alq₃: δ-Alq₃ and the γ-phase.

Blue luminescent Alq₃ obtained by train sublimation as described in the previous section and by annealing showed the same behavior with respect to its solubility as well as its properties in PL, DSC, and IR measurements, confirming that in both cases the δ-phase of Alq₃ was obtained. In the sublimation tube the different phases were separated due to the temperature gradient. Since δ-Alq₃ and the other high temperature phase (most likely γ-Alq₃) are formed in a relatively narrow temperature region, the separation of the two phases by train sublimation is difficult and a certain ratio of γ-Alq₃ is still present in the samples of δ-Alq₃, as indicated by the small shoulder at $2\theta{=}7.05°$ in the X-ray spectrum. On the other hand, under appropriate annealing conditions it is possible to obtain pure δ-phase without any visible admixtures of other phases, as demonstrated in curve (III) of Figure 11. A further advantage of this simple annealing process compared to temperature gradient sublimation is the possibility of obtaining large amounts (several grams) of pure δ-Alq₃ in a well-controlled process.

The relative content of δ- and γ- Alq₃ very critically depends on the preparation conditions (e.g. vacuum/atmosphere and temperature) as can also be seen in Figure 11 and can thus be tuned at will choosing suitable parameters. The samples measured in Figure 11 (III) and Figure 12 consist of more than 98% of δ- Alq₃, as can be derived from our X-ray data [67].

Chemical reactions during the annealing process can be excluded because the usual yellowish-green Alq$_3$ (α-phase) and the blue luminescent δ-Alq$_3$ can be easily converted into each other. Annealing yellowish-green Alq$_3$ at temperatures higher than 380 °C results in δ-Alq$_3$, while heating δ-Alq$_3$ above the melting point and cooling the melt down slowly results in yellowish-green powder again. Another method of reconverting blue Alq$_3$ into yellowish-green Alq$_3$ is to evaporate the material or to dissolve it in any appropriate solvent (e.g. chloroform). The same holds for the glassy state of Alq$_3$ obtained by quenching the melt. It is readily dissolved in chloroform and films of good quality can be cast from such solutions. The PL spectrum of such films is the same as for evaporated films of Alq$_3$. By annealing material in the glassy state, it is possible to obtain both the yellowish-green α-Alq$_3$ and the blue δ-Alq$_3$, depending on the temperature. In all cases pure Alq$_3$ with no visible contaminating material is obtained. The possibility of transforming Alq$_3$ from one phase into the other implies that even at these high temperatures there is no decomposition or chemical reaction of the material. So it is important to emphasize that for all temperatures up to 425 °C we are dealing with Alq$_3$, in agreement with ^{1}H NMR and FT-IR analysis of Alq$_3$ annealed at 422 °C, where no decomposition products have been found [34]. By excluding chemical reactions the difference in the phases must be of physical and structural origin.

4.4
The Molecular Structure of δ-Alq$_3$

4.4.1
High resolution powder diffraction using synchrotron radiation

In the previous sections a new phase of Alq$_3$, the δ-phase, which exhibits major differences to all other phases, was introduced and characterized. Based on the observed blue-shift of the PL by almost 0.2eV and the quantum chemical calculations of Curioni et al., which predicted a difference in the energy gap of the two isomers in that range [28], it could be supposed that the δ-phase contains the facial isomer of Alq$_3$. However, to prove this hypothesis it was necessary to resolve the crystal structure of the new phase, including the structure of the constituting molecules. The problem in determining the structure of organic molecular crystals is mainly due to the large number of atoms (104 for Alq$_3$) in the unit cell. Standard laboratory equipment requires single crystals to solve the structure of a new phase of a material; however, so far single crystals large enough for a full analysis of the structure have only been available for the β-phase of Alq$_3$ [14]. On the other hand, due to the use of high brilliance synchrotron radiation sources powder diffraction methods have progressed substantially in recent years, allowing very reliable determination of the structure from powder material without the need for larger single crystals. For this, high quality experimental data and specialized software for the analysis of the structure are required. These methods are very sensitive and unambiguous results are only to be expected if samples of one uniform crystal phase are measured. As the δ-phase can be isolated and δ-Alq₃ is easily obtained as a fine polycrystalline powder, these are good preconditions for this method.

In the case of a molecular crystal like Alq₃ it is necessary to start the simulation of the spectrum with an assumed configuration of the molecules within the unit cell in order to achieve convergence within a reasonable calculation time. Therefore we assumed a molecular configuration on the basis of the known connectivity of the molecule. The ligands were assumed to be planar and were randomly moved within a range of ±20° by a simulated annealing procedure until a minimal difference to the measured spectrum was obtained. After this, the position of the atoms was optimized by Rietveld refinements [39]. The accuracy of the structure obtained is given by the R-values and the goodness of fit χ. More information on the experimental procedure and analysis is found in Ref. [16] as well as in the literature [40, 43-45].

The following analysis of the data of the δ-phase of Alq₃ was made on the assumption that one of the two isomers is the constituent of this phase. First the results for the facial isomer are given, followed by the results for the meridional isomer for comparison.

Figure 12 shows the spectrum observed together with the best Rietveld-fit profiles for the assumption of the facial isomer. The enlarged difference curve between observed and calculated profiles is given in an additional window below. Indexing of this very well resolved powder spectrum with the ITO routine [46] led to a primitive

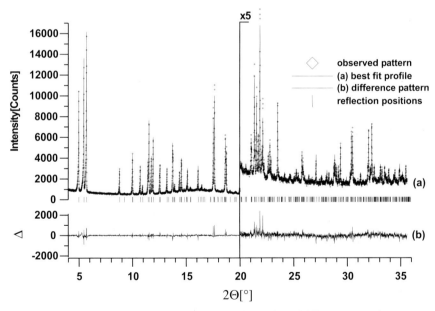

Figure 12 Scattered X-ray intensity for δ-Alq₃ under ambient conditions as a function of diffraction angle 2Θ. Shown are the observed patterns (diamonds), the best Rietveld-fit profiles on the assumption of a facial isomer (line) and the enlarged difference curves between observed and calculated profiles in an additional window below. The high angle part is enlarged by a factor of 5, starting at 20°. The wavelength was $\lambda = 1.15$ Å.

triclinic unit cell for Alq$_3$ with lattice parameters given in Table 2. The number of formula units per unit cell could be determined as Z=2 from packing considerations and density measurements. P-1 was selected as the most probable space group, which was confirmed by Rietveld refinements. The high quality of the refinement becomes obvious from the excellent differential pattern in particular at high diffraction angles (corresponding to small distances in real space), the R$_{wp}$ value of 6.5%, and the Bragg R value R-F^2 of 10.5%. Crystallographic data for δ-Alq$_3$ are listed in Table 2.

Table 2 Crystallographic data for δ-Alq$_3$. R$_p$, R$_{wp}$, and R-F^2 refer to the Rietveld criteria of the fit for profile and weighted profile respectively, defined by Langford and Louer [40].

Formula	Al(C$_9$H$_6$NO)$_3$	ρ-calc [g/cm^3]	1.423
Temperature [K]	295	2Θ range [°]	4-35.7
Formula weight [g/mol]	918.88	Step size [°2Θ]	0.005
Space group	P-1	Wavelength [Å]	1.14982(2)
Z	2	μ [1/cm]	2.48
a [Å]	13.2415(1)	Capillary diameter	0.7
b [Å]	14.4253(1)	R$_p$ [%]	5.0
c [Å]	6.17727(5)	R$_{wp}$ [%]	6.5
α [°]	88.5542(8)	R-F^2 [%]	10.5
β [°]	95.9258(7)	Reduced χ^2	1.6
γ [°]	113.9360(6)	No. of reflections	337
V [Å^3]	1072.52(2)	No. of variables	115

The molecular structure of δ-Alq$_3$ obtained from these measurements is shown in Figure 13. Compared to the idealized isolated facial Alq$_3$ isomer, the molecule is only slightly distorted, which reduces its symmetry only negligibly, and the planes defined by the O- and N-atoms, respectively, are parallel. The molecules form linear stacks in the c-direction of the crystal. The projection along the c-axis as well as the projection perpendicular to the planes of the hydroxyquinoline ligands, which shows the overlap between ligands of neighboring Alq$_3$ molecules, are shown in Figure 14.

The data was also evaluated on the assumption of the meridional isomer. The best fit obtained for this case is plotted in Figure 15 together with the differential curve. A comparison with Figure 12 clearly shows that the fit assuming the meridional isomer is far worse than the result for the facial isomer. Refinement resulted in a distorted meridional molecule, whereby the distance for one coordination bond (Al-N) was elongated more than 10% compared to the others (ligand A and B: ca. 2.1Å, ligand C: 2.39Å) and a Bragg R value R-F^2 of 19.4% was obtained. R-Values, tables and a picture of the distorted meridional molecule are given in Ref.[16] and Ref. [52].

The most important outcome of these refinements is that the δ-phase of Alq$_3$ consists of the facial isomer. For a long time it was believed that the facial isomer is unstable and would not exist. Thus the results shown here give clear evidence for the existence of this facial isomer [16, 47]. The simulations assuming the facial isomer closely match the measured spectrum, as can be seen in the differential spectrum in Figure 12, which is much better than the differential spectrum in Figure 15

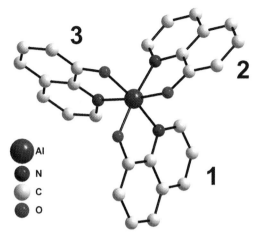

Figure 13 Facial Alq₃ molecule of the δ-phase with the three hydroxyquinoline ligands labeled by 1, 2 and 3. H-atoms are omitted for simplicity.

of the best possible fit for the meridional isomer. For the meridional isomer the molecule is distorted and a substantially higher Bragg R value (by 9%) was obtained compared to the facial isomer (R-F²= 10.5% facial, 19.4% meridional). The R values for the facial isomer indicate a high quality of the refinement, resulting in a very high probability that the δ-phase consists of this isomer. Furthermore, the high quality of the fit and the very well resolved spectrum suggests that the samples of δ-Alq₃ are an almost pure phase, confirming the results in the previous section. Therefore it can be concluded that the δ-Alq₃ samples prepared under defined annealing conditions as described above are a pure phase without significant admixtures of other phases and that δ-Alq₃ consists of the facial isomer. Thus, as a result of the preparation of δ-Alq₃, we have for the first time successfully isolated the long sought-after facial isomer of Alq₃.

This assignment of the facial isomer as being the only constituent of the δ-phase of Alq₃ was also very recently confirmed by NMR measurements [50] where the different electric field gradient tensors for the two isomers give characteristic fingerprints for their identification in solid state Al-NMR spectra. Moreover, the group at Eastman Kodak has recently grown single crystals of δ-Alq₃ large enough for a single crystal structure analysis. An excellent confirmation of the structure and the packing presented here was obtained [51].

The data also gives information about distance and orientation of the molecules and thus about molecular packing in the crystal. It is noteworthy that the molecules are arranged in a manner minimizing the possible overlap of the π-orbitals between pairs of hydroxyquinoline ligands belonging to neighboring Alq₃ molecules, as shown in Figure 14. As demonstrated by Brinkmann et al., the orbital overlap influences the optical properties and can explain shifts in the photoluminescence spectra of different phases of Alq₃ [14]. In δ-Alq₃ the pyridine rings of antiparallel ligands 1

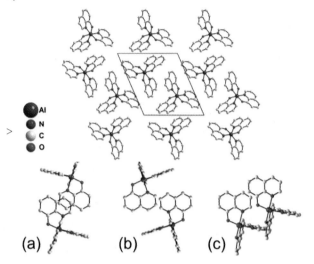

Figure 14 Crystal structure of δ-Alq₃ in a projection along the c-axis. (a), (b), and (c) are projections perpendicular to the planes of the hydroxyquinoline ligands 1, 2, and 3, respectively, showing the overlap between ligands of neighboring Alq₃ molecules

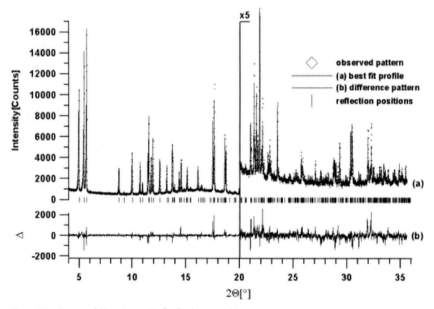

Figure 15 Scattered X-ray intensity for δ-Alq₃ at ambient conditions. Shown are the observed patterns (diamonds), the best Rietveld-fit profiles on the assumption of a meridional isomer (line) and the enlarged difference curves between observed and calculated profiles in an additional window below. Best values obtained for R_p, R_{wp} and R-F^2 are 7.3%, 9.4% and 19.4%, respectively.

face each other with an interligand distance of 3.4Å (Figure 14 (a)). The partial overlap of the rings is smaller than in the other known phases, and the atoms are slightly displaced, further reducing the overlap of the π-orbitals. Figure 14 (b) and (c) show the projection perpendicular to the planes of ligand 2 and ligand 3, respectively. The interligand distance is about 3.45 Å and these ligands do not overlap at all. Thus a strongly reduced π-orbital overlap of neighboring ligands is found in δ-Alq$_3$ as compared to the α- and β-phase. As only one ligand of each molecule overlaps with a neighboring molecule, there are no π-π links generating an extended one-dimensional chain, as reported for the β-phase [14]. In view of this, both the packing effect with reduced intermolecular interaction and the changed symmetry of the molecule are likely to be responsible for the large blue-shift of the photoluminescence by 0.2eV, which is in the same range as predicted theoretically by Curioni et al. for the two isomers [28].

For transformation from the meridional isomer to the facial isomer one ligand, namely ligand C in Figure 2, has to flip by 180°. From our results the facial isomer is formed at temperatures above 380 °C; thus the question is of interest whether this transition is energetically allowed for this molecule. Amati et al. made theoretical calculations for several possible transition processes between the geometric isomers of Alq$_3$ and its stereoisomers, and they found that thermal conversion from the meridional isomer to the facial isomer is energetically possible [48]. Very recently Utz et al. reported on NMR measurements of solutions demonstrating an internally mobile nature of the Alq$_3$ complex [49]. They found a high probability of ligands flipping by 180° and suggested that this process takes place on a time scale of about $5s^{-1}$ at room temperature in solution. In these measurements they were only able to determine the meridional isomer for two reasons: First, the facial isomer is predicted to be less stable by about 17kJ/mol for the isolated molecule [28, 48], thereby reducing its lifetime in solution; second, only the flip of ligand C may result in the facial isomer, giving a lower probability for this process, and thus the expected concentration of this isomer in solution is likely to be too small to be measured [49]. These measurements and the theoretical work of Amati et al. demonstrate that the transformation from the meridional isomer to the facial isomer at elevated temperature is possible, as was carried out for the δ-phase.

Brinkman et al. used the preparation method that is described in this review to obtain different phases of Tris(8-hydroxyquinoline) Gallium(III) [33]. It is highly remarkable, that they observed the same behavior as we describe here for Alq$_3$. Thus it is not simply related to the Alq$_3$ molecule but seems to be a general property of this class of chelate complexes.

4.2
Vibrational Analysis

Due to the different molecular symmetry of the meridional and facial isomers (C_1 versus C_3), vibrational analysis using infrared (IR) spectroscopy should be another possible method to differentiate between them. In particular, the first coordination sphere or central part of the molecule AlO$_3$N$_3$ should show characteristic vibrational properties for each isomer (Al-O and Al-N modes located below 600 cm^{-1}, as calcu-lated by Kushto et al. [25]). Furthermore, there is a weak coupling of the three li-gands via the central part, and movements around the central aluminum atom are involved in most of the molecular vibrations below 1700 cm^{-1}. This coupling depends on the relative positions of the oxygen atoms of the ligands (compare Fig. 2). For the facial isomer each oxygen atom faces a nitrogen atom, and thus the cou-pling via the Al atom is identical for all ligands, whereas for the meridional isomer one can clearly distinguish between the ligands labeled by A, B and C in Figure 2. For the meridional isomer, the coupling mainly affects the ligands B and C, where the oxygen atoms face each other, and to a lesser extent the A and B ligands, which have the oxygen and nitrogen atoms opposite. The coupling mechanism of ligand A and C is mainly characterized by the modes of the two opposite nitrogen atoms. This means that due to the lower symmetry of the meridional molecule each vibra-tional mode has a slightly different energy for the three ligands.

For the δ-phase of Alq$_3$ it was shown above by structural investigations to consist of the facial isomer. We can therefore use the IR spectra to identify characteristic differences in the vibrational properties of the two isomers. Figure 16 shows a com-parison of the FT-IR-spectra of δ-Alq$_3$, α-Alq$_3$, and the ligand hydroxyquinoline (8-Hq) alone. In principle one has to distinguish between two regions, above and below 600 cm^{-1}. The lines above 600 cm^{-1} are mainly related to vibrations within the ligands, as one can see from comparison with the 8-Hq spectrum. Due to the differ-ent symmetry of the isomers there is a different interaction of the ligands via the Al-atom leading to small differences in this region. The region below 600 cm^{-1} is dominated by the modes of the first coordination sphere or central fragment around the Al-atom. A detailed discussion of these spectra and individual lines as well as a discussion of the influence of crystallinity of the sample can be found in Refs [47, 52]. In this review we only give the main results and exemplify the discussion by the Al-N and Al-O stretching modes that are marked with arrows in Figure 16.

If we consider the central fragment AlO$_3$N$_3$, the local symmetry for each isomer is C$_{2v}$ and C$_{3v}$, respectively, as shown in Figure 17. The separation of the central part from the ligands is justified by the different and well-separated vibrational energies belonging to these groups, as observed in the comparison of Alq$_3$ with the hydroxy-quinoline parent of the ligands in Figure 16. Particular focus is on the stretching vibrations of this central part. For α-Alq$_3$, which consists of the meridional isomer (C$_{2v}$), six stretching vibrations are expected, three involving Al-N and three involving Al-O modes (see Figure 17). As they are all dipole-allowed, they are observable by IR-spectroscopy. According to Kushto et al. the following assignments for α-Alq$_3$ are made: Al-N stretching: 396 cm^{-1}, 405 cm^{-1}, 418 cm^{-1}, Al-O stretching: 522 cm^{-1}, 542 cm^{-1}, 549 cm^{-1}. By contrast δ-Alq$_3$ shows a total of only four bands in this re-gion (397 cm^{-1}, 423 cm^{-1}, 531 cm^{-1}, 548 cm^{-1}). As the AlO$_3$N$_3$ fragment of the facial isomer belongs to symmetry C$_{3v}$, six stretching vibrations are expected here too, but four of them belong to two degenerate vibrational states and therefore only four bands should be observed in IR-spectroscopy, as is the case for δ-Alq$_3$. The Al-N stretching is found at 397 cm^{-1} and 423 cm^{-1}, the Al-O stretching at 531 cm^{-1} and 548 cm^{-1}. The degeneracy of the first and last band is not present in the α-phase of Alq$_3$ (see Figure 16

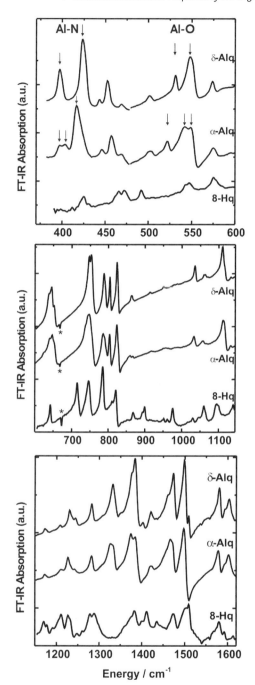

Figure 16 Comparison of the FTIR-spectra of δ-Alq₃ (upper trace), α-Alq₃ (middle trace) and hydroxyquinoline (8-Hq, lower trace) in the range from 350 cm⁻¹ to 1650 cm⁻¹

(a)

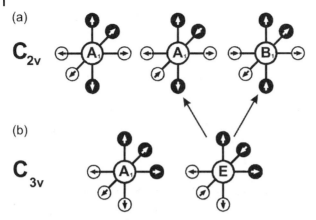

(b)

Figure 17 Schematic picture of the central part of the meridional (a) and the facial (b) isomer of Alq₃. Hollow and filled circles around the central Al-atom represent oxygen and nitrogen atoms, respectively. The three stretching modes of the meridional molecule (C_{2v}-symmetry) and the two for the facial molecule (C_{3v}-symmetry, one is degenerated) are marked by arrows in the O and N atoms.

and Figure 17), which consists of the meridional isomer. Two lines are observed at $400\,cm^{-1}$ and at $550\,cm^{-1}$ in α-Alq₃, in agreement with theoretical calculations of Kushto et al.[25] From this and the discussion in Ref [47] it can be seen, that the analysis of the IR-spectra for the region above as well as below $600\,cm^{-1}$ confirms the presence of the meridional and facial isomer in α-Alq₃ and δ-Alq₃, respectively. Furthermore, the specific fingerprints of the two isomers obtained by IR-spectroscopy may help to identify the isomers in other Alq₃-samples [47].

4.5
Population and properties of the electronic excited triplet state

4.5.1
Population of the triplet states

In the previous part of this review structural investigations and properties of the molecule in the electronic ground state were discussed, giving evidence for the existence of the two different geometric isomers. However, not only the electronic ground state should be different for the two isomers, but also the excited states are expected to have different properties due to the different geometry of the molecule. Two types of photoexcited states are distinguished: the singlet state and the triplet state. In the singlet state the total spin quantum number of the unpaired electrons S=0, whereas in the triplet state the total spin quantum number is S=1.

As in most cases the S_0-S_1 transition is an allowed transition, the lifetime of the S_1-state is very short. For Alq₃ it was measured to be about 12ns [3, 29, 53, 54]. On the other hand, the S_0-T_1 transition is a so-called forbidden transition and thus the lifetime of the T_1 state is expected to be several orders of magnitude larger. However, so far very few experimental data on the triplet state of Alq$_3$ have been available, and thus this section includes the first direct measurements of the triplet state. First, a method to investigate the population of the triplet states due to intersystem crossing (ISC) is introduced and applied to discuss the properties of the different phases. Then we briefly discuss very recent results related to the characterization of the electronic excited triplet state in Alq$_3$.

The triplet state T_1 is populated due to intersystem crossing, as schematically shown in Figure 18. In reality the triplet state splits into three levels $|x>$, $|y>$ and $|z>$. Their energetic distance is characterized by the zero field splitting parameters E and D. To simplify the following discussion this splitting of the triplet state is neglected and only T_1 is given in the schematic Figure 18. Due to photoexcitation by the absorption of incident laser light, mainly the singlet states S_n are excited ($S_0 \rightarrow S_n$) and relax to the lowest excited singlet state S_1 (process a). The excited singlet state S_1 can relax to the ground state ($S_1 \rightarrow S_0$) by emission of a photon (process b) or simply relax thermally (process c). The triplet state is populated by intersystem crossing with the rate constant d and f is the rate constant for the $T_1 \rightarrow S_0$ transition. In the literature f is often denoted as k_T.

Under constant photoexcitation and for long periods of time ($t \rightarrow \infty$) there is a dynamic equilibrium of the $S_1 \rightarrow T_1$ and $T_1 \rightarrow S_0$ transitions, resulting in a constant concentration of the triplet states $[T_1]^\infty$. The molecules which are in the long-lived triplet state are not able to emit fluorescent light and, at high excitation density, this leads to a decrease in fluorescence intensity. Therefore the process of intersystem crossing can be investigated by transient PL measurements and as a result the ratio of molecules in the triplet state can be estimated. The time dependence of the population process and the concentration of the triplet states $[T_1]^\infty$ is obtained from the rate equations (see Figure 18):

$$\frac{d[S_0]}{dt} = -a[S_0] + b[S_1] + c[S_1] + f[T_1] \tag{1}$$

$$\frac{d[S_1]}{dt} = a[S_0] -, b[S_1] -, c[S_1] - d[S_1] \tag{2}$$

$$\frac{d[T_1]}{dt} = d[S_1] - f[T_1] \tag{3}$$

These equations were solved by Sveshnikov, and Smirnov et al. [55, 56]. As the rate constants *b* and *c* cannot be distinguished experimentally here, they can be replaced by b′=b+c. Further, if we bear in mind both that the lifetime of the triplet state is much longer than the lifetime of the singlet state and that the rate of intersystem crossing is much higher than the rate of triplet decay (*b′≫f, d≫f*), the solutions are

$$[S_1] = \frac{af[S_0]^0}{(b'+d)f+da} + \frac{Aa^2[S_0]^0}{(b'+d)f+da}e^{-\frac{1}{\tau_1}t} - \frac{a[S_0]^0}{b'+d+Ba}e^{-\frac{1}{\tau_2}t} \tag{4}$$

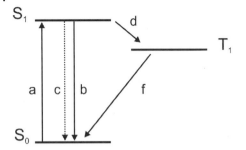

Figure 18 Schematic diagram of the electron levels and the transitions between the levels in an organic molecule. S_0, and S_1 are the non-excited ground state and the first excited singlet level; T_1 is the lowest triplet level. The coefficient a is proportional to the intensity of the exciting light and the probability of excitation of the molecule. b, c, d and f are the corresponding rate constants. The transition d is the population of the triplet state due to intersystem crossing.

$$[T_1] = \frac{ad[S_0]^0}{(b'+d)f+da}\left(1 - e^{-\frac{1}{\tau_1}t}\right)$$ (5)

with

$$f + \frac{d}{b'+d}a = f + Aa = \frac{1}{\tau_1}$$ (6)

and

$$b' + d + \frac{b'}{b'+d}a = b' + d + Ba = \frac{1}{\tau_2}.$$ (7)

From these equations it is evident that τ_1 is the characteristic time for the accumulation of molecules in the triplet state. For t→∞ (stationary regime) the concentration of molecules in the triplet state is given by

$$[T_1]^\infty = \frac{ad[S_0]^0}{(b'+d)f+da} = Aa[S_0]^0\tau_1$$ (8)

and finally with equation (6) one can express $[T_1]^\infty$ by the characteristic accumulation time τ_1 and the lifetime of the molecules in the triplet state $\tau_0 = 1/f$.

$$[T_1]^\infty = [S_0]^0\left(1 - \frac{\tau_1}{\tau_0}\right)$$ (9)

As a result it is possible to estimate the ratio of the molecules excited in the triplet state $[T_1]^\infty$ from the lifetime τ_0, determined from delayed fluorescence measurements that will be discussed below, and the characteristic accumulation time τ_1, which can be measured using transient PL studies.

These measurements for the Alq$_3$ phases as well as for an evaporated amorphous film are shown in Figure 19. Instantaneously with the turning-on of the excitation light the fluorescence is observed, which subsequently decreases with a decay time

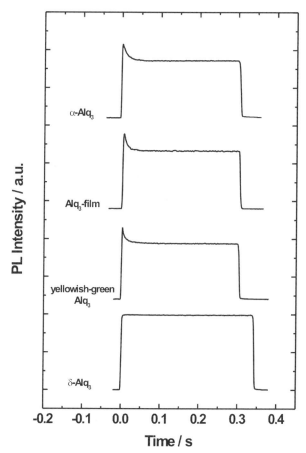

Figure 19 Time dependence of the PL-intensity during an optical excitation pulse for polycrystalline Alq₃ phases and an evaporated film. The measurements were performed at 1.3K by using excitation at 363.8nm. The signal was detected at 2.64eV (470nm) for δ-Alq₃ and at 2.48eV (500nm) for all other samples.

τ_1 to an equilibrium value. α-Alq₃, yellowish-green Alq₃ and the evaporated amorphous film behave in a similar fashion. Their decay time τ_1 is 11ms, 11ms, and 7ms, and the triplet lifetime τ_0 at that temperature (1.3K) was measured as 15ms, 14ms, and 9ms respectively. Therefore in these samples about 20% to 30% of the molecules are in the triplet state. This similarity between these samples of Alq₃ containing the meridional isomer is remarkable, because it clearly demonstrates that the morphology and thus intermolecular interactions seem to have no significant influence on the intersystem crossing process in Alq₃. However, for δ-Alq₃ there is only a very small decrease in the PL intensity and the equilibrium value remains at 98%. Due to the small decay and the noise of the measurement, the error for determination of τ_1 is too large, but from the decrease in intensity one may roughly estimate that only about 2% of the molecules in δ-Alq$_3$ are in the triplet state. From the independence of the morphology of the samples it can be concluded that the low population of the triplet state due to strongly reduced intersystem crossing is a molecular property of the facial isomer in δ-Alq$_3$. This has also been confirmed recently by Amati et al. using quantum chemical calculations [57].

4.2
Phosphorescence of Alq$_3$

The measurement of the transient PL discussed above gives information about intersystem crossing to the triplet state but not about its energetic position and lifetime. Until recently, the determination of the triplet properties has been based on theoretical calculations [24] and only very few experimental data were available so far: for example, the lifetime of the triplet state was derived from measurements of the diffusion length at room temperature using a phosphorescent sensing layer on the assumption of nondispersive transport, and the triplet energy was inferred from other metal-chelate complexes [61]. These experimental methods were necessary because no radiative triplet emission of Alq$_3$ (no phosphorescence) has been found. Only very recently direct observations of the electronic excited triplets have been reported: H.D. Burrows et al. published a phosphorescence spectrum of Alq$_3$ in an ethyl iodide glass matrix [62]. We measured the electro-phosphorescence of Alq$_3$-based OLEDs by using delayed electroluminescence [59], further, phosphorescence was also shown for all crystalline phases and for the evaporated amorphous film [52,60,63]. Here we summarize the results that were mainly obtained by delayed fluorescence as well as by phosphorescence of Alq$_3$.

The zero-field splitting parameters E and D are characteristic values for the triplet state. They were determined by ODMR measurements at zero field [60]. The measured values of the zero-field splitting parameters of $|E|=0.0114$ cm^{-1} and $|D|= 0.0630$ cm^{-1} are in the same range for the crystalline phases (α- and yellowish-green Alq$_3$) and for evaporated amorphous films, indicating a weak influence of the morphology. Calculations of the zero-field splitting parameters of the meridional Alq$_3$ molecule, starting from the D and E values of the isolated ligands, seem to support a mini-exciton-like behavior of the triplet state on the three ligands of the Alq$_3$ molecule [60].

In order to learn more about the properties of the triplet state, measurements of the transient PL in the millisecond time range were taken. From these it is possible to obtain information about the lifetime and the population of the long-lived triplet state due to intersystem crossing. The principle of these measurements is shown in Figure 20. The sample is excited by a rectangular laser pulse (dotted line), and as soon as the excitation light is turned on the spontaneous Alq$_3$ fluorescence is observed, which subsequently decreases to an equilibrium value with a decay time of about 10 ms. This decay is related to the population of the triplet states by ISC as discussed in detail above.

After the laser is turned off, the intensity of spontaneous PL from the singlet states decreases very fast due to their short lifetime of about 10 ns [3,29,53,54] and

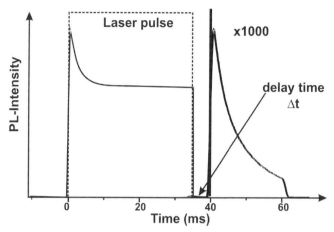

Figure 20 Principle of the transient PL measurements shown for α-Alq$_3$ measured at 20 K. By applying an intense rectangular laser pulse, PL from the sample is observed instantaneously. During the first 10-15 ms the PL-signal decreases due to intersystem crossing to a constant equilibrium value. After the laser is turned off, the spontaneous PL decreases within less than 1μs. The fluorescence observed after the laser has been turned off is called delayed fluorescence, with an intensity of about three orders of magnitude less than the spontaneous PL. In our experiments this delayed fluorescence was measured after a delay time Δt of several milliseconds. The cutoff of the delayed fluorescence, here at about 60 ms, is due to the chopper system of the setup.

only the triplet state whose lifetime is often in the range of several milliseconds, is still populated; thus after just 1 μs no spontaneous fluorescence is present any more. However, even after a delay time Δt of 4 ms, a weak PL is still observed. It is about three orders of magnitude less intense than the spontaneous PL and shows a slow decay rate. This process is known as delayed fluorescence (DF) in the literature. It occurs due to collision of two triplet excitations and is therefore a bimolecular process, which has a probability proportional to the square of the density of the triplet states [T$_1$]. The collision process is also often referred to as triplet exciton fusion or sometimes as T-T annihilation. If the energy of the lowest excited singlet state is less than the sum of the energies of the colliding triplet excitons, the fusion reaction may yield triplet and singlet states. The generated singlet exciton S$_1$ then decays, as would be the case for a directly excited singlet exciton state. As the DF originates from the S$_1$→S$_0$ transition, its spectrum is the same as observed in usual PL measurements, only significantly less intense than the instantaneous PL spectrum (in our case the difference is about three orders of magnitude). The distinguishing feature of the fluorescence from the S$_1$-state generated by the fusion process is that its apparent lifetime is determined by the triplet excitons and hence is much longer than the lifetime of the spontaneous fluorescence.

Figure 21 shows the delayed PL spectrum of α-Alq$_3$ at 10 K measured with a delay time Δt of 4 ms and the spectrum was integrated over about 20 ms. It shows two distinct bands, one at about 500 nm, similar to spontaneous PL, and an additional

Table 3 Lifetimes obtained from the transient measurements of I_{DF} of the different polycrystalline Alq$_3$ phases and amorphous films at a temperature of about 20 K. τ_{DF} is the time constant for the exponential decay of I_{DF} and from this the triplet lifetime τ_0 was obtained. τ_{700} is the time constant of the exponential decay of the luminescence intensity measured at 730 nm.

	τ_{DF}	τ_0 (=$2\tau_{DF}$)	τ_{700}	τ_{700}/τ_{DF}
α-Alq$_3$	6.6±0.5	13.2±1	13.6±0.5	2.05
Yellowish-green	7.8±0.5	15.6±1	16.2±0.5	2.08
δ-Alq$_3$	6.2±0.5	12.4±1	13.2±0.5	2.13
Film	4.33±0.5	8.66±1	9.3±0.5	2.15

band at about 700 nm. For α-Alq$_3$, yellowish-green Alq$_3$ and the evaporated film the position of the bands is approximately the same, whereas for δ-Alq$_3$ the bands are slightly blue-shifted. The relative intensity of the two bands is different for the different phases and temperature-dependent [64]. As the excitation wavelength of 442 nm is located below the absorption edge of δ-Alq$_3$, the obtained density of excited states is significantly lower for δ-Alq$_3$. Under these experimental conditions only a very weak band at 500 nm was observed for δ-Alq$_3$ due to the low density of triplet states, but the band at 700 nm was still clearly resolved [52,60,64].

The vibronic progressions on the high energy side of the band at 700 nm are clearly resolved. By subtraction of the usual PL spectrum from the delayed PL spectrum bands at 700 nm with vibronic progressions are obtained as shown for α-Alq in Figure 21. These are at approximately the same positions for α-Alq$_3$ and yellowish-green Alq$_3$. The vibronic progressions for α-Alq$_3$ and yellowish-green Alq$_3$ are

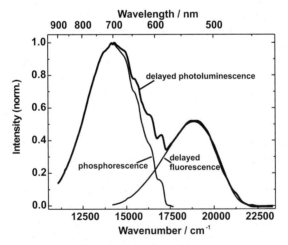

Figure 21 Delayed PL spectrum of α-Alq$_3$ at 10 K measured with a delay time Δt of 4ms. Excitation wavelength was 442 nm. The spectrum shows two distinct bands: the typical PL spectrum and a new additional band at about 700 nm (1.77 eV). The band at about 525nm is the delayed fluorescence (DF), the band at about 700nm the phosphorescence of Alq$_3$.

located at about 586 nm, 606 nm, 627 nm, 645 nm and 668 nm (17065 cm^{-1}, 16502 cm^{-1}, 15950 cm^{-1}, 15504 cm^{-1} and 14970 cm^{-1}), and for δ-Alq$_3$ at about 574nm, 594 nm, 613 nm and 635 nm (17422 cm^{-1}, 16835 cm^{-1}, 16313 cm^{-1} and 15748 cm^{-1}), and thus have an average distance of about 550 cm^{-1}, similar to the vibronic progression observed for the PL [15]. For the amorphous film the distance between the vibrational modes are similar but their position seems to be slightly red-shifted as reported in Refs [59,63]. The vibrational modes of the new band at about 700 nm are due to the vibrational modes of the Alq$_3$ molecule in its electronic ground state.

Later on it is shown, that the new band at about 700 nm is the $T_1 \rightarrow S_0$ transition. As this phosphorescence spectrum shows well-resolved vibronic progressions, one can directly determine the triplet energy by assignment of the lowest resolved vibronic band to the 0-0 transition. Hence the triplet energy for the meridional isomer (in α-Alq$_3$) can be determined as 2.11±0.1 eV and for the facial isomer (in δ-Alq$_3$) as 2.16±0.1 eV. For the evaporated film the lowest resolved vibronic progression seems to be slightly red-shifted [60,63] These experimental values are similar to the values roughly estimated by Baldo et al. (2 eV)[61], are very close to the theoretical value calculated by Martin et al. for an isolated molecule (2.13 eV) [24] and also close to the triplet energy of 2.17±0.1 eV that has been estimated from the phosphorescence spectrum of Alq$_3$ in an ethyl iodine glass matrix [62].

The transient PL intensity of both bands was also investigated. By measuring the decay of the intensity of the DF (I_{DF}), the lifetime of the triplet states can be determined [52,60,65]. For this one has to distinguish between two regimes, namely high and low triplet concentration. At very high triplet densities the T-T-annihilation directly influences the lifetime of the triplets in the sample and therefore the regime of low concentration has to be chosen to measure the correct triplet lifetime τ_0. This is equivalent to $(k_T[T_1] \gg \gamma_{tot}[T_1]^2)$, where k_T is correlated with the triplet lifetime τ_0 by $(\tau_0 = 1/k_T)$ and γ_{tot} is the total bimolecular annihilation (fusion) rate constant. In this case I_{DF} decays according to a monoexponential law $I_{DF}(t) \sim e^{-2k_T t}$ The decay time of I_{DF} is half of the correlated triplet lifetime τ_1, and thus it is possible to determine the lifetime of the triplet state by transient measurements of I_{DF}. By choosing the delay time Δt to be at least 4 ms we obtained very good monoexponential decays of the DF detected at 500 nm, confirming that the measurements were in the correct regime.

Figure 22 shows the intensity decay of the delayed luminescence of polycrystalline samples detected at 500 nm and of the additional band detected at 730 nm and measured at a temperature of 20 K. The measured apparent lifetimes of the delayed fluorescence detected at 500 nm are 6.6 ±0.5 ms, 7.8±0.5 ms and 6.2±0.5 ms, resulting in triplet lifetimes of 13.2±1 ms, 15.6±1 ms and 12.4±1 ms for α-Alq$_3$, yellowish-green Alq$_3$ and δ-Alq$_3$, respectively. The values are summarized in Table 3. Although in the film the triplet lifetime is about 60% - 70% of that in the polycrystalline samples, all values are in the same range and thus the morphology of the samples seems to have only little influence on the lifetime of the triplet states.

The decay of the band at about 700 nm is also shown in Figure 22. All polycrystalline phases show a similar monoexponential decay, which is significantly slower than that detected at 500 nm. Measured monoexponential decay times are also given

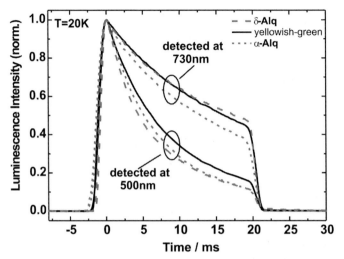

Figure 22 Transient intensity of the delayed luminescence shown in Figure 5 detected at 500 nm and at 730 nm, respectively. The delay time Δt was 4 ms in all cases. The steep edge at 20 ms is due to the experimental setup. The temperature was 20 K.

in Table 3. Within the accuracy of the measurement these values are about a factor of 2 higher than the values for the band at 500 nm and thus almost identical to the triplet lifetimes obtained.

The temperature dependence of the decay of both bands was investigated for all phases of Alq_3 including amorphous thin films, and as a result it became clear that both bands are directly correlated [52,60,64]. In principle the lifetime of the delayed fluorescence and the band at 700 nm increases with decreasing temperature, as shown in Figure 23 for yellowish-green Alq_3 powder. A local minimum is observed at about 50 K. This is due to a reversible phase transition that we also observed in temperature-dependent PL quantum efficiency measurements and ESR measurements [66]. Similar behavior was observed for δ-Alq_3 and amorphous films. For the evaporated amorphous film the local minimum is located at about 100 K (Figure 24). Measurements of the delayed fluorescence and phosphorescence of Alq_3-based OLEDs gave similar values and are also included for comparison [59]. As shown in Figure 25 for yellowish-green Alq_3-powder and in Figure 26 for the evaporated film, the lifetime of the band located at about 700 nm and the apparent lifetime of the delayed fluorescence always differ by a factor of 2 within the accuracy of the measurements, and thus the lifetime of this band is identified to be the lifetime of the triplet state. Therefore it is obvious that the band at 700 nm is directly linked with the triplet state of Alq_3. This justifies the assignment of this band to the $T_1 \rightarrow S_0$ transition and thus this spectrum is the phosphorescence spectrum of Alq_3.

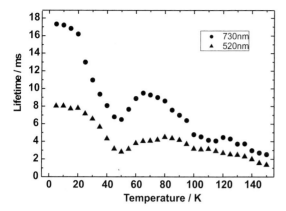

Figure 23 Lifetimes of both bands of the delayed PL of yellowish-green Alq$_3$-powder shown in Figure 2 in a temperature range between 6 K and 150 K. The lifetimes of both bands increase with decreasing temperature but show a local minimum at about 50K that is assigned to a phase transition [60,64].

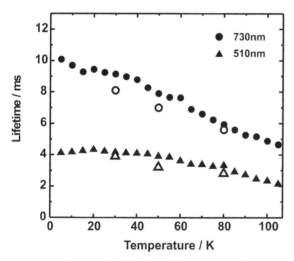

Figure 24 Lifetimes of both bands of the delayed PL of an evaporated amorphous film (thickness 5μm) over a temperature range of 100 K. Delay time Δt was 4 ms, the detection wavelengths were 510 nm and 730 nm, respectively. Hollow symbols indicate values measured by delayed electroluminescence of an Alq$_3$-LED which lead approximately to the same results as measurements using photoexcitation [59,63].

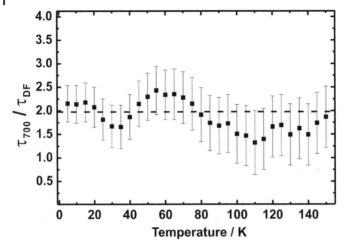

Figure 25 Temperature dependence of the factor (τ_{700}/τ_{DF}) between the decay times measured for the DF at 500 nm (τ_{DF}) and for the band at 700 nm (τ_{700}) of yellowish-green Alq$_3$-powder.

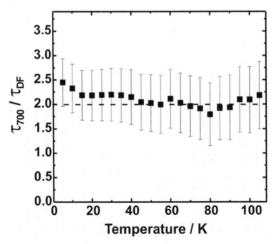

Figure 26 Factor (τ_{700}/τ_{DF}) between the decay times measured for the delayed fluorescence at 500 nm (τ_{DF}) and for the band at 700 nm (τ_{700}) of the Alq$_3$ film. The dashed line at the factor (τ_{700}/τ_{DF})=2 serves as a guide to the eye. The lifetimes of both bands are correlated by a factor of 2. The dashed line at the factor (τ_{700}/τ_{DF})=2 serves as a guide to the eye. From theory the lifetime of the DF should be half of the lifetime of the phosphorescence. Within the accuracy of the measurement τ_{700} and τ_{DF} are very well correlated by a factor of 2 over a wide range of temperatures, which confirms the assignment of the band at 700 nm to the phosphorescence of Alq$_3$.

4.6
Summary

In this review we have presented detailed investigations on polycrystalline Alq$_3$-samples prepared using sublimation in a horizontal glass tube with a temperature gradient or by annealing. The thermal, structural and photophysical properties were investigated by using differential scanning calorimetry (DSC), X-ray diffraction, infrared spectroscopy, and photoluminescence measurements, with the focus on the comparison of the different phases.

It was shown that different fractions of Alq$_3$ can be separated in the sublimation tube; they differ in the shape of the crystals, their color, their solubility, and their fluorescence. An important outcome was the discovery of a new blue luminescent crystalline phase of Alq$_3$ (δ-Alq$_3$), which has significantly different properties compared to all other known phases (α, β, γ). Furthermore, it was demonstrated that the material commonly used for the evaporation of thin films in OLEDs (here named yellowish-green Alq$_3$) mainly consists of α-Alq$_3$ with some small admixtures of δ-Alq$_3$.

As temperature has a strong influence on the formation of these polycrystalline phases, the formation conditions of the different phases of Alq$_3$ were investigated using differential scanning calorimetry measurements in combination with structural and optical characterization. As a result a phase transition at about 380 °C was identified where blue luminescent Alq$_3$ is formed. From detailed investigations of the processes in the temperature region between 385 °C and 410 °C, the two high-temperature phases δ-Alq$_3$ and γ-Alq$_3$ were identified. In addition, an efficient method was developed to prepare large amounts (several grams) of pure blue luminescent δ-Alq$_3$. It was also shown that all phases can easily be transformed into each other; thus chemical reactions could be excluded.

Well-defined preparation of pure δ-Alq$_3$ was the prerequisite for high resolution X-ray measurements on δ-Alq$_3$ powder including structural refinements. The high quality of these refinements gave convincing evidence that the facial isomer constitutes the δ-phase of Alq$_3$. The data also provided information about distance and orientation of the molecules and thus about molecular packing in the crystal. Compared to the α- and β-phase, both consisting of the meridional isomer, a strongly reduced π-orbital overlap of hydroxyquinoline ligands belonging to neighboring Alq$_3$ molecules was found in δ-Alq$_3$. Therefore both the packing effect with reduced intermolecular interaction and the changed symmetry of the molecule are likely to be responsible for the large blue-shift of the photoluminescence.

Using infrared spectroscopy it was demonstrated how the isomerism of the Alq$_3$ molecule is manifested in the vibrational properties. Comparison of the experimental results with theoretical calculations provided further evidence of the presence of the facial isomer in the blue luminescent δ-phase of Alq$_3$.

Experiments on other chelate complexes [33] suggest that the structural and thermal properties described for Alq$_3$ in this review are general properties of this class of chelate complexes.

Finally this article presented first results of the electronic excited triplet state in Alq_3. The intersystem crossing behavior was investigated, the phosphorescence spectrum was measured and it was found that the reduced population of the triplet states in δ-Alq_3 due to intersystem crossing is a characteristic property of the facial isomer.

Although Alq_3 is a so called singlet-emitter it is possible to characterize its triplet state and to measure the radiative $T_1 \rightarrow S_0$ transition, the phosphorescence. This was shown for different morphologies: crystalline phases, amorphous films and of diluted material in a frozen glass matrix. The delayed luminescence of the crystalline phases and the amorphous films shows two distinct bands: the delayed fluorescence and the phosphorescence of Alq_3. The data obtained for the amorphous films are in very good agreement with the data reported for Alq_3-based OLEDs. Well-resolved vibrational progressions for all samples allowed a precise determination of the triplet energy, namely 2.11±0.1 eV and 2.16±0.1eV for the meridional isomer (in α-Alq_3) and facial isomer (δ-Alq_3), respectively. Temperature-dependent investigations of the delayed luminescence in the range between 6K and 150K confirmed the identification of the phosphorescence band since the lifetime of the delayed fluorescence and of the phosphorescence always differ by a factor of two, as expected from theory. The triplet lifetime shows a clear temperature dependence from about 17 ms at 6 K to 4 ms at 120 K for yellowish-green Alq_3. These measurements also indicated a phase transition of Alq_3 at about 50 K. This direct characterization of the triplet state in Alq_3 is highly relevant for applications such as Alq_3-based OLEDs as most of the excitons formed in OLEDs are triplet excitons.

With regard to the application of Alq_3 in OLEDs, it is important to note that the simple annealing process can be used to obtain thin films of blue luminescent Alq_3. In preliminary experiments evaporated amorphous thin films with thicknesses from 300nm to 15µm, were encapsulated and converted at 390 °C into thin films showing blue luminescence. In addition it was possible to evaporate blue luminescent thin films directly onto heated glass substrates. At present, these films are still comparatively thick (several microns), have polycrystalline structure, and need further optimization. This would then allow OLEDs to be manufactured with blue luminescent Alq_3. Further work to characterize such films and their application in OLEDs is in progress.

Acknowledgements

The authors would like to thank all those that made this work possible. In particular we would like to thank Jürgen Gmeiner for the preparation of the Alq_3-samples, Markus Braun, Oliver Wendland and Jost-Ulrich von Schütz for transient PL-measurements, Falko D. Meyer, Wolfgang Milius and Harald Hillebrecht for X-ray analysis, Robert E. Dinnebier for performing the measurements and analysis of the synchrotron data, Stefan Forero-Lenger for IR-measurements, Marian Tzolov for Raman measurements that confirmed the IR-data, Christoph Gärditz for measurements of the delayed luminescence and, last but not least, Thomas Stübinger, Anton G. Mückl and Markus Schwoerer for helpful discussions.

References

[1] W. Ohnesorge, and L. Rogers, Spectrochim. Acta, Part A **15**, 27 (1959).

[2] C. Tang, and S. VanSlyke, Appl. Phys. Lett. **51**, 913 (1987).

[3] C. Tang, and S. VanSlyke, J. Appl. Phys. **65**, 3610 (1989).

[4] J. Shi, and C. Tang, Appl. Phys. Lett. **70**, 1665 (1997).

[5] L. Hung, C. Tang, and M. Mason, Appl. Phys. Lett. **70**, 152 (1997).

[6] H. Aziz, Z. Popovic, N.-X. Hu, A. Hor, and G. Xu, Science **283**, 1900 (1999).

[7] H. Kubota, S. Miyaguchi, S. Ishizuka, T. Wakimoto, J. Funaki, Y. Fukuda, T. Watanabe, H. Ochi, T. Sakamoto, T. Miyake, M. Tsuchida, I. Ohshita, and T. Tohma, J. Lumin. **87-89**, 56 (2000).

[8] T. Tsutsui, M. Yang, M. Yahiro, K. Nakamura, T. Watanabe, T. Tsuji, Y. Fukuda, T. Wakimoto, and S. Miyaguchi, Jpn. J. Appl. Phys. 2 **38**, 1502 (1999).

[9] P. Burrows, Z. Shen, V. Bulovic, D. McCarty, S. Forrest, J. Cronin, and M. Thompson, J. Appl. Phys. **79**, 7991 (1996).

[10] S. Barth, U. Wolf, H. Bässler, P. Müller, H. Riel, H. Vestweber, P. Seidler, and W. Rieß, Phys. Rev. B **60**, 8791 (1999).

[11] M. Stößel, J. Staudigel, F. Steuber, J. Blässing, J. Simmerer, and A. Winnacker, Appl. Phys. Lett. **76**, 115 (2000).

[12] W. Brütting, S. Berleb, and A. Mückl, Organic Electronics **2**, 1 (2001).

[13] S. Berleb, and W. Brütting, Phys. Rev. Lett. **89**, 286601 (2002).

[14] M. Brinkmann, G. Gadret, M. Muccini, C. Taliani, N. Masciocchi, and A. Sironi, J. Am. Chem. Soc. **122**, 5147 (2000).

[15] M. Braun, J. Gmeiner, M. Tzolov, M. Cölle, F. Meyer, W. Milius, H. Hillebrecht, O. Wendland, J. von Schütz, and W. Brütting, J. Chem. Phys. **114**, 9625 (2001).

[16] M. Cölle, R.E. Dinnebier, and W. Brütting, Chem. Comm. **23**, 2908 (2002).

[17] M. Cölle, J. Gmeiner, W. Milius, H. Hillebrecht, and W. Brütting, Adv. Funct. Mater. **13**, 108 (2003).

[18] G. Kauffmann, Coord. Chem. Rev. **12**, 105 (1974).

[19] R. Larsson, and O. Eskilsson, Acta Chem. Scand. **22**, 1067 (1968).

[20] B.C. Baker, and D.T. Sawyer, Anal. Chem. **40**, 1945 (1968).

[21] J. Majer, and M. Reade, Chem. Comm. **1**, 58 (1970).

[22] M. Halls, and R. Aroca, Can. J. Chem. **76**, 1730 (1998).

[23] N. Johansson, T. Osada, S. Stafström, W. Salaneck, V. Parente., D. dos Santos, X. Crispin, and J. Brédas J. Chem. Phys. **111**, 2157 (1999).

[24] R. Martin, J. Kress, I. Campbell, and D. Smith, Phys. Rev. B **61**, 15804 (2000).

[25] G. Kushto, Y. Iizumi, J. Kido, and Z. Kafafi, J. Phys. Chem. A **104**, 3670 (2000).

[26] M. Ichikawa, H. Yanagi, Y. Shimizu, S. Hotta, N. Suganuma, T. Koyama, and Y. Taniguchi, Adv. Mater. **14**, 1272 (2002).

[27] R.J. Curry, W.P. Gillin, J. Clarkson, and D.N. Batchelder, J. Appl. Phys. **92**, 1902 (2002).

[28] A. Curioni, M. Boero, and W. Andreoni, Chem. Phys. Lett. **294**, 263 (1998).

[29] W. Humbs, H. Zhangand M. Glasbeek, Chem. Phys. **254**, 319 (2000).

[30] J. Steiger, R. Schmechel, and H. von Seggern, Synth. Met. **129**, 1 (2002).

[31] E. Ito, Y. Washizu, N. Hayashi, H. Ishii, N. Matsuie, K. Tsuboi, Y. Ouchi, Y. Harima, K. Yamashita, and K. Seki, J. Appl. Phys. **92**, 7306 (2002).

[32] M. Amati, and F. Lelj, J. Phys. Chem. A. **107**, 2560 (2003).

[33] M. Brinkmann, B. Fite, S. Pratontep, and C. Chaumont, Chem. Mat. **16**, 4627 (2004).

[34] L.S. Sapochak, A. Padmaperuma, N. Washton, F. Endrino, G.T. Schmett, J. Marshall, D. Fogarty, and S.R. Forrest, J. Am. Chem. Soc. **123**, 6300 (2001).

[35] Mettler-Toledo, User Com **11**, 4 (2000).

[36] B. Wunderlich. (1990). *Thermal Analysis*. Academic Press, San Diego.

[37] J. Ford, and P. Timmins, (1989). *Pharmaceutical Thermal Analysis*.

[38] K. Naito, and A. Miura., J. Phys. Chem. **97**, 6240 (1993).

[39] H. Rietveld, J. Appl. Cryst. **2**, 65 (1969).

[40] L. Langford, and D. Louer, Rep. Prog. Phys. **59**, 131 (1996).

[41] R. Young, and D. Wiles J. Appl. Cryst. **15**, 430 (1982).

[42] R.J. Hill, and R.X. Fischer, J. Appl. Cryst. **23**, 462 (1990).

[43] W. Massa, (2000). *Crystal structure determination*. Springer, Berlin.

[44] R. Young (1993). *The Rietveld Method*. Oxford University Press, New York.

[45] P. Coppens, (1992). *Synchrotron Radiation Crystallography*. Academic Press, London.

[46] J.W. Visser., J. Appl. Cryst. **2**, 89 (1969).

[47] M. Cölle, S. Forero-Lenger, J. Gmeiner, and W. Brütting, Phys. Chem. Chem. Phys. **5**, 2958 (2003).

[48] M. Amati, and F. Lelj, Chem. Phys. Lett. **363**, 451 (2002).

[49] M. Utz, C. Chen, M. Morton, and F. Papadimitrakopoulos, J. Am. Chem. Soc. **125**, 1371 (2003).

[50] M. Utz, M. Nandagopal, M. Mathai, and F. Papadimitrakopoulos, Appl. Phys. Lett. **83**, 4023 (2003).

[51] M. Rajeswaran, T.N. Blanton and K.P. Klubek, Z.Kristallogr. NCS **218**, 439 (2003).

[52] M. Cölle, (2004). *The Electroluminescent Material Alq$_3$*, Logos-Verlag, Berlin.

[53] T. Mori, K. Obata, K. Miyachi, T. Mizutani, and Y. Kawakami, Jpn. J. Appl. Phys. 1 **36**, 7239 (1997).

[54] Y. Kawasumi, I. Akai, and T. Karasawa, Int. J. Mod. Phys. B **15**, 3825 (2001).

[55] B. Sveshnikov, Z. Eksp. Teo. Fiz. **18**, 878 (1948).

[56] V. Smirnov, and M. Alfimov, Kinetika i Kataliz **7**, 583 (1966).

[57] M. Amati and F. Lelj, Chem. Phys. Lett., **358**, 144 (2002)

[58] C.E. Pope and M. Swenberg, (1982), *Electronic processes in organic crystals*. Clarendon Press, Oxford.

[59] M. Cölle, and C. Gärditz, Appl. Phys. Lett., **84**, 3160, (2004).

[60] M. Cölle, C. Gärditz and M. Braun, J. Appl. Phys., **96**, 6133 (2004).

[61] M. A. Baldo and S. R. Forrest, Phys. Rev. B **62**, 10958 (2000).

[62] H.D. Burrows, M. Fernandes, J.S. de Melo, A.P. Monkman, and S. Navaratnam, J. Am. Chem. Soc. **125**, 15310 (2003).

[63] M. Cölle and C. Gärditz, Synth. Met., **147**, 97 (2004).

[64] M. Cölle and C. Gärditz, J. Lumin., **110**, 200 (2004).

[65] W. Kao and K. Hwang, *Electrical Transport in Solids*, Pergamon Press, 1981, Oxford.

[66] M. Cölle *et. al.* manuscript in preparation.

[67] X-ray data can be downloaded from the homepage of R. Dinnebier: http://www.fkf.mpg.de/xray/html/isomers.html

Part II
Photophysics

Physics of Organic Semiconductors. Edited by W. Brütting
Copyright © 2005 WILEY-VCH Verlag GmbH & Co. KGaA, Weinheim
ISBN 3-527-40550-X

5
Ultrafast Photophysics in Conjugated Polymers

G. Lanzani, G. Cerullo, D. Polli, A. Gambetta, M. Zavelani-Rossi, C. Gadermaier

5.1
Introduction

Conjugated polymers (CP) are chains of carbon atoms similar to natural quantum wires, having π-electron density delocalized along the backbone in a quasi 1-D geometry, with lateral confinement of about 0.5 nm. This peculiarity yields a distinctive character to the electronic structure and dynamics of CP, which are somewhat in between large organic molecules and low dimensional semiconductors. In spite of the validity of the molecular approach for describing their excitations, which is based on the framework used for large organic molecules [1], one should keep in mind that CP have unique features. Theory suggests that long enough chains can support quasi-particles [2], which in turn implies they have rich intra chain dynamics (self-trapping, singlet fission, charge separation, triplet pairs, soliton, polarons, bi-polarons). Basic excitations are neutral singlet, neutral triplet and charged states (in terms of spin quantum number S=0, S=1 and S=1/2, 3/2, etc. respectively). All those can be labelled, independently from theoretical models, as S_k, T_k and D_k, where S stays for singlet, T for triplet, D for doublet (these are charges, for we neglect at present the case of spin-less charge carriers [3]) and k is the enumerative integer index giving the order in energy. In the solid, molecular states can acquire an inter-chain character, be mobile or trapped at different sites. This makes the description rather complicated. Usually CP are amorphous, or have very short correlation length in space. The solid state properties are then closely related to those of conventional, disordered organic solids, while collective excitations (excitons) are rarely encountered. Yet conjugated segments on the same chain, separated by some kind of kink, defect or conjugation break, are a special case.

This paper is an overview on organic CP photophysics focussed on the early events following photo-excitation, as it can be deduced by ultrafast spectroscopy studies. The development of sub-ps and fs lasers has paralleled the discovery and growth of CP, and any significant advance in the technology of short pulse generation has found, from the very beginning, application to CP basic science. The reason is that CPs display extremely fast, often large responses to optical probes and have interesting photonic properties, such as large non-linear coefficients, stimulated emission, lasing, etc. The paper is organized as follows: in section (2) the laser sys-

Physics of Organic Semiconductors. Edited by W. Brütting
Copyright © 2005 WILEY-VCH Verlag GmbH & Co. KGaA, Weinheim
ISBN 3-527-40550-X

tems and experimental set-ups used are described; in section (3) the experimental technique is introduced; In section (4) we describe vibrational coherence; in section (5) we deal with vibrational relaxation, in section (6) with electronic relaxation.

5.2
Femtosecond laser systems

Femtosecond spectroscopy is a very powerful tool for investigating CP photophysics. The dramatic advances in femtosecond laser technology that have taken place in the last decade have greatly improved their stability, reliability and ease of operation and allow a routine investigation of many different CP materials with very high productivity.

The ideal spectroscopic system should be able to excite a molecule in resonance with an electronic transition and probe, with a delayed pulse, the transmission changes induced by the excitation pulse over the broadest possible wavelength range; in addition the temporal resolution, given by the cross-correlation between pump and probe pulses, should be as high as possible. Often the beneficial features (broad tunability of pump and probe pulses, high temporal resolution) are obtained at the expense of a greater complication and difficulty of operation of the experimental setup. In this section we will describe two femtosecond laser systems used by our research group for ultrafast spectroscopy of CPs. The first system that we describe is quite easy to handle and is widely used in many ultrafast spectroscopy laboratories: it provides a temporal resolution of $\approx$ 200 fs and allows to probe continuously the wavelength range from 400 to 1500 nm. In our laboratories it is used for the initial characterization of an unknown material. The second, more advanced system is based on a Noncollinear Optical Parametric Amplifier (NOPA) and allows to improve the temporal resolution by over an order of magnitude, down to 10-20 fs: it enables to capture the very early events of energy relaxation in CPs. An additional, unique feature enabled by the very short pulsewidths is the excitation and direct time-domain observation of coherent nuclear vibrations of the CP backbone.

The first spectroscopic system we use is based on an amplified Ti:sapphire laser (model CPA1 from Clark Instrumentation) producing 500 µJ, 140-fs pulses at 1 kHz and 780 nm wavelength (see figure 1).

The output of the system is divided by a beam splitter into pump and probe beams. Since most CPs have absorption bands in the UV, in most cases the pump beam can be obtained simply by frequency doubling the fundamental wavelength beam; using a 1-mm-thick LiB_3O_5 (LBO) crystal, 180 fs pulses at 390 nm wavelength with energies up to a few µJ can be generated. Broadband probe pulses can be obtained by a white light supercontinuum generated focusing 1-2 µJ of the 780 nm beam in a 2-mm-thick sapphire plate. By suitably choosing the pulse energy and the position of the sapphire plate around the focus, a very stable single-filament white-light continuum can be produced, extending from 400 to 1500 nm; typical energies are 10 pJ per 10 nm bandwidth, more than enough for probing purposes. The pump and probe pulses are both focused onto the sample after passing the pump beam

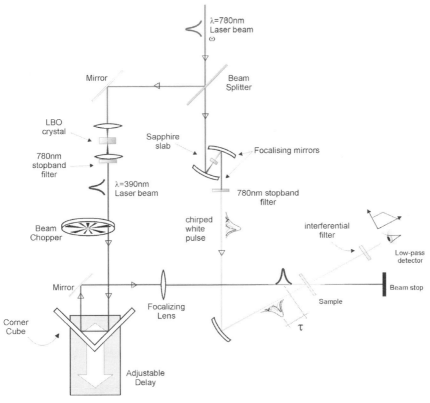

Figure 1 Experimental set-up for pump-probe spectroscopy.
All the components are labelled in the figure.

through a suitable computer-controlled delay line. The pump beam is focused to a spot of 80 μm diameter and the excitation fluence is between 0.3 and 12 mJ/cm^2 per pulse. Mirrors are used to collect the supercontinuum pulse and to focus it on the sample in order to minimize frequency chirp effects. The following measurements are carried out: i) The whole white light pulse is spectrally analyzed after travelling through the sample using a monochromator and a silicon diode array (OMA). Transmission difference spectra are obtained by subtracting pump-on and pump-off data. ii) Temporal evolution of the differential transmission is recorded at selected wavelengths (using interference filters of 10 nm bandwidth) by means of a standard lock-in technique.

The second experimental system used is based on a novel NOPA source, the experimental layout of which is shown in Fig. 2. The system starts again with the amplified system described above. Both the pump and the white light seed beams are focused into a β-barium borate (BBO) crystal, where the parametric amplification takes place. Using a non-collinear geometry with a suitable pump-seed angle (≈3.7°), the phase-matching angle becomes essentially independent of wavelength over most of the visible, so that very broad gain bandwidths can be obtained. The

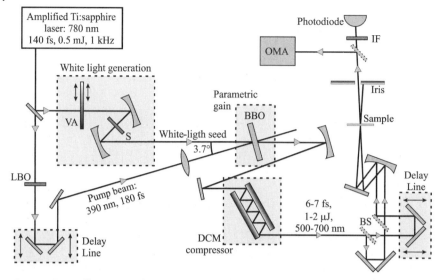

Figure 2 Non-collinear Optical Parametric amplifier, compression stage and pump-probe set-up. DCM=Double Chirp Mirrors, BS=Beam Splitters; S=Sapphire plate; OMA=Optical Multichannel Analyzer; IF=Interference Filters.

amplified pulses have bandwidths broader than 150 THz, extending from 500 nm to 700 nm. The pulses have energy of 1-2 µJ, which is more than enough for spectroscopic applications, and display good amplitude stability, with peak-to-peak fluctuations of less than 7%. The amplified pulses are then sent to a pulse compressor, which provides the negative dispersion necessary to cancel the positive frequency chirp introduced by the continuum generation and the amplification processes. The compressor consists of suitably designed chirped dielectric mirrors, i.e. mirrors providing a frequency-dependent group delay; after 10 bounces on the mirrors, the group delay of the pulses is essentially constant over their bandwidth and their duration approaches the transform-limited value of 6-7 fs. The compressed pulses are sent to a standard setup for pump-probe experiments. Given the very short pulsewidths, special care must be taken to minimize frequency chirping, that would cause a pulse lengthening; therefore only reflective optics are used to steer the pump pulses and focus them onto the sample and special 100-µm-thick inconel coated broadband beam splitters are employed. A stepper motor with time resolution of 0.6 fs controls the delay of the probe beam; the probe beam is selected with an iris and its differential transmission is measured with a photodiode followed by a lock-in amplifier. Passing the probe beam through a 10-nm-bandwidth interference filter after it traverses the sample makes measurements at selected wavelengths; this bandwidth reduction, if performed after the sample, does not cause any loss of temporal resolution.

5.3
The pump-probe experiment

In the basic experiment a first pulse, named pump, excites the sample and a second, weaker and delayed pulse, named probe, interrogates the excited region. The pump-induced relative transmission change (differential transmission) ΔT is measured by a lock-in amplifier referenced to the chopped pump pulse train and normalized to yield $\Delta T/T$. Upon re-excitation of the excited states created by the pump pulse with a second pulse called push, the transmission assumes a different value $(\Delta T/T)_{push}$. By chopping the push pulse train one obtains the push-induced variation of $\Delta T/T$ (double-differential transmission) $\Delta^2 T/T = (\Delta T/T)_{push} - \Delta T/T$.

On first base the interpretation of this experiment can be done by disregarding coherent effects (those require the full third order non linear response function in the field interaction picture), focussing on the population dynamics. In simple terms, let us consider an electronic transition between two levels, "i" and "j". The absorption coefficient is $\alpha_{ij}(\omega) = \sigma_{ij}(\omega)(N_i - N_j)$, where $\sigma_{ij}(\omega)$ is the cross-section (in cm^2), containing the dipole moment of the transition and the line-shape, while N_i, N_j are the populations of the initial and final states respectively. The convention is α positive for absorption and negative for gain (when $N_j > N_i$). This expression can be extended to a set of levels as

$$\alpha = \sum_{i,j} \sigma_{ij}(\omega)(N_i - N_j) = \sum_j \left(\sum_i \sigma_{ij}(\omega) N_j \right) \tag{1}$$

In the last sum we take σ positive for upwards transitions and negative for downwards ones.

The pump pulse acts on the sample by changing level occupation, $N \rightarrow N + \Delta N$. As a consequence some excited states will acquire population at the expense of the ground state, which gets depleted. In the small signal limit the measured quantity, i.e. the change in probe transmission, is:

$$\frac{\Delta T}{T} = -\sum_{i,j} \sigma_{ij}(\omega) \Delta N_j d \tag{2}$$

where d is the sample thickness and the expression is derived from the Lambert-Beer relation within the small signal approximation. "j" describes all possible excited states, irrespective of their charge or spin state, including each vibrational replica of the bare electronic ones. Time dependence is embodied in $\Delta N(t)$ where t is the pump-probe delay. If $\Delta N(t)$ already contains the pump pulse shape, the correct measured quantity and kinetics is obtained by correlation with the probe pulse profile.

Out of a huge number of states, only those transitions between dipole-coupled states have non-negligible cross section, thus the expression usually contains a limited number of terms. In addition it is of interest only for the frequency range actually explored by the probe pulse, so few transitions have to be practically considered. For j=0, ground state depletion is depicted, named bleaching. Photobleaching (PB) corresponds to positive ΔT and has the spectral shape of ground state absorption when thermalization, which is usually very fast, is over. For all the other levels (j > 0) both upwards and downwards transitions are possible. In particular, transitions from a level populated by the pump pulse to a higher lying one give rise to photoinduced absorption (PA). One should keep in mind that real systems are in this respect simpler. Internal conversion is very fast (about 100 fs), thus in most cases one probes the thermalized sample with occupation of the lowest states only. When the lowest singlet excitation (S_1) is dipole-coupled to S_0, i.e. the system is luminescent, stimulated emission (SE) is taking place. Yet this does not mean that positive ΔT is observed in the region of emission, for it depends on the spectral overlap with the absorbing transitions and on the relative cross sections involved. Often SE does not appear at all.

The population change ΔN induced by the pump can be expressed within the density matrix formalism by using the perturbation expansion to the second order in the pump interaction. This "jump" or window-state term [4] contains no electronic coherence, given the even number of interactions, but it has vibrational coherence, which represents impulsive excitation of the nuclear motion. The latter contributes an oscillation component to the pump-probe signal. While the theory is well established and will not be discussed here, in the following section we provide a qualitative picture of this phenomenon.

5.4
Coherent vibrational spectroscopy

The potential energy surfaces (PES) with respect to a nuclear configuration coordinate Q, for the electronic ground state and an electronic excited state, are shown in Figure 3. If the duration of the pump pulse is short compared to the period of nuclear vibration, the transition between these two electronic states is vertical in this picture. The excited molecule is left in a non-equilibrium position in the excited state PES and starts to oscillate around its new equilibrium position, much like a compressed spring. In the eigenstate description, the ground state vibrational wavefunction projected up to the excited state is a linear combination of many different vibrational eigenfunctions of the excited state with a certain phase relation to each other, separated by the oscillator energy hv. This wavepacket oscillates on the excited state PES [5,6,7], leaving and returning to the Franck-Condon (FC) region where it has been created. If the duration of the pump pulse is short enough, the excited molecules in the ensemble oscillate coherently, which results in a periodic modulation of the photoinduced absorption and stimulated emission signals that we observe. The interaction of the stationary ground state with the electromagnetic field of the pump pulse, which creates the wavepacket in the excited state, is called the first field interaction. Since the pump pulse has a finite duration, it also interacts with the excited state, and a fraction of the vibrational wavefunction can be projected down to the ground state. This is called the second field interaction. The further the wavepacket propagates between the first and second field interaction, the more the vibrational wavefunction on the ground state PES will be different from a stationary wavefunction and hence the stronger coherent oscillations on the ground state PES will be excited by the second field interaction. The relative contributions of excited and ground state to the oscillatory pattern are strongly dependent on the ratio between pulse length and oscillation period. For very short pump pulses, the pump-probe signal is dominated by excited state dynamics, because the polarization wave-packet created by the first field does not move significantly from the FC region upon arrival of the second field. As the pump pulse length increases, the polarization wave-packet has time to propagate further out of the FC region within the pump pulse, thus increasing the amplitude of ground state oscillations.

Observation of vibrational coherence provides a number of information, complementary to those of conventional Raman. Excited state modes can be detected, their frequency, dephasing and electronic coupling; Dephasing of ground state modes is measured, concerning coherence loss of the 0-1 as well higher vibrational levels. Having in mind the relation between coherence loss (T_2) and vibrational lifetime (T_1), $1/T_2=1/(2T_1)+\gamma$, these measurements provide a lower limit estimate for T_1 as well.

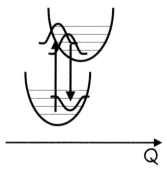

Figure 3 Schematics of wavepacket generation on the excited and ground state potential energy surfaces.

5.5
Vibrational relaxation

Upon optical excitation in CP, energy is distributed between the electronic and vibrational degrees of freedom, according to the molecular picture for vibronic transitions. Very few optically active (Franck-Condon) modes are initially populated. Their energy is quickly redistributed into a combination of modes (thermalization) by anharmonic coupling. Simple statistics predict that after this process most of the excess vibrational energy be stored into uncoupled modes, i.e. the electronic transition appears as starting from the zero-phonon level. Detailed studies of these processes in CPs do not exist, what we know is based on some indirect evidence. Except a few observations in solution[8], there are no reports on photoluminescence or stimulated emission build-up that could be assigned to vibrational relaxation in the excited state. This is usually interpreted as evidence for ultrafast vibrational relaxation, $T_1 < 100$ fs. The subject is however contradictory. We carried out pump-probe experiments in β-carotene in solution, finding that upon internal conversion the "hot" S_1 state thermalizes in about 2 ps [9]. Figure 4 shows the time evolution of the transient PA signal, S_1-S_n. The left panel reports experimental spectra, which clearly display a change in shape and a blue-shift of the peak. The right panel shows numerical simulation obtained according to a simple model for vibrational population relaxation. It is assumed that internal conversion places a fraction of population into the vibrational level *2* of the electronic state S_1. Three vibrational levels (j=0,1,2) were considered for each electronic state (S_1 and S_n). The vibrational populations of the electronic state S_1 are accounted for by the equations:

$$\frac{dN_{S_1,2}}{dt} = G_{IC} - k_{2,1} N_{S_1,2} \tag{3a}$$

$$\frac{dN_{S_1,1}}{dt} = k_{2,1} N_{S_1,2} - k_{1,0} N_{S_1,1} \tag{3b}$$

$$\frac{dN_{S_1,0}}{dt} = k_{1,0} N_{S_1,1} - R_{S_{1-0}} \tag{3c}$$

where $k_{x,y}$ is a vibrational relaxation rate ($1/T_1$), G_{IC} is the internal conversion rate of population transfer from some other electronic state and R is the lifetime of the relaxed S_1 state. The absorption spectrum is worked out considering the Frank-Condon factors for reasonable electron-phonon coupling of all possible vibronic transitions to $S_{n,w}$, with w=0,1,2, and introducing a gaussian broadening. Fitting the rate constants k_{21} and k_{10} (see Fig. 4), we find that vibrational relaxation slows down approaching the bottom of the potential well, as expressed by the ratio k_{21}/k_{10}=20. This result, which is confirmed by time-resolved Raman experiments [10], may be an "anomaly" in the vibronic structure of β-carotene. However, by measuring vibrational coherence dynamics in PPV, both film and solution, we estimated the vibrational lifetime $T_1 \geq 150$ fs [11], suggesting the finding has some generality. So it

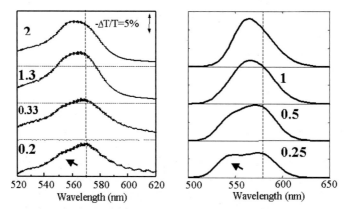

Figure 4 Left panel: transient spectra measured in β-carotene in solution at different pump-probe delays (labels) in ps; Right Panel: Numerical simulation (see text). The arrows points out a feature of the lineshape.

looks like direct probes of vibrational amplitude detect long phonon lifetime, while population dynamics suggest ultrashort lifetime.

The situation is clearer for the ground state, where some more direct indications exist about the time scale of the process. In Fig. 5 we show vibrational coherence damping in PDA, excited and probed by sub-10-fs pulses generated by the previously described NOPA. The observed coherence is assigned to ground state modes, given that the excited state undergoes ultrafast relaxation caused by singlet fission (see below), losing coherence. Fourier transform shows two main frequencies, assigned to double and triple stretching of the carbon backbone. From the data we extract the time scale for dephasing of the double stretching mode as $T_2 \approx 600$ *fs*, consistent with Raman linewidth and typical value for CPs, indicating that the vibrational life-time is $T_1 \geq 300$ fs in the ground state.

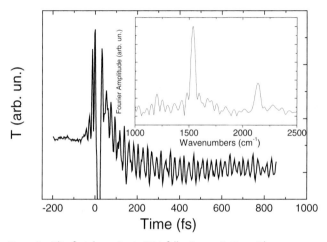

Figure 5 Ultrafast dynamics in PDA following excitation with sub-10 fs pulses as measured monitoring the 620 nm spectral component of the probe pulse band. Structures at negative delay are due to coherent artefacts, oscillations at positive delay reflect coherent nuclear motion. The inset shows Fourier transform of the oscillating signal at positive delays.

5.6
Electronic relaxation

Polyenes and carotenoids are good model compounds for understanding CP electronic structure, having a similar π-electron density delocalised along linear segments. Yet their photophysics turned out to be quite complex. The fact that they are not luminescent can be rationalised by the scheme depicted in the left Figure 6. Optical excitation induces a dipole transition from the ground state S_0 to the state S_2, which relaxes vibrationally towards S_1 much faster than it could radiatively relax towards S_0. The transition from S_1 to S_0 is dipole-forbidden [12] and hence can only occur via non-radiative, vibrational processes. This suggests the classification of electronic excited states into emissive ones and dark ones. Assuming the molecular structure bears inversion symmetry, which is approximately valid in most conjugated polymers, this distinction is equivalent to a classification into states of odd (and hence optically allowed towards the even ground sate) and even (and hence optically forbidden) parity. However, the recent identification of an intermediate state S_X in the S_2-S_1 transition in carotenoids [13], which is emissive but very short-lived, arouses the need for further subcategories (see right Figure. 6).

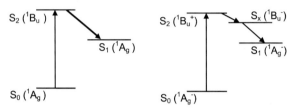

Figure 6   Electronic levels in β-carotene.

A second important classification of electronic states is based on their probability to dissociate into pairs of charges (polarons) or triplet states [14]. In the following we briefly review the case of singlet fission into triplet pairs for polydiacetylene (PDA), discuss the connection to polyacetylene and more extensively discuss charge generation, especially for ladder type poly-(paraphenylene) and fluorene polymers.

5.6.1
Triplet and soliton pairs

For linear, non-emitting conjugated systems theory predicts that the lowest covalent excitation is actually a spin wave, i.e. it can be obtained from the valence-bond ground state configuration by double spin flip. For what matters to us, this state configuration, with overall singlet character, can be described as a pair of triplets [2] or solitons. The latter are topological excitations [3], which require the reverse dimerization phase between the soliton and the antisoliton, so that they can be supported only in degenerate ground state systems (See Figure 7). For short chains this picture does not have any physical consequence, but for long enough chains, with large density of states, that's dramatic. The singlet state breaks into the internal components, two separated triplets for non-degenerate ground state. The process has distinct features. Triplet generation is ultrafast (triplet-triplet absorption build-up ~ 100 fs), circumventing the well-known limit due to spin flip induced by spin-orbit coupling. Second, the lifetime of the triplet states so produced is relatively short (~ 400 ps), due to bi-molecular annihilation (or geminate pair recombination)[16]. Third the process has a threshold, for $h\nu > 2E_T$, where $h\nu$ is the energy of the exciting photon, and E_T the energy of the triplet state (as measured from the ground state). In order to explore the possibility of state fission in CP we investigated with pump-probe spectroscopy substituted PDA dissolved in benzene [15]. Such a sample displays isolated chains, so it gives access to intrachain physics, and has a "carotenoid-like" enegy level ordering, with lower lying covalent state. Last but not least PDAs have technological interest, for instance in photonics.

Photoexcitation with sub-10 fs pulses creates a wave packet in the singlet manifold, which contains optically-allowed states. The spectral signature is optical gain and PB of the ground state transition. The initial wavepacket reaches a conical intersection and undergoes a branching into two states: the covalent state and the relaxed ionic state, responsible for emission. Fast PB recovery and vibrational dephasing are tentatively assigned to this process, completed in less than 100 fs.

a   b

Figure 7   (a) Schematic picture of the lowest covalent state. Circles points out spin inversion. (b) A segment of trans-polyacetylene showing the dimerization kink, neutral soliton S_0. A second kink (anti-soliton $\bar{S}_0$) will happen at some distance on the chain, reverting the dimerization to the initial one.

The energy located in the covalent state quickly funnels into the triplet manifold by singlet fission. The build-up of an absorption band in the gain region is followed in time, providing the dynamics of the process, which is completed in about 200 fs (see Figure 5). At this point the triplet-triplet absorption band is formed, peaking at 900 nm. The relaxed ionic state decays with lifetime in the 10-100 ps region, partially radiatively, while triplet pairs decay on a slightly longer time domain [16]. An estimate can be made for the time scale of the *intrachain* fissioning process, by evaluating the actual time needed for distinguishing the two triplet states. We depict fission as a separation in real space of the two triplets, which are seen as initially overlapped in the singlet configuration. Triplets may move along the chain at sound speed V_s. This can be estimated from the average phonon frequency, ω_0 about *1500 cm-1* for PDA, as $V_s \sim \omega_0 a/2 = 21$ *nm/ps*, where a is the average lattice constant. The triplet wavefunction extension, computed at the AM1 level in PDA oligomer, is 2.5 repeating unit [17]. So two triplets which are 5 repeating unit far apart can be distinguished. The time needed turns out to be $\sim$ *125 fs*.

The case of polyacetylene, with degenerate ground state with respect to dimerization, has been investigated in great details a decade or longer ago. To summarize, PA bands around 0.5 eV and 1.3-1.4 eV are assigned to charged and neutral solitons [18,19] respectively. Ultrafast spectroscopy has shown that such PA bands appear within 150 fs following photoexcitation [20] above the optical gap. A process of equilibration has been found too, with time constant of about 160 fs [21], interestingly matching the fissioning time estimated above. All this points to the ultrafast separation of a soliton-antisoliton pair on the chain. The initial interpretation was essentially based on the tight binding approach including electron-phonon coupling [22]. No electron-electron correlation was considered at the early stage. While the band picture seems nowadays to be unsuitable for reproducing the wealth of phenomena occurring in CP, if not incorrect, the soliton generation can still be understood within the picture presented above. Polyacetylene is the "infinite" limit of the polyene chain, it is non emissive and non-linear spectroscopy has identified a dark state below the optically active one [23,24]. Photoexcitation populates the optical state, and then the lowest lying dark state, by ultrafast internal conversion. This is a covalent excitation, which undergoes fission into soliton pairs and not triplet pairs, due to degeneracy. To conclude this section, it is worth mentioning that in di-substituted CHx, a form of air stable polyacetylene, photo excitation comprises both the "ionic" channel, leading to emission, and the "covalent" channel leading to separation into neutral soliton [25]. This system is then the counterpart, with degenerate ground state, of polyDCHD-HS. Both are emissive, but possess a closer lying (slightly above) dark state. Internal conversion from above reaches a branching point (possibly a conical intersection) bringing some population to the *covalent* state and some to the *ionic* state. In ring containing, photoluminescent CPs, there is evidence for the existence of fissioning states too, albeit at higher energy. For instance triplet photogeneration quantum efficiency in a ladder-type polymer [26] displays a jump around 3.2 eV, about 0.5 eV above S_1. This is assigned to the onset for singlet fission probably due to the presence of a covalent state. The process has small efficiency with respect to PDA, because of the competition with internal conversion, which brings population down to the lowest-lying ionic state.

5.6.2
Charge state dynamics

Charge dynamics, including generation, transport and recombination are of utmost importance in the understanding of CPs, for they are at the heart of most technological applications and characterize the materials as semiconductors. The early interpretation was based on the well-established band theory, the very same scheme adopted for inorganic semiconductors. In the last twenty years however, amid of spectacular development of devices and applications, the picture has changed, to be now quite far from that of a conventional semiconductor. In the following we present a short review of the ideas that have been put forward before moving to the more recent interpretation.

5.6.2.1 A brief overview about charge photogeneration
The first theoretical description of photoexcitation dynamics and charge carrier generation has been given in the framework of the Su-Schrieffer-Heeger (SSH) model [22]. This model treats the polymer chain as a quasi one-dimensional semiconductor, where photoexcitation directly creates a delocalised e-h pair with a low binding energy, a few kT at room temperature. However, there is ample experimental evidence from transient absorption, electroreflectance [27], electroabsorption [28,29], third-harmonic generation spectroscopy [30], scanning tunnelling spectroscopy [31], and inverse photoelectron spectroscopy [32], that singlet neutral states are the primary photoexcitations in CPs and pins their binding energy at 0.3–1.4 eV. For

instance in transient absorption experiments most CPs show a SE band with clear vibronic structures, which forms instantaneously upon photoexcitation. Note that in earlier experiments SE was not detected, and this was attributed to an instantaneous branching between neutral singlet states and charge carrier pairs [33]. It has later been found however that this is true only in photo-oxidised samples [34] and that the earlier experiments had been unintentionally been carried out in such condition.

In light of the shortcomings of the SSH model the effort to explain charge carrier photogeneration was turned towards the molecular framework, where singlet, localized excited states, are precursors to charged ones. In materials with low charge carrier mobility, which comprise CPs, charge carrier photogeneration is traditionally modelled within the Onsager theory of geminate charge recombination [35,36]. Here the initially hot excited electron performs a Brownian random walk during which its excess energy thermalises and it reaches a distance r from its geminate hole. In this configuration it can either recombine to a bound molecular state or dissociate into a pair of free charge carriers. It should be noted however that the Onsager model is essentially dealing with charge recombination, fully neglecting the nature of the very initial process of charge separation. The Onsager model is consistent with the frequently observed increase of photocurrent with increasing photon energy. CPs however do not show a universal behaviour in this regard, and often present a peak at the absorption edge or a jump at some higher energy. It was found that the electrical field could enhance singlet dissociation probability in CP as it is well known to occur in molecular solids [37]. This gave a handle to modulate the dissociation probability and hence study the dissociation mechanism. The first contribution came from studies of field-induced fluorescence quenching [38]. The large field applied in these experiments, in the order of MV/cm, and the relatively small fraction of dissociated states (up to 10%) is an important indication on the neutral state binding energy, which must largely exceed kT. The so-called Poole-Frenkel mechanism [39,40] can be convenient for describing the process if one focuses on the intra-chain process. Accordingly the formation of the charge-transfer state, intermediate to the formation of free carriers, is a tunnelling process across the Coulumb barrier, which is lowered by the electric field. In CPs however the dynamics of dissociation may quite reasonably involve inter-chain mechanisms, via neutral state migration and inter-chain separation. The difficulty here arising of separating the different contribution, in particular between interchain and intrachain charge separation, is clear. Very recent measurements on CPs in films and frozen solution might indicate the presence of an intra-chain process [41]. Vissenberg and de Jong [42] derived an analytic model that describes the interplay between migration, recombination and dissociation of neutral states. The basic idea is that as long as the neutral state is mobile it can reach sites where dissociation can take place, because of the favouring field orientation or local environment. Yet another crucial issue is that of the on-site energy, e.g. vibrational energy, which can assist dissociation. A detailed study of the temperature- field- and excitation energy dependence of the photocurrent of an m-LPPP photodiode [43,44] was interpreted in terms of the vibrational excess energy constituting a vibrational heat bath. This model has been challenged for its description as a quasi-equilibrium heat bath of what is actually a highly non-equilibrium situation. Therefore an alternative approach has been undertaken by Basko and Conwell [45], who assumed that the excess energy is initially stored in a low number of discrete vibrational modes (2 phonons of 0.18 eV and one of 0.08 eV for m-LPPP excited at 3.18 eV). The calculated dissociation time and hence the time scale of the loss of vibrational coherence according to this model is 30 fs. Coherent oscillations have not yet been observed in m-LPPP for lack of a sufficiently short pulse in the spectral range suitable for exciting m-LPPP. However, the decoherence time measured in PPV [11] (see above) was found to be one order of magnitude higher, questioning the assignment for m-LPPP.

5.6.2.2 Field-assisted pump-probe spectroscopy

The experimental data discussed so far provides only indirect evidence of charge carrier generation (fluorescence quenching rests on the assumption that all or a constant fraction of quenched singlets forms charge carriers) or detects only a fraction of the charges created (only mobile carriers contribute to the photocurrent, trapped carriers go unnoticed, only long-lived carriers are detected in slow photomodulation experiments). More direct evidence is obtained by optically detecting the charge carriers via their transient absorption signature. Field-modulated pump-probe mea-

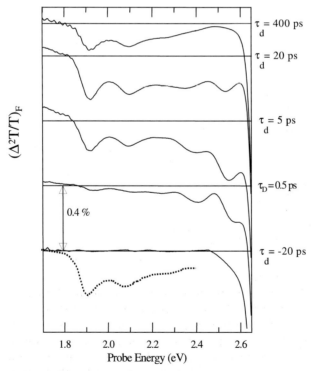

Figure 8 Electric Field assisted Pump-probe spectra for m-LPPP films. The dashed line shows the absorption spectrum of charged states in m-LPPP measured by chemical doping.

surements have been carried out on m-LPPP [46]. The field leads to the formation of additional polarons at the expense of singlets (Figure 8); note that these results have been exactly reproduced also recently, confirming their validity [47]. The temporal evolution of the singlet and polaron signals is independent of both the electric field amplitude and the pump intensity. The rate of dissociation of singlets into polaron pairs, $\gamma(t)$, shows a fast initial decrease which significantly slows down after 2-3 ps (Fig. 9), reaching a plateau. The initially higher dissociation rate points to a phenomenon favouring charge separation and fading off in time. There are two interpretations, without clear-cut separation: migration and on chain thermalization. Both represent re-equilibration following photo-excitation and have similar time scale [48], albeit the latter one is not precisely known. Likely they are both involved. The long time scale process is less straightforward to be interpreted due to the possible screening of the electric field by charges, a phenomenon which has been address in ref. [47]. In m-LPPP these experiments indicate that singlet states in presence of an electric field can dissociate for their whole lifetime, i.e. even after relaxation. Possibly this finding has to be understood as coming from the inter-chain process. In the next paragraph we will mention polyfluorene, which seems to behave in a different way.

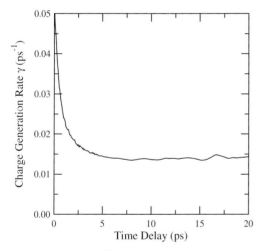

Figure 9   Singlet-state breaking rate γ versus pump-probe delay in presence of an external electric field in m-LPPP.

5.6.2.3   The role of higher-lying states

To conclude this discussion about charge generation we now consider the role of higher-lying excited states as intermediate in the process, so re-connecting to the ideas about fissioning expressed in the beginning. In photoconductivity experiments on a PPV derivative a large "jump" in the response was found at 4.7 eV, indicating the presence of a distinct electronic state at this energy [49]. The jump is due to a sudden increase in the charge generation once this state is reached [50]. Double-excitation experiments on PPV hint towards higher electronic states with a significant probability of dissociating into charge pairs [51].

In light of these findings, double-excitation techniques have been systematically used as a tool to study higher lying electronic states and their role in charge generation. In such an experiment a first (pump) pulse excites the transition S_0-S_1 (See Fig. 10). At a defined delay a second pulse (which we call push) excites the transition S_1-S_n. In the experiments discussed below two ultrafast deactivation pathways have been identified for S_n: relaxation towards S_1 and dissociation into a charge pair. Hence the modulation of the push pulse train gives a handle on the charge generation efficiency similar to the electric field discussed above. Two experimental configurations have been realised: photocurrent cross-correlation, where the sample is a photodiode whose photocurrent variation upon double excitation is measured, and pump-push-probe, where an optical detection via a third pulse, analogous to the probe in a pump-probe experiment, is applied. The optical detection allows monitoring of the temporal evolution of all involved electronic states while the photocurrent detection is selective to mobile charges. Moreover, in the optical detection the sample is a neat film without any electrode or interface effects.

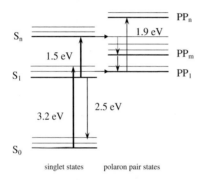

Figure 10 Electronic states involved in double excitation techniques. Thick arrows: first and second excitation pulses (photon energy to their left side), thin vertical arrows: detection via probe pulse (photon energy to their right side), other arrows: spontaneous transitions.

Photocurrent cross-correlation measurements [52] on m-LPPP have found a photocurrent increase upon excitation from S_1 to S_n. This photocurrent enhancement decreases with a double-exponential dependence on the pump-push delay. The slower component of this double-exponential behaviour corresponds to the singlet lifetime and hence has a straightforward explanation: the less S_1 states there are upon re-excitation, the less S_n states and consequently charges can be generated. The faster time constant, about 0.5 ps, is consistent with the time scale of the thermalisation processes of the vibrationally excited S_1 state. It indicates that re-excitation of S_1 when still out of equilibrium, creates S_n states with a higher dissociation probability than exciting a vibrationally and migrationally relaxed S_1. Upon tuning the push energy to 1.9 eV, a longer-lived photocurrent increase has been found [53]. The authors of that work ascribed this behaviour to a long-lived charge precursor state, formed by bound polaron pairs, which separate into free polarons [54].

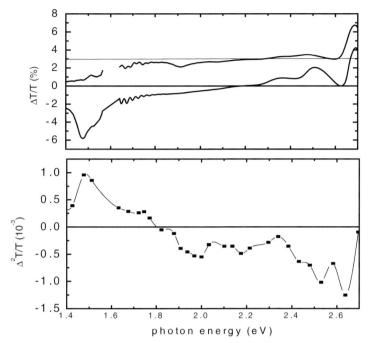

Figure 11 Upper panel: Pump-probe spectrum of m-LPPP at room temperature at 0 ps (lower curve) and 400 ps (upper curve) delay. Lower panel: Push-induced change in the Pump-induced transmission ($\Lambda T^2/T$) spectrum for 300 fs pump-push delay. $\Delta T^2 > 0$ at 1.5 eV is reduced absorption, $\Delta T^2 < 0$ between 1.9-2.3 eV is enhanced absorption from charges, $\Delta T^2 < 0$ above 2.3 eV is reduced emission

Via optical detection it has been directly observed that in ladder-type phenylene based polymers re-excitation of S_1 to S_n leads to charge generation [55]. Figure 11 shows the push-induced modification of the pump induced transmission difference spectrum for m-LPPP films. Compared with the regular pump–probe spectrum, also reported in the figure, one can see that re-excitation gives rise to increased absorption in the polaron region (around 2 eV) at the expense of a reduction of the singlet features (the broad absorption peak around 1.5 eV and the stimulated emission around 2.5 eV). The time scale for the process is limited by internal conversion, which is quite fast having a time constant of about 55 fs [56]. Figure 12 shows the recovery of S_1 and the concomitant rise of polaron population as measured probing at 2 eV, where the singlet and polaron absorptions overlap. The pump-push-probe experiment is "bi-dimensional" in the time domain, because both the push-probe and the pump-push delay can be varied. Taking advantage of this we observed a decrease in the efficiency of charge generation for increasing the pump-push delay, the process being negligible after 3 ps. This time scale again points to re-equilibration in the photo-excited polymer.

In summary, there exist hard evidence for charge generation occurring from high lying excited states, yet without definitive statement on intrachain fission, for the process may well be inter-chain too. From our data it is also clearly seen that the efficiency of this process strongly depends on the "age" of the starting singlet, an observation that is correlated to the re-equilibration process.

Finally we briefly mention the case of polyfluorene (PFO), for which the role of higher lying states seems dominant [57]. Here field assisted pump-probe indicates that singlets get dissociated only during the pump pulse, contrary to the finding in m-LPPP. Interpretation is that only higher lying states can be dissociated, and the later are only reached by two photon transition during pumping [58]. Experiments of double excitation photoconductivity shed additional light on the phenomenon. They were carried out on single layer diode structures kept under vacuum at room temperature. The pump pulse at 3.2 eV (390 nm) excites the polymer, reaching S_1.

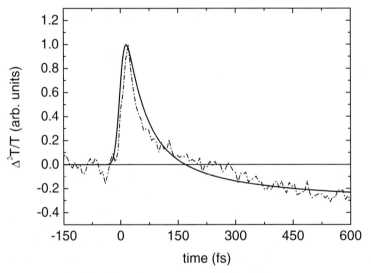

Figure 12 Push induced change in transmission difference for m-LPPP film measured at 2 eV (dashed) . Re-excitation occurs 300 fs after the pump. The solid line is a fit according to a system of rate equations discussed in detail in Ref. 56, and describes both singlet recovery and charge population build up.

The second, delayed pulse at 1.6 eV (800 nm) (push) re-excites S_1 to the higher lying S_n state (symmetry selection rules require S_n to be even parity). Photocurrent increase (ΔPC) is detected following second excitation as a function of time delay. We find that ΔPC(t) is closely correlated to the decay of S_1 population, which is monitored by transient transmission at 680 nm, in the S_1-S_n absorption band as shown in Fig. 13. This clearly shows that re-excitation generates charges. The coincident time scale implies S_1 can be re-excited for its whole lifetime, with constant dissociation efficiency into charges, while no other species are involved. Contrary to the case of m-LPPP, there is no evidence of an initial higher efficiency, a puzzle still to be resolved.

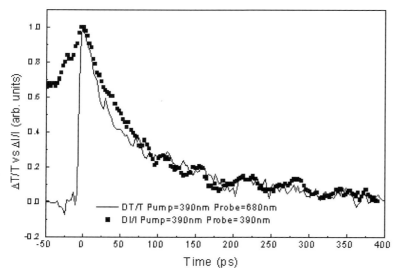

Figure 13 Double excitation change in photocurrent (solid squares) and singlet population dynamics (solid line) measured by pump-probe

5.7
Conclusion

The broad range of results and phenomena here discussed shows that ultrafast spectroscopy is a distinguished tool for understanding CP photophysics, yet many other techniques should be considered if one wishes to achieve a comprehensive picture. Not only this, the material under study must be well characterized in order to skip the well known chaos generated by unintentional doping, oxidation, inhomogeneous broadening, irreproducible synthesis and so on. This let alone the intrinsic difficulty of having hundreds of different CP structures and morphologies, not always under full control. While large improvement has been done in this direction, and virtually any existing material science probe has been applied to CPs, a conclusive picture is still far from available. Technological development and applications however are a strong push in this research, and guarantee that the interest from the scientific community will last for a long time. Likely this will bring us to a universal, broadly accepted, comprehensive picture of CP photophysics.

Acknowledgements

C.G. acknowledges the support from the European Community Access to Research Infrastructure action of the improving Human Potential Programme, Contract n. HPRI-CT-2001-00148 (Center For Ultrafast Science and Biomedical Optics, CUSBO).

References

[1] J. B. Birks "*Photophysics of Aromatic Molecules*" Wiley-Interscience, London 1970

[2] P. Tavan, K. Schulten, Phys. Rev. **B36**, 4337(1987)

[3] A. J. Heeger, S. Kivelson, J. R. Schrieffer, W. P. Su, Rev. Mod. Phys. **60**, 781(1988).

[4] S. Mukamel, *Principles of nonlinear optical spectroscopy*, Oxford University Press, New York, 1995.

[5] For a review, see R. Merlin, Solid State Commun. 102, 207 (1997), C.J. Bardeen, Q. Wang, and C.V. Shank, Phys. Rev. Lett. 75, 3410 (1995)

[6] H. L. Fragnito, J.Y. Bigot, P.C. Becker, and C.V. Shank, Chem. Phys. Lett. 160, 101 (1989).

[7] W. T. Pollard, H.L. Fragnito, J.-Y. Bigot, C.V. Shank, and R.A. Mathies, Chem. Phys. Lett. 168, 239 (1990).

[8] O. J. Korovyanko, R. Österbacka, X. M. Jiang, Z. V. Vardeny, and R. A. J. Janssen, Phys. Rev. B **64**, 235122 (2001).

[9] G. Cerullo, G. Lanzani, M. Zavelani-Rossi, and S. De Silvestri Phys. Rev. B **63**, 241104 (2001).

[10] M. Yoshizawa, H. Aoki, H. Hashimoto, Phys. Rev. B63, 180301(2001).

[11] G. Lanzani, G. Cerullo, C. Brabec, and N. S. Sariciftci Phys. Rev. Lett. **90**, 047402 (2003).

[12] B.E. Koheler in "*Conjugated polymers: the Novel science and technology of conducting and non linear optically active materials*" edited by J. L. Bredas and R. Silbey, Kluwer Press, Dodrecht, (1991).

[13] G. Cerullo *et al.* Science298, 2395(2002), G. Cerullo *et al.* Phys. Rev. **B**(R)63, 241104(2001).

[14] A. Shukla, H. Ghosh and S. Mazumdar, Phys. Rev. **B** 67, 245203 (2003).

[15] G. Lanzani et al. Phys. Rev. Lett **87**, 187402(2001).

[16] B. Kraabel *et al.,* Chem. Phys. **227**, 83 (1998); G. Lanzani *et al.,* Chem. Phys. Lett. **313**, 525 (1999).

[17] G. F. Musso et al. Synth. Met. 102, 1414(1999).

[18] N. Suzuki et al. Phys. Rev. Lett. **45**, 1209(1980).

[19] Z. V. Vardeny et al. Phys. Rev. Lett. **57**,2995(1986).

[20] Z. V. Vardeny et al. Phys. Rev. Lett. **49**, 1656(1982) ; L. Rothberg et al. Phys. Rev. Lett. **57**, 3229(1986).

[21] C. V. Shank et al. Phys. Rev. Lett. **28**, 6095(1983).

[22] W. P. Su, J. R. Schrieffer, A. J. Heeger, Phys. Rev. Lett. **42**,1698(1979).

[23] T. Nishioka et al. , Jpn. J. Appl. Phys. 36, 1099)(1997).

[24] W. –S. Fann et al. **62**, 1492(1989).

[25] I. Gontia et al. Phys. Rev. Lett. 82, 4058(1999). O.J. Korovyanko et al. Phys. Rev. B **67**, 035114 (2003)

[26] M. Wohlgenannt, W. Graupner, G. Leising, Z. V. Vardeny Phys. Rev. Lett.**82**, 3344(1999).

[27] G. Weiser, Phys. Rev. B 45, 14076 (1992).

[28] L. M. Blinov, S. P. Palto, G. Ruani, C. Taliani, A. A. Tevesov, S. G. Yudin, R. Zamboni, Chem. Phys. Lett 232, 401 (1995).

[29] S. J. Martin, H. Mellor, D. D. C. Bradley, P. L. Burn, Opt. Mater. 9, 88 (1998).

[30] A. Mathy, K. Ueberhofen, R. Schenk, H. Gregorius, R. Garay, K. Müllen, and C. Bubek, Phy. Rev. B 53, 4367 (1996).

[31] S. F. Alvarado, P. F. Seidler, D. G. Lidzey, and D. D. C. Bradley, Phys. Rev. Lett. 81, 1082 (1998).

[32] I. G. Hill, A. Kahn, Z. G. Soos, R. A. Pascal Jr, Chem. Phys. Lett. 327, 181 (2000)

[33] M. Yan, L. J. Rothberg, F. Papadimitrakopoulos, M. E. Galvin, and T. M. Miller, Phys. Rev. Lett. 72, 1104 (1994).

[34] M. Yan, L. J. Rothberg, E. W. Kwock, and T. M. Miller, Phys. Rev. Lett. 75, 1992 (1995).

[35] L. Onsager, J. Chem. Phys. 2, 599 (1934).

[36] L. Onsager, Phys. Rev. 54, 554 (1938).

[37] J. Kalinowski, W. Stampor, P.G. Di Marco, J. Chem. Phys. 96, 4136(1992).

[38] R. Kersting, U. Lemmer, M. Deussen, H. J. Bakker, R. F. Mahrt, H. Kurz, V. I. Arkhipov, H. Bässler, and E. O. Göbel, Phys. Rev. Lett. 73, 1440 (1994).

[39] J. Frenkel, Zh. Tekh. Fiz. 5, 685 (1938).

[40] J. Frenkel, Phys. Rev. 54, 647 (1938).

[41] A. Hayer, H. Baessler, B. Falk, S. Schrader, J. Phys. Chem. A 106, 11045(2002).

[42] M. C. J. M. Vissenberg and M. J. M. de Jong, Phys. Rev. Lett. 77, 4820 (1996).

[43] V. I. Arkhipov, E. V. Emilianova and
      H. Bässler, Phys. Rev. Lett. 82, 1321 (1999).
[44] V. I. Arkhipov, E. V. Emilanova, S. Barth,
      and H. Bässler, Phys. Rev. B 61, 8207
      (2000).
[45] D. M. Basko and E. M. Conwell, Phys. Rev.
      B 66, 155210 (2002).
[46] W. Graupner, G. Cerullo, G. Lanzani,
      M. Nisoli, E. J. W. List, G. Leising, and
      S. De Silvestri, Phys. Rev. Lett. 81, 3259
      (1998).
[47] V. Gulbinas, Yu. Zaushitsyn, V. Sundström,
      D. Hertel, H. Bässler, and A. Yartsev Phys.
      Rev. Lett. 89, 107401 (2002)
[48] G. Cerullo, S. Stagira, M. Nisoli,
      S. De Silvestri, G. Lanzani, G. Kranzelbin-
      der, W. Graupner and G. Leising, Phys. Rev.
      B 57, 12806 (1998).
[49] A. Kohler et al. Nature392, 903(1998).
[50] J. L. Bredas *et al*. Acc. Chem. Res. 32,
      267(1999).

[51] S. V. Frolov, Z. Bao, M. Wohlgenannt, and
      Z. V. Vardeny, Phys. Rev. Lett. 85, 2196,
      (2000).
[52] C. Zenz, G. Lanzani, G. Cerullo,
      W. Graupner, G. Leising, S. De Silvestri,
      Chem. Phys. Lett. 341, 63 (2001).
[53] J. G. Müller, U. Lemmer, J. Feldmann, and
      U. Scherf, Phys. Rev. Lett. 88, 147401
      (2002).
[54] E. Frankevich, Y. Nishihara, A. Fujii,
      M. Ozaki, K. Yoshino Phys. Rev. **B66**,
      155203(2002).
[55] C. Gadermaier,.G. Cerullo, G. Sansone,
      G. Leising, U. Scherf, and G. Lanzani, Phys.
      Rev. Lett. 89, 117402 (2002)
[56] C. Gadermaier,.G. Cerullo, C. Manzoni, U.
      Scherf, E.J.W.List, G. Lanzani Chem. Phys.
      Lett.384,251(2004).
[57] C. Silva *et al*. Phys. Rev. **B 64**, 125211
      (2001).
[58] T. Virgili, G. Cerullo, L. Lüer, G. Lanzani,
      C. Gadermaier, D. D. C. Bradley Phys. Rev.
      Lett. 90, 247402, (2003).

6

The Origin of the Green Emission Band in Polyfluorene Type Polymers

S. Gamerith, C. Gadermaier, U. Scherf, E. J. W. List

6.1
Introduction

Among polyphenylene-based materials highly emissive polyfluorenes (PFs) have received particular attention during the last decade as a promising class of conjugated polymers which can be used as blue light emitting materials in polymer light emitting diodes (PLEDs). The excellent optical and electronic properties of 9,9-substituted polyfluorenes, polyindenofluorenes [1], and fluorene copolymers [2,3] have brought this class of materials in the focus of scientific and industrial interest.

The photophysics of PF-type polymers has been intensively studied and impressive improvements concerning colour purity and device stability of PF-based devices have been made. Still, the strict requirements for commercialization, demanding tens of thousands of hours of operation, are an unreached goal for blue emitting PLEDs.

The first report on the synthesis of soluble poly(9,9-dialkylfluorenes) from the corresponding fluorene monomers with akyl chain lengths of up to $n = 12$ using $FeCl_3$ as the oxidative coupling agent was given by Yoshino and co-workers in 1989 [4], and first blue electroluminescence (EL) from such poly(9,9-dialkylfluorene) was shown by the same group two years later [5]. After these initial experiments, a broad variety of efficient synthesis of polyfluorenes has been developed, especially by Suzuki- and Yamamoto-type aryl-aryl couplings [8,11]

Being crucial for the fabrication of efficient and stable electroluminescent devices, the interplay of the physical structure and photophysics of polyfluorenes has been studied in detail [6]. Such photophysical investigations showed that poly(9,9-bis(2-ethylhexyl') fluorene), a PF type material with an optical energy gap of around 3 eV, exhibits a phosphorescence emission from a triplet T_1 state some 0.7 eV below the lowest lying excited singlet state [7].

Investigations of the solid state morphology and structure revealed that poly(9,9-dioctylfluorene) (PFO) forms a well defined thermotropic liquid crystalline (LC) state above its melting point (ca. 160 °C), but well below the temperature for thermal decomposition. The fact that this nematic LC state can be quenched into a glass or crystallized [8,9] allows the fabrication of PLEDs which emit highly polarized light making them potential candidates for backlight illumination in LC displays [10,11].

Physics of Organic Semiconductors. Edited by W. Brütting
Copyright © 2005 WILEY-VCH Verlag GmbH & Co. KGaA, Weinheim
ISBN 3-527-40550-X

Upon different physical treatment protocols such as dissolving the polymer in toluene, spin coating a film, quenching the heated film from 200 °C and crystallizing the film in nitrogen, the luminescence spectrum of PFO is continuously red-shifted as expected for such a series of increased densification of the material and related increase of effective conjugation length. An even further red shift of the PL spectrum of ~100 meV is found for thermally cycled PFO-films compared to as spin coated films. This PL red shift is always observed in conjunction with an additional absorption feature at 2.86 eV (the optical band gap of pristine PFO films is 3 eV) and is attributed to the appearance of a higher ordered solid phase of the polymer, the so called β-phase, which shows a more extended conjugation of the planarized PF-chains than the glassy α-phase [12].

On the device-physics side a variety of low-onset, highly efficient, blue-emissive devices have been realised with PF-type polymers. PLEDs comprising additional hole- and electron transport layers (N,N′-diphenyl-N,N′-(3-methylphenyl)-1,1′-biphenyl-4,4′-diamine; (TPD), and 2-(4-biphenyl)-5-(4-tert-butylphenyl)-1,3,4-oxadiazole; (PBD), respectively) with vapour deposited poly (9,9-dihexylfluorene) as blue emitter showed increased emission efficiency as compared to single layer devices [13].

Blending PF with low molecular weight hole transporting molecules (triaryl-amines) with an ionisation potential of ~4.9 eV which act as hole traps was reported to increase the efficiency of electron-injection limited PF devices [14]. Triphenylamine incorporation into the PF –chain (alternating copolymers, endcapping, or sidechain substitution) can change the oxidation potential of PF from 5.8 eV to values as low as 4.98 eV and significantly improves the hole injection from the ITO electrode thus enabling the fabrication of diode structures with an ohmic contact at the anode and trap free, space-charge limited current characteristics [2,3,15,16]. Time of flight measurement of such (co)polymers revealed non-dispersive hole transport with mobilities up to $3 \times 10^{-3} cm^2/Vs$, about one order of magnitude higher than for the PFO homopolymer [17,18,].

In contrast to the anode of a PF-PLED, where a high work function electrode such as O_2–plasma treated ITO or polyethylenedioxithiophene doped with poly(styrene sulfonate) (PEDOT/PSS) is necessary to match the HOMO level of the PF-type polymer, the realization of a good electron injecting contact requires the use of low work function materials. Such materials as Ca or alkali metals have their Fermi levels close to the PF LUMO level and, therefore, no or only minor energy barriers have to be overcome when electrons are injected. Electrode/polymer interfaces of PLEDs made of PF-type emitters and low work function metals have been thoroughly investigated by photoelectron spectroscopy (XPS and UPS) [19-22]. For PFO, a significant change of the electronic structure of the polymer with a broadening of the HOMO and LUMO as well as bipolaron states at the interface with Ca or alkali metal electrodes have been detected [19,20]. The thickness of the calcium layer and the calcium concentration within the interface region, as investigated by time of flight secondary ion mass spectroscopy (TOF SIMS), strongly influence the EL efficiency via Ca-induced quenching of excitons [21]. Regarding the influence of oxygen at the Ca/PFO interface small amounts of oxygen are described to reduce the number of gap states while a high amount of oxygen alters the original LUMO position due to changes of the chemical structure of the polymer [22].

One of the main reasons which have so far prevented a breakthrough of PF-type polymers as *the* material for the realization of blue light emitting devices is their poor long term stability. Upon photo-oxidative or thermal degradation and during device operation the formation of a low energy emission band around 530 nm (ca. 2.3 eV) is frequently observed in PF-type polymers. The origin of these low energy photo- and electroluminescence peaks was firstly attributed to an aggregate or excimer emission [9,23,24,25,26,27,28]. This primary assignment led to various synthetic efforts efforts, such as copolymer generation, attachment of dendronic substituents [29,30,31], endcapping, and spiro-linking [32] in order to modify and stabilize the bulk emission properties. Frequently, the observed improvements of the bulk PL properties have then been related to the variations of the chemical structure of single polymer chains.

However, recently the low energy emission bands in PF-type polymers were identified as the emission from an exciton and/or charge trapping on-chain keto defect [33,34,35,36,37]. It was shown that by incorporating keto defects into the polymer backbone the emissive characteristics of oxidatively degraded polyfluorenes, especially the appearance of the 530 nm band (ca. 2.3 eV) can be faithfully reproduced [38]. The absence of any concentration dependence of the 530 nm bands' intensity in solution as well as the appearance of a vibronic structure in the 530 nm-band at low temperatures support the picture of an emissive on-chain defect as the origin of this particular spectral feature [39]. Efficient energy migration to these defects and the localization of the emissive state which is strongly confined to the fluorenone unit lead to a significant modification of the optical properties of PF films already at low defect concentrations [40]. This behaviour causes the need for polyfluorenes with high operational EL stability for an use in electroluminescent devices. Mono-alkylated fluorene units have been identified as preferred sites for oxidative degradation of PF- type polymers. A significant improvement of the oxidative stability was obtained by a careful purification of the fluorene monomers prior to polymerization [41].

Given the research effort which has already been undertaken and the very encouraging results of the last years, PF-type materials certainly must be considered for future optoelectronic applications.

6.2
Photo-physical Properties of Polyfluorene-type Polymers

Pristine polyfluorene-type polymers and oligomers display a rich and attractive variety of optical and electronic properties. In the following we will focus on the basic properties which are inherent to the molecular structure of this type of polymers. Furthermore we will also discuss the so called β-phase formation in the solid state, an effect altering the electronic structure of some PF-type polymers.

6.2.1
Optical properties of pristine polyfluorenes

Independently of the substitution in 9-position most PF-type polymers (see Fig. 1 for chemical structure) in dilute solution exhibit an unstructured absorption with its maximum at ca. 3.25 eV and a photoluminescence spectrum with the 0-0 π^*-π transition peaking at ca 2.9 eV as depicted in Fig. 2. In contrast to the absorption, the emission spectra exhibit a well-resolved vibronic progression with an energetic spacing of ca. 180 meV, which is related to the stretching mode of C=C-C=C substructures of the polymer backbone. Very high photoluminescence quantum yields of > 50% have been reported for polyfluorenes in the solid state [12,42]. When spin or drop cast from solution, pristine PF films typically display a solid-state emission spectrum which is similar in shape and position to the spectrum in dilute solution. In some cases, however, a slight bathochromic shift of up to 100 meV is observed. The absorption spectra in the solid state are generally slightly broadened by 100-250 meV as result of small local variations of the π-overlap due to some conformational disorder of the polymer chains in the solid state.

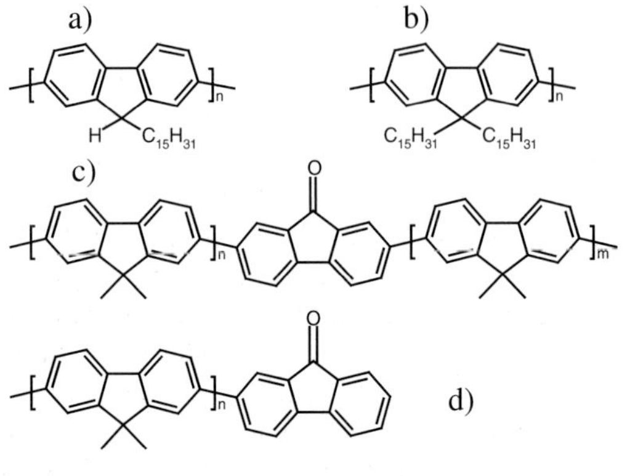

Figure 1 Chemical structure of a) 9-monoalkylated PF (MA-PF), b) 9,9-dialkylated PF (DA-PF), and fluorene-fluorenone copolymers with c) on-chain, and d) chain-end fluorenone moieties.

As depicted in Fig. 3 the photoinduced absorption (PA) spectrum of most PF-type polymers at low temperatures shows only one dominant band with a maximum at ca. 1.5 eV. From a comparison with conjugated polymers of similar chemical structure such as methyl-substituted ladder-type poly (para-phenylene) (m-LPPP) in optically detected magnetic resonance experiments [43] this band has been assigned to a transition from the lowest triplet state 1^3B_u to an excited m^3A_g state. Remarkably, most PF-type polymers do not exhibit significant polaronic absorption bands in steady state PA measurements at low temperatures, which is a) a consequence of the rather low density of electronic traps and low energetic disorder of the PF bulk polymers, and b) due to the masking of the polaronic absorption by the triplet absorption. The latter becomes evident when one compares PA at low temperature

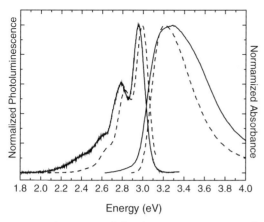

Figure 2 Absorption and photoluminescence emission of DA-PF in dilute solution and in the solid state (full line). Figure modified from [33]

with room temperature measurements on the same polymer. In such a case the dominant triplet feature at 1.5 eV is drastically reduced and the polaronic absorption centred at approximately 2.1 eV becomes clearly visible [44]. Note that the polaron band is normally identified from a comparison with doping induced absorption measurements done on the same polymer. The low concentration of polaronic species and the small overlap of the PA with the PL emission make PF-type derivatives promising candidates for organic solid-state lasers since the absence of a PA/PL overlap strongly prevents both polaron self-absorption processes as well as exciton quenching via energy-transfer onto the species giving rise to PA [45]. The low energetic disorder is also reflected in the rather high charge carrier mobility of polyfluorenes of up to 4×10^{-4} cm^2/Vs (for holes) combined with the observation of a non-dispersive charge carrier transport from TOF measurements [17]. The energetic spacing between the ground state 1^1A_g and the lowest triplet exciton state 1^3B_u has been determined from pulsed radiolysis measurements to be 2.1 eV, which is in excellent agreement with the observed phosphorescence energy of poly [9,9-bis(2-ethylhexyl)fluorene] (PF2/6) of ca. 2.1 eV [7]. Also in the solid state PF-type polymers exhibit very high PL quantum yields, which have been found to be in the order of 50-60 % [12,42].

6.2.2
Optical properties of polyfluorenes in the β-phase

Among the differently substituted PFs, *n*-alkyl-substitued PFs can undergo conformational changes in the solid state upon thermal treatment [12], upon exposing the films to solvent vapours [6,9], or in appropriate solvent/non-solvent mixtures [16]. It has been shown that the conformational changes are strongly related to the alkyl side-chains in 9-position. For such samples, in addition to the regular, glassy (α-)

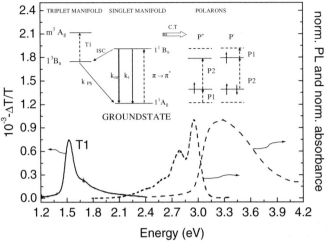

Figure 3 Absorption (dashed line), photoluminescence (PL, dotted line), and photoinduced absorption (PA, solid line) spectra of a PF film. The inset shows the schematic Jablonski diagram for this PF-material, depicted are the energy levels and optical transitions associated with singlet, triplet, and charged excitations. The 1^1A_g and the 1^3B_u states are the ground state and lowest triplet exciton (TE) states, respectively. The 1^1B_u is the lowest allowed excited singlet exciton (SE) state. k_r and k_{nr} are the radiative and nonradiative rate constants for the 1^1B_u singlet exciton (SE) decay. T1 denotes the first allowed triplet absorption from the 1^3B_u to the lowest excited m^3A_g state as observed in photoinduced absorption (PA), while k_{PH} is the radiative phosphorescent decay of triplets to the ground state. p^+ and p^- represent polarons created upon charge transfer (CT), while P1 and P2 are the optical transitions within the polaronic states. Figure reproduced from [34]

phase, a second, the so-called β-phase, has been identified and thoughtfully studied [6,12]. It has been shown that the β -phase formation is a result of the crystallisation of the *n*-alkyl side-chains forcing the polymer main chains into a more coplanar arrangement of the fluorene building blocks resulting in an increased π-conjugation along the polyfluorene main-chain.

As shown in Fig. 4 essentially this effect can be also observed for poly(9,9-dioctyl-fluorene) (PFO) in (chloroform/methanol) of increasing non-solvent (methanol) content which gives rise to an agglomeration of individual polymer chains [16]. The agglomeration (particle formation) is detectable e.g. by light scattering and microfiltration. As shown for PFO the agglomeration of single macromolecules is accompanied by the occurrence of a series of novel, red-shifted absorption peaks at ca. 2.85 eV, 3.04 eV and 3.22 eV, as indication of the β-phase formation. The relative intensity of the new β-phase-related bands increases for increasing nonsolvent content, and simultaneously the intensity of the broad unstructured PF absorption decreases.

Simultaneously to the changes in absorption the PL spectrum of PFO in chloroform/methanol mixtures also dramatically changes with increasing non-solvent content. In addition to the emission bands of "isolated" (molecularly dissolved) PF chains a series of novel emission bands at lower energies (2.81 eV, 2.65 eV, and 2.49 eV) arises, which is, however, diminished at higher methanol contents. Further-

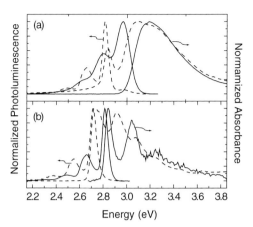

Figure 4 Absorption and photoluminescence spectra of poly (9,9–dioctylfluorene) (PFO) in dilute solution (chloroform, solid lines) and after partial agglomeration (β-phase formation) in a chloroform/methanol mixture (v/v: 75/25, dashed lines); b) absorption and photoluminescence spectra of the so-called β-phase of polyfluorene (solid lines, as derived from a numerical subtraction of the absorption and emission spectra in chloroform and in chloroform/methanol mixture –75/25-, resp.); for comparison the absorption and photoluminescence spectra of ladder-type poly(para-phenylene) LPPP (R_1: -hexyl, R_2: -4-decylphenyl, R_3: -methyl) is given (dilute solution, toluene, dashed lines). Figure reproduced from [34]

more, the newly emerging red-shifted absorption and emission bands of β-phase PFO possess, in contrast to "isolated" PF chains, a vibronic progression both in absorption and emission.

From a comparison to the absorption and emission properties of a fully planarized PPP-type ladder polymer (LPPP) a clearer picture for the observations in the partly agglomerated (β-phase) and "isolated" PF sample can be derived. As can be clearly observed both materials possess nearly identical spectral characteristics, which only differ in a slight red-shift of ca. 100 meV. Both the β-Phase PF and LPPP display an identical vibronic progression with a spacing of 180 meV. The observed vibronic structure of β-phase PF in absorption is fully consistent with the interpretation of a side-chain driven planarization of the PF backbone. LPPP and β-phase PF show a similar geometrical main-chain arrangement and hence similar ground and excited state potential energy surfaces. In contrast, the absence of a well-resolved vibronic structure in the absorption of "isolated" PF chains is indicative for a distinct geometry change during transition from the ground state into the excited state which is documented by rather differently shaped ground and excited state potential energy surfaces, as observed e.g. also for biphenyl [53].

In addition to the remarkable similarities between the absorption and emission spectra of LPPP and β-phase PF a comparable similarity is found for the photoinduced spectral features such as triplet excitons [12,43] and polarons which occur at nearly identical energies both for β-phase PF and LPPP (P1@ 0.5 eV and P2@ 1.9 eV. For the definition of the triplet and polaronic states please see the inset and caption of Fig. 3. As shown by Cadby et al., in films composed of both "regular"

(α-phase) and β-phase PF a dramatic increase in the polaron steady state density occurs as a consequence of the increase of the energetic disorder in the bulk material [12]. Since a high steady state polaron density represents a serious drawback in OLED and solid-state laser applications due to non-radiative quenching of singlet excitons at polarons [45] PF derivatives without β-phase formation (e.g. with branched alkyl side-chains in 9-position) have to be favoured [46].

6.2.3
Fluorescent on chain defects in polyfluorene-type polymers

As it has been laid out in the introduction of this paper there has been considerable effort in both academic and industrial research during the past five years towards improving the overall stability of blue light emitting polymers of the polyfluorene family. Yet, most of the attempts to improve in particular the colour stability, i.e. to avoid the formation of the low energy emission band at 2.2-2.3 eV, for quite some time have been based on the assumption that the PF chains can form aggregates, excimer or even exciplexes. However, as described in detail in the following chapters it is very unlikely that aggregation, excimer, or exciplex formation can be held responsible for the colour instability of PF-based PLEDs.

In the following it will be shown that a) on-chain emissive keto-type chemical defects can be formed already during the synthesis, or during thermal-, photo- or electro-degradation, and b) that these keto type chemical defects when incorporated into a PF backbone lead to an efficient low energy emission band at 2.2-2.3 eV. Even at low defect concentrations this emission band can completely dominate the solid state PL spectrum due to energy transfer processes.

6.2.3.1 Formation of keto-type defects in PF-type polymers
The formation of keto-type chemical defects may be best understood when correlating the formation of low energy emission bands to the generation of such keto defects, which can be monitored by infra-red spectrometry. Such correlated experiments can be carried out by different means, as now discussed for thermal and photo-oxidative degradation. In addition, a possible mechanism for the formation of the keto defects during the polymer synthesis is discussed.

6.2.3.2 The formation of keto defects during the synthesis
The formation of keto-type defects during the synthesis of the polymer can be best illustrated by a comparison of two derivatives of PF. a) a 9-monoalkylated (MA-PF) and the 9,9-dialkylated derivative (DA-PF), which can be both synthesized from the corresponding 2,7-dibromofluorene monomers in a reductive aryl-aryl coupling according to Yamamoto [47]. From our investigations we find that MA-PF degrades much more rapidly than DA-PF which can be attributed to reduced chemical stability of MA-PF due to the presence of the relatively weakly bound, acidic methylene bridge hydrogen. Oxidative transformation of the 9-alkylfluorene moiety leads to 9-fluorenone defect sites as depicted in Fig. 5. The highly active Ni[0]-species used in the reductive coupling of the dibromo monomers reduce a certain amount of the

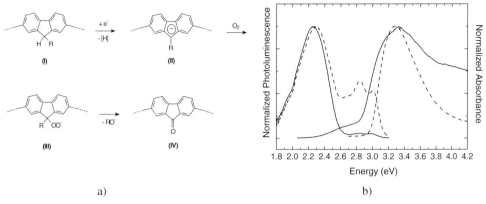

a)   b)

Figure 5 Proposed mechanism for the generation of keto-defect sites in MA-PF (for explanations see text). Figure reproduced from [34] b) Absorption and photoluminescence emission of MA-PF in dilute solution (dotted line) and in the solid state (full line). Figure modified from [33]

9-monoalkylated fluorene building blocks (I) to (aromatic) fluorenyl anions (II) under formation of hydrogen. These anions can form hydroperoxide anions (III) with atmospheric oxygen during the work-up of the reaction mixture. The hydroperoxide anions then undergo a final rearrangement to fluorenone moieties (IV).

As depicted in Fig. 5 b) the incorporation of the 9-fluorenone defect sites in MA-PF dramatically changes the emission properties in photoluminescence of the polymer backbone as compared to the photoluminescence of defect free pristine DA-PF.

In dilute solution MA-PF exhibits a broad absorption peak at 3.2 eV, the PL emission spectrum shows maxima at 2.95 eV and 2.8 eV as well as the low energy band peaking at 2.3 eV. In the solid state the absorption of MA-PF becomes much broader and an additional weak contribution centred at ca. 2.8 eV occurs. The solid-state PL spectrum of the MA-PF is dominated by the low energy emission peak at ca. 2.28 eV. In the infra-red absorption spectrum of MA-PF one observes an additional IR band at ca. 1721 cm^{-1} already in pristine samples, while no such signal is detectable in pristine DA-PF as depicted in Fig. 7 b). Since this band is an IR fingerprint of a carbonyl stretching mode (>C=O) of the fluorenone building block [48] this demonstrates the clear correlation between the presence of 9-fluorenone defects and the presence of the low energy emission band at 2.3 eV. [49]

6.2.3.3 Formation of keto defects during thermal degradation

To further clarify the formation of keto-type defects, especially in defect free DA-PFs, thermal degradation measurements have been carried out. Stability and morphology related issues will be discussed in the following. Note that this method has the advantage to be less dependent upon the particular film thickness than photo-oxidation (where preferably the top layer of the film is affected) and device-degradation experiments. Since such experiments can be easily carried out in vacuum, air, or any other atmosphere they allow for a simple investigation of the influence of atmospheric components such as O_2 and H_2O as well as of structural changes upon heating above the glass transition temperature T_g of the polymer.

In Fig. 6 a) a detailed study of the absorption and PL emission properties of pristine DA-PF (R: -3,7,11-trimethyldodecyl) films heated in air for one hour at 66 °C, 100 °C, 125 °C, 150 °C, 175 °C, and 200 °C, respectively, is presented. While the blue emission at 2.96 eV and 2.78 eV initially increases, which we attribute to the removal of residual solvent or ordering effects, it decreases significantly after heating the sample to higher temperatures (175 and 200 °C, respectively). At the same

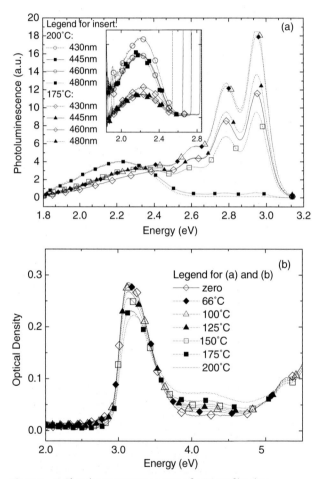

Figure 6 a) Photoluminescence spectra of a DA-PF film thermally degraded in air (1 h at all given temperatures, times cumulative), excitation wavelength 390 nm. The insert shows the site-selective PL spectra obtained for excitation wavelengths as given in the legend after annealing at 175 °C and 200 °C. b) Optical density of the degraded film.

time the broad, low energy emission band around 2.3 eV evolves. As shown in the insert of this figure the low energy emission can be directly excited when an excitation energy distinctly lower than the band gap energy (ca. 2.8 eV) is used. This not only allows us to exclude excimer emission as explained in the discussion below, but in fact it also reveals that there is a true increase of emissive keto defects in the sample (by about a factor of three during the 1h - 200 °C heating step). When exciting the sample at 390 nm the latter information is masked by a significant quenching of excitons due to their migration to non-emissive chemical defects, which are also created in the degradation experiment. Upon heating in air, the broad emission band around 2.3 eV of DA-PF is also found when the material is incorporated in a polystyrene (PS) matrix at a very low concentration, where any interchain interactions are hindered by the matrix material. This observation also rules out the formation of fluorene or fluorenone excimers and also fluorene-fluorenone exciplexes as the source of the 2.3 eV emission band. (Fig. 9).

In absorption a blue-shift of the π-π* peak from 3.14 to 3.21 eV is observed, which is caused by the lowering of the effective conjugation length as it has already been observed for other polymers (e.g. PPV) under similar conditions [50]. The overall decrease of the main absorption is accompanied by a broadening of the spectrum on the high energy flank upon thermal degradation in air. This supports the picture of a process where one introduces a larger number of chemical defects, which break the conjugated backbone into shorter conjugated segments. Note, that in contrast to the absorption of fluorene-fluorenone copolymers (see Fig. 10), a subgap absorption band at 2,6-2,9 eV due to keto defects cannot be observed in the degraded samples. This is, however, not surprising as the amount of keto defects is comparably small

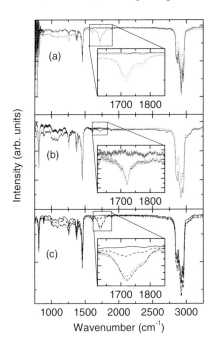

Figure 7 a) FTIR spectra of DA-PF film thermally degraded in air; "pristine film" (full line) after 1 hour at 66 °C (short dash), 100 °C (short dot), 133 °C (dash dot), 166 °C (dashed line) and 200 °C (dotted line), times cumulative. b) "pristine" MA-PF (dashed line) and "pristine" DA-PF (full line) on Si. c) "pristine" DA-PF (full line), and this film after photooxidation (1000 W xenon lamp under air) for 2 (dashed line), 4 (dotted line), and 6 min illumination (dash-dotted line). The insets show the region of the >C=O stretching mode around 1721 cm^{-1}. (b) and (c) modified from [34]

in our degradation experiment, which makes them undetectable in the optical absorption measurement.

The correlating IR transmission spectra of the PF-film thermally stressed in air are shown in Fig. 7. The observed bands can be grouped into three main regions; the region around 3000 cm^{-1}, which is associated with C-H stretching vibrations, the region of C-H scissoring vibrations below 1500 cm^{-1}, and the region between 1600 cm^{-1} and 1800 cm^{-1} where two features evolve by heating in air. The strongest band appears at around 1718 cm^{-1}, which has been assigned to the carbonyl stretching mode of the fluorenone building block [48]. Moreover, a second band at 1609 cm^{-1}, which has been interpreted as a stretching mode of an asymmetrically

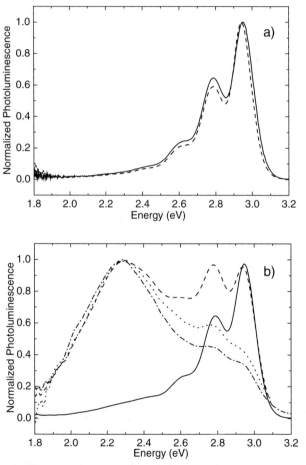

Figure 8 a) PL spectra of a DA-PF film before and after heating in vacuum at 200 °C for 4.5 hours. b) PL emission spectrum of a "pristine" DA-PF film (full line), and after photooxidation (1000 W xenon lamp under air) for 2 (dashed line), 4 (dotted line), and 6 min illumination (dash-dotted line). Figure reproduced from [33]

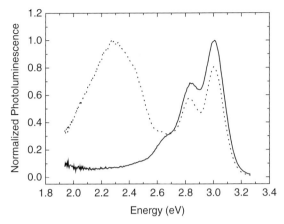

Figure 9 PL spectra of DA-PF (0,4 wt %)in a Polystyrene (PS) matrix. Full line: Pristine film; Dashed line: after 2 h at 180 °C in air

substituted benzene ring [28], and a third band at 1770 cm^{-1}, which we have not yet been able to assign to a certain molecular vibration, appear. These features have been observed before for different PFs and also for many other polymers after thermal and photo-oxidative degradation [28,34,51] as well as for fluorene-fluorenone co-polymers (see below). Note, however, that the bands observed for the copolymers are much narrower and that the 1770 cm^{-1} feature is absent there [52], which indicates that upon degradation also other chemical defects are created. The new IR-bands clearly appear in the DA-PF film after heating to temperatures ≥ 166 °C for one hour. Further heating to 200 °C only slightly increases the peak height of the 1718 cm^{-1} band.

Comparing the PL and the FTIR thermal degradation data, a clear correlation between the appearance of the keto (carbonyl) absorption in the IR spectra and the broad green emission peak in PL is found. In contrast, heating in vacuum (200 °C for 4.5 hours) does not significantly alter the PL spectra of the polymer under investigation as shown in Fig. 8 a). Obviously, the threshold for the thermal instability (> 133 °C) is not related to T_g (or T_m) values as the used DA-PF (R: -3,7,11-trimethyl-dodecyl) is a low T_g derivative ($T_g < 50$ °C) [37].

6.2.3.4 Formation of keto defects due to photo-oxidative degradation

The illumination with UV light from a high power halogen lamp of a spin cast DA-PF film in the presence of oxygen leads to a rapid degradation of the layer accompanied by spectral changes of the PL as shown in Fig. 8 b). Photooxidation of the sample for 6 min gives a dramatic decrease of the integral PL quantum yield to approximately 10 % of the initial value. Simultaneously, the low energy emission band at 2.3 eV emerges and increases relative to the initial PF emission during ongoing photooxidation of the DA-PF film. IR spectra recorded simultaneously (depicted in Fig. 7 c) display the appearance of a fluorenone-related signal (>C=O stretching mode) at 1721 cm^{-1}, which gradually increases. These findings give evidence of the formation of 9-fluorenone defects also upon photo-degradation. In summary, the experimental results of this subchapter clearly demonstrate that chemical, photo- or thermal degradation processes, lead to the formation of low energy emission bands at 2.2-2.3 eV coupled with the occurrence of characteristic IR-features due to the formation of 9-fluorenone defect sites.

6.2.4
The origin of the green emission band at 2.2-2.3 eV

The experimental results outlined above do not fully explain the origin of the green emission band around 2.2-2.3 eV, they only show the clear correlation between the observation of the low energy emission band and occurrence of >C=O -related IR bands. To further clarify the origin of the green emission we have synthesized and studied model polymers for degraded poly(fluorene), namely dialkylfluorene/fluorenone copolymers. -They show very similar properties as oxidatively degraded PFs, however, bearing the advantage that the 9-fluorenone content of the copolymers can be exactly controlled. In particular, the model copolymers have been invoked in a number of experiments which allow to unambiguously determine whether the excited species in polyfluorenes emitting in the green spectral region is an excimer, an aggregate or an on-chain defect [39].

Fig. 10 shows the absorption and photoluminescence spectra in toluene solution of the model copolymers of varying fluorenone content (ranging between 0.5 and 25% of fluorenone; y: 0.005 to 0.25 units in the fluorene-fluorenone co-polymer). A weak absorption shoulder at 2.6-3.1 eV is observed on the low energy side of the π-π* transition of polyfluorene (ca. 3.3 eV). This band is assigned to the absorption of charge transfer excited states of the poly(fluorene-*co*-fluorenone) chains by quantum chemical calculations [40]. The oscillator strength of this band increases with increasing fluorenone content. Excitation within the polyfluorene π-π* band results in emission spectra composed of the "regular" blue poly(fluorene) feature and a low energy emission peaking at ca 2.2 eV, which increases with increasing fluorenone content in the copolymer. From a comparison with MA-PF, where degradation and 9-fluorenone formation already take place during synthesis (as discussed above), one can estimate that pristine MA-PF contains 5-10% fluorenone units within the polymer backbone.

Isolated fluorenone molecules show a weak photoluminescence band at ca. 2.3 eV [49], very similar to the emission from fluorenones incorporated into the polymer backbone. Yet, despite this coincidence, one might argue that the oscillator strength (emission yield) of isolated fluorenone molecules is not high enough to account for the spectroscopic properties of the fluorene-fluorenone copolymers. However, quantum-chemical calculations performed on oligofluorenes containing 9-fluorenone moieties show that a dramatic increase of the 9-fluorenone oscillator strength occurs when the ketone is incorporates into a conjugated oligofluorene. This leads to the conclusion that the 9-fluorenone unit in such oligomers can be regarded as an on-chain emissive guest defect "accidentally" incorporated into the π-conjugated backbone by synthetic design or during oxidative degradation. [40]

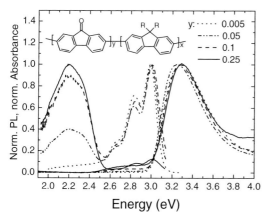

Figure 10 Absorption spectra of poly(fluorene-*co*-fluorenone) in toluene solution. For the fluorenone content (y) in the copolymers see legend. Photoluminescence (PL) spectra of the copolymers were recorded in very dilute toluene solution (<1µg/ml). The inset shows the chemical structure of the model copolymers. Figure reproduced from [39]

In order to support this assignment and to finally rule out excimers and aggregates we will now discuss a series of experiments which are typically used to study these excimer and aggregate related phenomena. An excimer [53], per definition, is an excited state dimer with a dissociative ground state. Since it is only formed in the excited state, its excitation spectrum is that of its constituents. Excimer emission is characterized by a broad featureless PL spectrum with an emission energy below that of single molecules. Also for aggregates, a relatively broad emission spectrum is expected as the inter-molecular distance should vary due to disorder. It has been shown that only slight variations of the inter-chain distance result in significant shifts of the emission energy [54]. Since both aggregates and excimers are intermolecular phenomena, their emission should display a pronounced concentration dependence in solution. In the experiments discussed in the following, we, however, observe that (a) the low energy emission band can be directly excited ruling out excimers, (b) the PL spectra do not show any significant concentration dependence, (c) a distinct solvatochromism is observed as predicted by theory [40], (d) the low energy emission band displays a pronounced vibronic structure at low temperatures, and (e) the low energy emission behaves like that of a guest molecule imbedded in a host of higher emission energy regarding excitation energy migration. These findings are direct evidence against an excimer or aggregate nature of the low energy emission band in polyfluorene-type materials. The observation of the low energy emission also in the highly diluted solutions counts as additional evidence against any physical dimerization (i.e. aggregation) effects. In the following, the individual observations will be discussed:

(a) Direct excitation of highly diluted solutions of the model copolymers within the low energy absorption band as shown in the inset of Fig. 11 results in an emission band at ca. 2.2 eV, similar to the low energy emission feature observed for degraded PF. This observation is contradictory to the formation of aggregates (the emission is observed also for dilute solutions) as well as to the formation of excimers (direct excitation of the low energy band is possible).

(b) Fig. 11, shows the concentration dependence of the photoluminescence spectra of model copolymers. We did not observe any significant difference of the spectra as well as of the relative intensities of the "regular" polyfluorene and the low energy emission bands. This is in contradiction to the behaviour of excimers, where the relative intensity of the excimer band strongly increases with increasing concentration of the material in the solvent as observed e.g. for pyrene [55]. We remark that also aggregation-related effects should increase with increasing concentration, since both phenomena arise only between molecules which are sufficiently close to each other.

(c) The emission spectra, however, show a distinct solvatochromism. This is shown in Fig. 12, where the photoluminescence spectra (for excitation within the π-π* band) are shown for dilute solutions of a fluorene-fluorenone copolymer (y: 0.25) in solvents of different polarity. While the spectral position of the blue emission band of polyfluorene is hardly affected by the solvent, the low energy emission band shows a strong bathochromic shift with increasing solvent polarity and a decrease of the relative intensity of the low energy emission peak. The most drastic effect is observed in chloroform, where the low energy emission band is very weak.

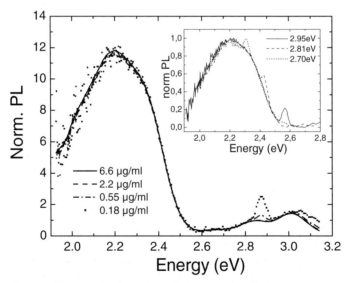

Figure 11 Concentration dependence of the photoluminescence of poly(fluorene-*co*-fluorenone) (y: 0.25) in toluene solution normalized to the blue emission component. Note that the additional peak at ca. 2.85 eV is due to Raman scattering by the solvent. The inset shows the photoluminescence spectra of the copolymer excited at 2.95, 2.81, and 2.70 eV respectively in dilute toluene solution (Note that the peaks appearing at 2.56 eV, 2.43 eV, and 2.31 eV are due to Raman scattering by the solvent). Figure reproduced from [39]

The bathochromic shift of the emission is indicative of a strongly polar emissive state. This is consistent with the high static dipole moment of 3.7 Debye calculated for the emissive charge-transfer excited state in oligofluorenes with a central 9-fluorenone moiety (following the methodology outlined in [40]). To rule out that the increased solubility of the copolymers in more polar solvents decreases the low energy emission band due to de-aggregation, we have measured the (relative) quantum yields of different co-polymers in the various solvents, as shown in Fig. 12.

For polyfluorene (y: 0) the PL quantum yields in toluene and dichloromethane are approximately equal and the PL quantum yield in chloroform is somewhat increased. The picture is strongly different for the fluorene-fluorenone copolymers. Here, the quantum yield has its maximum value in toluene and decreases with solvent polarity. This decrease is more pronounced in copolymers with a higher fraction of fluorenone (y: 0.25). The low energy emission band is weaker where the quantum yield is lower. Obviously the fluorenone containing segments are still populated either directly or via an intramolecular energy transfer, possibly due to a

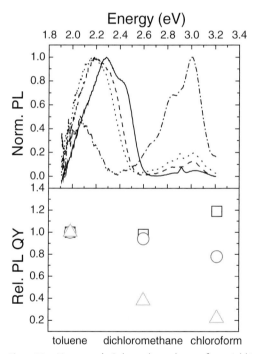

Figure 12 *Upper graph*: Solvent dependence of the fluorescence of poly(fluorene-*co*-fluorenone) with y: 0.25. Emission spectra in cyclohexane (solid line), toluene (dashed line), dichloromethane (dotted line), and chloroform (dash-dotted line). *Lower graph*: solvent dependence of the relative photoluminescence quantum yield (reference value is the PL quantum yield in toluene solution) for different copolymers: polyfluorene (squares), poly(fluorene-*co*-fluorenone) with y: 0.05 (circles), and poly(fluorene-*co*-fluorenone) with y: 0.25 (triangles). The absolute quantum yield of the materials is found to decrease strongly with increasing fluorenone content. Figure reproduced from [39]

reordering of the excited states by the solvent polarity (n-π^* and CT-$\pi\pi^*$; with n-π^* being non-emissive according to [40]), but the luminescence is efficiently quenched due to interaction with the polar solvent. On the contrary, if aggregates or excimers were the cause of the 2.3 eV emission, the quantum yield should increase with increasing solubility (i.e. solvent polarity), as observed for defect–free polyfluorene.

(d) To study the emission properties at low temperature as well as the effect of excited state migration, the photoluminescence of thin solid films of two model co-polymers (both for excitation within the polyfluorene π-π^* band, as well as within the low energy absorption band) were studied as a function of temperature. The results are shown in Fig. 13. For both fluorenone concentrations (y: 0.005 and 0.25) the previously broad, unstructured emission band displays a clear vibronic structure at low temperatures. The peakwidths decrease with decreasing temperature. In order to ensure that the double-peak structure is not caused by different chemical species with different positions of the fluorenone unit within the polymer chain (central, terminal), we have performed site selective PL measurements with the co-polymer (y: 0.25) in the solid state as shown in Fig. 14. The fact that the relative intensity of the two peaks remains constant, independently of the excitation energy indicates that both peaks are caused by the same chromophore, thus confirming the assignment of the double peak to a vibronic progression. Direct excitation of the sample at low temperatures probes the specific sites only, and due to the strong lo-calization of the excited states [40] at low temperature the migration [56] of the excited state is strongly hindered. Thus the observed emission stems only from the probed sites and no broadening caused by e.g. the exact position of the fluorenone unit on the polymer backbone [57] can be observed. As discussed below and depicted in Fig. 14 direct excitation into the fluorenone absorption band results in an even more pronounced vibronic structure at low temperatures.

The findings (a)-(d) obtained on solutions of fluorene-fluorenone copolymers clearly exclude excimers as the source of the low energy emission bands.

(e) While the results reported above for the solution measurements in non-polar solvents qualitatively hold true for the solid state, the relative intensity of the low energy band drastically increases (as seen from a comparison of Fig. 10 and Fig. 13) for the fluorene-fluorenone copolymers as well as for MA-PF (Fig. 5 b).

In the copolymer with y: 0.005, for which both blue polyfluorene and green low energy emission bands are observed simultaneously, the relative intensity of the low energy emission band gradually decreases with decreasing temperature. This find-ing is similar to observations made for the temperature dependence of a polymer guest-host system as investigated in [56]. It was shown that this behaviour can be explained by temperature assisted excited state migration and is inconsistent with an excimer formation for which the relative intensity of the low energy (the excimer) peak is expected to increase with decreasing temperature as observed else-where [58].

In addition, quantum chemical calculations and related experiments have revealed that the on- chain migration rate in conjugated polymers is almost two orders of magnitude lower than the interchain migration rate [59], Therefore, the 9-fluorenone emission is predominantly observed in the solid state at rather high

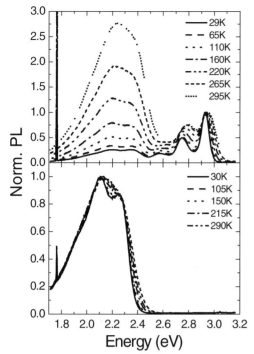

Figure 13 *Upper graph:* Temperature dependence of the PL spectra of poly(fluorene-*co*-fluorenone) with y: 0.005 in the solid state. *Lower graph:* Temperature dependence of the PL spectra of poly(fluorene-*co*-fluorenone) with y: 0.25 in the solid state. The sharp spikes occurring below 1.8eV are due to second order scattering of the laser beam (Ar-ion laser operated in multi-line UV mode) used for the PL excitation. Figure reproduced from [39]

9-fluorenone concentrations and strongly diminished in isolated molecules in solid state solution or polymers with large interchain distances e.g. PFs with dendronic sidechains [29, 31, 60]. As shown for a series of fluorene-9-fluorenone copolymers, for isolated polymer chains in solution it was found that an average of at least 1.5 % 9-fluorenone moieties need to be incorporated into the PF backbone to observe the low energy emission band in solution of a proper solvent.

In summary, fluorenone containing polyfluorenes can be regarded at as a guest–host system where the excitation energy can be transferred from the host (polyfluorene) to the guest (fluorenone). In order to further confirm these explanations and to enlarge our knowledge of the optical and electronic properties of PF-type polymers, we have studied the photoexcitation dynamics of DA-PF and MA-PF as well as of fluorene-fluorenone copolymers with defined concentrations of keto-defects via differential transmission spectroscopy on the fs- to ms- timescale.

The photoinduced absorption spectra for a modulation frequency of 17 Hz are depicted in Fig. 15. DA-PF shows a triplet absorption peak at 1.5 eV typical for PF-type materials [12]. This peak is completely absent in MA-PF as well as in the fluorene-fluorenone copolymer with y: 0.02 which, concerning its absorption and PL

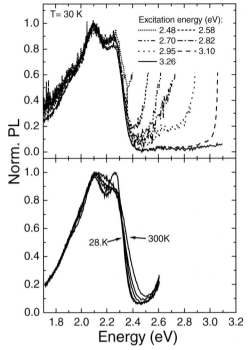

Figure 14 *Upper graph:* Site selective photoluminescence spectra of a thin solid film of poly(fluorene-co-fluorenone) with y: 0.25 at 30K, the photon energies of the excitation light are shown in the inset. *Lower graph:* Temperature dependence of the fluorescence of a thin solid film of poly(fluorene-co-fluorenone) with y: 0.25 excited at 2.71eV. Reproduced from [39]

spectra, mimics the behaviour of MA-PF rather accurately. These two polymers show two broad, unstructured polaron signatures peaking at 1.3 and 2.2 eV. The fluorene-fluorenone copolymer with y: 0.002 displays an intermediate behaviour between that of MA-PF and DA-PF. The evaluated decay times are ca. 5 ms for the triplet absorption of DA-PF and ca. 1 ms for the polarons of MA-PF and the fluorene-fluorenone copolymer with y: 0.002 [61]. The triplet signal is already significantly reduced in the fluorene-fluorenone copolymer with y: 0.002 and only one fluorenone for 500 fluorene units.

In order to clarify why mainly triplets are observed in the pristine PF and mainly polarons in the PFs containing fluorenone defects, a closer look at the generation of these species, hence at the photoexcitation dynamics immediately after excitation, is helpful. These processes can be studied by time-resolved pump-probe measurements. The transient differential transmission spectrum of DA-PF is shown in Fig. 16 for various delay times after excitation. Two main features originating from singlet excitons are observed [62]: stimulated emission (StE) in the energy range from 2.5 to 3.0 eV and a broad absorption of excited singlet excitons peaking at 1.6 eV (PA$_1$). Both features show an analogous temporal evolution and decay to zero in the monitored time interval of 400 ps. On the high energy side of the singlet absorption

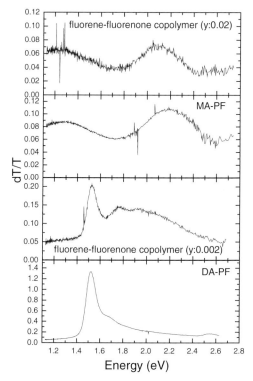

Figure 15 Steady–state photo induced absorption spectra of DA-PF, MA-PF, and the poly(fluorene-co-fluorenone)s with y: 0.002 and 0.02, respectively

peak a second broader absorption feature is observed in the range of 2.0–2.5 eV (PA₂), which we ascribe to polarons, consistently with the assignment of the PA spectra. In Fig. 17 the transient differential transmission spectra for MA-PF are displayed. The StE is immediately overlapped by a broad long living absorption. The PA spectra suggest that the intense polaronic absorption (PA₂) dominates the spectrum and no StE can be observed. Such a behaviour has also been found for other poly(para-phenylene)s and accounts for the fast disappearance of the StE signal [63]. The intense PA₂ signal dominates the spectrum at all times. As in the steady state measurements a much stronger polaron signal is found for the fluorene-fluorenone copolymers. Since the polaron population is higher at all times of observation, this increased polaron content should not be a lifetime-related effect but indicates a higher polaron generation yield, probably via charge transfer to the keto-defect sites. This confirms the assumption that the keto-emission is stronger in electroluminescence than in photoluminescence due to charge carrier trapping [33].

Fig. 18 compares the temporal evolution of the pump-and-probe-signal of DA-PF and MA-PF at selected wavelengths. For MA-PF the stimulated emission signal monitored at 2.75 eV changes its sign after 300 fs and is then dominated by the

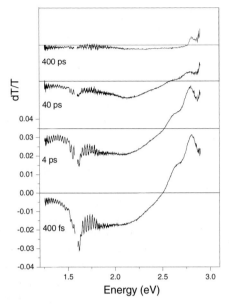

Figure 16   Pump and probe spectrum of DA-PF for different time delays.

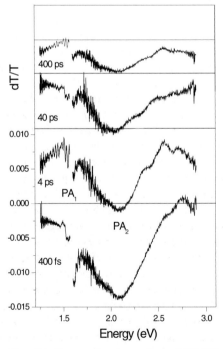

Figure 17   Pump and probe spectrum of MA-PF for different time delays.

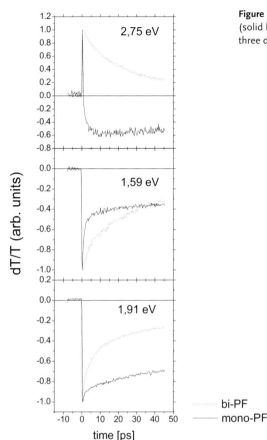

Figure 18 dT/T delay curves for of MA-PF (solid line) and DA-PF (dotted line) films for three different observation energies.

polaron absorption. The stimulated emission of DA-PF is barely biased by a counteracting absorption feature, the decay of the 2.75 eV feature is very similar to the decay of the singlet absorption at 1.59 eV. The decay of the polaron signal monitored at 1.91 eV is initially slower for MA-PF, since the polarons are repopulated via charge transfer to keto-defects, which counteracts their decay at short times.

From the ultrafast together with the steady-state photoinduced absorption data we can derive a straightforward picture for the photoexcitation kinetics in fluorenone-free and fluorenone-containing polyfluorenes both on the femtosecond and the millisecond time scale. In DA-PF a small fraction of polarons is formed via dissociation of singlets. A further small fraction of triplets is formed via intersystem crossing (ISC) during the whole singlet lifetime or non-geminate coalescence of polarons. The major fraction of the singlets decays towards the ground state via radiative recombination (PL) or internal conversion. In MA-PF a large fraction of the singlets reaches the fluorenone sites via single- or multi-step Förster-type energy transfer where the probability to dissociate into a pair of polarons is much higher. The formation of triplets via ISC is not significant in the fluorenone-containing polymers.

We are currently performing a more detailed study [64] to clarify the role of the charge transfer states predicted by quantum mechanical calculations [40].

6.2.5
Degradation in polyfluorene-based electroluminescent devices

While the electroluminescence spectra of unstressed PF-based devices show very similar bands as observed in photoluminescence, prolonged operation of the devices frequently leads to the appearance of low energy emission bands (Fig. 19). In case of PF-type polymers containing keto defects (e.g. MA-PF), but also for a variety of presumably more stable DA-PFs the low energy emission appears as a broad feature peaking at around 2.3 eV which was assigned to bulk keto-defects as discussed above. The intensity of the defect emission usually increases with ongoing operation time and is influenced by a variety of device- and fabrication-related parameters. Careful exclusion of oxygen and water during device preparation, i.e by annealing the polymer layers in vacuum before processing and the preparation of the devices in an inert atmosphere, results in less spectral degradation. Especially for the polymers containing an intrinsic amount of keto-defects (MA-PF, Fluorene-fluorenone copolymers) the relative spectral contribution of the keto emission in EL is considerably stronger than in PL. This demonstrates that the keto defects act as a trap for charge carriers thus leading to increased exciton localization at the defect sites. Devices made from PFs of higher structural regularity do not exhibit the undesired green emission even after extensive electrical stress [41]. PF derivatives with bulky sidechains, aggravating the exciton migration to the defect sites, have also been reported to significantly improve the colour stability [31].

An additional low energy spectral feature of PF-based PLEDs is often observed upon device operation which is energetically located between the "regular" blue emission bands at 2.9 eV and the broad keto-related emission at 2.3 eV, i.e. between 2.45 and 2.6 eV (Fig. 20) [65]. These emission features appear most strongly for devices with Ca electrodes. By investigating the voltage dependence of the EL emission intensity as well as PL spectra of very thin, electrode-covered films, these additional bands were identified as originating from interface-defects close to the low work-function cathode of the PLED [66]. The parameters of electrode deposition such as the base pressure in the evaporation chamber and the metal electrode deposition rate strongly determine the structure and quality of the interface and thus the intensity of the defect emission at 2.45 and 2.6 eV. Residual oxygen incorporated into the interface region caused by low deposition rates or an insufficient vacuum may play a crucial role in the formation of the interface-related defect. Thermal degradation experiments of devices affected by this interface defect demonstrate a different spectral signature as compared to that of bulk keto-defects. This is shown in Fig. 20 which depicts the EL spectra of a pristine DA-PF PLED, a DA-PF PLED after operation in argon atmosphere, this DA-PF PLED after subsequent heating in air for 45 minutes at 150 °C and a DA-PF PLED prepared by a different route where less care was taken to exclude oxygen during the preparation. While the device operated in argon shows only additional emission of the interface- related defect, the annealing of the PLED in air adds a contribution from bulk keto-defects, this EL component becomes dominant in the device prepared without proper exclusion of oxygen.

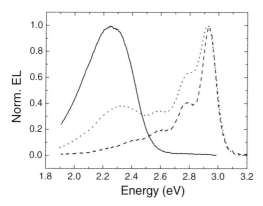

Figure 19 Electroluminescence spectrum of an ITO/MA-PF/Al device (solid line), an ITO/DA-PF/Al device (dashed line), and this ITO/DA-PF/Al device (dotted line) after 30 min continuous operation under air (electrooxidative degeneration). Figure reproduced from [33]

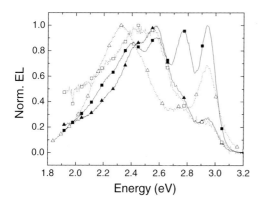

Figure 20 Spectral signature of the interface-related and the bulk-keto defect in DA-PF. Emission of an "as prepared" device (full squares). After operation in argon (filled triangles), occurrence of interface defects. After heating the device in air (open squares), occurrence of interface and bulk defects. Emission from device prepared in the presence of oxygen (open triangles), occurrence of bulk keto defects.

6.3
Conclusion

In conclusion, the now available broad variety of experimental and theoretical results for polyfluorene-based materials yield a comprehensive picture of the (photo-)physics of this promising class of semiconducting materials. Understanding the still existing problems is crucial for the fabrication of efficient and stable PF-based optoelectronic devices; solutions to overcome a large number of these problems have been developed and applied successfully.

In this article we have reviewed the basic optical properties of PF-type materials including morphology related properties (β-phase formation). As a result of side chain crystallization and subsequent planarization of the polyfluorene backbone several polyfluorenes tend to form a highly ordered solid state phase.

Keto-defects (9 fluorenone building blocks) have been unambiguously identified as the origin of the (usually unwanted) green emission component in PF type materials by a variety of experimental and theoretical findings: i) The correlation between the appearance of the characteristic carbonyl stretching absorption of the fluorenone unit in the IR spectrum and the low energy emission bands in degraded polyfluorenes as well as in fluorenone-containing model systems. ii) The finding that the green emission can be directly excited with an excitation energy well below the bandgap energy of polyfluorene. iii) The lack of any concentration-dependence of the defect emission intensity in solution. iv) The distinct influence of the solvent polarity on the low-energy emission (solvatochromism). v) The emission properties at low temperatures and vi) Results from quantum chemical calculations showing a dramatic increase of the PL quantum yield of 9 fluorenone when it is incorporated into a conjugated oligofluorene backbone.

The formation of the fluorenone defects can either occur already during synthesis of the polymer or during thermal, photo- or electrochemical oxidation. Studies of the photoexcitation dynamics in mono- and dialkylated PFs, show a large difference in their tendency to form fluorenone units. Model studies of fluorene-fluorenone copolymers further support the picture of the degraded PFs as a guest-host system where the excitation energy is to a large part transferred from the host (PF) to the guest (keto defect) which emits at a lower energy. Charge trapping and subsequent exciton localisation at the defect sites lead to considerable green emission components already for low defect concentrations. This means that the availability of extremely defect-poor and stable polymers is crucial for future commercial EL applications comprising PFs.

An additional issue in electroluminescent devices is a good control of the low work function cathode/polymer interface in order to avoid chemical degradation of the polymer at the interface which can also lead to unwanted low-energy emission components.

Summing up a large amount of work towards a commercialization of PF-based optoelectonic devices has already been done. However, further fine tuning will certainly be necessary for any PF- based industrial products especially concerning stability-related issues.

Acknowledgements

The presented experiments have been sponsored and supported by the Christian Doppler Forschungsgesellschaft, Sonderforschungsbereich "Elektroaktive Stoffe", the Fonds zur Förderung der wissenschaftlichen Forschung – Austria (P 12806-PHY), SONY International (Europe), Stuttgart, Germany, StiftungVolkswagenwerk, Hannover, Germany, and the Max-Planck-Gesellschaft, München, Germany. CDL-AFM is an important part of the long term research strategies of AT&S. We would like to thank the team of co-workers and collaborators which have been involved in the interdisciplinary research projects on semiconducting polyfluorenes, especially Lorenz Romaner, Martin Gaal, Heinz-Georg Nothofer, Roland Güntner, Michael Forster, Patricia Scandiucci de Freitas, Günther Lieser, Akio Yasuda, Gabi Nelles, Dieter Neher, Tzenka Miteva, Dessislava Sainova, Alexander F. Pogantsch, Franz P. Wenzl, Andrew C. Grimsdale, Egbert Zojer, Günther Leising, and Jean Luc Brédas. We also wish to thank Klaus Müllen, Mainz, Giulio Cerullo and Guglielmo Lanzani, Milano, for the cooperation as well as for their continuous and generous support of our investigations.

References

[1] A.C. Grimsdale, P. Leclère, R.Lazzaroni. J. D. Mackenzie. C. Murphy, S. Setayesh, C. Silva, R.H. Friend, K. Müllen, Adv. Funct. Mater. **12**, 729 (2002)

[2] A.J. Campbell, D.D.C. Bradley, H. Antoniadis, M. Inbasekaran, W.W. Wu, E.P. Wu, Appl. Phys. Lett. **76**, 1734 (2000)

[3] A.J. Campbell, D.D.C. Bradley, H. Antoniadis, J. Appl. Phys. **89**, 3343 (2001)

[4] M. Fukuda, K. Sawada, K. Yoshino, Jpn. J. Appl. Phys. **28**, L1433 (1989)

[5] Y. Ohmori, M. Uchida, K. Muro, K. Yoshino, Jpn. J. Appl. Phys. **11B**, L1941 (1991)

[6] M. Ariu, M. Sims, M.D. Rahn, J. Hill, A.M. Fox, D.G. Lidezy, Phys. Rev. B **67**, 1953333 (2003) and references therein

[7] D. Hertel, S. Setayesh, H.-G. Nothofer, U. Scherf, K. Müllen, H. Bässler, Adv. Mater. **13**, 65 (2001)

[8] M. Grell, D.D.C. Bradley, M. Inbasekaran, E.P. Woo, Adv. Mater. **9**, 798 (1997)

[9] M. Grell, D.D.C. Bradley, G. Ungar, J. Hill, K.S. Whitehead, Macromolecules **32**, 5810 (1999)

[10] K.S. Whitehead, M. Grell, D.D.C. Bradley, M. Jandke, P. Strohriegl, Appl. Phys. Lett. **76**, 2946 (2000)

[11] M.Grell, W. Knoll, D. Lupo, A. Meisel, T. Miteva, D. Neher, H.-G. Nothofer, U. Scherf, A. Yasuda, Adv. Mater. **11**, 671 (1999)

[12] A.J. Cadby, P.A. Lane, H. Mellor, S.J. Martin, M. Grell, C. Giebler, D.D.C. Bradley, Phys. Rev. B, **62**, 15604 (2000)

[13] Y. Ohmori, M. Uchida, C. Moroshima, A. Fujii, K. Yoshino, Jpn. J. Appl. Phys. **32**, L1663 (1993)

[14] D. Sainova, T. Miteva, H.G. Nothofer, U. Scherf, I. Glowacki, J. Ulanski, H. Fujikawa, D. Neher, Appl. Phys. Lett. **76**, 1810 (2000)

[15] C. Ego, A.C. Grimsdale, F. Uckert, G. Yu, G. Srdanov, K. Müllen, Adv. Mater. **14**, 809 (2002)

[16] H.-G. Nothofer, Ph.D. Thesis, Universität Potsdam, (Logos Verlag, Berlin, Germany, 2001)

[17] M. Redecker, D.D.C. Bradley, M. Inbasekaran, W.W. Wu, E.P.Woo, Adv. Mater. **11**, 241, (1999)

[18] M. Redecker, D.D.C. Bradley, M. Inbasekaran, E.P.Woo, Appl. Phys. Lett. **73**, 1565 (1998)

[19] L.S. Liao, L.F. Cheng, M. K. Fung, C.S. Lee, M. Inbasekaran, E. P. Woo, W. W. Wu, Chem. Phys. Lett. **325**, 405 (2000)

[20] M. K. Fung, S. L. Liai, S. N. Bao, C. S. Lee, S. T. Lee, W. W. Wu, M. Inbasekaren, J. J. O'Brian. J. Vac. Sci. Technol. **20**, 911 (2002)

[21] M. Stoessel G. Wittmann, J. Staudigel, F. Steuber, J. Blässing, W. Roth, H. Klausmann, W. Rogler, J. Simmerer, A. Winnacker, M. Inbasekaran, e.P. Woo, J. Appl. Phys. **87**, 4467 (2000)

[22] L. S. Liao, L.F. Cheng, M. K. Fung, C.S. Lee, S. T. Lee, M. Inbasekaran, E. P. Woo, W.W. Wu; Phys. Rev. B. **62**, 10004 (2000)

[23] U.Lemmer, S.Heun, R.F. Mahrt, U. Scherf, M. Hopmeier, U. Siegner, E.O. Göbel, K. Müllen, H. Bässler, Chem. Phys. Lett. **240**, 373 (1995)

[24] J.I. Lee, G. Klärner, R.D. Miller, Chem. Mater. 1999, 11, 1083.

[25] E. Conwell, Trends Polym. Sci. **5**, 218 (1997)

[26] V. Cimrova, U. Scherf, D. Neher, Appl. Phys. Lett. **69**, 608 (1996)

[27] K.-H. Weinfurter, H. Fujikawa, S. Tokito, Y. Taga, Appl. Phys. Lett. **76**, 2502 (2000)

[28] V.N. Bliznyuk, S.A. Carter, J.C. Scott, G. Klärner, R.D. Miller, Macromolecules **32**, 361 (1999)

[29] S. Setayesh, A.C. Grimsdale, T. Weil, V. Enkelmann, K. Müllen, F. Meghdadi, E.J.W. List, G. Leising, J. Am. Chem. Soc. **123**, 946 (2001)

[30] J.M. Lupton, P. Schouwink, P.E. Keivanidis, A.C Grimsdale, K. Müllen, Adv. Funct. Mat. **13**, 154 (2003)

[31] A. Pogantsch, F.P. Wenzl, E.J.W. List, G. Leising, A.C. Grimsdale, K. Müllen, Adv. Mater. **14**, 1061 (2002)

[32] Y.K. Nakazawa, S.A. Carter, H.-G. Nothofer, U. Scherf, V.Y. Lee, R.D. Miller, J.C. Scott, Appl. Phys. Lett. **80**, 3832 (2002)

[33] E. J. W.List, R. Güntner, P. Scandiucci de Freitas, U. Scherf, Adv. Mater. **14**, 374 (2002)

[34] U. Scherf, E. J. W. List, Adv. Mater. **14**, 477 (2002)

[35] J.M. Lupton, M.R. Craig, E.W. Meijer, Appl. Phys. Lett. **80**, 4489 (2002)

[36] D. Sainova, D. Neher, E. Dobruchowska, B. Luszczynska, I. Glowacki, J. Ulanski, H.-G. Nothofer, U. Scherf, Chem. Phys. Lett. 371, 15 (2003)

[37] M. Gaal, E.J.W. List, U. Scherf, Macromolecules **36**, 4236 (2003)

[38] P. Scandiucci de Freitas, U. Scherf, M. Collon, E.J.W. List, e-polymers 009 (2002)

[39] L. Romaner, A. Pogantsch, P. Scandiucci de Freitas, U. Scherf, M. Gaal, E. Zojer, E.J.W. List, Adv. Funct. Mat. **13**, 597 (2003)

[40] E. Zojer, A.Pogantsch, E. Hennebicq, D. Beljonne, J.-L. Bredas, P. Scanducci de Freitas, U. Scherf, E.J.W. List, J. Chem. Phys. **117**, 6794 (2002)

[41] M.R. Craig, M.M. de Kok, J.W. Hofstraat, A.P.H.J. Shenning, E.W. Meijer, J. Mater. Chem. **13**, 2861 (2003)

[42] M. Ariu, D.G. Lidzey, M. Sims, A.J. Cadby, P.A. Lane, D.D.C. Bradley, J. Phys.: Condens. Matter. **14**, 9975 (2002)

[43] E.J.W. List, J. Partee, J. Shinar, U. Scherf, K. Müllen, W. Graupner, K. Petritsch, E. Zojer, G. Leising, Phys. Rev. B **61**, 10807 (2000)

[44] A. Pogantsch, F.P. Wenzl, U. Scherf, A.C. Grimsdale, K. Müllen, E.J.W. List, J. Chem. Phys. **119**, 6904 (2003)

[45] E.J.W. List, C.-H. Kim, A.K. Naik, U. Scherf, G. Leising, W. Graupner, J. Shinar Phys. Rev. B **64**, 155204 (2001)

[46] G. Fytas, H.-G. Nothofer, U. Scherf, D. Vlassopoulos, G. Meier, Macromolecules **35**, 481 (2002)

[47] H.-G. Nothofer, A. Meisel, T. Miteva, D. Neher, M. Forster, M. Oda, G. Lieser, D. Sainova, A. Yasuda, D. Lupo, W. Knoll, U. Scherf, Macromol. Symp. **154**, 139 (2000)

[48] R.M. Silverstein, G.C. Bassler, T.C. Morrill, Spectroscopic Identification of Organic Compounds, 4th ed. (Wiley, New York, 1981)

[49] A.R.G. Ilharco, J. Lopes da Silva, M. João Lemos, L.F. Vieira Ferreira, Langmuir **13**, 3787 (1997)

[50] L.J. Rothberg, M. Yan, F. Papadimitrakopoulos, M.E. Galvin, E.W. Kwock, T.M. Miller, Synth. Met. **80**, 41 (1996)

[51] H. Zweifel, Stabilization of Polymeric Materials, (Springer-Verlag, Berlin/Heidelberg, 1998)

[52] S. Panozzo, J.-C. Vial, Y. Kervella, O. Stéphan, J. Appl. Phys. **92**, 7 (2002)

[53] M. Pope and C.E. Swenberg, Electronic processes in Organic Crystals and Polymers,

(Oxford University Press, New York 1999), ch. 1

[54] J. Cornil, A. J. Heeger, and J. L. Bredas, Chem. Phys. Lett. **272**, 463 (1997)

[55] J.B. Birks and L.G. Christophorou, Spectrochim. Acta **19**, 401 (1963)

[56] E.J.W. List, C. Creely, G. Leising, N. Schulte, A.D. Schlüter, U. Scherf, K. Müllen, W. Graupner. Chem. Phys. Lett. **325**,132 (2000)

[57] As experimental observations [24] as well as correlated quantum chemical calculations [40] show, the actual emission energy in a fluorenone-fluorene co-polymer does depend on the exact position of the fluorenone unit versus the fluorene units. Differences in the range of 0.15 eV have been observed.

[58] J.B. Birks and A.A. Kazzaz, Proc. Roy. Soc. **A 304**, 291 (1968)

[59] D. Beljonne, G. Pourtois, C. Silva, E. Hennebicq, L. M. Herz, R. H. Friend, G. D. Scholes, S. Setayesh, K.Müllen and

J. L. Bredas, Proc. Natl. Acad. Sci. USA 99, 10982 (2002)

[60] E. J. W. List, C. H. Kim, J. Shinar, M. A. Loi, G. Dongiovanni, A. Mura, S. Setayesh, A. C. Grimsdale, T. Weil, V. Enkelmann, U. Scherf, K. Müllen and G. Leising, Mat. Res. Soc. Symp. Proc. 665, C5.47.1–6, (2001)

[61] L. Romaner, T. Piok, C. Gadermaier, R. Guentner, P. Scandiucci de Freitas, U. Scherf, G. Cerullo, G. Lanzani and E. J. W. List, Synth. Met. **139**, 851 (2003)

[62] T. Virgili, D.G. Lidzey, D.D.C. Bradley, G. Cerullo, S. Stagira, S. De Silvestri, Appl. Phys. Lett. **74**, 2767 (1999)

[63] C. Gadermaier et al., Phys. Rev. B **66**, 125203 (2002)

[64] C. Gadermaier, in preparation.

[65] X. Gong, P.K. Iyer, D. Moses, G.C. Bazan, A.J. Hegger, S.S. Xiao, Adv. Funct. Mater. **13**, 325 (2003)

[66] S. Gamerith. H.-G. Nothofer, U. Scherf, E.J.W. List, Jpn. Journ. Appl. Phys. **43**, L891 (2004)

7

Exciton Energy Relaxation and Dissociation in Pristine and Doped Conjugated Polymers

V. I. Arkhipov and H. Bässler

7.1
Introduction

The enormous potential of organic semiconductors comes from the fact that they combine basic electronic properties of the traditional inorganic crystalline semiconductors with easy processing of plastics [1,2]. This unique combination of properties paves the way for the design of new large area and/or low cost, and even flexible optoelectronic devices such as solar cells, light-emitting diodes, and lasers. As far as device applications are concerned, organic and inorganic semiconductors reveal rather similar photoelectrical properties. Both types of semiconductors are capable of efficient electrical power conversion into visible or infrared optical emission in light emitting diodes. The inverse process, i.e. electrical power generation upon illumination is used in photovoltaic devices. However, the underlying mechanisms are entirely different in these groups of materials.

An important characteristic feature of both radiative carrier recombination and carrier photoproduction in crystalline inorganic semiconductors is that these are *direct* processes, which proceed without formation of any intermediate excited electronic states [3]. Radiative recombination of free electrons and holes, occupying extended states in the conduction and valence bands, respectively, is responsible for electroluminesce in such materials. Free carriers are immediately produced upon photon absorption in crystalline inorganic semiconductors. Therefore, both radiative recombination and photogeneration of charge carriers can be simply characterized by respective cross-sections that are virtually independent of the applied electric field, carrier densities, dopant and trap concentrations, etc. The cross-section of charge photogeneration is also temperature-independent while the probability of radiative recombination often strongly decreases with increasing temperature due to enhanced energy dissipation via phonons at higher temperatures.

The situation is somewhat different in amorphous inorganic semiconductors [4]. Due to strong carrier localization, photogenerated charges cannot escape from their common Coulomb potential well before both carriers are thermalized and trapped by localized states. Therefore, Coulombically bound geminate pairs rather than free carriers should be considered as primary photoexcitations in disordered inorganic materials. Such pairs can subsequently either recombine or dissociate into free car-

Physics of Organic Semiconductors. Edited by W. Brütting
Copyright © 2005 WILEY-VCH Verlag GmbH & Co. KGaA, Weinheim
ISBN 3-527-40550-X

riers as described by the Onsager theory of geminate recombination [5]. This theory predicts very strong dependences of the dissociation probability upon both external electric field and temperature in good agreement with experimentally observed field and temperature dependences of the charge photogeneration yield in *a*-Se and chalcogenide glasses [6].

In organic semiconductors including conjugated polymers, both photogeneration and radiative recombination of charge carriers proceed via formation of strongly bound Frenkel-type molecular excitons [7]. There is abundant evidence in favor of this concept [8-14]. It offers a plausible explanation for electroluminescent properties of organic materials especially for the notorious similarity of their electro- and photoluminescence spectra as well as for a weak temperature dependence of their yields. However, the question remains why and how a strongly bound molecular exciton can dissociate into free charge carriers and give rise to dc photocurrent. Rather extended excited states in conjugated polymers tend to diminish the energy difference between neutral singlet excitons and charge transfer states, i.e. geminate electron-hole pairs. Nonetheless, there is good reason to believe that the energy of charge transfer states is, on average, above that of the singlet excitons. Our review is focused on the description and analysis of various processes related to exciton dissociation into geminate pairs and further into free carriers in conjugated polymers.

In these materials, a super-excited Frank-Condon state can autoionize directly or via local heating when the excess optical energy is funneled into vibrations coupled to the hot exciton [15]. This conjecture is supported by non-congruent absorption and photocurrent action spectra observed in those materials. The excess energy needed to dissociate an already relaxed singlet exciton can also be supplied by a sufficiently strong electic field. It can promote an electron from the excited to an adjacent molecule or chain segment and form an intermediate charge transfer state, i.e. a geminately bound electron-hole pair. The field dependence of the fluorescence quenching is a measure of the primary dissociation rate. Since this initial step is field sensitive it is obvious that an Onsager-like description for the photoconduction is inappropriate. However, relaxed singlet excitons can somehow dissociate into free carriers even at moderate electric fields although the dissociation yield is very small in pristine materials. This dissociation can be caused by defects having higher electron affinity/lower ionization energy than that of the bulk material. There is experimental evidence that chemical acceptors can, in fact, sensitize dissociation of an exciton [16-18] but one cannot discard the possibility that physical defect acts similarly. Sensitization generates ionized donor-acceptor pairs that can subsequently dissociate into free charges although with low probability [16,17].

Strong Coulomb interaction in a material with the dielectric constant of about 3 combined with weak inter-chain coupling implies strong inelastic scattering of carriers before they can leave their mutual coulombic potential well. The majority of charges, generated upon exciton dissociation, form metastable geminate pairs of which only some fraction can further separate into free carriers. Rigorous tests for the existence of geminate pairs are e.g. delayed field collection of charges [18], delayed fluorescence from geminate pair recombination [19,20], and field quenching of delayed fluorescence [21,22]. In a strongly disordered material, the lifetime of some pairs can be up to hours especially at low temperatures [23] at variance with molecular crystals in which charge transfer states are rather short lived. Since the absorption spectra of radical cations/anions, i.e. polarons, overlap with the fluorescence from neutral chain segments this will be a major handicap for electrically pumped lasing in such systems.

In conjugated polymers, heavily doped with electron scavengers, i.e. in polymer/acceptor blends, the yield of exciton dissociation into free charges can increase up to 60...70% provided that the morphological demands are met [24-27]. It implies that excitons can somehow almost immediately dissociate into free carriers in polymer/acceptor blends. This notion is supported by the fact that no geminate pairs can be detected upon photoexcitation of heavily doped conjugated polymers [28]. There must be, therefore, an efficient way for making profitable use of the energy gained upon exciton dissociation at a polymer/acceptor interface. The Coulomb electric field of the electron, trapped on the acceptor side of the interface, together with the external electric field forms both a potential well and a potential barrier for the on-chain hole on the polymer side. The kinetic energy of zero-point oscillations of the hole within the potential well facilitates dissociation of geminate pairs. Under certain conditions, an effectively repulsive potential of the electron-hole interaction can be formed at a polymer/acceptor interface.

7.2
Field-induced exciton dissociation probed by luminescence quenching

7.2.1
Steady state fluorescence quenching

In inorganic semiconductors, such as GaAs, the binding energy E_b of an exciton, i.e. the energy needed to dissociate an exciton into a coulombically unbound electron-hole pair, is amenable to direct absorption spectroscopy. At low temperatures the Wannier-type exciton yields a sharp absorption feature, which is slightly but noticeably offset from the onset of the valence-to-conduction band transition that determines the bandgap. Typically E_b is not larger than 10 meV in those materials. In organic solids, the situation is different because both the valence and conduction bands are narrow, typically < 0.1 eV [7], and the dielectric constants are 3...4 implying a coulombic capture radius of an electron hole pair of (roughly) 20 nm at room temperature. As a consequence, the dominant absorption is of excitonic nature and the oscillator strength for a transition from the neutral state to a fully ionized state is extremely weak. It escapes spectroscopic probing except in perfect poly-diacetylene (PDA) crystals. By applying the electroreflection technique, a Franz-Keldysh feature can be observed some 0.5 eV above the exciton transition [29]. This energy difference has to be identified with the exciton binding energy. In view of their similar dielectric responses this value should be representative for other π-conjugated polymers as well. It is gratifying that calculations, made by solving the Bethe-Salpeter equation for a two body Green function of the electron-hole (eh) pair, yielded a value of

$E_b = 0.43$ eV for PDA. The same computational method predicts E_b-values of 0.54 eV for poly-phenylenevinylene (PPV), 0.64 eV for the ladder-type poly-para-phenylene (LPPP), 0.43 eV for poly-acetylene, and 0.61 eV for poly-thiophene [11]. There is abundant experimental evidence that E_b is, indeed, of this order of magnitude although the reasoning is indirect.

The above listed values of E_b may be compared to the coulombic energy of 0.4 eV of an eh-pair at a distance of 1 nm in a medium with a dielectric constant of 3.5. However, associating that energy with E_b would imply that no energy is needed to split an on-chain singlet exciton of a π-conjugated polymer into an off-chain eh-pair localized on adjacent chains. A similar problem has been encountered in classic molecular crystals already. The calculations by van der Horst et al. indeed show that transfer of an electron from an LPPP chain in its first excited singlet state (S_1) to an adjacent chain does cost an energy of 0.1...0.2 eV. Therefore, E_b can be considered as the sum of the energy needed to create a charge transfer (CT) state from the S_1 state and the binding energy of that eh-pair. The fact that the energy difference $E(CT) - E(S_1)$ is, on average, a positive quantity is supported by the fluorescence spectroscopy. If it were negative a polymer film would not emit fluorescence from the on-chain singlet state but feature exclusively excimer emission instead. Although the experimentally observed fluorescence spectra may contain an excimer component, caused by energy transfer to topological defects, it is known that most of the emission comes from on-chain singlet excitons.

The simplest experimental test whether or not extra energy is required to break up an on-chain singlet excitation into a pair of charges, if coulombically bound, is to measure the fluorescence intensity upon applying an electric field to the sample [21,30]. Note that the drop of the electrostatic energy upon displacing a charge by 1 nm in an electric field of 2×10^6 V/cm is 0.2 eV. A field of that magnitude can, therefore, compensate for the energy needed to create an eh-pair on adjacent chains from an on-chain S_1 exciton and, concomitantly, quench fluorescence. Recall that in inorganic semiconductors like GaAs or CdS fields of a few V/cm or even less are sufficient to ionize a Wannier exciton [31].

On a variety of systems steady state fluorescence quenching was measured as a function of an electric field applied to typically 100 nm thick polymer films, such as phenyl-substituted PPV (PPPV) [32], the methyl-substituted planarized poly-para-phenylene (MeLPPP) [19,33], and a copolymer of an alkoxy-substituted poly-phenylenevinylene (PhPPV) [34]. The samples were sandwiched between an indium-tin-oxide (ITO) electrode and an evaporated Al electrode. In order to prevent spurious fluorescence quenching by charge carriers injected from the electrodes all experiment were done under reverse bias, i.e. with negatively biased ITO contact. Figure 1 summarizes data on the fluorescence quenching efficiency defined as the fluorescence decrement normalized to the fluorescence intensity at zero electric field,

$$Q(F) = \frac{I(0) - I(F)}{I(0)} \tag{1}$$

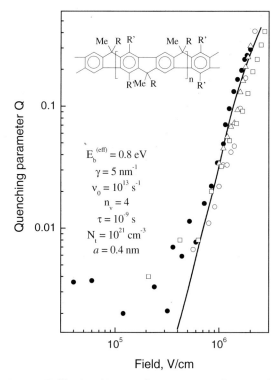

Figure 1 Field-induced PL quenching in MeLPPP (●) [33] and (○) [19], PPPV (□) [32], and PhPPV (△) [34]. The solid line is calculated from Eq. (10). The inset shows the chemical structure of MeLPPP.

In all these materials, a significant fluorescence quenching was observed featuring a superlinear field dependence at fields exceeding 10^6 V/cm. It turned out that in PPPV, $Q(F)$ is by a factor of 2 lower than in MeLPPP. It is worth noting that, as revealed in the absorption spectra, the inhomogeneous broadening of the $S_1 \leftarrow S_0$ transition in PPPV is about a factor of 5 larger than in MeLPPP, the latter being the least disordered of all non-crystalline π-conjugated polymers due to the rigidity of its polymer backbone. This discrepancy clearly indicates that disorder is not the distinguishing property that controls the efficiency of fluorescence quenching.

These experiments demonstrate that a roughly 500 times larger electric field is typically required to quench the photoluminescence in conjugated polymer films as compared to inorganic semiconductors. This supports the notion that the exciton binding energies are similarly different in these two classes of materials. An electric field of several MV/cm is obviously required for preventing an electron-hole pair to recombine and repopulate a fluorescent S_1 state. An important question is whether or not an initially hot Franck-Condon state, created via exciting into the manifold of vibronic states above the optical absorption threshold, is more liable to dissociation. This issue was raised by the Rothberg group [35,36] arguing that an excited state decides early on, i.e. prior to dissipation of the vibrational excess energy, if it chooses the fluorescence or dissociation channel for its subsequent decay. Earlier experiments on spectrally resolved fluorescence quenching in PPPV and recent data for MeLPPP (see Fig. 2) demonstrate that the fluorescence quenching at selected electric fields is constant within a spectral range of 2.7 eV< $h\nu$ < 3.45 eV [34]. At higher photon energies, a weak tendency towards more efficient quenching is noted. Although this result cannot be taken as an evidence against the notion that no photodissociation occurs from a hot Franck-Condon state, it indicates that at least the majority of the vibronic S_1 states with excess energy < 0.7 eV first relax to the vibrational cold S_1 state and only afterwards can be quenched by electric field.

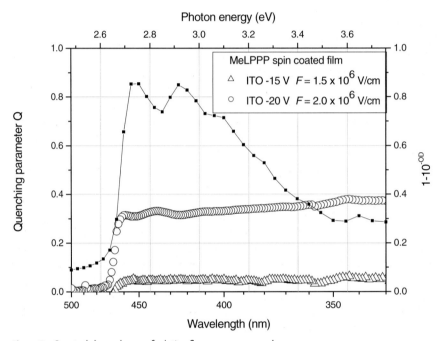

Figure 2 Spectral dependence of relative fluorescence quenching and optical density in a MeLPPP film at different applied voltages. The film thickness was 100 nm. (C.Im, unpublished results.)

7.2.2
Time resolved fluorescence quenching

Another way to delineate exciton dissociation is to monitor the evolution of the transient optical absorption due to the generated charge. Upon adding or removing a charge carrier to or from a molecule a radical anion or cation is generated whose absorption spectrum is red-shifted relative to the absorption of the neutral parent molecule. In conjugated polymers, associated optical transitions are typically in the vicinity of 2 eV. The photoinduced absorption is easily detectable by pump-probe spectroscopy. Such experiments were performed by Graupner *et al.* [37] on MeLPPP upon applying an electric field. A transient absorption feature was indeed observed at 1.9 eV and has been ascribed to charge carriers that are normally identified as polarons although that optical transition gives no information on the relaxation of the polymer chain upon ionization. It has been concluded that (i) charges are generated directly from excitons without any additional intermediate states and (ii) the process occurs on the time scale of 10 ps, i.e. much less than the lifetime of singlet excitons suggesting that they are reactive before vibronic relaxation and thermalization towards the bottom states of the excitonic density-of-states (DOS) is completed.

Analogous experiments were performed recently by Gulbinas *et al.* [33] the only difference being the photon dose applied upon excitation. The experiments were done using 100 fs laser pulses at $h\nu = 3.1$ eV with the dose of 14 µJ/(cm^2×pulse), i.e. almost a factor of 100 less than that used in Ref. [37]. At so high excitation intensities the rate of bimolecular exciton annihilation can limit their lifetime and, concomitantly, the characteristic time of exciton dissociation. At an incident intensities of around 14 µJ/(cm^2×pulse) this effect is absent as indicated by the fact that the fluorescence decay is governed by the intrinsic exciton lifetime of 300 ps.

Figure 3 shows the electromodulated differential absorption (EDA) spectra at different delay times after excitation at a field of 2.2×10^6 V/cm. Positive signals correspond to quenching of stimulated emission and to electro-stimulated induced absorption by polarons at 2.1 eV and 1.9 eV. The top EDA spectrum, measured without optical excitation, is due to a Stark shift of the excitonic absorption edge. In the following analysis, the Stark-shift contribution was subtracted from the spectra measured at positive delays. The EDA spectrum at 1 ps delay displays a sizeable positive contribution in the stimulated emission region but no signal has been detected at energies corresponding to absorption from excitons (≤ 1.85 eV, not shown) and polarons. At 20 ps delay, induced absorption with pronounced bands at 1.9 eV and 2.1 eV and a negative signal at the low energy part is observed. At 400 ps the new absorption bands are even more pronounced, whereas the negative signal below 1.85 eV has disappeared. The new induced absorption has a spectrum very similar to that reported in Ref. [37] and is unambiguously assigned to the charged species, i.e. polarons. Most remarkably, their concentration grows within the entire exciton lifetime.

From the analysis of the transient absorption the rate, at which singlet excitons dissociate into charged species, i.e. the exciton breaking rate γ_b can be inferred (Fig. 4). It is time dependent and can be fitted by an algebraic law, $\gamma_b(t) \propto (t_0/t)^{0.4}$, with a cut-off value of 2.2×10^{10} s^{-1} at short times. Such time dependences of rate constants is often found in random systems [7] and, in this case, can be explained by a distribution of charge transfer rates in a disordered solid MeLPPP polymer film. The dissociation rate has its maximum at short times when the nearest-neighbor geminate pair formation dominates but at longer times jumps to more distant sites may prevail depending upon the electric field and the DOS distribution.

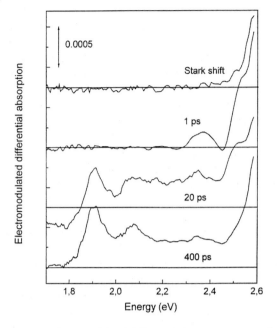

Figure 3 Electromodulated differential absorption spectra of MeLPPP obtained at different delay times after excitation. The upper curve shows the electric field induced differential absorption, measured without excitation. The dashed line is EDA spectrum corrected for the Stark effect.

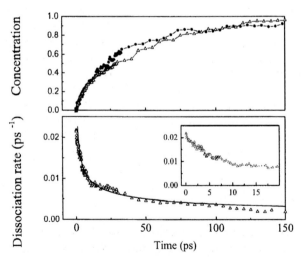

Figure 4 (*a*) Normalized concentration of broken excitons (open triangles) and polarons (solid circles) in MeLPPP as a function of time; (*b*) the rate of exciton dissociation as a function of time (open triangles). The inset shows the signal at short times. The full curve is a fit based upon an algebraic $t^{-0.4}$ law.

Field-assisted exciton breaking occurs if the energy difference between a nearest neighbor inter-chain geminately bound eh-pair and a neutral excitation is counterbalanced by the electrostatic potential. The fact that photogeneration becomes efficient at $F \geq 10^6$ proves that transfer of an electron from the excited chromophore to an adjacent (neutral) chain with an inter-chain separation of 1 nm requires an energy of no less than ca 0.1 eV, in agreement with the calculations in Ref. [11]. It is gratifying to note that the field dependences derived from transient probing of polaron generation at a delay time of 400 ps and from cw-fluorescence quenching experiments (see section 2.1) are virtually identical as shown in Fig. 1.

From experiments on both delayed [22] and thermally stimulated [23] fluorescence it is known that there can be some exciton dissociation already at zero applied field. The field modulation method allows detection only of the field-induced exciton breaking. However, as concluded from the lack of temporal evolution in the exciton/polaron spectral region of the field-free transient absorption and from the measured dynamic range of the field-assisted process (see Fig. 4) it can be concluded that field-free exciton breaking is not strong (less than 1%).

It is a straightforward conjecture that transfer of an electron from an excited chromophore is facilitated if the latter is a Franck-Condon state with excess vibronic energy. Very recent experiments by Gulbinas et al. [38] indeed showed that, upon exciting a MeLPPP film with 4.6 eV photons, onset of the polaron signal occurs on a time scale of $\cong 1$ ps, comparable with the time resolution of the experimental setup. This is in agreement with the works of Gadermaier et al. [39] and Silva et al. [40] who showed that excitation with 4.8 eV photons generates charges in the absence of an electric field. Remarkably, though, that this process is fast yet not instantaneous. Employing fs probing the authors of Ref. [39] were able to demonstrate that 'hot' exciton dissociation occurs on a time scale of 1 ps. This is in agreement with the model of Arkhipov et al. [41,42] who suggested that hot exciton dissociation occurs from a vibrationally hot chain segment after funneling the excess photon energy into the local vibrational heat bath. Recent studies of the intrachain vibrational dynamics supported this notion [43,44].

In summary, the EDA experiments on MeLPPP showed that (i) the photodissociation of singlet excitons into geminately bound electron-hole pairs is facilitated by external electric field and (ii) even at an excess photon energy of 0.4 eV above the $S_1 \leftarrow S_0$ (0–0) transition, the dissociation predominantly proceeds via vibrationally relaxed S_1 excitons during their entire lifetime with a rate decreasing as $t^{-0.4}$.

7.2.3
A model of field-assisted dissociation of optical excitations

In order to produce a geminate pair, an optical excitation must dissociate before it decays radiatively, i.e. at times normally shorter than 1 ns in conjugated polymers. Two different scenarios of dissociation can be envisaged.

In a conjugated polymer consisting of segments of typically 5 nm in length, an optical excitation can dissociate into an on-chain geminate pair if the gain of the electrostatic energy within the segment is sufficient to stabilize the carriers on the opposite ends of a segment and to prevent them from geminate recombination [45]. In other words, the external field must be strong enough to provide for the on-chain potential barrier separating the carriers within the same segment. Under these conditions, the external field must also assist further dissociation of on-chain geminate pairs by stimulating further carrier jumps to other segments. Therefore, most of the on-chain geminate pairs must eventually dissociate into free carriers and contribute to photoconductivity. However, the yield of free carrier photogeneration is always much smaller than the yield of field-induced photoluminescence quenching, implying that only a small fraction of geminate pairs can subsequently dissociate into free carriers [46]. This argues against on-chain carrier separation as the main mechanism of non-radiative field-assisted relaxation of optical excitations in conjugated polymers and implies that off-chain dissociation has to be more important.

Two molecules which belong either to different polymer chains or to different segments of the same chain may be sufficiently close to each other such that the Coulomb binding energy E_C of an electron–hole pair occupying these molecules exceeds the exciton binding energy E_b in either of the molecules. This pair of molecules will then serve as a charge transfer center at which an exciton can dissociate into a pair of charges localized in these molecules. The distance between the molecules must be short enough to comply with the following conditions. Firstly, the binding energy of the off-chain geminate pair must exceed the binding energy of the on-chain exciton. Secondly, either an electron or a hole must make a tunneling jump from the excited molecule into its neighbor before the excitation decayed. Since the energy of the eh-pair is less than the exciton energy the pair will be stabilized against radiative decay as well as against dissociation into free carriers. Although the carriers may execute a short random walk within the Coulomb potential well most of them must sooner or later recombine non-radiatively.

Dissociation of an optical excitation into a metastable coulombically bound eh-pair is energetically feasible if the binding energy of the letter, $E_b^{(GP)}$, is larger than that of the former, E_b. Both these energies are eigenvalues of the corresponding Hamiltonians and they can be represented as sums of the potential, $E_p^{(GP)}$, E_p, and kinetic, $E_k^{(GP)}$, E_k, energies, respectively. Since very little is known about relative contribution of those energies to the total binding energy of a geminate pair, in the following treatment we consider the kinetic energy $E_k^{(GP)}$ as a parameter. The dissociation is energetically possible if the distance between carriers in a geminate pair is such that the following inequality is valid,

$$E_b \leq \frac{e^2}{4\pi\varepsilon_0\varepsilon r} + eFrz - E_k^{(GP)}, \tag{2}$$

or, equivalently,

$$E_b^{(eff)} \leq \frac{e^2}{4\pi\varepsilon_0\varepsilon r} + eFrz, \tag{3}$$

where r is the distance between carriers in a geminate pair, $z = \cos\vartheta$ with ϑ being the angle between the external field F and the direction of the carrier jump over the distance r, and the effective binding energy of the exciton $E_b^{(eff)}$ is defined as,

$$E_b^{(eff)} = E_b + E_k^{(GP)}. \tag{4}$$

The use of $E_b^{(eff)}$ instead of E_b allows to account only for the potential energy of geminate pairs as far as the probability of exciton dissociation is concerned. Equation (3) determines the region in space within which the sites, available for dissociation of excitons into geminate pairs, can be found.

Another limiting factor is the rate of exciton dissociation. The rate v of energetically downward tunneling carrier jumps over the distance r is given by,

$$v(r) = v_0 \exp(-2\gamma r), \tag{5}$$

where γ is the inverse localization radius and v_0 the attempt-to-jump frequency. The probability w that, during its lifetime τ on a given segment, an exciton will dissociate into a geminate pair of radius r is given by the Poisson distribution of probability as,

$$w(r) = 1 - \exp[-v_0 \tau \exp(-2\gamma r)]. \tag{6}$$

The average number of dissociation sites around an excited molecule, $<n_{diss}>$, can be evaluated by the integration over the dissociation volume of the density of neighboring sites, $N_n(r)$, weighted with the dissociation probability $w(r)$. The result reads,

$$\langle n_{diss} \rangle (F) = \int_{V_{diss}} d\mathbf{r}\, w(r)\, N_n(r). \tag{7}$$

The coordinate dependence of the function $N_n(r)$ accounts for possible correlations between positions of neighboring sites. The probability W that no sites are located within the dissociation volume is also determined by the Poisson distribution as,

$$W = \exp[-\langle n_{diss} \rangle (F)]. \tag{8}$$

The dissociation yield η_{diss}, i.e. the probability to find a dissociation site around at least one of n_v sites, visited by an exciton during its lifetime, is then given by,

$$\eta_{diss} = 1 - \exp[-n_v \langle n_{diss} \rangle (F)]. \tag{9}$$

It is worth noting that the number of visited sites can either be taken as an input parameter or calculated as described in section 3.2. Since the quenched PL intensity is proportional to the dissociation probability η_{diss} the use of Eq. (1) leads to the following expression for the quenching parameter Q,

$$Q(F) = 1 - \exp\{-n_v[\langle n_{diss} \rangle (F) - \langle n_{diss} \rangle (0)]\}. \tag{10}$$

Field dependences of the quenching parameter are plotted in Fig. 5 parametric in the effective exciton binding energy for a completely disordered system without any correlation between positions of nearest sites. At weak and moderate fields all these dependences remarkably well feature a square field dependence of Q. At strong fields the functions $Q(F)$ approach unity and saturate. If one accounts for positional correlations the quenching parameter remains practically zero at weak fields and steeply increases at moderate and strong fields with subsequent saturation as illustrated in Figure 6. The data shown in this figure are calculated under the assumption that a given site has no neighbors at distances shorter than a and that there is a constant probability density of finding a neighbor at longer distances.

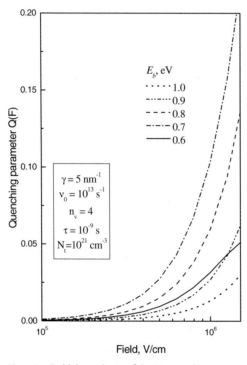

Figure 5   Field dependence of the PL quenching parameter calculated for different values of the exciton binding energy in a material with uncorrelated positions of neighbouring localized states.

Figure 1 illustrates the comparison between the calculated field dependence of the PL quenching and experimental data. At strong fields the theoretical curve does reproduce the observed field dependence of the quenching parameter. The deviation at weaker fields can be due to exciton quenching by defects and/or inadvertent impurities. This quenching mode will be considered in the following chapter.

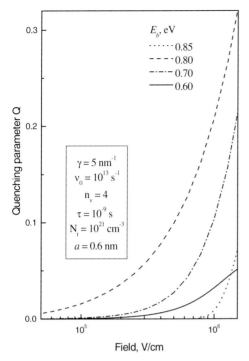

Figure 6 Field dependence of the PL quenching parameter calculated for different values of E_b in a material with correlated positions of neighboring localized states.

7.3
Exciton dissociation in doped conjugated polymers

7.3.1
Experiments on photoluminescence quenching in doped polymers

Exciton quenching can occur whenever the energy of an eh-pair, if coulombically bound, is less than that of its excitonic precursor. In the absence of an external electric field this requires the presence of a sensitizing dopant whose HOMO (highest occupied molecular orbital) or LUMO (lowest unoccupied molecular orbital) fulfills the criterion that the energy of the exciplex (or, synonymously, the charge transfer state or geminately bound eh-pair (GP)) be less than the energy of the lowest singlet state of either the host or the dopant,

$$E_{LUMO}^{(dopant)} - E_{HOMO}^{(host)} - E_b^{(GP)} < E_{S_1}^{(host/dopant)}. \tag{11}$$

where $E_b^{(GP)}$ is the binding energy of the geminate pair. Since conjugated polymers are typically p-type materials whose HOMOs are located near –5.5 +/–0.5 eV relative to the vacuum level this condition can be fulfilled if the dopant is a strong electron acceptor such as trinitrofluorene (TNF). Its LUMO is at −3.9 eV and the absorption threshold is in the UV-region. A typical example studied recently is PhPPV doped with TNF [16]. The intention of that work was to delineate the effect of charge accepting impurities/dopants and to extrapolate it to the behavior of the pristine material. Figure 7 shows the 295 K and 80 K absorption and cw-fluorescence spectra of both a pristine PhPPV film and a film doped with 1 wt% of TNF excited at $hv_{exc} =$ 2.76 eV. It turns out that the absorption spectra are unaffected by doping while the latter causes significant fluorescence quenching. In passing it was noted that in the neat film the fluorescence spectrum is the superposition of two PhPPV emitters with slightly different morphology. Their intrinsic lifetimes, measured on dilute solution at 295 K, are 650 ns and 750 ns, respectively. This is irrelevant in the present context, however. The effect of doping was measured via both cw-fluorescence and time-resolved Streak camera experiments. The results are summarized in Fig. 8.

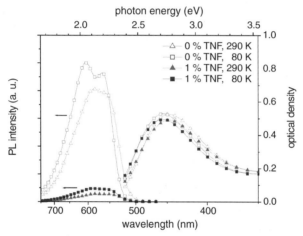

Figure 7 Steady state photoluminescence and absorption spectra of spin-coated PhPPV films with and without TNF at 290 and 80 K.

Consistent with both types of experiments, there is an almost hyperbolic decrease of the fluorescence efficiency above a TNF concentration of 0.1 wt%. By combining these data with those of time resolved measurements in dilute solution one recognizes, though, that in the film there is additional quenching by unidentified dopants/impurities with a concentration of 0.04%. The Streak camera experiments also revealed the non-exponential character of the fluorescence decay in the doped film. It can be quantified by plotting the decay function in terms of Kohlrausch-Williams-Watts' (KWW) stretched exponential law,

$$I(t) = I_0 \exp\left[-(t/t_0)^\beta\right],$$ (12)

as shown in Fig. 9. In such a plot an exponential decay would be reproduced as a straight line with slope 1. The fact that the KWW plots of the fluorescence intensity

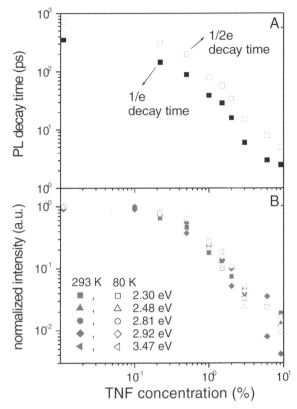

Figure 8 (A) TNF concentration dependences of PL decay time in a PhPPV film as 1/*e* (full rectangle) and 1/2*e* (open rectangle). (B) TNF concentration dependences of cw-PL intensity at 295 K and 80 K with various excitation energies.

from both doped and neat PhPPV films approach a straight line with $\beta = 0.65 \pm 0.05$ asymptotically while the kinetics of fluorescence from an isolated PhPPV chain in solution is exponential indicates that it is not due to the Förster-type energy transfer to randomly dispersed acceptors, that would lead to $\beta = 1/2$, but due to intra- as well as interchain exciton transport, which ultimately leads to the formation of a $(\text{TNF})^{-}$ and $(\text{PhPPV})^{+}$ geminate pairs via short range electron transfer. It is important, though, to recognize that the same functional KWW dependence is observed in pristine PhPPV films. The implication is that this behavior is also due to a small yet finite concentration of charge acceptors of unidentified origin that quench singlet excitons. They may either be inadvertent chemical impurities or physical dimers at which the coulombic energy of a GP exceeds the binding energy of an on-chain singlet state and which also are the origin of the ubiquitous photoconductivity response in the spectral regime of the $S_1 \leftarrow S_0$ transition at moderate electric fields (see chapter 4). In the following section a theoretical framework will be developed in order to rationalize the experimental observations.

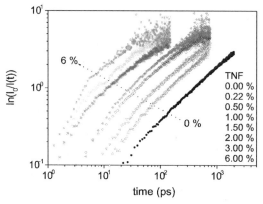

Figure 9 KWW plots for PhPPV films with various TNF concentrations.

7.3.2
Exciton energy relaxation and quenching by Förster energy transfer

Excitons can be quenched either via dissociation into geminate pairs of charges at charge transfer centers or by Förster energy transfer to exciton acceptors. The former process requires carrier tunneling which is feasible only if an electron acceptor is located close to the excited molecule or segment. The latter quenching mechanism requires only energy transfer, which implies a much larger possible quenching range. Below, we consider these quenching modes separately and start with the Förster quenching.

In this case, both energy relaxation and quenching of excitons proceed similarly via Förster-type energy transfer either within intrinsic exciton density-of-states (EDOS) or between an excited host molecule and a guest exciton acceptor. This similarity implies that jumps of excitons to quenching centers can be described in the same way as exciton jumps between host molecules and, concomitantly, the exciton acceptors can be incorporated in the EDOS distribution as 'deep traps' distributed below all intrinsic states.

One should expect that, in a conjugated polymer, the energy transfer rate ν depends upon both the distance r between occupied and target segments and their mutual orientation. Below, we ignore the latter dependence and describe the rate of exciton jumps from an occupied chromophore of the energy E to a target chromophore of the energy E' by the Förster formula as

$$\nu(E, E', r) = \frac{1}{\tau_0} \left[\frac{r_F(E,E')}{r} \right]^6 \times \begin{cases} \exp\left(-\dfrac{E'-E}{kT}\right), & E' > E \\ 1, & E' < E \end{cases} \tag{13}$$

where r_F is the Förster radius that depends upon the energy mismatch between emitter and acceptor chromophores, τ_0 the exciton lifetime, T the temperature, and k the Boltzmann constant. Since the widths of both the intrinsic and exciton acceptor EDOS distributions are normally considerably smaller than the energy difference between them, it is possible to approximate the energy dependence of the Förster radius by two constants. The first one, r_i, describes energy transfer between chromophores within the intrinsic EDOS while the second, r_d, corresponds to exciton jumps from intrinsic states to traps.

Before the energy relaxation is finished, every exciton can normally find a close molecule of a smaller energy as a target site for the next jump. Since jumps to such molecules do not require thermal excitation an exciton of an energy E will, most probably, jump to a molecule in which its energy will be smaller than E. The density of such easily accessible target sites will decrease after every exciton jump and, therefore, the distance to a next deeper site will, on average, increase with time. The density of target molecules, accessible for an exciton of an energy E, $N(E)$, is fully determined by the EDOS distribution $g(E)$:

$$N(E) = \int_{-\infty}^{E} dE' g(E').$$ (14)

As one can see from Eq. (13), the Förster jump rate strongly decreases with increasing distance between an emitter (starting site) and an acceptor (target site). Therefore, excitons will normally jump to nearest accessible target sites. If an energy of an excited molecule is E, the probability densities $w_i(E,E',r)$ and $w_d(E,E',r)$ of having such an either intrinsic or dopant target site, respectively, of the energy $E' < E$ over the distance r is given by the Poisson distribution as

$$w_{i(d)}(E, E', r) = 4\pi r^2 g_{i(d)}(E') \exp\left[-\frac{4\pi}{3} r^3 N_{i(d)}(E)\right], E' < E,$$ (15)

where N_i and N_d are total densities of intrinsic, $g_i(E)$, and dopant, $g_d(E)$, sites of energies below E. If an exciton does have a nearest accessible intrinsic (dopant) target site over the distance r, the probability $p_{i(d)}(r,t)$ that it has not yet jumped to this site until the time t is also described by the Poisson formula:

$$p_{i(d)}(r, t) = \exp\left[-v_{i(d)}(r)t\right] = \exp\left[-\frac{t}{\tau_0}\left(\frac{r_{i(d)}}{r}\right)^6\right].$$ (16)

An exciton still occupies an intrinsic starting molecule of the energy E at the time t if it jumped to neither host nor dopant molecules until this time. The respective probability, $W_i(E,t)$, can, therefore, be calculated as a product of the probabilities of having jumped to neither host nor guest molecules as

$$W_i(E, t) = \exp\left\{-\frac{4\pi}{3}\left(\frac{t}{\tau_0}\right)^{1/2}\left[r_i^3 N_i(E) + r_d^3 N_d(E)\right]\right\}.$$ (17)

where we have accounted for the very steep coordinate dependence of the Förster jump rate that can be written as

$$\exp\left[-\frac{t}{\tau_0}\left(\frac{r_{i(d)}}{r}\right)^6\right] \approx \begin{cases} 0, r < r_{i(d)}\left(\frac{t}{\tau_0}\right)^{1/6}, \\ 1, r < r_{i(d)}\left(\frac{t}{\tau_0}\right)^{1/6}. \end{cases}$$ (18)

An exciton, occupying a host molecule, can jump to both intrinsic and dopant target sites independently. If this exciton has jumped to another host molecule, it is liable to further energy relaxation. However, if it has jumped to a dopant molecule it is quenched and cannot return to a host molecule. Although energy relaxation can further proceed within the manifold of dopans, its rate and, concomitantly, the value of the corresponding Förster radius do not affect the kinetics of exciton quenching. In the following we simply assume that all dopants are deeper in energy than any host molecule and disregard energy transfer between dopant chromophores. Under these conditions, Eq. (17) reduces to

$$W_i(E,t) = \exp\left[-\frac{4\pi}{3}r_d^3 N_d \left(\frac{t}{\tau_0}\right)^{1/2}\right]\exp\left[-\frac{4\pi}{3}r_i^3 N_i(E)\left(\frac{t}{\tau_0}\right)^{1/2}\right], \tag{19}$$

In order to obtain the energy distribution function of excitons, occupying the intrinsic sites, $f_i(E,t)$, one has to account for contributions from both excitons that still occupy molecules in which they have been initially generated and excitons that made one or more jumps within the manifold of intrinsic sites. The former contribution, $p_0(t)$, can be calculated by multiplying Eq. (19) by the intrinsic EDOS distribution and integrating the product over energy. The result reads:

$$p_0(t) = \left(\frac{3}{4\pi r_i^3 N_t}\right)\left(\frac{t}{\tau_0}\right)^{-1/2}\exp\left[-\frac{4\pi}{3}r_d^3 N_d\left(\frac{t}{\tau_0}\right)^{1/2}\right]\left\{1-\exp\left[-\frac{4\pi}{3}r_i^3 N_t\left(\frac{t}{\tau_0}\right)^{1/2}\right]\right\}, \tag{20}$$

where N_t is the total (energy-integrated) density of intrinsic sites. An exciton, currently occupying a host molecule, can jump to either a dopant molecule or a deeper intrinsic site. As one can see from Eq. (19) the relative probability of such jumps is determined by the products of the respective densities of states and Förster radii. Therefore, the occupational probability for intrinsic sites, occupied by excitons that made at least one jump by the time t, $f_i(E,t)$, can be written as

$$f_i(E,t) = [1 - p_0(t)]g_i(E)\exp\left[-\frac{4\pi}{3}r_d^3 N_d\left(\frac{t}{\tau_0}\right)^{1/2}\right]\exp\left[-\frac{4\pi}{3}r_i^3 N_i(E)\left(\frac{t}{\tau_0}\right)^{1/2}\right]$$

$$\times\left(\frac{3}{4\pi r_i^3}\right)^{1/2}\left(\frac{t}{\tau_0}\right)^{1/2}\exp\left[-\frac{4\pi}{3}r_d^3 N_d\left(\frac{t}{\tau_0}\right)^{1/2}\right]\left\{1-\exp\left[-\frac{4\pi}{3}r_i^3 N_t\left(\frac{t}{\tau_0}\right)^{1/2}\right]\right\}+\left(\frac{r_d}{r_i}\right)^3 N_d\right)^{-1}$$

$$\tag{21}$$

Integrating this equation over energy, adding the density of excitons that still reside in 'parent' host molecules, and accounting for the radiative decay of excitons yields the total density $p_i(t)$ of excitons occupying intrinsic sites at the time t:

$$p_i(t) = \exp\left(-\frac{t}{\tau_0}\right)p_0(t) + \exp\left(-\frac{t}{\tau_0}\right)[1 - p_0(t)]$$

$$\times\left(1+\frac{4\pi}{3}r_d^3 N_d\left(\frac{t}{\tau_0}\right)^{1/2}\exp\left[\frac{4\pi}{3}r_d^3 N_d\left(\frac{t}{\tau_0}\right)^{1/2}\right]\left\{1-\exp\left[-\frac{4\pi}{3}r_i^3 N_t\left(\frac{t}{\tau_0}\right)^{1/2}\right]\right\}^{-1}\right)^{-1}, \tag{22}$$

In the absence of the exciton energy diffusion within the intrinsic EDOS distribution, i.e. with $r_i = 0$, Eq. (22) reduces to the $(t/\tau_0)^{1/2}$ dependence obtained by Klafter and Blumen [47] for the Förster-type donor photoluminescence (PL) quenching by randomly distributed exciton acceptors. Equation (22) indicates that the time dependence of the exciton quenching rate is not sensitive to the specific shape of the EDOS distribution. It rather depends upon the total density of hopping sites that is typical for low-temperature energy relaxation of both excitons and charge carriers [48].

The role of exciton energy relaxation in the PL quenching is illustrated in Fig. 10. In this figure, $p_i(t)$ curves are plotted for different values of the intrinsic Förster radius r_i. These curves clearly indicate that energy relaxation within the intrinsic EDOS considerably enhances the rate of PL quenching. The origin of this enhancement is obvious: the energy relaxation of excitons proceeds in parallel with their random walk in space which strongly increases the probability for an exciton to encounter an acceptor in the proximity of the visited host molecules. The effect of the density of exciton acceptors on the PL quenching kinetics is shown in Fig. 11. Each pair of curves, plotted in this figure, is calculated for different acceptor densities with (solid lines) and without (dashed lines) account for exciton energy relaxation. As one can see from this figure, the energy relaxation always facilitates PL quenching. Intuitively, one would expect a stronger effect of the energy relaxation (and the related random walk) on the PL quenching at lower doping levels, when excitons have to migrate to attain a quenching center. However, one should bear in mind that, to some extent, the effect of exciton diffusion is equivalent to increasing Förster radius of quenchers and the same increase of r_d leads to a stronger quenching at higher doping levels.

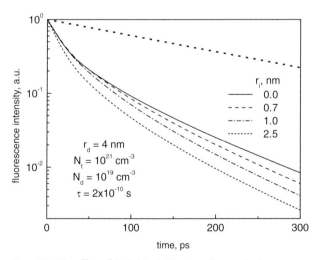

Figure 10 The effect of the intrinsic Förster radius on the time dependence of the fluorescence intensity controlled by the energy transfer from donor host molecules to exciton acceptors. The dot line indicates the fluorescence kinetics in the absence of exciton quenching.

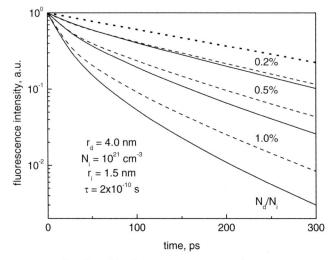

Figure 11 The effect of the dopant concentration on the time dependence of the fluorescence intensity controlled by the energy transfer from donor host molecules to exciton acceptors. Dashed lines are calculated disregarding energy relaxation of excitons within the intrinsic EDOS, i.e. for $r_i = 0$. The dot line indicates the fluorescence kinetics in the absence of exciton quenching.

It is worth noting that PL quenching controlled by low-temperature energy relaxation of excitons is an intermediate regime in between single-step quenching [47] and quenching controlled by exciton diffusion. The latter requires equilibration of excitons at times shorter then their lifetime, which is possible only at higher temperatures. Experimentally, the transition from exciton energy relaxation to equilibrium exciton diffusion is manifested by the change of PL kinetics from stretch-exponential to purely exponential [49].

Fluorescence quenching in films of poly(phenylphenylene vinylene) (PPPV) doped with 4-dicyanomethylene-2-methyl-6-p-dimethylaminostyryl-4H-pyran (DCM) molecules has been studied at both low (10 K) and room temperature by the authors of Ref. [49]. Upon photoexcitation by a laser pulse at room temperature, the fluorescence intensity decayed exponentially, both the total fluorescence intensity and the decay time strongly decreasing with dopant concentration. At the low temperature, stretched-exponential fluorescence decay kinetics was observed in both pristine and doped films as shown in Fig. 12. The exciton lifetime, measured in the pristine film, turned out to be much smaller than the intrinsic radiative lifetime [49], which suggests exciton quenching at unidentified inadvertent impurities and/or structural defects. Since Förster quenching requires rather specific molecular dopants it seems reasonable to assume that excitons in the pristine film were quenched via charge transfer at electron scavengers [50]. This quenching mechanism will be considered in the following section.

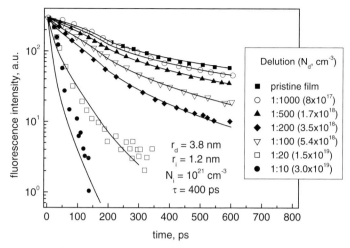

Figure 12 Fluorescence transients in PPPV films doped with DCM [49]. The lines are fitting curves calculated from Eq. (22) – see text for further details. The inset shows experimental DCM:PPPV ratios and the dopant concentrations used in the calculations.

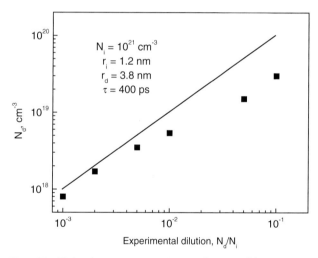

Figure 13 Fitting dopant concentration as a function of the experimental DCM:PPPV ratio.

Solid lines in Fig. 12 were calculated from Eq. (22) with the fitting values of the dopant concentration shown in Fig. 13. The time dependence of fluorescence decay in the pristine film has been used in the calculations as a reference curve instead of the exponential decay characterized by the radiative exciton lifetime. It is worth noting that the values of the dopant concentration, used for the data fitting, are somewhat smaller than they should be according to the relative number of the guest

(DCM) molecules diluted in the host (PPPV) matrix as indicated by the solid line in Fig. 13. Since this discrepancy increases with increasing dopant concentration clustering of dopant molecules can be suggested as a possible explanation.

It is worth noting that the above consideration has been based on the assumption that both host and guest molecules can be treated as point-like as far as exciton jumps between them are concerned. This approximation seems to be justifiable for the relatively long-range Förster-type exciton jumps. Further support for this approximation comes from the fact that fitting the data, obtained on a conjugated polymer, was possible with both intrinsic and dopant Förster radii close to their spectroscopic values [49]. However, a similar approach could not be fully applied to the problem of exciton quenching at charge transfer centers. The reason is that charge transfer requires tunneling of particles with non-zero masses, which is a short-range process in disordered organics characterized by very small charge carrier localization radii. Although the model of point-like sites can still be used for modeling exciton energy relaxation one has to account for a finite length of conjugated molecular segments in the treatment of charge transfer from excitons to electron scavengers. A model of exciton quenching at charge-transfer centers, described in the following section, is based on this notion.

7.3.3
Exciton quenching at charge transfer centers

An exciton can dissociate into a geminate pair of charge carriers if a deep trap (usually for electrons) is located next to a molecule or segment visited by the exciton in the course of its energy relaxation. Since the spatial distribution of traps is random and does not correlate with energies of the host molecules, the probability for an exciton to encounter a charge transfer center is fully determined by the number of sites visited by this exciton. In the absence of exciton acceptors ($N_d = 0$), the exciton distribution function can be found from Eqs. (15) and (16) as

$$f(E, t, r) = r^2 g_i(E) N_i(E) \exp\left[-\frac{4\pi}{3} r^3 N_i(E)\right] \exp\left[-\frac{t}{\tau_0}\left(\frac{r_i}{r}\right)^6\right]$$

$$\times \left(\int_0^\infty dr r^2 \exp\left[-\frac{t}{\tau_0}\left(\frac{r_i}{r}\right)^6\right] \int_{-\infty}^\infty dE g_i(E) N_i(E) \exp\left[-\frac{4\pi}{3} r^3 N_i(E)\right]\right)^{-1}. \quad (23)$$

Since electron scavengers are distributed randomly there is a constant probability for an exciton to encounter a charge transfer center upon every next jump. Consequently, the rate of quenching must be proportional to the average exciton jump rate $\nu(t)$. In order to calculate the latter one should multiply the distribution function, given by Eq. (23), by the intrinsic Förster jump rate and by the radiative-decay exponential, $\exp(-t/\tau_0)$, and integrate the product over energy and jump distance. The result reads

$$v(t) = \frac{r_i^6}{\tau_0} \exp\left(-\frac{t}{\tau_0}\right)$$

$$\times \left\{ \int_0^\infty \frac{dr}{r^4} \left[1 - \exp\left(-\frac{4\pi N_t}{3} r^3\right) - \frac{4\pi N_t}{3} r^3 \exp\left(-\frac{4\pi N_t}{3} r^3\right) \right] \exp\left[-\frac{t}{\tau_0} \left(\frac{r_i}{r}\right)^6 \right] \right\}^{-1}. \quad (24)$$

$$\times \int_0^\infty \frac{dr}{r^{10}} \left[1 - \exp\left(-\frac{4\pi N_t}{3} r^3\right) - \frac{4\pi N_t}{3} r^3 \exp\left(-\frac{4\pi N_t}{3} r^3\right) \right] \exp\left[-\frac{t}{\tau_0} \left(\frac{r_i}{r}\right)^6 \right]$$

Integrating Eq. (24) over time yields the average number of jumps, n, to be made by an exciton after the time t:

$$n(t) = \frac{r_i^6}{\tau_0} \int_t^\infty dt' \exp\left(-\frac{t'}{\tau_0}\right)$$

$$\times \left\{ \int_0^\infty \frac{dr}{r^4} \left[1 - \exp\left(-\frac{4\pi N_t}{3} r^3\right) - \frac{4\pi N_t}{3} r^3 \exp\left(-\frac{4\pi N_t}{3} r^3\right) \right] \exp\left[-\frac{t'}{\tau_0} \left(\frac{r_i}{r}\right)^6 \right] \right\}^{-1}.$$

$$\times \int_0^\infty \frac{dr}{r^{10}} \left[1 - \exp\left(-\frac{4\pi N_t}{3} r^3\right) - \frac{4\pi N_t}{3} r^3 \exp\left(-\frac{4\pi N_t}{3} r^3\right) \right] \exp\left[-\frac{t'}{\tau_0} \left(\frac{r_i}{r}\right)^6 \right] \quad (25)$$

Note that the value $n(0)$ gives the average number of jumps made by an exiton during its entire lifetime. Similar to Eq. (22), the results given by Eqs. (24) and (25) are not sensitive to the specific shape of the EDOS distribution. Two cautious notions are pertinent to this conclusion: (i) one should bear in mind that Eqs. (24) and (25) have been derived under the assumption of the Förster radius independent upon the energies of both emitter and acceptor molecules within the intrinsic EDOS distribution and (ii) the possibility of thermally assisted jumps of excitons to sites of higher energies is disregarded implying that the time scale of energy relaxation is longer than the exciton lifetime unless most excitons were generated within the deep tail of EDOS.

In fact, Eq. (25) determines the exciton dissociation potential, i.e. the probability that an exciton, that has survived to the time t, will make a jump that can result in dissociation after t. Time dependence of the exciton dissociation potential is illustrated in Fig. 14 for different values of the Förster radius in comparison with the exciton density decay. It is assumed that dissociation occurs once an exciton jumps onto a segment that has an electron scavenger as its neighbor. For excitations, which survived until the time t, the probability to meet charge transfer centers and, thereby, to dissociate into pairs of free carriers is proportional to the average number of segments that these excitations will visit during the rest of their lifetimes. This number decreases for two reasons: (i) the density of excitations decreases with time due to both intrinsic decay and quenching by defects and impurities other than charge transfer centers, and (ii) the exciton jump rate decreases with time in the course of energy relaxation when excitations are getting stronger localized on segments that belong to a deep tail of the excitonic DOS distribution.

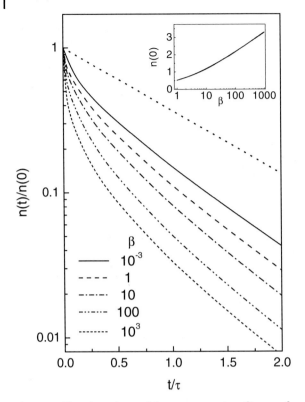

Figure 14 Time dependence of the average number of jumps of an exciton with intrinsic lifetime τ to be made after the time t. The dotted line shows an exponential decay of the exciton density, i.e. the PF intensity in the absence of spectral relaxation. The parameter β is determined as $\beta = [(4\pi/3)N_t r_i^3]^2$.

In two-pulse experiments [51], the number of singlets, quenched by the second depleting pulse, is proportional to the density of excitations generated by the first inducing pulse and survived until the delay time t_d. Therefore, the quenching of prompt fluorescence is proportional to the density of excitations at the time t_d while the photocurrent quenching gives a direct measure for the exciton dissociation potential. Exactly this time dependence is described by the function $n(t_d)$ as illustrated in Fig. 15. The experimental points shown in this figure were taken from Ref. [51] and the solid line illustrates the fit of these data by Eq. (25) with the exciton lifetime $\tau_0 = 180$ ps determined in Ref. [51] from the delay-time dependence of the photoluminescence quenching.

After every intermolecular jump an exciton can find itself in a quencher, i.e. in a molecule that has a deep (electron) trap in its close neighbourhood. The relation between the density of deep electron traps N_a and the concentration of quenchers depends upon the molecular configuration. In the following, we consider exciton quenching in conjugated polymers, in which excitons are delocalized within conju-

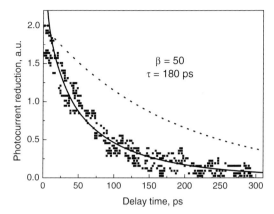

$\beta = 50$
$\tau = 180$ ps

Figure 15 The decrease of the photocurrent in a MeLPPP film due to the second depleting laser pulse in the two-pulse experiment of Ref. [14]. The solid line is calculated from Eq. (25) with the exciton lifetime $\tau_0 = 180$ ps, as determined in Ref. [14] from the PF decay, and $\beta = 50$ that corresponds to e.g. $N_t = 10^{21}$ cm^{-3} and $r_i = 1.2$ nm. The dotted line shows for comparison an exponential decay of the exciton density with the intrinsic lifetime of 180 ps.

gated molecular segments. Provided that deep traps are distributed homogeneously, the probability w_q that a given segment is an exciton quencher, i.e. that it has a deep trap as a close neighbor, is again determined by the Poisson distribution as

$$w_q = 1 - \exp(-\pi r_q^2 l N_a), \tag{26}$$

where l is the conjugation length and r_q the maximum distance between a segment and a deep trap which still allows for quenching. An exciton, occupying a quencher, can still avoid quenching if it jumps to another accessible segment before it is quenched. The probability to be indeed quenched at a quencher, W_q, is given by

$$W_q = \frac{\tau_j}{\tau_q + \tau_j}, \tag{27}$$

where τ_q and τ_j are the quenching and jump times, respectively. In order to decay radiatively and to contribute to the photoluminescence an exciton must avoid quenching at all segments visited during its entire lifetime including the segment on which the exciton has been generated. Estimating the exciton jump time as $\tau_j = \tau/n$ and using the Poisson distribution of probabilities yields the following expression for the probability η that an exciton has not been quenched and eventually decayed radiatively

$$\eta = 1 - Q(\infty) = \left\{ \frac{n\tau_q + \tau_0 \exp(-\pi r_q^2 l N_a)}{n\tau_q + \tau_0} \right\}^{n+1}. \tag{28}$$

The dependence of the radiative yield η upon the concentration of deep traps N_d is shown in Fig. 16 together with experimental data obtained on an alkoxy-substituted poly-phenylenevinylene (PhPPV) doped by trinitrofluorene (TNF) [16]. The yield, calculated for realistic values of the model parameters, quantitatively fits the data within the entire experimental range of the trap concentrations. However, one should bear in mind that the average number of exciton jumps has been calculated disregarding possible correlation between energies and positions of exciton hopping sites. In the presence of such correlation, the average dissipated energy per single carrier jump decreases, which leads to increasing average number of molecules or segment visited by an exciton in the course of energy relaxation. Since the correlations should not affect the spatial distribution of traps, increasing number of visited sites results in increasing quenching probability. One should, therefore, expect stronger quenching of excitons by impurities in materials with correlated energies and positions of exciton hopping sites.

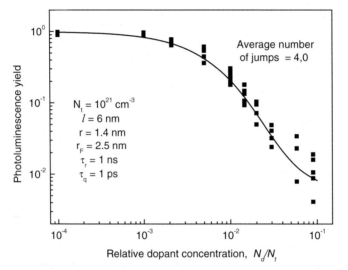

Figure 16 Dopant concentration dependence of the PL intensity in PhPPV films. Experimental data are taken from Fig. 8 B, the solid line was calculated from Eq. (25).

7.4
Photoconductivity in pristine and weakly doped polymers

The ubiquitous onset of photoconductivity at the absorption threshold of π-conjugated polymers, except for poly-diacetylenes, has been taken as evidence in favor of the notion that they resemble inorganic semiconductors regarding their optoelectronic properties. This would imply that (i) this threshold marks the valence-to-conduction band transition and (ii) coulombic and electron-electron corrections are negligible. Meanwhile there is abundant experimental as well as theoretical evidence against this conjecture such as (i) the recognition that the results of linear and non-linear spectroscopy of poly-acetylene indicate the importance of electron-electron correlations [52], (ii) the energy splitting between the first excited singlet and triplet states is about 0.7 eV implying strong exchange interaction [53-55], (iii) efficient fluorescence quenching requires an electric field on the order of several MV/cm [21,32-34], and (iv) the value of the exciton binding energy, calculated by the use of various theoretical approaches, is ~ 0.5 eV [11].

The fact that the action spectra of polydiacetylenes do not start until about 0.5 eV above their absorption spectra is also consistent with the excitonic origin of optical absorption in π-conjugated polymers. It is known that in polydiacetylenes the lifetime of singlet excitons is less than 1 ps. Therefore, they can hardly contribute to photoconductivity. It is similarly well known that in classic organic crystals photoconductivity usually starts right at the excitonic $S_1 \leftarrow S_0$ transition unless special precautions are made in order to eliminate charge photoinjection from the electrode(s). The origin of this extrinsic photogeneration is related to exciton dissociation at an appropriate electrode. In the early days of photoconductivity studies in organic crystals, the spectral dependence of this extrinsic photocurrent has, in fact, been taken as a measure of the exciton diffusion length. Experiments on steady state photoconductivity in MeLPPP and a derivative of PPV proved that this effect plays an important role in conjugated polymers as well [46,56]. These experiments showed that the introduction of a thin SiO_x layer between the sample and an ITO electrode greatly reduces the photocurrent within the spectral range of the $S_1 \leftarrow S_0$ 0-0 transition. Electrode-sensitized photoinjection can also be avoided by illuminating through a semi-transparent metal electrode because excitons are quenched efficiently at a metal via dipole-allowed energy transfer. On the other hand, there is unambiguous evidence that (i) there is finite photogeneration in the bulk of a sample within the spectral range of the $S_1 \leftarrow S_0$ transition and (ii) it increases at higher photon energies. In this chapter, representative examples for intrinsic as well as dopant-assisted photogeneragtion will be described in greater detail. They will illustrate the inherent problems encountered when trying to distinguish between intrinsic and external effects and to separate primary and secondary steps in photoionization.

As outlined in the introduction and in the beginning of chapter 2, photoionization in organic solids is, in general, a multi-step process. Initially a more or less tightly bound geminate eh-pair is generated upon exciton dissociation. This pair can subsequently either recombine or fully dissociate into free carriers. In organic crystals the latter process is tractable in term of Onsager's theory of geminate recombination based on the concept of diffusive escape of carriers from their mutual coulombic potential. This theory predicts similarly strong field and temperature dependences of the photogeneration yield. However, a notorious feature of the yield in conjugated polymers is that it reveals a very weak temperature dependence while its field dependence is as strong as predicted by the Onsager theory. Below, we describe alternative approach to explain intrinsic charge photogeneration in intentionally and/or accidentally doped conjugated polymers.

7.4.1
Intrinsic photogeneration

Experiments on transient photogeneration in MeLPPP, which is an exceptionally pure material with a low degree of disorder, will be described first [57]. These experiments employ the techniques of delayed and instantaneous collection of charge carriers excited by a laser flash combined with studies of fluorescence quenching and transient absorption. In the delayed field collection experiment, the sample is biased towards a variable external field F_{app} and illuminated by a laser flash. After a delay time τ_{del} of 100 ns a rectangular field pulse F_{coll} of 10^6 V/cm is applied that is sufficient to collect all charge carriers still alive after τ_{del}. The field dependence of the collected charge is shown in Fig. 17. Simultaneously the amount of field-induced fluorescence quenching is measured. The latter results are included in Fig. 1. It turns out that the number of collected charges as a function of F_{app} is more or less proportional to the number of the quenched excitons suggesting that the primary field-induced exciton dissociation is the rate-limiting step in the process of charge photogeneration. At low fields, the ratio of collected charges and quenched excitons increases with the field because then the defect-sensitized photoconduction is likely to take over. Experiments on conventional transient photoconduction in the same material confirm the above results.

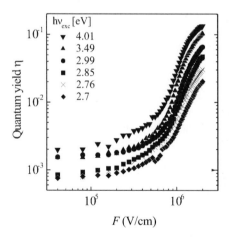

Figure 17   Field dependence of the charge carrier photogeneration yield in MeLPPP for different photon energies of photoexcitation.

Figure 18 shows a conventionally recorded photocurrent in a 100 nm thick MeLPPP film at 295 K at a incident photon dose of 40 µJ/pulse, equivalent to an absorbed photon dose of 6×10^{12} photons, and at different external electric fields. The rise time is determined by the RC-time constant of the circuit that was about 30 ns. At times shorter than 1 µs and at low electric fields the current features a plateau in a double logarithmic representation. At higher fields an algebraic decay followed by an almost exponential decay at $t > 1$ μs is observed. At $t > 10$ μs a weak algebraic tail is noticeable. The number of collected charges Q_{coll} was calculated by integration of the current. A plot of Q_{coll} as a function electric field at different photon doses is presented in Fig. 19. Experiments within the temperature range from 310 to 150 K showed that generation of mobile positive charge carriers is temperature insensitive. This concurs with previous experiments on cw photoconductivity [56].

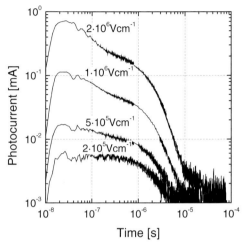

Figure 18 Transient photocurrents measured in a MeLPPP film at different external electric field and at room temperature.

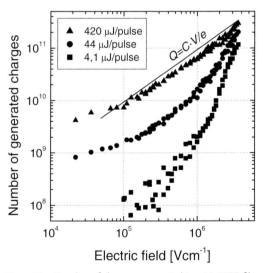

Figure 19 Number of charges, generated in a MeLPPP film, as a function of the applied electric field parametric in the excitation intensity. The straight line indicates the dependence of the corresponding capacitance charge on the electric field.

A transient current can be either transport or generation limited. From previous results, it is known that in freshly prepared MeLPPP films hole transport is trap-free, time-of-flight signals yielding a hole mobility of 2×10^{-3} cm^{2}V^{-1}s^{-1} with an exceptionally weak temperature and field dependence [58]. This is a consequence of the low degree of structural disorder. On the other hand, electron transport has never been observed in a time-of-flight experiment indicating that electrons are trapped in this material. At the lowest electric field, employed for the recording of the signals in Fig. 18, the expected transit time of holes is 25 ns, which is comparable to the RC-time of the circuit, i.e., about three orders of magnitude shorter than the duration of the photocurrent pulse. Obviously, the observed transient photocurrent must be controlled by generation of charge carriers rather than by their transport. This conclusion is supported by the fact that the duration of the signal is virtually independent of the electric field, at variance with the notion of it being determined by transport. It also indicates that this signal cannot be controlled by electron transport. Detrapping of the negative space charge must occur on the longer time scale.

The obvious source of the transient photocurrent is the final dissociation of meta-stable geminately bound eh-pairs whose existence has been well established by (i) delayed field collection experiment [19], (ii) delayed fluorescence [22], (iii) thermally stimulated luminescence [23] and (iv) two-color photoconduction [59]. At a given time t the photocurrent is determined by the product of the number of GPs, N_{GP}, which survived until time t and the rate constant of their subsequent escape k_{esc} from the coulombic potential well

$$j(t) = ek_{esc}N_{GP}(t). \tag{29}$$

The total collected charge Q_{coll} can be evaluated by integrating the current as

$$Q_{coll} = \int_0^\infty dt j(t) = e\eta_{diss}\eta_{esc}N_{ph}, \tag{30}$$

where η_{diss} is the initial yield of dissociation into GPs of N_{ph} singlet excitations, produced by the laser, and η_{esc} is the fraction of GPs which escape geminate recombination. In principle, both η_{diss} and η_{esc} can be functions of the applied electric field and temperature. In the delayed field experiment, this product has been disentangled by applying a collecting field pulse after a delay time of 100 ns after the generation pulse. In the transient photogeneration study, generated charge carriers are collected and their number is compared to the number of generated GPs as monitored by the transient absorption within time domain of 200 fs to 500 ps which is too short for their subsequent escape from the coulombic potential [38]. This procedure is superior to that employed in [19] where the difference between the generation and collection field vanished at high electric fields. Moreover, this experimental approach was extended towards lower photon dose thus avoiding any possible effects, which bimolecular annihilation of singlet excitons during the laser pulse might have.

In Fig. 20, field dependence of the yield of collected holes is compared with the yield of GP generation by field-assisted dissociation of relaxed singlet excitons as monitored by transient pump–probe experiments [38]. Note that the yield rather

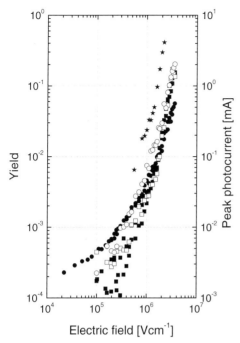

Figure 20 Charge generation efficiency (full symbols, left coordinate) and the photo-current peak (open symbols, right coordinate) in a MeLPPP film versus applied electric field, parametric in the excitation energy (squares: 4.1 µJ/pulse, circles: 44 µJ /pulse). The stars indicate the field dependent evolution of the transient absorption of geminate pairs generated by the dissociation of singlet excitons at photon dose of 14 µJ /cm².

than the number of collected charges is shown in this figure. The proportionality of both dependences is striking. It confirms that (i) the rate limiting field-assisted step is the dissociation of relaxed singlet excitons into GPs rather than their subsequent escape from the coulombic potential well, (ii) the former process proceeds on a 200 fs to 0.5 ns time scale while the dwell time of GPs extends to the microsecond regime, and (iii) approximately 10% of initially generated GP are liable to complete dissociation at a field of 3 MV/cm.

The field dependence of the yield of fully dissociated GPs is weaker at pump intensity beyond the threshold at which bimolecular annihilation becomes the dominant channel for the decay of singlet excitons. The corresponding plot is included in Fig. 20. At low electric fields the annihilation-assisted yield exceeds the value measured at low intensity while at high fields both dependences intersect due to the onset of bimolecular recombination of mobile holes and trapped electrons, which shortens the hole range. The former effect is generic and confirms the notion that if a highly excited singlet state has been generated by singlet–singlet annihilation, its dissociation is more efficient [40] and requires less stimulation by an electric field.

The above conclusion can be cast into a schematic diagram for photoionization shown in Fig. 21. Depending upon the applied electric field, an exciton can transfer one of its charges to a polymer segment within the coulombic well. At zero external electric field only very few excitons can form tightly bound GPs unless the initial dissociation event involves bimolecular collision of two singlet excitons. Under the latter circumstance delayed fluorescence, arising from geminate pair recombination, can be observed. Upon increasing the electric field some excitons can dissociate into metastable GPs. Most of them recombine nonradiatively because the external potential overcompensates the energy difference between the GP and singlet exciton. However, such GPs can be revitalized and emit delayed fluorescence when the electric field is turned off [22].

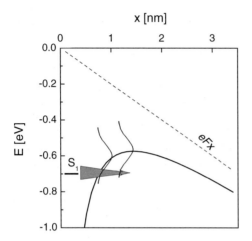

Figure 21 Schematic diagram of the photoionization in organic hoping system with a exciton binding energy of 0.7 eV and a Gaussian width of the density of states distribution of 0.05 eV in an external electric field of 2 MV/cm and assuming a dielectric constant of 3.5.

In summary the comparative study of charge photoproduction and field-induced exciton quenching prove that the field dependence of the intrinsic photogeneration in the π-conjugated polymer MeLPPP is controlled by the initial step of the dissociation of a relaxed singlet exciton into a geminate pair rather than its subsequent escape from the coulombic potential. Therefore it is illegitimate to analyze the field (and temperature) dependence of the yield in terms of Onsager's theory although the functional forms of η can accidentally be similar. This conclusion is not specific for a conjugated polymer but is generic for disordered organic solids in which the energy gap between the geminate electron–hole pair state and the parent singlet state is narrow so that it can be compensated by a strong electric field.

A similarly ubiquitous phenomenon as the onset of photoconductivity at the $S_1 \leftarrow S_0$ threshold is the observation that the yield of photogeneration increases considerably at a photon energy of 0.7...1.0 eV in excess of that threshold energy. In the pioneering work of the Vardeny group [8] it has been assigned to intrinsic photogeneration above the genuine electrical gap between valence and conduction manifolds. Phenomenologically, the effect appears to be analogous to autoionization in conventional molecular crystals [7]. It is generally believed that, in these systems, a superexcited, i.e. vibrationally hot Franck-Condon state ejects an electron into surrounding medium. After thermalization it forms a geminately bound eh-pair that subsequently can escape from the coulombic well in the course of a temperature and field assisted random walk. An eh-pair might also be excited directly via a weak charge transfer transition that overlaps with the dominant vibronic exciton transitions.

An alternative model of field-assisted ultrafast on-chain dissociation of optical excitations in conjugated polymers [41,42] suggests that the excess photon energy above the $S_1 \leftarrow S_0$ 0-0 transition is transferred into intrasegmental vibronic energy and establishes, while dissipating, an important additional source of energy required for carriers in order to cross the potential barrier and to separate within the segment. Similar to Onsager's theory, the external electric field lowers the potential barrier for carrier separation and, thus, assists the dissociation of optical excitations. Different from Onsager's treatment, the rate-determining step is of the interchain type, not involving diffusive motion inside the Coulombic potential.

The process of photogeneration has to become more complicated if the length of a conjugated segment decreases, for instance due to either excessive disorder or poor π-overlap among the repeat units. In this case a metastable electron-hole pair at the ends of a segment can no longer be established. Instead, dissociation must involve interchain coupling inside the Coulombic well. Asymptotically, the classic Onsager treatment should be obeyed as in the case of a dense array of small molecules. Dissociation due to the excess energy of optical excitation is also involved in that case, but the thermalization process is completed before the geminate pair escapes from its mutual Coulombic potential. Onsager-type photogeneration and hot exciton dissociation in an elongated conjugated polymer are, thus, complementary limiting processes depending upon the size of the uninterrupted π-system. Real materials will often fall into an intermediate range and escape rigorous analytical description.

7.4.2
Dopant assisted photogeneration

An alternative to field-assisted intrinsic exciton dissociation in a neat polymer film is dissociation at inadvertent or deliberately added dopants that can act as electron/hole acceptors depending upon the position of the mutual HOMO/LUMO levels as described in section 3.1. In order to explore this possibility steady state photocurrents in PhPPV films doped with various amounts of TNF were further measured under low intensity irradiation of the sample with a xenon lamp. Note that, under these conditions, the photocurrent is solely a measure of the generation rate of mobile charge carriers irrespective of their transport velocity. Figure 22 shows photocurrents in a neat PhPPV film and a film doped with 1 wt% of TNF at either positive or negative bias of the irradiated ITO electrode while the counterelectrode was Al.

The fact that the photocurrent is about one order of magnitude higher if ITO was positive illustrates the importance of photoinjection of holes from ITO. This effect vanishes upon replacing ITO by Al because of non-radiative decay of excitons near a metal.

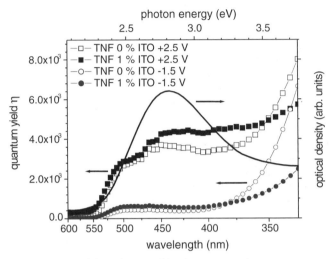

Figure 22 Spectral dependencies of the charge carrier photogeneration quantum yield in ITO/PhPPV/Al configuration at 293 K. The electric fields for both bias directions were approximately 2.5×10⁵ V/cm.

Figure 23 shows that, upon adding TNF at a concentration between 0.1 and ~9 wt.%, the photocurrent does increase but very slowly. The simultaneous measurement of the fluorescence quenching provides a straightforward explanation of this phenomenon [16]. Operationally, the simplest way to quantify the effect of doping is to analyze the reduction of the fluorescence decay time after pulsed excitation as a function of the dopant concentration. The rate equation for the concentration of singlet excitations is

$$\frac{d[S]}{dt} = -(k_r + k_{nr} + k_{q0} + k_q c)[S],$$
(31)

where k_r and k_{nr} are the rate constants of intrinsic radiative and nonradiative decay, k_{q0} and k_q the rate constants for quenching by unidentified quenchers and deliberately added dopants at concentration c, and $(k_{q0} + k_{qc})[S]$ is the rate of singlet quenching. From the experimentally determined fluorescence lifetime τ of a PhPPV as a function of concentration, shown in Ref. [16], it is known that τ does not extrapolate to the intrinsic lifetime τ_0 at zero concentration of TNF. Instead, there is a residual quenching effect at $c = 0$ due to an unidentified quencher present at a concentration of <0.04%. Provided that the light intensity is below a critical level beyond which nonlinear terms, e.g., exciton–exciton annihilation, become important the relevant terms in the rate equation for cw excitation are the same. Under this premise the relative number of quenching events can be inferred from the relative loss of fluorescence,

$$Q = \frac{I_0 - I_c}{I_0} = \frac{k_{q0} + k_q c}{\sum_i k_i}.$$ (32)

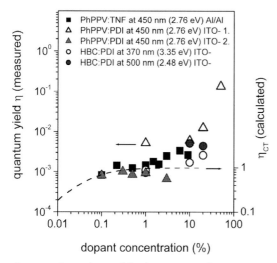

dopant concentration (%)

Figure 23 Dependence of the charge carrier photogeneration quantum yield at 293 K on the dopant concentration for various sample configurations.

Provided that the number of quenching events is proportional to the photogeneration yield, concentration dependence of both the yield and quenching should be proportional. If $k_{q0} = 0$, Q should decrease linearly with decreasing concentration of the dopant. However, if about 50% of the singlets are already quenched by accidental doping, the total defect sensitized photogeneration can increase by a factor of 2 only at complete singlet quenching. Considering the uncertainly regarding absolute photogeneration yields this effect can hardly be distinguished. This is the reason why the photogeneration yield is almost constant even at low dopant concentration.

The yield of complete dissociation of a strongly coulombically bound eh-pair is much less than unity and strongly depends on external electric field [5,6]. However, its temperature dependence is weak without noticeable spectral dependence except at large excess photon energies (Fig. 22). The facts that (i) in the pristine sample the photogeneration yield is close to that of a sample of $c < 0.1\%$ and (ii) the concentration of inadvertent dopants, which cause fluorescence quenching (see Ref. [16]) in the neat material, is also comparable suggest that both effects have the same origin, i.e., that at moderate electric fields the photocurrent in the spectral range of the $S_1 \leftarrow S_0$ transition is defect sensitized. It is still an open question whether the defects are of chemical or physical origin. The latter type of defects could be a pair of chains, which are tightly bound so that an eh-pair is more strongly bound than a

S_1 excitation. The present experiments do not allow differentiating between the two possibilities.

The polarity independent increase of photogeneration at higher photon energies has been explained in terms of the model for dissociation of vibrationally hot excitons. The action spectrum of intrinsic photogeneration in undoped PhPPV (Fig. 22) is no different from that of other conjugated polymers and features an onset at 3 eV. It is remarkable, though, that at zero-order approximation, the magnitude of the yield and its spectral dependencies are basically retained upon doping by TNF. Since doping decreases the lifetime of S_1 excitons, this is unambiguous proof that the intrinsic contribution to photogeneration at higher $h\nu$ cannot be due to "cold" S_1 excitations. Instead, it must either be due to autoionization from the primarily excited Franck–Condon state [56] or dissociation of electronically "cold" S_1 excitations, or, rather coulombically strongly bound on-chain eh-pairs coupled to a temporally hot vibrational bath [41,42]. However, there are subtle differences. Plotting the photogeneration yield at $h\nu = 4.13$ eV as a function of concentration shows that the yield drops by a factor of 3 upon increasing the TNF concentration features a shallow minimum and increases again for $c > 2\%$. Such a behavior is at variance with the ballistic dissociation concept [60] because the lifetime of the initially excited Franck–Condon state must be unaffected by its future decay. The concept of hot exciton dissociation provides a plausible explanation, though. If the exciton is excited next to a TNF molecule it can transfer its electron on a time scale of <1 ps. Since donor and acceptor form a quantum mechanically coupled system, part of the excess energy will be transferred to the TNF. Therefore the lifetime of the excess vibrational bath decreases and, concomitantly, the dissociation yield will decrease.

Complementary to the above experiments, transient photocurrents (see section 4.1) were measured on a neat PhPPV film and a PhPPV film doped with 1 wt% of TNF [61]. At variance with the experiments on MeLPPP, the measured photoresponse must involve dissociation after pulse excitation and the dissociating entity has to be a metastable electron-hole pair produced from a short lived primary optical excitation. However, it is open to conjecture what the rate-determining step for generation of mobile carrier is. Is it the primary field-dependent dissociation of a singlet exciton into a coulombically bound geminate electron-hole pair or its subsequent escape from its initial potential? Based upon the field dependence of the generation yield alone (Fig. 24) no unambiguous assignment can be made. The reason is that any experimental superlinear field dependence of Q_{coll} could be fitted by the conventional Onsager model in its either 1D [62] or 3D [5] version by choosing an empirical value for the so-called thermalization distance of the GP regardless whether or not the conceptual basis of the data analysis is valid. The temperature dependence of Q_{coll} at selected values of electric field is presented in Fig. 25. At $F = 2.2 \times 10^5$ V/cm Q_{coll} is weakly temperature dependent featuring activation energy of about 0.07 eV for $T > 200$ K. At lower temperatures Q_{coll} tends to saturate.

In order to distinguish between the above possibilities one needs additional experimental information. In the case of MeLPPP this was the simultaneous measurement of field-modulated picosecond transient absorption of charge carriers and excitons. By comparing the field dependence of the evolution of the transient

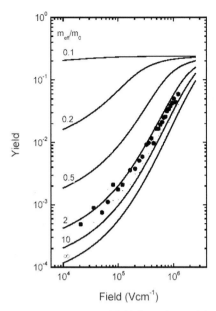

Figure 24 Experimental field dependence of the quantum yield in a PhPPV film. Solid lines are calculated from Eq. (40) parametric in the on-chain effective carrier mass. The following set of material parameters was used for calculations: $l = 6$ nm, $r_{min} = 0.6$ nm, $\tau = 1$ ns $\nu_0 = 10^{12}$ s^{-1}, $1/\gamma = 0.2$ nm, $\varepsilon = 3$.

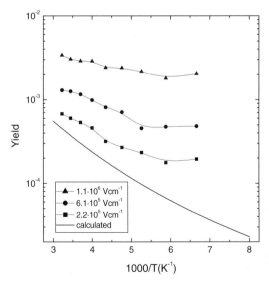

Figure 25 Temperature dependence of the quantum yield in a PhPPV film at different electric field as a function of temperature.

absorption of geminately bound positive polarons with the microsecond transient photocurent signal we could ascertain that it must be the field-assisted initial dissociation event which is rate controlling [57]. In this case dissociation creates more loosely bound GPs that can subsequently ionize quite easily without requiring too much of assistance by either electric field or temperature.

In the case of PhPPV the situation is different. Previous spectroscopic [16] and cw-photoconduction [17] work on PhPPV with and without controlled doping by trinitrofluorene showed that neat PhPPV contains about 0.04 % of dopants which already quench about 50% of excitations that, upon quenching, form electron-hole pairs on TNF and PhPPV. Intentional doping by 1% TNF can therefore increase the photogeneration of charge carriers by about a factor of 2 or less only, in agreement with Fig. 23. This proves that in this case the field-assisted step for photogeneration must be the subsequent escape of the pair from its coulombic well while the efficiency of the initial event, i.e. of exciton diffusion towards the sensitizes and following charge transfer, is close to unity. Since exciton quenching by dopants is exothermic it does not require an electric field.

At first glance, one would expect that the field dependence of Q_{coll} is tractable in terms of Brown's adaptation [63] of Onsager's theory for the dissociation/recombination of a geminate pair with a finite lifetime. However, in this theory both the field and temperature dependences of the dissociation yield are related via the value of the initial intra-pair separation. Since the measured activation energy of the yield at the lower electric fields is as low as 0.07 eV that value should be 6 nm. This is incompatible with the fact that there is no indication that the photocurrent saturates even at the field of 2.5×10^6 V/cm. At the same time, the extrapolated values of the yield at infinite either temperature or field are different whereas conventional Onsager theory predicts $\lim_{T \to \infty} \eta = \lim_{F \to \infty} \eta = \eta_0$, where η_0 is the initial yield for GP formation from an optical excitation. This problem has been recognized for some time already [56] and has been taken as evidence that classic Onsager-type theories are inappropriate for conjugated polymers. In the following section, an alternative model of charge photogeneration in intentionally or accidentally doped conjugated polymers will be outlined that is able to account for the relevant experimental results.

7.4.3
A model of dopant-assisted charge photogeneration

In Chapter 3 we showed that doping facilitates dissociation of relaxed excitons into geminate pairs. However, the Coulomb binding energy of such pairs of typically 0.5 eV is still much higher than the experimentally observed activation energies of photoconductivity [46,56,61]. Although some energy must be released upon carrier trapping, it is not clear how this energy can be transferred to the twin carrier that still occupies the polymer segment. In the present section we describe a model of exciton dissociation on a CT center, which consists of a conjugated segment and an adjacent acceptor molecule or a deep electron trap (electron scavenger). An external electric field and the Coulomb field of the trapped carrier form a potential well for the on-chain twin carrier. Since this carrier is delocalized within the segment its minimum energy must also include the energy of zero-point quantum oscillations. The latter effectively lowers the energy barrier for full carrier separation and, therefore, facilitates dissociation of the GP into free charge carriers.

Exciton dissociation at a charge transfer center is possible if the electron affinity of the electron scavenger is large enough in order to compensate the exciton binding energy. In the following we assume that this condition is always fulfilled. In the course of energy relaxation an exciton can visit several conjugated segments before it either decays radiatively or dissociates into a geminate pair at a CT center. At low dopant concentrations, the dissociation probability linearly increases with the exciton dwell time on a given segment. Since the energy relaxation is a dispersive process an excitation, on average, spends the longest time on the last visited segment implying that most excitations will dissociate at such segments. This reasoning may be not correct at large concentrations of traps but, under such circumstances, the exciton dissociation probability is close to one anyhow. The exciton dissociation probability, w_d, is determined by the relative rates of the radiative decay and the electron tunneling to a scavenger located at the distance r from the polymer chain. The former is characterized by the radiative exciton lifetime τ and the latter by the tunneling rate $v_0 \exp(-2\gamma r)$ where v_0 is the attempt-to-jump frequency and γ the inverse localization radius. The result then reads:

$$w_d = \left[1 + (v_0\tau)^{-1}\exp(2\gamma r)\right]^{-1}, \tag{33}$$

Since off-chain exciton dissociation rate exponentially decreases with increasing r carrier trapping by the nearest available electron scavenger is most feasible at low and moderate concentrations of traps. In a random system, the probability density $P(r)$ to find a nearest deep trap over the distance r from a conjugated segment is determined by the Poisson distribution as

$$P(r) = 2\pi r l N_d \exp\left[-\pi l N_d \left(r^2 - r_{min}^2\right)\right], \tag{34}$$

where l is the conjugation length, r_{min} the minimum possible distance from a deep trap to a polymer chain, and N_d the concentration of deep traps in the space between polymer chains. It is worth noting that Eq. (34) slightly underestimates the trap density for the following reasons: (i) this equation disregards the possibility of exciton dissociation at any visited segment except the last one and (ii) the value of N_d is determined as the trap density in free space between chains ($r > r_{min}$) rather than as the true average density of traps.

After exciton dissociation the electron is localized in a deep trap at the distance r from the polymer chain. The Coulomb field of this electron and the external electric field F determine the potential energy distribution, $U(x)$, for the hole that still occupies the polymer chain,

$$U(x) = -eFzx - \frac{e^2}{4\pi\varepsilon_0\varepsilon\left(x^2 + r^2\right)^{1/2}}, \tag{35}$$

where $z = \cos\theta$ and θ is the angle between the external field and the chain direction, e the elementary charge, ε_0 the dielectric permittivity, and ε the dielectric constant. At strong external fields and/or large values of r this potential monotonously decreases along the chain and, therefore the on-chain carrier becomes free immediately after exciton dissociation. Otherwise, the potential distribution given by Eq. (35) has a minimum at some $x = x_{min}$. Disregarding non-harmonic terms in the power-series expansion the potential well for on-chain carrier oscillations U_{osc} can be written as,

$$U_{osc}(x) = -eF_x x_{min} - \frac{e^2}{4\pi\varepsilon_0\varepsilon\left(x_{min}^2+r^2\right)^{1/2}} + \frac{e^2\left(r^2-2x_{min}^2\right)}{8\pi\varepsilon_0\varepsilon\left(x_{min}^2+r^2\right)^{5/2}}(x-x_{min})^2. \tag{36}$$

The third term in the right-hand side of Eq. (36) directly determines the oscillator strengths for quantum carrier oscillations within the potential well. Evaluating the minimum carrier energy E_{min} as a sum of the minimum potential energy and the energy of the zero-point oscillations yields,

$$E_{min} = U_{osc}(x_{min}) + \frac{1}{2}\hbar\omega = -eF_x x_{min} - \frac{e^2}{4\pi\varepsilon_0\varepsilon\left(x_{min}^2+r^2\right)^{1/2}}$$

$$+\hbar\left[\frac{e^2\left(r^2-2x_{min}^2\right)}{16\pi\varepsilon_0\varepsilon\left(x_{min}^2+r^2\right)^{5/2}m_{eff}}\right]^{1/2}, \tag{37}$$

where $\hbar$ is the Planck constant, ω the frequency of oscillations, and m_{eff} the effective carrier mass.

The height of the barrier for full on-chain dissociation is determined by the difference between the maximum potential energy, $U_{max} = U(x_{max})$, and E_{min}. If either the potential energy monotonously decreases along the chain or the potential well is present but $E_{min} > U_{max}$ full separation of carriers happens instantaneously. Since the depth of the potential well increases with decreasing r and z full carrier separation is granted for $r > r_0(z)$ where $r_0(z)$ can be found from the condition $E_{min}(r_0, z) = U_{max}(r_0, z)$. Otherwise, if $r < r_0(z)$ carrier escape from the potential well requires further thermal activation and the probability of full separation is determined by the condition that the hole must escape from the on-chain potential well before it recombines with the trapped sibling carrier. Since the recombination rate, v_{rec}, is controlled by the off-chain tunneling it can be written as $v_{rec} = v_0\exp(-2\gamma r)$. The escape rate, v_{esc}, is given by the rate of over-barrier jumps:

$$v_{esc}(r) = v_0\exp\left[-\frac{U_{max}(r)-E_{min}(r)}{kT}\right], \tag{38}$$

Combining Eq. (33) with expressions for v_{rec} and v_{esc} yields the total probability of exciton dissociation into free carriers w as a function of r and z,

$$w = \left[1+(v_0\tau)^{-1}\exp(2\gamma r)\right]^{-1}\left\{1+\exp\left[-2\gamma r+\frac{U_{max}(r,z)-E_{min}(r,z)}{kT}\right]\right\}^{-1}. \tag{39}$$

Averaging Eq. (39) over z and r with the probability density $P(r)$, given by Eq. (34), leads to the following expression for the quantum yield η of the dopant assisted carrier photogeneration:

$$
\eta = 2\pi l N_d \int_0^1 dz \left(\int_{\max\{r_{min},r_0(z)\}}^{\infty} dr\, r \exp\left[-\pi l N_d \left(r^2 - r_{min}^2\right)\right] \left[1 + (v_0\tau)^{-1}\exp(2\gamma\, r)\right]^{-1}\right.
$$

$$
+ \int_{r_{min}}^{\max\{r_{min},r_0(z)\}} dr\, r \exp\left[-\pi l N_d \left(r^2 - r_{min}^2\right)\right] \left[1 + (v_0\tau)^{-1}\exp(2\gamma\, r)\right]^{-1}
$$

$$
\left. \times \left\{1 + \exp\left[-2\gamma\, r + \frac{U_{max}(r,z) - E_{min}(r,z)}{kT}\right]\right\}^{-1}\right)
$$

$$. \quad (40)$$

The field dependence of the photogeneration yield is illustrated in Fig. 24 parametric in the effective carrier mass at a relatively high trap density, $N_d = 10^{20}$ cm^{-3}. Recent quantum-chemical calculations [11] showed that m_{eff} in conjugated polymers is around $(0.03...0.08)m_e$, where m_e is the free electron mass. At small values of m_{eff} the yield remains practically independent of both the field and carrier effective mass indicating that virtually every exciton-dissociation event results in producing a pair of free carriers. The situation changes at the value of m_{eff} around $0.1m_e$. Heavier carriers have smaller energy of zero-point oscillations and such carriers are localized within on-chain potential wells. Therefore, field-controlled height of on-chain barriers essentially determines the probability of full dissociation and the field dependence of the free carrier yield becomes stronger with increasing m_{eff}. The contribution from zero-point oscillations decreases with increasing m_{eff} and practically vanishes at $m_{eff} > 2m_e$. Concomitantly, the field dependences of the yield for heavier carriers practically merge into a universal curve.

The almost complete lack of the field and temperature dependences of the yield in materials with 'light' carriers may resemble direct photoexcitation of free charge carriers as described by the 1D semiconductor model of conjugated polymers. The results of the present work suggest another explanation based on the concept of strongly bound excitons as primary optical excitations. This explanation does not involve excess photon energy and, therefore, accounts for coincidence of the absorption edge and the onset of photoconductivity.

Figures 24 and 25 show the $Q_{coll}(F)$ and $Q_{coll}(T)$ data for the photogeneration yield for the PhPPV film in comparison with the prediction of theory under the premise that all primary excitations will migrate to a dopant. It turns out that a perfect fit is obtained for $m_{eff} = 2m_e$. Such a value of m_{eff} for a π-conjugated polymer is at variance with the results of quantum chemical calculations [11] and electro-reflection experiments on single crystalline poly-diacetylenes featuring a Franz-Keldysh effect [64]. On the other hand, there is experimental evidence that even the on-chain charge carrier mobility, probed via microwave techniques [65] is of the order of 0.1 cm^{2}V^{-1}s^{-1}, i.e. comparable with that of molecular crystals at 295 K and implying that m_{eff}/m_e must be of the order of unity.

In summary, it is a well-known phenomenon that donor or acceptor doping of organic semiconductors can sensitize photoconduction. The novel aspect regarding conjugated polymers is that in these systems one of the sibling carriers is delocalized within a chain segment comprising several repeat units. This carrier executes zero-point oscillations, which effectively lowers the barrier for complete dissociation of the GP into free carriers and explains why the field and temperature dependences can be weak although the absolute yield is much smaller than unity.

7.5
Photoconductivity in polymer donor/acceptor blends

7.5.1
Enhanced exciton dissociation at high concentration of electron acceptors

Nearly all π-conjugated polymers exhibit photoconduction within the spectral range of the $S_1 \leftarrow S_0$ transition, albeit with quantum efficiency as low as 10^{-4} to 10^{-3} depending on electric field. As discussed in the previous chapter, this weak photogeneration is due to the presence of inadvertent impurities that can sensitize photodissociation by virtue of low reductive/oxidative potentials. The yield of such a process is low because the dissociation product is a coulombically bound eh-pair confined as a donor–acceptor couple that requires further activation in order to escape from its mutually attractive potential.

However, it is well known that exciton dissociation into free charge carriers can be highly efficient in two-component donor/acceptor blends, as evidenced by the photovoltaic response of several such systems. A notable example is a hexa-*peri*hexabenzocoronene (HBC)/perylenediimide (PdI) blend system. The obvious conclusion is that the dissociation must strongly depend on the concentration of the sensitizing, i.e., electron accepting moiety. In order to bridge the gap between experiments on cw-photoconductivity in weakly doped samples and on the photovoltaic response in donor–acceptor blends one has got to study photoconductivity in donor-acceptor systems with a broad concentration range of the electron accepting moiety as it has been done in Ref. [27]. In that study, the electron donor was either PhPPV or a hexa-*peri*-hexabenzocoronene derivative (HBC-PhC12) and the acceptor was PdI – see structures in Fig. 26.

In both studied systems the quantum efficiency η was found to be field dependent with a characteristic variation upon the acceptor concentration (Fig. 27). As the concentration of the electron acceptor increases, η tends toward saturation at lower electric fields. Essential information is contained in Fig. 23 that shows quantum efficiency for both systems as a function of the relative acceptor concentration. Previous data for the PhPPV/TNF system [17] have been also included for comparison. Remarkably, within the concentration range $0.1\% < c < 10\%$, the yield is virtually independent of the kind of donor.

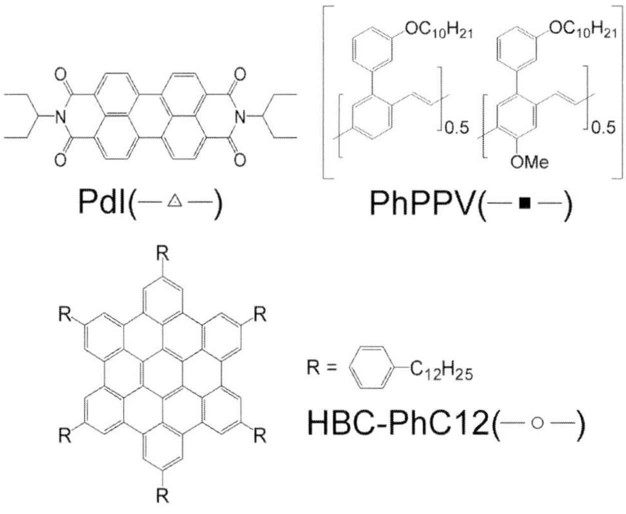

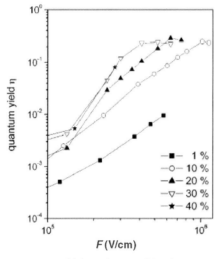

PdI(—△—)   PhPPV(—■—)

R = ⬡—C$_{12}$H$_{25}$

HBC-PhC12(—○—)

Figure 26   Chemical structures of PhPPV, HBC-PhC12 and PdI.

Figure 27   Field dependencies of the charge carrier photogeneration quantum yield in the ITO/PhPPV:PdI/Al configuration at 293 K for various PdI dopant concentrations. Excitation wavelength was 550 nm (2.25 eV).

An essential observation is that the yield of photogeneration increases steeply above roughly 10% acceptor concentration. Since the yield of exciton quenching at the acceptors is already saturated, this increase has to be attributed to increasing yield η_{esc} of escape of the eh-pair from the mutual coulombic potential. In terms of the rate constants for escape, k_{esc}, and geminate recombination, k_{rec}, the escape yield can be expressed as

$$\eta_{esc} = \frac{k_{esc}}{k_{esc} + k_{rec}}. \tag{41}$$

It is straightforward to conjecture that at high acceptor concentration, k_{esc} should increase above the percolation threshold for electron motion across the ensemble of acceptors. This process becomes efficient at concentrations above 10% [27]. However, one still has to wonder about the nature of the rate-limiting factor for geminate pair dissociation. Is it the escape of the eh-pair from the coulombic potential or the transport velocity of the escaping electron? An answer is provided by the field dependence of the free carrier yield, which has entirely to do with the field dependence of η_{esc}. Figure 27 shows that η_{esc} increases with electric field and tends to saturate at higher fields. Importantly, the saturation of the yield occurs at lower fields with increasing concentration. Qualitatively, the saturation electric field, F_{sat} must be a measure of how strongly the geminate eh-pair is bound. Therefore, decreasing F_{sat} suggests that the coulombic field is somehow screened progressively at higher concentrations. Concomitantly, the rate-limiting step must be the primary dissociation event. A theoretical model of the enhanced geminate pair dissociation at a donor/acceptor interface is described in the following section. Percolative electron motion above a critical dopant concentration will, of course, further enhance escape of the electron from the potential well.

It is remarkable that at a given electric field, the photoconductive yield as a function of acceptor concentration is virtually independent of the systems studied so far although the increase at the highest dopant concentrations has not been yet measured on the PhPPV/TNF system. This suggests that, at least at moderate doping, the dissociation is independent of the chemical constitution but is rather controlled by the coulombic binding energy of the eh-pair whose internal separation and dielectric environment do not change substantially upon changing the system, except at high doping when the morphology of the donor and acceptor phases next to their interface, e.g., regioregularity, becomes important. In the PhPPV/PdI system used in the current study the morphology is unlikely to be optimized. The 30% yield measured with PhPPV/PdI still exceeds the yield in a system with a polyfluorene copolymer backbone carrying perylene substituents as electron acceptors (16%), but is only about 1/3 of that measured in a blend containing regioregular polythiophene ($\eta \sim 100\%$). This illustrates the crucial role of improved packing within the array of donors and acceptors for the advancement of eh-pair separation and subsequent charge carrier motion.

7.5.2
A model of efficient exciton dissociation at a donor/acceptor interface

It has been shown in the previous section that, in conjugated polymers strongly doped with electron acceptors, i.e. in polymer donor/acceptor blends, the yield of intrinsic charge carrier photogeneration can increase up to almost 100% provided that the morphological demands are met [26,27,66]. Moreover, such efficient exciton dissociation can be achieved even at moderate electric field and low temperatures.

This indicates a really efficient mechanism preventing geminate pairs from recombination at a polymer/acceptor interface. A strong difference between exciton dissociation in weakly and strongly doped polymers is also supported by the fact that eh-pairs can be detected at small concentrations of acceptors while in polymer/acceptor blends mostly free charges and/or loosely bound long pairs have been observed [28]. The fast and efficient dissociation of geminate pairs suggests that a bottleneck for geminate recombination must exist at a polymer/acceptor interface. Below, we describe a model of interfacial exciton dissociation in donor/acceptor blends that shows how this bottleneck can be formed.

The model rests on the notion that a dipolar layer can exist at a polymer/acceptor interface as testified by the observation of a vacuum level shift of 0.25 eV at the interface between the five-ring conjugated oligomer p-bis [(p-styryl)styryl]benzene and C_{60} [67]. An optical excitation, generated in a conjugated polymer, can diffuse towards the interface and dissociate into an eh-pair. Immediately after dissociation, the electron will be localized at the electron acceptor molecule while the hole remains on the polymer chain next to the interface. The potential energy of the on-chain hole, E_p, is determined by its interaction with the negatively charged nearest electron acceptor and partial charges of dark dipoles as shown in Fig. 28. By neglecting contributions of all partial charges except for four nearest ones, the coordinate dependence of the potential energy along the chain can be estimated as

$$E_p(x) = \frac{e^2}{4\pi\varepsilon_0\varepsilon} \left(-\frac{1}{\sqrt{x^2+b^2}} + \frac{\alpha}{a+x} + \frac{\alpha}{a-x} - \frac{\alpha}{\sqrt{(a+x)^2+b^2}} - \frac{\alpha}{\sqrt{(a-x)^2+b^2}} \right),$$

(42)

where a is the distance between electron acceptors, b the distance between the chain and electron acceptors, and α the share of elementary charge transferred from the chain to an acceptor in the dark.

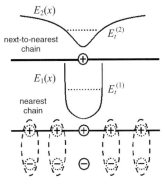

Figure 28 Potential energy distribution and total energy of a hole occupying a polymer chain either nearest or next-to-nearest to a polymer/acceptor interface. The potential distributions are caused by dark interfacial dipoles and the sibling electron trapped at an acceptor next to the interface.

Although an on-chain hole is supposed to occupy an extended state it cannot move freely along the chain mainly because of positive partial dipole charges residing on the same chain to the right and to the left from the hole. Together with negatively charged acceptors, these charges form an on-chain potential well for the hole and restrict its otherwise free motion along the chain. In order to estimate the energy of zero-point oscillations within this potential well one can approximate the real potential distribution given by Eq. (42) by the parabolic oscillatory potential as

$$E(x) \approx \frac{e^2}{4\pi\varepsilon_0\varepsilon}\left(-\frac{1}{b}+\frac{2\alpha}{a}-\frac{2\alpha}{\sqrt{a^2+b^2}}\right)+\frac{e^2}{8\pi\varepsilon_0\varepsilon}\left[\frac{1}{b^3}+\frac{4\alpha}{a^3}-\frac{6\alpha a^2}{(a^2+b^2)^{5/2}}\right]x^2, \quad (43)$$

The total energy $E_t^{(1)}$ of a hole, occupying the ground state on the chain nearest to the interface, is a sum of the minimum potential energy $E_p(x=0)$ and the kinetic energy of zero-point oscillations

$$E_t^{(1)} = \frac{e^2}{4\pi\varepsilon_0\varepsilon}\left(-\frac{1}{b}+\frac{2\alpha}{a}-\frac{2\alpha}{\sqrt{a^2+b^2}}\right)+\frac{\hbar}{2}\sqrt{\frac{e^2}{4\pi\varepsilon_0\varepsilon m_{eff}}\left[\frac{1}{b^3}+\frac{4\alpha}{a^3}+\frac{2\alpha}{(a^2+b^2)^{3/2}}-\frac{6\alpha a^2}{(a^2+b^2)^{5/2}}\right]}, \quad (44)$$

If the hole has been transferred from the chain nearest to the interface to the next nearest one, its coulombic potential energy increases but, at the same time, its kinetic energy drops significantly because the caging effect is eroded. The calculation, similar to the described above, leads to the following expression for the total energy $E_t^{(2)}$ of such a hole

$$E_t^{(2)} = \frac{e^2}{4\pi\varepsilon_0\varepsilon}\left[-\frac{1}{(b+d)}+\frac{2\alpha}{\sqrt{a^2+d^2}}-\frac{2\alpha}{\sqrt{a^2+(b+d)^2}}\right]$$

$$+\frac{\hbar}{2}\sqrt{\frac{e^2}{4\pi\varepsilon_0\varepsilon m_{eff}}\left\{\frac{1}{(b+d)^3}-\frac{2\alpha}{(a^2+d^2)^{3/2}}+\frac{6\alpha a^2}{(a^2+d^2)^{5/2}}+\frac{2\alpha}{\left[a^2+(b+d)^2\right]^{3/2}}-\frac{6\alpha a^2}{\left[a^2+(b+d)^2\right]^{5/2}}\right\}}. \quad (45)$$

where d is the distance between polymer chains. The energy difference $E_t^{(1)} - E_t^{(2)}$ is shown in Fig. 29 as a function of the effective carrier mass for different values of the dark charge transfer parameter α. Remarkably, this energy difference is positive at sufficiently small values of the effective carrier mass, i.e. a repulsive potential barrier separating a negatively charged acceptor and a hole, occupying a segment of a chain next nearest to the interface, is established. This prevents geminate recombination and stabilizes the interfacial geminate pairs provided that the effective mass is ~ 0.2 free electron mass or less. However, even if on-chain holes are heavier and the total energy of holes increases upon their transfer from nearest to next nearest chains, this increase is weaker than it would be without the cage effect. From the Onsager theory of geminate recombination it is known that even a small reduction of the

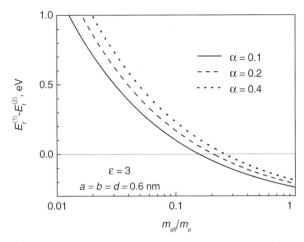

Figure 29 Dependence of the repulsive barrier height upon the effective on-chain carrier mass for different values of the dark-charge-transfer parameter α.

attraction potential, especially at short distances, leads to a strong increase in the dissociation probability. Therefore, the cage effect should strongly enhance dissociation of geminate pairs into free carriers also at larger values of the effective carrier mass.

Since the model requires polymer chains being fully parallel to the interface within, at least, two molecular layers nearest to the interface and the array of acceptor molecules being ordered it readily explains why the molecular ordering near the internal donor/acceptor interface combined with a sufficiently high concentration of acceptors is crucial for attaining a high photovoltaic efficiency. Because of the importance of the on-chain zero-point oscillations it is also obvious why polymeric photovoltaic cells are normally more efficient than small molecular devices. The only exception from the rule are cells containing a discotic donor, such as derivatives of benzocoronene [68]. In the context of the present theory this is remarkable because it is known that lowest excited states in these materials are excitons with efficient charge oscillations within a pair of molecules, which is equivalent to a low effective mass of the on-chain charge.

The model explains the increase of the photogeneration efficiency at higher acceptor concentrations in terms of the increase of the GP dissociation rates. Alternatively, one could invoke the onset of percolative motion of the electron across the array of acceptor molecules. However, any change of the carrier diffusivity without an independent increase of the dissociation rate constant could not affect the trade-off between dissociation and recombination of geminate pairs. The model also has several practical implications. (i) The interface morphology must strongly affect the exciton dissociation yield as experimentally verified. Any structural disorder at the interface is counterproductive because it will diminish the energy of zero-point oscillations and, therefore, destroy the bottleneck for geminate recombination. (ii) The existence of a dipolar layer at the donor/acceptor interface facilitates dissociation. Although, as evidenced by the data plotted in Fig. 29, the degree of the dark charge transfer has a relatively minor effect on the exciton dissociation yield the latter must increase with increasing α especially at weak electric field. (iii) Efficient dissociation of excitons is possible at moderate fields only if carriers within eh-pairs are separated by potential barriers, which prevents geminate recombination. This is not the case for so-called 'double-cable' polymers [69]. Therefore, the present model of interfacial dissociation predicts a relatively low efficiency of photovoltaic devices based on double cables.

7.6
Conclusions

Independent counting of the number of singlet excitons quenched upon applying an external electric field to selected pristine and doped π-conjugated polymers by employing fluorescence and transient absorption techniques and by measuring the number of charge carriers collected in the course of a photoconduction experiment proves that exciton dissociation is a sequential process. Initially a geminately bound electron hole pair is formed that can subsequently dissociate fully. In a neat, chemically pure system the initial step is strongly field dependent because the formation of the eh-pair requires additional Coulomb energy. From the field dependence of fluorescence quenching as well as from transient absorption one can conclude on the magnitude of the exciton binding energy. It turns out to be no less than ~0.5 eV. Since the rate limiting step is the formation of the eh-pair that field dependence cannot be analyzed in terms of Onsager's theory of geminate pair dissociation because it rests upon the assumption that the initial step be field-independent. Instead, an alternative theory for exciton quenching has to be invoked. At high photon doses, when non-linear processes such as singlet-singlet annihilation and tandem excitation [40] become important the pair yield increases and can approach 10 % [70].

In a doped system eh-pair formation is controlled by exothermic, time-dependent exciton migration followed by charge (usually electron) transfer to the scavenger. In that case the field dependence of photogeneration reflects the dissociation of the charged donor-acceptor pair. In both cases, though, the fraction of collected charges can – dependent on the applied electric field – be much less than unity because of the geminately bound eh-pairs recombine non-radiatively. The recombination features a power law on time implying long lifetimes of the pairs. In donor-acceptor blends the yield of pair dissociation can increase significantly if a dipole layer at the internal interface between the two components established that prevents pair recombination or if the LUMO energies of the components are appropriately matched in order to facilitate pair dissociation [71].

Acknowledgements

The authors are indebted to M. Deussen, E. V. Emelianova, V. Gulbinas, D. Hertel, C. Im, and M. Weiter for their contributions to the present review. We thank U. Scherf, and the Covion Organic Semiconductors company for supplying the polymers used in this work. This work was supported by the Optodynamics Research Center of the Philipps University of Marburg, the Deutsche Forschungsgemeinschaft, and EU-RTN project LAMINATE.

References

[1] R. H. Friend, R. W. Gymer, A. B. Holmes, J. H. Burroughes, R. N. Marks, C. Taliani, D. D. C. Bradley, D. A. Dos Santos, J. L. Bredas, M. Lögdlund, and W. R. Salaneck, Nature **397**, 121(1999).

[2] G. Hadziioannou and P. F. van Hutten (eds.), Semiconducting Polymers: Chemistry, Physics and Engineering (Wiley VCH, Weinheim, 2000).

[3] S. O. Kasap, Optoelectronics and Photonics: Principles and Practices (Prentice Hall, Harlow, 2001).

[4] N. F. Mott and E. A. Davis, Electronic Processes in Non-Crystalline Materials, 2nd edn. (Clarendon, Oxford, 1979).

[5] L. Onsager, Phys. Rev. **54**, 554 (1938).

[6] D. M. Pai and R. C. Enck, Phys. Rev. B **11**, 5163 (1975).

[7] H. Pope and C. E. Swenberg, Electronic Processes in Organic Crystals and Polymers, 2nd ed. (Oxford, Univ. Press, 1999).

[8] M. Chandross, S. Mazumdar, S. Jeglinski, X. Wei, Z. V. Vardeny, E. W. Kwock, and T. M. Miller, Phys. Rev. B **50**, 14702 (1994).

[9] H. Bässler, in "Primary photoexcitations in conjugated polymers: Molecular excitons versus semiconductor band model", N. S. Sariciftci, (ed.) (World Scientific, Singapore, 1997).

[10] P. G. Dacosta and E. M. Conwell, Phys. Rev. B **48**, 1993 (1993),

[11] J.-W. van der Horst, P. A. Bobbert, M. A. J. Michels, and H. Bässler, J. Chem. Phys. **114**, 6950 (2001).

[12] M. Rohlfing and S. G. Louie, Phys. Rev. Lett. **82**, 1959 (1999).

[13] S. F Alvarado, S. Barth, H. Bässler, U. Scherf, J.-W. van der Horst, P. A. Bobbert,

and M. A. J. Michels, Adv. Funct. Mater. **12**, 117 (2002).

[14] I. D. W. Samuel, G. Rumbles, and R. F. Friend, in "Primary photoexcitations in conjugated polymers: Molecular excitons versus semiconductor band model", N. S. Sariciftci, (ed.) (World Scientific, Singapore, 1997), p. 140.

[15] N. Geacintov and M. Pope, J. Chem. Phys. **47**, 1194 (1967).

[16] C. Im, J. Lupton, P. Schouwink, S. Heun, H. Becker, and H. Bässler, J. Chem. Phys. **117**, 1395 (2002).

[17] C. Im, E. V. Emelianova, H. Bässler, H. Spreitzer, and H. Becker, J. Chem. Phys. **117**, 2961 (2002).

[18] V. I. Arkhipov, E. V. Emelianova, and H. Bässler, Chem. Phys. Lett. **372**, 886 (2003).

[19] D. Hertel, E. V. Soh, H. Bässler, and L. J. Rothberg, Chem. Phys. Lett. **361**, 99 (2002).

[20] D. Hertel, Y. V. Romanovskii, B. Schweitzer, U. Scherf, H. Bässler, Synth. Met. **116**, 139 (2001).

[21] B. Schweitzer, V. I. Arkhipov, and H. Bässler, Chem. Phys. Lett. **304**, 365 (1999).

[22] B. Schweitzer, V. I. Arkhipov, U. Scherf, and H. Bässler, Chem. Phys. Lett. **313**, 57 (1999).

[23] A. Kadashchuk, Y. Skryshevskii, A. Vakhnin, N. Ostapenko, V. I. Arkhipov, E. V. Emelianova, and H. Bässler, Phys. Rev. B **63**, 115205 (2001).

[24] N. S. Sariciftci and A. J. Heeger, Int. J. Mod. Phys. B **8**, 237 (1994).

[25] J. J. M. Halls, C. A. Walsh, N. C. Greenham, E. A. Marseglia, R. H. Friend, S. C. Moratti, and A. B. Holmes, Nature **376**, 498 (1995).

[26] S. E. Shaheen, C. J. Brabec, N. S. Sariciftci, F. Padinger, T. Fromherz, and J. C. Hummelen, Appl. Phys. Lett. **78**, 841 (2001).

[27] C. Im, W. Tian, H. Bässler, A. Fechtenkötter, M. D. Watson, and K. Mullen, J. Chem. Phys. **119**, 3952 (2003).

[28] M. C. Scharber, N. A. Schultz, N. S. Sariciftci, and C. J. Brabec, Phys. Rev. B **67**, 085202 (2003).

[29] A. Horvath, G. Weiser, C. Lapersonne-Meyer, M. Schott, and S. Spagnoli, Phys. Rev. B **53**, 13507 (1996).

[30] U. Lemmer and E. O. Göbel, in "Primary photoexcitations in conjugated polymers: Molecular excitons versus semiconductor band model", N. S. Sariciftci, (ed.) (World Scientific, Singapore, 1997), p. 211.

[31] W. Bludau and E. Wagner, Phys. Rev. B **13**, 5410 (1976).

[32] M. Deussen, M. Scheidler, and H. Bässler, Synth. Met. **73**, 123 (1995).

[33] V. Gulbinas, Y. Zaushitsyn, V. Sundström, D. Hertel, H. Bässler, and A. Yartsev, Phys. Rev. Lett. **89**, 107401 (2002).

[34] C. Im, unpublished results.

[35] M. Yan, L. J. Rothberg, F. Papadimitrako-poulos, M. E. Galvin, and T. M. Miller, Phys. Rev. Lett. **72**, 1104 (1994).

[36] L. J. Rothberg, in "Primary photoexcitations in conjugated polymers: Molecular excitons versus semiconductor band model", N. S. Sariciftci, (ed.) (World Scientific, Singapore, 1997), p. 129.

[37] W. Graupner, G. Cerullo, G. Lanzani, M. Nisoli, E. J. W. List, G. Leising, and S. De Silvestri, Phys. Rev. Lett. **81**, 3259 (1998).

[38] V. Gulbinas, Y. Zaushitsyn, H. Bässler, A. Yartsev, and V. Sundström, submitted.

[39] C. Gadermaier, G. Cerullo, G. Sansone, G. Leising, U. Scherf, and G. Lanzani, Phys. Rev. Lett. **89**, 117402 (2002).

[40] C. Silva, A. S. Dhoot, D. M. Russell, M. A. Stevens, A. C. Arias, J. D. MacKenzie, N. C. Greenham, R. H. Friend, S. Setayesh, and K. Müllen, Phys. Rev. B **64**, 125211 (2001).

[41] V. I. Arkhipov, E. V. Emelianova, and H. Bässler, Phys. Rev. Lett. **82**, 1321 (1999).

[42] V. I. Arkhipov, E. V. Emelianova, S. Barth, and H. Bässler, Phys. Rev. B **61**, 8207 (2000).

[43] G. Lanzani, G. Cerullo, C. Brabec, and N. S. Sariciftci, Phys. Rev. Lett. **90**, 047402 (2003).

[44] D. M. Basko and E. M. Conwell, Phys. Rev. B **66**, 155210 (2002).

[45] M. C. J. M. Vissenberg and M. J. M. de Jong, Phys. Rev. Lett. **77**, 4820 (1996).

[46] S. Barth, H. Bässler, H. Rost, and H. H. Hörhold, Phys. Rev. B **56**, 3844 (1997).

[47] J. Klafter and A. Blumen, Chem. Phys. **119**, 377 (1985).

[48] B. Movaghar, B. Ries, and M. Grünewald, Phys. Rev. B **34**, 5574 (1986).

[49] U. Lemmer, A. Ochse, M. Deussen, R. F. Mahrt, E. O. Göbel, H. Bässler, P. Haring Bolivar, G. Wegmann, and H. Kurz, Synth. Metals **78**, 289 (1996).

[50] V. I. Arkhipov, E. V. Emelianova, and H. Bässler, Chem. Phys. Lett. **383**, 166 (2004).

[51] J. G. Müller, U. Lemmer, J. Feldmann, and U. Scherf, Phys. Rev. Lett. **88**, 147401 (2002).

[52] B. E. Kohler, C. Spangler, and C. Westerfield, J. Chem. Phys. **89**, 5422 (1988).

[53] A. P. Monkman, H. D. Burrows, M. D. Miguel, I. Hamblett, and S. Navaratnam, Chem. Phys. Lett. **307**, 303 (1999).

[54] Y. V. Romanovskii, A. Gerhard, B. Schweitzer, U. Scherf, R. I. Personov, and H. Bässler, Phys. Rev. Lett. **84, 1027 (2000).**

[55] Y. V. Romanovskii and H. Bässler, Chem. Phys. Lett. **326**, 51 (2000).

[56] S. Barth, H. Bässler, U. Scherf, and K. Müllen, Chem. Phys. Lett. **288**, 147 (1998).

[57] M. Weiter, H. Bässler, V. Gulbinas, and U. Scherf, Chem. Phys. Lett. **379**, 177 (2003).

[58] D. Hertel, H. Bässler, U. Scherf, and H. H. Horhold, J. Chem. Phys. **110**, 9214 (1999).

[59] J.G.Müller, U. Scherf, and U. Lemmer, Synth. Met. **119**, 395 (2001).

[60] V. I. Arkhipov, E. V. Emelianova, and H. Bässler, Chem. Phys. Lett. **296**, 452 (1998).

[61] M. Weiter, V. I. Arkhipov, and H. Bässler, Synth. Met. **141**, 165 (2004).

[62] R. Haberkorn and M. E. Michel, Chem. Phys. Lett. **23**, 128 (1973).

[63] C. L. Braun, J. Chem. Phys. **80**, 4157 (1984).

[64] G. Weiser, A. Horvàth, Chem. Phys. **227**, 153 (1998).

[65] A. M. van de Craats, J. M. Warman, P. Schlichting, U. Rohr, Y. Geerts and K. Müllen, Synth. Met. **102**, 1550 (1999).

[66] D. M. Russell, A. C. Arias, R. H. Friend, C. Silva, C. Ego, A. C. Grimsdale, and K. Mullen, Appl. Phys. Lett. **80**, 2204 (2002).

[67] S. C. Veenstra and H. T. Jonkman, J. Polym. Sci. B: Polym. Phys. **41**, 2549 (2003).

[68] L. Schmidt-Mende, A. Fechtenkötter, K. Müllen, E. Moons, R. H. Friend, and J. D. MacKenzie, Science **293**, 1119 (2001).

[69] A. Cravino and N. S. Sariciftci, J. Mater. Chem. 12, 1931 (2002).

[70] D. Moses, A. Dogariu, and A. J. Heeger, Chem. Phys. Lett. **316**, 356 (2000).

[71] A. C. Morteani, A. S. Dhoot, J. S. Kim, C. Silva, N. C. Greenham, C. Murphy, E. Moons, S. Cina, J. H. Burroughes, and R. H. Friend, Adv. Mat. **15**, 1708 (2003).

8

Polarons in π-conjugated Semiconductors: Absorption Spectroscopy and Spin-dependent Recombination

M. Wohlgenannt

8.1
Introduction

Since the first report of metallic conductivities in 'doped' polyacetylene in 1977 [1], the science of electrically conducting, π-conjugated polymers has advanced very rapidly. As high-purity polymers have become available, a range of semiconductor devices have been investigated; these include transistors [2–6], photodiodes [7, 8], and organic light-emitting diodes (OLEDs) [9–13]. It is well-known that chemical doping or electrical injection of charge carriers into organic semiconductors results in the formation of polarons [14]. The nature of the charge carriers is therefore quite different than in inorganic semiconductors and their study is of fundamental scientific interest, and also of high practical significance. Many questions about polarons remain: Are they described satisfactorily by simple phenomenological models that are extensively used in computer simulations [15]? How do defects in real samples influence the behavior of polarons [16]? Do interchain-interaction and solid state effects in films significantly alter the nature of polarons [17, 18]? What distinguishes high-mobility films from lower mobility organic films [19]? Is the recombination of polarons spin-dependent, i.e. do singlet excitons and triplet excitons form with different rates and probabilities in OLEDs? We think that the last question is especially intriguing and important since it is directly related to the maximum achievable efficiency of OLEDs, as we will discuss below.

8.1.1
Polarons in π-conjugated polymers and oligomers

Fig. 1 shows a comparison between different models that have been used for describing polarons. Panel a) depicts the electron-phonon or Su-Schrieffer-Heeger (SSH) model [14, 20, 21]. It was originally developed for the degenerate groundstate π-conjugated polymer, trans-polyacetylene, but it was later extended to include also non-degenerate polymers [22]. It predicts that the electron-phonon coupling causes a forbidden gap between valence and conduction band, respectively (Peierls instability). In addition, two localized polaron levels appear inside the forbidden gap in the singly charged polymer. Experimentally one finds two optical transitions [23] upon

Physics of Organic Semiconductors. Edited by W. Brütting
Copyright © 2005 WILEY-VCH Verlag GmbH & Co. KGaA, Weinheim
ISBN 3-527-40550-X

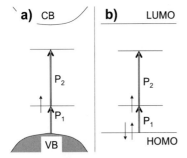

Figure 1 Comparison between different models for the positive polaron referring to the discussion in the text. a) Electron-phonon (SSH) model. b) Molecular orbital picture.

(photo)doping that are interpreted as the P_1 and P_2 transitions depicted in Fig. 1. The molecular orbital picture is illustrated in panel b). It relates to quantum-chemical calculations that yield a spectrum of molecular orbitals. In the singly charged molecule, electron phonon coupling gives rise to two localized polaron levels located between the highest occupied and lowest unoccupied molecular orbitals (HOMO and LUMO, respectively) [24–26].

Oligomers are often used as model compounds instead of polymers because they can be obtained with high purity and a well-defined chemical structure and conjugation-length. Although the molecular weight of polymers is typically much larger than that of oligomers, nevertheless it is established that the polymer should be viewed as a string of effectively independent segments, separated by chemical or physical defects. These segments are called conjugation segments, their (average) length is the (effective) conjugation-length (CL).

8.1.2
Organic electroluminescence

Π-conjugated organic semiconducting materials, both small molecules and polymers, have attracted much scientific and commercial interest. Commercial interest is to a large extend motivated by the electroluminescent property of many π-conjugated compounds. Research in the use of π-conjugated compounds as the active semiconductors in light-emitting diodes has advanced rapidly (for a review see Ref. [27]). OLEDs are now approaching efficiencies and lifetimes sufficient for the commercial market for e.g. large-area flexible displays or panel lighting. Organic electroluminescence (EL) was first reported for anthracene single crystals in the 1960s [28, 29]. Efficient EL was demonstrated in the 1980s by Tang and Van Slyke in two-layer sublimed molecular film devices [30]. These devices consisted of a hole-transporting layer of an aromatic diamine and an emissive layer of 8-hydroxyquinoline aluminium (Alq_3) (see Fig. 2). Indium tin oxide (ITO) is used as the transparent hole-injecting electrode, through which the electroluminescence is harvested, and a magnesium-silver alloy as the electron-injecting electrode. A large number of other

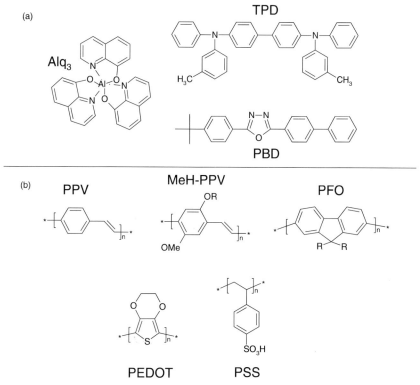

Figure 2 (a) Structures of some molecular semiconductors that have been used in thin-film electroluminescent devices. Alq$_3$ is used as an electron transport and emissive layer, TPD is used as a hole transport layer, and PBD is used as an electron transport layer. (b) Polymers used in electroluminescent diodes. The prototypical (green) fluorescent polymer is poly(p-phenylene vinylene) (PPV). One of the best known (orange-red) solution processable conjugated polymers is MEH-PPV. Poly(dialkyl-fluorene)s (PFO) are blue-emitting, high-purity polymers, which show high luminescence efficiencies. 'Doped' polymers such as poly(dioxy-ethylene thienylene), PEDOT, doped with poly-styrenesulphonic acid, PSS, are widely used as hole-injection layers.

molecular materials have been used as the charge-transporting or emissive layer, respectively in OLEDs.

EL from conjugated polymers was first reported in 1990 [9], using poly-p-phenylene vinylene, PPV (PPV and other π-conjugated polymers used in OLEDs are shown in Fig. 2). The main advantage of using polymers, compared to small π-conjugated molecules, is their potential for large scale, low cost production using solution processing techniques.

The processes responsible for EL have been established in the early studies: Positive and negative polarons are first injected from the electrodes into the luminescent layer. The polarons then migrate through this layer and can form neutral, bound excitons, either spin singlet or triplet. The electroluminescence quantum efficiency,

η_{EL} can therefore be defined as a product of three factors [27, 31]: $\eta_{EL} = \eta_1\eta_2\eta_3$. η_1 is the fraction of emitted photons per optically active exciton formed, η_2 is the fraction of optically active excitons per total number of excitons and η_3 is the fraction of excitons formed per charge carrier flowing in the external circuit. In hydrocarbon materials usually only singlet states luminesce, whereas triplet states are usually non-emissive [32]. The maximum possible efficiency of fluorescence-based (only singlet states are emissive) OLEDs is therefore determined by the fraction of injected electrons and holes that recombine to form emissive spin-singlet excitons, rather than non-emissive triplet excitons.

An assumption often employed until recently is that excitons are formed in the ratio one singlet to three triplets, since only one singlet combination can be formed from the addition of two spin 1/2 charge carriers, however the triplet state is three-fold degenerate (see Fig. 3). This is correct only if the polaron recombination process by which these excitons form were spin-independent, and then the maximum quantum efficiency, η_{max} of OLEDs would be limited to 25% [27]. The validity of this 25% spin-degeneracy statistics was tested in an Alq$_3$ device, and a singlet fraction of $(22\pm3)\%$ was found [33], in good agreement with expectations based on spin degeneracy statistics. Much research has therefore focused on finding schemes where both spin states, singlet and triplet excitons, can be used for light-emission in OLEDs. Such schemes generally involve materials, primarily organometallic complexes, having significant spin-orbit coupling to promote the spin-flip needed for optical emission. The most prominent schemes for harvesting triplet excitons involve either using phosphorescence emission directly or using Förster transfer from the triplet state of the "phosphorescent sensitizer" to the singlet state of a fluorescent guest complex [34]. Recent reports by Adachi et al. have demonstrated an impressive internal phosphorescence efficiency of 87% [35], in such devices the external efficiency is then mostly determined by the light out-coupling efficiency of the device.

Spin-dependent Exciton Formation
(e.g. in OLED):

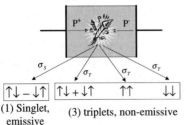

(1) Singlet,   (3) triplets, non-emissive
emissive

Figure 3 Spin dependent exciton formation in OLEDs: The OLED device is schematically drawn as a single layer sandwich device. Upon application of forward bias, positive (P^+) and negative (P^-) polarons are injected. Polaron recombination is symbolized as a firework, the possible spin states for the resulting exciton are shown. Assigning different exciton formation cross-sections for the singlet (σ_S) and triplet (σ_T) channels allows for spin-dependent exciton formation.

The triplet harvesting scheme has proven very successful, however a surprising recent discovery indicates that for electrical excitation in OLEDs made from π-conjugated polymers a considerably higher number of singlet excitons are generated compared to what is expected from spin-degeneracy statistics (i.e. 25%), and triplet harvesting might not always be necessary. The original report gave a singlet fraction of ≈ 50% [36] in a PPV-based device, and this report was soon confirmed by another laboratory [37]. To explain the large singlet yield, the original report also speculated that either the exciton binding energy is weak or that singlet bound states are formed with higher probability than triplets. To our knowledge, all subsequent works consider the second alternative, since it is by now well established that the splitting between the lowest singlet and triplet excitons is typically 0.7 eV [32], so that at least the triplet exciton is strongly bound. The singlet exciton binding energy is therefore also expected to be relatively large. The exact value for the singlet exciton binding energy has, however remained a matter of intense discussion and ongoing research. (see e.g. [38] for a review)

Very recently, Baldo and coworkers [39] have used a novel technique employing reverse bias measurements of photoluminescent efficiency to determine the excitonic singlet-triplet formation statistics of electroluminescent organic thin films. Using this method, the singlet fractions in thin films of two organic emissive materials commonly used in organic light emitting devices, Alq_3 and poly[2-methoxy-5-(2-ethylhexyloxy)-1,4-phenylenevinylene] (MEH-PPV) (for their chemical structures see Fig. 2), were found to be (20 ± 1) % and (20 ± 4) %, respectively. These results are in apparent contradiction to reports by others claiming that the singlet fraction is much larger than 25% in polymer devices. We must therefore conclude that the current understanding of exciton formation statistics in polymeric and small molecular weight organic electroluminescent materials is still incomplete. We will discuss our contributions to this field in the light of this controversy.

8.1.2.1 Spin dependent exciton formation cross-sections

To accommodate the possibility that singlets and triplets form with different probabilities, Shuai et al. [31] have introduced the spin-dependent exciton formation cross-sections, σ_S and σ_T for singlet and triplet formation, respectively (see also Fig. 3). They also postulated that $\eta_2 = \sigma_S/(\sigma_S + 3\sigma_T)$; for $\sigma_S = \sigma_T$ we retrieve $\eta_2 = 25$ %, the spin-statistical limit. Since the original report, a number of recent reports have indicated that η_{max} in OLEDs ranges between 22% to 83%[33, 36, 37, 40–43], and the reason for this variation is under investigation. In our work we will assume that the exciton formation cross-sections are proportional to the exciton formation rates that we will measure experimentally.

8.2
Polaron absorption spectra

8.2.1
Experimental

We have used the continuous wave photoinduced absorption (PA) and the Fourier-transform infrared photoinduced absorption (FTIR-PA) spectroscopy to study the polaron optical transitions in films of π-conjugated polymers and oligomers. The PA spectroscopy has been widely used in π-conjugated materials for studying long-lived photoexcitations [44]. An Ar$^+$ laser beam modulated with a chopper was used as the pump beam. An incandescent tungsten-halogen lamp and a variety of diffraction gratings and solid-state detectors were used to span the probe transmission T in the spectral range $\hbar\omega$ between 0.3 and 3 eV. The PA spectrum $\Delta\alpha(\omega)$ was obtained

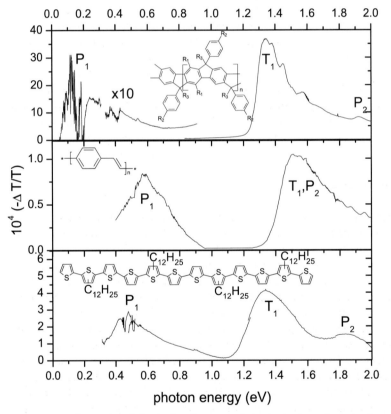

Figure 4 Typical examples of photoinduced absorption spectra of π-conjugated polymer and oligomer films , namely of a) mLPPP (see inset), b) PPV (see inset) and c) 12T (see inset). The spectra were measured at 80K under excitation by an Argon-Ion laser (typically 100mW) and modulated by an optical chopper (typically at 1kHz). The T_1 transition is due to triplet-triplet absorption, and P_1 and P_2 are due to polaron absorption.

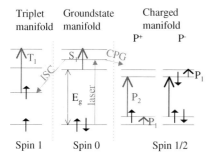

Figure 5 Schematic energy level diagram for photogenerated triplet excitons and polarons, for completeness the energy levels in the groundstate manifold (groundstate level together with neutral, spin singlet excitations) are also shown. The transitions important for cw photoinduced absorption spectroscopy and the intersystem crossing (ISC) and charge photogeneration (CPG) processes are assigned. E_g> is the optical gap.

by dividing $\Delta T/T$, where the modulated transmission ΔT was measured by a phase-sensitive technique. We extended the probe photon energy down to 0.05 eV using a Fourier Transform Infrared (FTIR) spectrometer in combination with a deuterated triglycine sulfide (DTGS) detector. To obtain the FTIR-PA spectrum, we measured the FTIR absorption spectrum with the pump beam on and, subsequently, with the pump beam blocked. We calculated the difference in transmission and repeated this alternation about 10 000 times to achieve a signal-to-noise ratio better than 10^5. Finally, the results were normalized by the sample transmission.

8.2.2
Experimental results

Fig. 4 shows the PA spectra of films of two π-conjugated polymers, namely methyl-substituted ladder-type poly-para-phenylene (mLPPP) (panel a)) and PPV (panel b)) and that of an oligothiophene, namely 12T (panel c)). All these spectra show three optical transitions (it is known that the high-energy transition in PPV (Fig. 4, panel b)) is actually composed of two transitions [46]). Thanks to a large number of previous PA studies the nature of the long-lived photoexcitations as well as their characteristic absorption bands inside the optical gap are well-understood and firmly established. In the spectra in Fig. 4 the characteristic two bands due to polarons (P_1 and P_2) are assigned, whereas the triplet exciton absorption is a single band (T_1). Although the majority of photoexcitations are expected to be singlet excitons [38], they usually do not show up in cw spectroscopy, because of their much shorter lifetime. All films we have studied show a PA spectrum similar to those of Fig. 4. The transitions important in PA spectroscopy together with a schematic energy-level diagram are shown in Fig. 5.

8.2.2.1 Polaron absorption spectra in doped oligomers; non-degenerate ground state systems

Fig. 6 shows the peak photon energies of the P_1 transition in a variety of π-conjugated oligomers, namely oligophenyls, oligophenylene-vinylenes, oligothiophenes and oligothienelyne-vinylenes plotted versus the oligomer-length, L [47]. We note that the data were taken from the literature [48–51]. Data on films of substituted oligothiophenes were measured in our laboratory using the PA spectroscopy and added to Fig. 6. The data in Fig 6 encompass unsubstituted and alkyl-substituted molecules in solution, anions and cations, and photodoped oligomer films and range from trimers to oligomers with 17 repeat units. From Fig. 6 we find that the P_1 transition in each of the four oligomer classes red-shifts as L increases; specifically $P_1 = P_{1,\infty} + const/L$. We note that this observation in itself is not surprising. In fact this scaling relation is ubiquitous in π-conjugated oligomers [24–26]: the optical gap (i.e. the singlet exciton energy) [52], triplet exciton energy [53] and also the P_2 transition (not shown here) obey such a scaling relationship. The novel observation is that if the P_1 data for the various oligomers are plotted versus L rather than the number of repeat-units (as it is usually done) and combined into a single graph, we find that the scaling law is universal, at least in the classes of materials studied here. To the best of our knowledge, this universal relationship has previously been overlooked. The universal scaling relationship does not hold true for neither singlet exciton or triplet exciton energies, or any other quantity, as far as we know. Secondly, we see in Fig. 6 that P_1 increases several times when going from long oligomers to the trimer, whereas singlet and triplet exciton energies only change by a fraction of their value for long chains. P_1 is therefore a universal and sensitive measure of the CL of oligomers.

8.2.2.2 Polaron absorption spectra in doped oligomers; degenerate ground state systems

In Fig. 6 we have limited ourselves to non-degenerate ground state oligomers, whereas Fig. 7 compares P_1 data measured in (degenerate ground state) polyenes to the "universal" scaling law that we found in Fig. 6. We note that the polyene P_1 data were taken from the literature [45] and that polarons are not stable in polyenes. It is energetically favorable for polarons to form solitons and antisolitons. Under certain experimental conditions it is however possible to observe polarons in polyenes [45]. It is seen in Fig. 7 that the polyene data do not follow the universal scaling law obtained for non-degenerate systems. Specifically, the scaling relation is somewhat steeper than that in non-degenerate ground state systems, and $P_{1,\infty} = 0.12eV$ rather than $P_{1,\infty} = 0.25eV$ found in Fig. 6. The steeper scaling relationship shows that polyenes are more sensitive to the boundary conditions at the chain ends and the resulting confinement. This indicates that polarons are more extended in degenerate ground state systems than in non-degenerate ones. This is expected based on the SSH model [21] where non-degeneracy leads to an additional confinement of the polaron, measured by the so-called confinement parameter.

In summary, the P_1 band shows a universal scaling law on L in a wide class of non-degenerate ground-state oligomers. This observation is then the basis for the

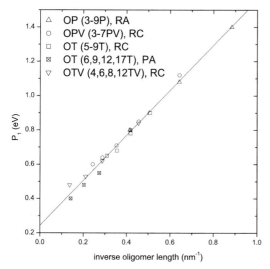

Figure 6 The peak photon energies of the P_1 polaron transition in a variety of oligomers, namely solutions of unsubstituted oligophenyls (OP, radical anion (RA)), alkyl-substituted (AS) oligophenylene-vinylenes (OPV, radical cation (RC)), end-capped oligothiophenes (OT, RC), films of AS OT (PA), AS oligothienylene-vinylenes (OTV, RC). The solid line is a linear fit to the data.

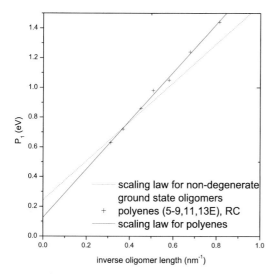

Figure 7 The peak photon energies of the P_1 transition in polyene oligomers. The data were taken from the literature [45] and are for radical cations in a Freon matrix. The solid line is a linear fit to the polyene data. The dotted line is the fit to the non-degenerate ground state oligomers shown in Fig. 6.

following generalization: P_1 is a universal, sensitive measure of the CL of both oligomer and polymer films (remember that the CL of a polymer is the length of the "equivalent" oligomer). In Ref. [54] we justify this generalization in more detail. Based on the universality of P_1, we may perform a linear fit to the data in Fig. 6 and the resulting function $P_1(L)$ is used to obtain the effective CL of polymer films using the following method: we measure P_1, then invert the function $P_1(L)$ and interpret L as the effective CL of the polymer film studied.

8.2.2.3 Polaron absorption spectra in photodoped polymer films

The procedure described above for measuring the CL is the basis for the presentation of the polymer P_1 data in Fig. 8 where crosses mark the data for a certain polymer (name is assigned). The P_1 value is the peak photon energy of low energy polaron transition in the PA spectrum measured in a thin film. The y-coordinate of the crosses is equal to the measured P_1 for the polymer, whereas the x-coordinate then yields the CL obtained by inversion of the fit function $P_1(L)$. We find that some polymer films (poly-phenylene-ethenylene (PPE), PPV and regio-random poly-3-hexyl-thiophene (r-ra P3HT)) have a short CL and behave rather like oligomers, whereas in MEH-PPV we find that the polarons are considerably more delocalized than in oligomer films. In addition, it is seen that this procedure does not yield a meaningful result for the CL for some polymer films for which $P_1 < P_{1,\infty}$, i.e. P_1 is lower in photon energy than the extrapolation of the oligomer data even to infinite L. These polymers are regio-regular hexyl-substituted polythiophene (r-re P3HT, its very close kinship to the oligothiophenes is obvious and its P_1 should therefore follow from

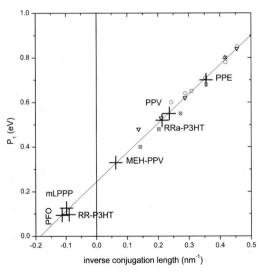

Figure 8 The peak photon energies of the P_1 transition in a variety of π-conjugated polymer films, their names are assigned. The crosses mark the intersect with the linear fit to the oligomerdata in Fig. 6; the y-coordinate therefore gives the measured P_1 peak photon energy, the x-coordinate gives the effective conjugation length (see text for a discussion).

extrapolation of the oligothiophene data) and mLPPP and PFO (that are closely related to the oligophenyls). It is clear that something interesting occurs in the these samples. We have shown elsewhere [54] that $P_1 < P_{1,\infty}$ is indicative of a destabilization of the polaron as a result of interchain interactions.

8.3
Is polaron recombination (exciton formation) spin-dependent?

In OLEDs the annihilation of positive and negative polarons to the ground state proceeds via two fundamental steps: (i) formation of a neutral exciton, either spin singlet or spin triplet, and (ii) the radiative or non-radiative decay of the exciton. In most π-conjugated materials only the spin singlet exciton is emissive, and this is why the spin-statistics of step (i) determines the maximum achievable η_{EL}. In this section we discuss the spectroscopic/magnetic resonance experiments we used [40, 42, 55, 56] to study the question whether polaron recombination is spin-dependent, i.e. $\sigma_S \neq \sigma_T$ (see Fig. 3). We performed such experiments in a large variety of π-conjugated polymer and oligomer thin films. Our technique uses a combination of the PA and PA-detected magnetic resonance spectroscopies.

8.3.0.1 Experimental
We used the spin 1/2 PA-detected magnetic resonance (PADMR) technique [57–61, 40, 42]. In PADMR we measure the changes, δT that are induced in the PA spectrum, ΔT by spin 1/2 magnetic resonance. δT is proportional to δn that is induced in the photoexcitation density, n that is probed by PA spectroscopy. Two types of PADMR spectra are possible: the H-PADMR spectrum where δT is measured at a fixed probe wavelength, λ as the magnetic field H is scanned, and the λ-PADMR spectrum where δT is measured at a resonant H while λ is scanned. A schematic drawing of the PADMR experimental setup used is shown in Fig. 9. For the detailed description see Ref. [62]

The photoluminescence-detected magnetic resonance (PLDMR) technique [63–67, 55] is closely related to PADMR: PLDMR measures changes, δPL induced in photoluminescence (PL) (rather than PA) upon magnetic resonance. We note that PA, PADMR and PLDMR can all be measured using the same setup and under identical conditions, allowing accurate comparison between results of the three methods. The term optically detected magnetic resonance (ODMR) is often used for both, PADMR and PLDMR. We note that ODMR experiments are performed at low temperature (typically 10K), mainly because of two reasons: (a) the polaron lifetime is strongly temperature dependent and is sufficiently long for study using cw spectroscopies only at temperatures below typically 100K; (b) the spin-lattice relaxation time is strongly temperature dependent and becomes very long at low temperatures, such that the spin alignment is conserved during the modulation period of the experiments.

In PA and spin-1/2 PADMR polaron recombination reactions occur between neighboring positive (P^+) and negative polarons (P^-); the products of polaron recom-

PADMR-setup

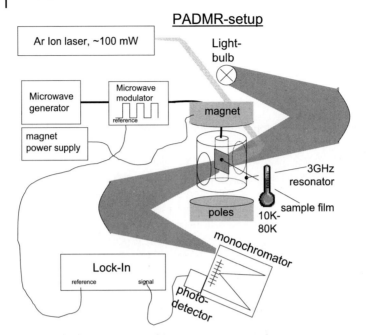

Figure 9 A schematic drawing of the PADMR experimental setup used to study spin-dependent polaron recombination and exciton formation.

bination are neutral excitons, either spin-singlet or triplet. The recombination reaction rate R_P between spin parallel pairs ($\uparrow\uparrow$, $\downarrow\downarrow$) is proportional to $2\sigma_T$, whereas the rate R_{AP} between spin antiparallel pairs ($\uparrow\downarrow$, $\downarrow\uparrow$) is proportional to ($\sigma_S + \sigma_T$), where the proportionality constant is the same in both cases [40]. These relations are obtained as follows (see Fig. 3): antiparallel pairs may form either the $\uparrow\downarrow - \downarrow\uparrow$ (total spin singlet state) or $\uparrow\downarrow + \downarrow\uparrow$ combination (total triplet), whereas parallel pairs form only total triplet pairs ($\uparrow\uparrow$ or $\downarrow\downarrow$).

8.3.0.2 Experimental Results

Using PADMR and PLDMR spectroscopy [40, 42, 55, 56] we have collected a large experimental body of evidence that the polaron recombination reaction to form excitons is spin-dependent, in particular $\sigma_S > \sigma_T$. In the following we present a detailed discussion of our main experimental observations and conclusions that can be drawn based on the experimental data.

Our claim that $\sigma_S > \sigma_T$ is based on a series of observations:

- Fig. 10 shows the PA spectrum of a 12T thin film together with the λ-PADMR spectrum at magnetic field H=1.05 kG corresponding to S=1/2 resonance (see inset in (b)). λ-PADMR spectra measured in other materials are very similar, Fig. 10 is therefore a typical example and the discussion that follows is equally valid in all other films we have studied. The spin

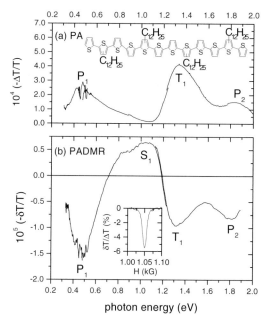

Figure 10 (a) The PA spectrum of 12T (inset); (b) the PADMR spectrum at magnetic field H=1.05 kG corresponding to S=1/2 resonance (see inset in (b)). Both spectra (a) and (b) show two bands (P_1 and P_2) due to polar- ons, T_1 is due to triplet absorption. S_1 is assigned to singlets. The PA was measured at 80 K, excitation by the 488 nm Ar^+ laser line (500 mW); the PADMR spectrum was measured at 10 K.

1/2 λ-PADMR spectrum clearly shows a negative magnetic resonance response at the polaron PA bands P_1 and P_2. The polaron density is therefore reduced upon spin-1/2 resonance. Importantly, this magnetic resonance response is unrelated to the thermal equilibrium spin-polarization. A back of the envelope calculation using Boltzmann statistics shows that the observed effect is much too large, in addition the effect does not scale with temperature in the way that Boltzmann statistics dictates. In fact, this situation is well-known from the study of amorphous silicon and it was concluded that this non-thermal-equilibrium magnetic resonance response is due to spin-dependent reaction rates [63]. We therefore interpret this simple observation as strong and direct evidence that polaron recombination is spin-dependent.

- The spin 1/2 PADMR spectrum (Fig. 10 (b)) clearly shows a negative magnetic resonance response at the T_1 triplet exciton PA band, in addition to P_1 and P_2 PADMR bands. As a matter of fact, the H-PADMR spectrum (not shown) at T_1 is identical to the spin-1/2 resonance, rather than the expected triplet resonance powder pattern (a weak triplet powder pattern is superimposed on the spin-1/2 resonance in some materials we studied, but is completely absent in other films). This observation is very intriguing, since it would at first appear improbable for spin-1/2 resonance to have any effect on the spin-1 triplet exciton population. However, an explanation may easily be found: Polarons can recombine to form triplet excitons. Since the polaron population is reduced by spin-1/2 resonance, then spin-1/2 resonance indirectly affects the triplet population via the exciton formation process.

- The observation of a negative T_1 spin-1/2 magnetic resonance indicates that the polaron recombination process studied is bimolecular and non-geminate in nature. In a geminate recombination process, on the other hand, the recombining pairs of polarons are prepared in a spin-antiparallel configuration, magnetic resonance (i.e. spin-flips) will therefore necessarily increase parallel alignment and triplet exciton formation [68], which would result in a *positive* triplet spin-1/2 resonance. We note however, that this conclusion is not unambiguous as Frankevich has recently pointed out [68]. PA and PLDMR measurements to be discussed now will however confirm our conclusion that we are dealing with bimolecular, non-geminate polaron recombination. As will become clear to the reader shortly, it is of crucial importance to carefully determine the recombination kinetics of polarons (non-geminate or geminate) for the correct interpretation of our experimental data [68].

- Fig. 11 shows the PA, PL, and λ-PADMR at magnetic field H=1.05 kG that corresponds to S=1/2 resonance of PPV. Inset (i) shows the spin 1/2 H-PLDMR resonance, whereas inset (ii) shows spin-1/2 H-PADMR resonance. It is seen that the PLDMR and PADMR resonances are mirror images

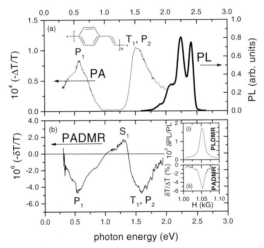

Figure 11 (a) The PA and PL spectra of PPV (see inset); (b) the PADMR spectrum at magnetic field H=1.05 kG that corresponds to S=1/2 resonance. Inset (i) shows the spin 1/2 PLDMR resonance, whereas inset (ii) shows spin-1/2 PADMR resonance. Both spectra (a) and (b) show two main bands (P_1 and (T_1, P_2) (see text)). P_1 and P_2 are due to polar-ons, T_1 is due to triplet absorption. The positive PADMR band S_1 is assigned to singlet absorption in agreement with the positive H-PLDMR response (see inset (i) in (b)). The PA and PL (PADMR and PLDMR) were measured at 80 K (10 K), excitation was 457 nm Ar$^+$ laser line (300 mW), modulated at 1 kHz.

of each other. Again, it would at first appear improbable for spin-1/2 resonance to influence spin-0 singlet excitations that give rise to the PL. Again, the interpretation comes to mind that polarons can recombine to form singlet excitons, and therefore spin-1/2 resonance indirectly influences the singlet population via the exciton formation process. However, this interpretation not unambiguous: whereas polaron recombination is necessarily an annihilation reaction between two oppositely charged polarons, singlet exciton recombination can be due to a multitude of reactions. These include radiative or non-radiative monomolecular recombination, singlet-polaron quenching and singlet-singlet annihilation. In particular, the singlet-polaron quenching reaction may also cause *positive* spin-1/2 PLDMR [67]. This is because a reduction in polaron density would then lead to a reduction in singlet-polaron quenching and therefore an increase in singlets (this ambiguity did not arise for the triplet exciton spin-1/2 resonance, since it is negative). In the following we therefore experimentally determine the origin of the spin-1/2 PLDMR.

- Fig. 12 shows the experimentally determined dependencies of PL, δPL and polaron PA band P_1 (where $-\Delta T/T \propto$ polaron density, N) on the laser intensity, Φ in a PPV film. We first discuss the polaron kinetics: Fig. 12 shows that the polaron PA signal scales as $\Phi^{1/2}$ at large Φ, which shows that polaron recombination follows a rate equation law with bimolecular recombination kinetics:

$$\frac{dN}{dt} = \eta\Phi - BN^2, \tag{1}$$

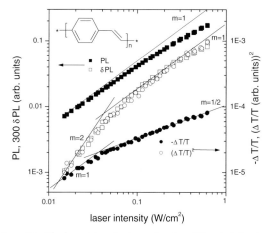

Figure 12   The laser intensity dependencies of the photoluminescence (PL, solid squares), the magnetic resonance effect on the photoluminescence (δPL, open squares), the polaron PA band measured at 0.55 eV (-$\Delta T/T$, solid circle) and its square (open circles, rescaled) in a PPV film measured at 10K. The modulation frequency was 1kHz.

where B is the bimolecular recombination constant. Such a rate equation yields the $\Phi^{1/2}$ law at large Φ [69]. We note that we found that the polaron PA signal scales as $\Phi^{1/2}$ at large Φ in all the films we studied (see also Ref. [70]). *This shows that polaron recombination is bimolecular and therefore non-geminate*, as was also concluded above.

In order to identify the mechanism responsible for the PLDMR signal, we have explored the relation between N and δPL. Importantly in Fig. 12 we see that $(\Delta T/T)^2$ coincides with high accuracy with the laser intensity dependence of δPL, i.e. $\delta PL \alpha N^2$. We have thus shown that the delayed PL is proportional to the recombination term in the polaron rate equation 1, and the proportionality between δPL and N^2 directly implies that δPL is a result of a magnetic resonance effect on the non-geminate polaron recombination. We note that the data presented in Fig. 12 for PPV does not mean that spin 1/2-PLDMR is due to non-geminate recombination in all materials under all possible experimental conditions. Indeed List et al. concluded that spin 1/2-PLDMR in mLPPP using X-band magnetic resonance is due to singlet-polaron quenching [67]. Similarly, our finding of non-geminate polaron recombination does not mean that polaron recombination is always non-geminate under all possible experimental conditions. Indeed, in ultrafast and picosecond experiments it is often found that polaron recombination is geminate [68]. Finally we note that we tentatively assign the positive PADMR-band S_1 in Fig. 10 b) as the spectroscopic signature of the resonantly enhanced singlet exciton population. We have been able to substantiate this assignment using ultrafast pump-and-probe spectroscopy [56].

• Having established the origin of the PLDMR and that polaron recombination is non-geminate, the positive sign of the H-PLDMR then shows that $\sigma_S > \sigma_T$. This may be understood as follows: Magnetic resonance leads to a randomization of the spin-alignment of the recombining pairs of polarons. At all times, the four possible spin states of the recombining polarons are equally populated, and each pair of polarons changes its spin state rapidly on the time scale of recombination (strictly speaking these statements are only true under saturated resonance conditions). This then leads to continuous competition between singlet and triplet formation. A straightforward rate equation calculation [55] then shows that the more efficient channel leads to a positive signal, whereas the less efficient channel leads to a negative signal. Our observation of a positive singlet signal and a negative triplet signal is therefore in striking agreement with this analysis.

• Frankevich has recently pointed out that the observation of the negative triplet spin-1/2 resonance signal does not necessarily imply that $\sigma_S > \sigma_T$ [68]. His analysis boils down to asking the question, whether it is possible that triplet formation is preferred (i.e. $\sigma_T > \sigma_S$) and still the triplet population is reduced upon spin-1/2 resonance. He shows that this is indeed possible, if the triplet lifetime is considerably different dependent on the spin-projection or magnetic quantum number, m of the triplets produced by polaron recombination. As a matter of fact, that this is true is well-known [44]. In particular, different lifetimes for triplets in different m states is considered the reason underlying the observation of spin-1 PADMR resonances. Such spin-1 resonances are observed as half-field resonances and full-field powder patterns in some of our samples. On the other hand, in some of the samples that we studied, namely PFO and mLPPP, such spin-1 resonances are entirely absent, but a strong negative spin-1/2 resonance at T_1 is nevertheless observed.

We have thus presented a large body of experimental evidence in support of the following conclusions:

- spin-1/2 PADMR studies the spin-dependent polaron recombination rates for forming either singlet or triplet excitons.
- Polaron recombination and exciton formation are spin-dependent.
- Polaron recombination is a bimolecular, non-geminate reaction under the present experimental conditions.
- $\sigma_S > \sigma_T$.

We have previously shown that PADMR experiments may be analyzed quantitatively under saturated magnetic resonance conditions to yield values for $r = \sigma_S/\sigma_T$. This quantitative interpretation goes beyond the scope of the present discussion, interested readers are referred to these references [42, 56]. In particular, we found evidence that r is mainly determined by the CL, in particular r increases with increasing CL. For completeness we simply state our main result, shown in Fig. 13. Fig. 13 shows the inverse ratio $r^{-1} = \sigma_T/\sigma_S$ as obtained from PADMR measurements in a large variety of π-conjugated materials vs. P_1. Since P_1 is a linear function of 1/CL we may actually plot r^{-1} vs 1/CL as is also shown in Fig. 13 (upper axis). This uncovers a universal behavior of r(CL), namely that r^{-1} depends linearly on 1/CL irrespective of the chain backbone structure, side groups or film morphology. The measured r values are apparently film morphology independent. This indicates that the recombination processes to form singlet and triplet excitons, respectively depend on morphology in a similar way and that this dependence cancels out in measurements of the ratio r. The Cambridge group have independently arrived at a similar conclusion, namely that r is mainly determined by the CL, based on fluorescence and phosphorescence emission efficiency measurements in OLEDs made from a Platinum-containing monomer and polymer, respectively [41].

It is important to emphasize that we studied recombination of *photogenerated* polarons in thin films rather than polarons injected into OLED devices. We note that in principle it is possible to use our technique directly in OLED devices, but it has been shown [71] that in such measurements electrode interface effects and transport (mobility) effects make a quantitative interpretation very involved at best. Measurements on films are also less time-consuming, easier and more general (also non-luminescent materials can be studied), and this allowed us to study a large number of materials. Baldo and coworkers [39] have recently pointed out that, based on our results, a large positive electroluminescence-detected magnetic resonance (ELDMR) signal is expected to occur when magnetic resonance measurements are

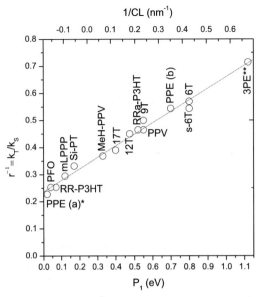

Figure 13 The ratio $r^{-1} = k_T/k_S$ of spin-dependent exciton formation cross sections in various polymers and oligomers as a function of the peak photon energy of the P_1 transition (lower x-axis). r^{-1} is also shown as a function of the inverse conjugation length 1/CL (upper x-axis), which was determined from P_1 (see text for discussion). The line through the data points is a linear fit. * The P_1 band of this polymer does not show a clear peak in the PA spectrum, the P_1 band extends to the longest wavelengths measured. ** The length of this oligomer was calculated. In addition to the chemical names defined in the text, 3PE stands for the PPE trimer, PPE for poly(phenylene-ethynylene), Si-PT for silicon bridged polythiophene. For details consult original publications.

performed directly in OLEDs. In particular, they calculated that this change in EL intensity is expected to be 10 to 50% for the range of σ_S/σ_T values determined by PADMR. However, the ELDMR signal of small molecular weight materials and polymers such as PPV actually shows a very slight increase [71] or even a decrease. These findings would imply that $\sigma_S \approx \sigma_T$, in stark contrast to our findings. We must therefore conclude that several important unanswered questions remain and that polaron recombination will continue to be an intriguing subject to study.

8.4
Conclusions

In summary, we studied polaron absorption spectra and (spin-dependent) polaron recombination. Revisiting doping induced absorption spectra of oxidized oligomer solutions measured by others, we found that the peak photon energy of the low-energy polaron transition is a universal and sensitive measure of the oligomer length. This observation provides us with a sensitive measure of the conjugation-length in polymer thin films using standard photoinduced absorption spectroscopy. We find that the wavefunction extend of polarons in polymer films may be similar to that in oligomers in some materials, but polarons are very much more extended in other polymer films, presumably depending on quality and purity of the polymer material.

We present strong and direct evidence that polaron recombination is spin-dependent and that singlet excitons form with a larger rate constant than triplet excitons. These claims are justified based on a large body of experimental data measured using photoinduced-absorption-detected and photoluminescence-detected magnetic resonance spectroscopy, respectively. Our findings are discussed in the context of works by other groups.

Acknowledgement

We thank X. M. Jiang and C. Yang for their invaluable contributions in the course of the measurements. I am greatly indebted to my postdoctoral and Ph.D. advisor, Prof. Z. V. Vardeny. This work was supported in part by DOE ER-45490 and NSF DMR-02-02790 [Utah], Carver 8 50152 00 and NSF ECS 04-23911 [Iowa] grants.

References

[1] C. K. Chiang, C. R. Fincher Jr., Y. W. Park, A. J. Heeger, H. Shirakawa, E. J. Louis, S. C. Gau, and A. G. MacDiarmid, "Electrical conductivity in doped polyacetylene," *Phys. Rev. Lett.*, vol. 39, pp. 1098–1101, 1977.

[2] F. Garnier, R. Hajlaoui, A. Yassar, and P. Srivastava, "All-polymer field-effect transistor realized by printing techniques," *Science*, vol. 265, pp. 1684–1686, 1994.

[3] L. Torsi, A. Dodabalapur, L. J. Rothberg, A. W. Fung, and H. E. Katz, "Intrinsic transport properties and performance limits of organic field-effect transistors," *Science*, vol. 272, p. 1462, 1996.

[4] Y. Yang and A. J. Heeger, "A new architecture for polymer transistors," *Nature*, vol. 372, pp. 344–346, 1994.

[5] A. R. Brown, A. Pomp, C. M. Hart, and D. M. Deleeuw, "Logic gates made from polymer transistors and their use in ring oscillators," *Science*, vol. 270, p. 972, 1995.

[6] H. Sirringhaus, N. Tessler, and R. H. Friend, "Integrated optoelectronic devices based on conjugated polymers," *Science*, vol. 280, pp. 1741–1744, 1998.

[7] G. Yu, J. Gao, J. C. Hummelen, F. Wudl, and A. J. Heeger, "Polymer photovoltaic cells: Enhanced efficiencies via a network of internal donor-acceptor heterojunctions," *Science*, vol. 270, p. 1789, 1995.

[8] M. Granström, K. Petritsch, A. C. Arias, A. Lux, M. R. Anderson, and R. H. Friend, "Laminated fabrication of polymeric photovoltaic diodes," *Nature*, vol. 395, pp. 257–260, 1998.

[9] J. H. Burroughes, D. D. C. Bradley, A. R. Brown, R. N. Marks, K. Mackay, R. H. Friend, P. L. Burn, and A. B. Holmes, "Light-emitting-diodes based on conjugated polymers," *Nature*, vol. 347, pp. 539–541, 1990.

[10] P. L. Burn, A. B. Holmes, A. Kraft, D. D. C. Bradley, A. R. Brown, R. H. Friend, and R. W. Gymer, "Chemical tuning of electroluminescent copolymers to improve emission efficiencies and allow patterning," *Nature*, vol. 356, pp. 47–49, 1992.

[11] N. C. Greenham, S. C. Moratti, D. D. C. Bradley, R. H. Friend, and A. B. Holmes, "Efficient light-emitting-

diodes based on polymers with high elec-tron-affinities," *Nature*, vol. 365, pp. 628–630, 1993.

[12] D. Braun and A. J. Heeger, "Visible-light emission from semiconducting polymer diodes," *Appl. Phys. Lett.*, vol. 58, pp. 1982–1984, 1991.

[13] G. Gustafsson, Y. Cao, G. M. Treacy, F. Klavetter, N. Colaneri, and A. J. Heeger, "Flexible light-emitting-diodes made from soluble conducting polymers," *Nature*, vol. 357, pp. 477–479, 1992.

[14] A. J. Heeger, S. Kivelson, J. R. Schrieffer, and W. P. Su, "Solitons in conducting poly-mers," *Rev. Mod. Phys*, vol. 60, pp. 781–850, 1988.

[15] A. Johansson and S. Stafstrom, "Modeling of the dynamics of charge separation in an excited poly(phenylene vinylene)/c-60 sys-tem," *Phys. Rev. B*, vol. 68, p. 035206, 2003.

[16] H. A. Mizes and E. M. Conwell, "Stability of polarons in conducting polymers," *Phys. Rev. Lett.*, vol. 70, pp. 1505–1508, 1993.

[17] J. A. Blackman and M. K. Sabra, "Interchain coupling and optical-absorption in degener-ate and nondegenerate polymers," *Phys. Rev. B*, vol. 47, pp. 15437–15448, 1993.

[18] R. Österbacka, C. P. An, X. M. Jiang, and Z. V. Vardeny, "Two-dimensional electronic excitations in self-assembled conjugated polymer nanocrystals," *Science*, vol. 287, pp. 839–842, 2000.

[19] H. Sirringhaus, P. J. Brown, R. H. Friend, M. M. Nielsen, K. Bechgaard, B. M. W. Langeveld-Voss, A. J. H. Spiering, R. A. J. Janssen, E. W. Meijer, P. Herwig, and D. M. de~Leeuw, "Two-dimensional charge transport in self-organized, high-mo-bility conjugated polymers," *Nature*, vol. 401, pp. 685–688, 1999.

[20] D. K. Campbell, A. R. Bishop, and K. Fesser, "Polarons in quasi-one-dimensional sys-tems," *Phys. Rev. B*, vol. 26, pp. 6862–6874, 1982.

[21] K. Fesser, A. R. Bishop, and D. K. Campbell, "Optical absorption from polarons in a model of polyacetylene," *Phys. Rev. B*, vol. 27, pp. 4804–4825, 1983.

[22] S. Brazosvkii and N. Kirova, "Excitons, polarons, and bipolarons in condcting poly-mers," *JETP Lett.*, vol. 33, p. 4, 1981.

[23] P. A. Lane, X. Wei, and Z. V. Vardeny, "Studies of charged excitations in pi-conju-gated oligomers and polymers by optical modulation," *Phys. Rev. Lett.*, vol. 77, pp. 1544–1547, 1996.

[24] E. Zojer, J. Cornil, G. Leising, and J. L. Brédas, "Theoretical investigation of the geometric and optical properties of neu-tral and charged oligophenylenes," *Phys. Rev. B*, vol. 59, pp. 7957–7967, 1999.

[25] J. Cornil, D. Beljonne, and J. L. Brédas, "Nature of optical-transitions in conjugated oligomers .1. theoretical characterization of neutral and doped oligo(phenylenevinyl-ene)s," *J. Chem. Phys.*, vol. 103, p. 834, 1995.

[26] J. Cornil, D. Beljonne, and J. L. Brédas, "Nature of optical-transitions in conjugated oligomers .1. theoretical characterization of neutral and doped oligothiophenes," *J. Chem. Phys.*, vol. 103, p. 842, 1995.

[27] R. H. Friend, R. W. Gymer, A. B. Holmes, J. H. Burroughes, R. N. Marks, C. Taliani, D. D. C. Bradley, D. A. D. Santos, J. L. Brédas, M. Löglund, and W. R. Sala-neck, "Electroluminescence in conjugated polymers," *Nature*, vol. 397, pp. 121–128, 1999.

[28] M. Pope, H. Kallmann, and P. Magnante *J. Chem. Phys.*, vol. 38, p. 2042, 1963.

[29] W. Helfrich and W. G. Schneider *Phys. Rev. Lett.*, vol. 14, p. 229, 1965.

[30] C. W. Tang and S. A. {Van Slyke, "Organic electroluminescent diodes," *Appl. Phys. Lett.*, vol. 14, pp. 913–915, 1987.

[31] Z. Shuai, D. Beljonne, R. J. Silbey, and J. L. Brédas, "Singlet and triplet exciton for-mation rates in conjugated polymer light-emitting diodes," *Phys. Rev. Lett.*, vol. 84, pp. 131–134, 2000.

[32] A. Köhler, J. S. Wilson, and R. H. Friend, "Fluorescence and phosphorescence in organic materials," *Adv. Mat.*, vol. 14, pp. 701–707, 2002.

[33] M. A. Baldo, D. F. O'Brien, M. E. Thompson, and S. R. Forrest, "Excitonic singlet-triplet ratio in a semiconducting organic thin film," *Phys. Rev. B*, vol. 60, pp. 14422–14428, 1999.

[34] M. A. Baldo, M. E. Thompson, and S. R. Forrest, "High-efficiency fluorescent organic light-emitting devices using a phos-phorescent sensitizer," *Nature*, vol. 403, pp. 750–753, 2000.

[35] C. Adachi, M. A. Baldo, M. E. Thompson, and S. R. Forrest, "Nearly 100% internal phosphorescence efficiency in an organic light-emitting device," *J. Appl. Phys.*, vol. 90, pp. 5048–5051, 2001.

[36] Y. Cao, I. D. Parker, G. YU, C. Zhang, and A. J. Heeger, "Improved quantum efficiency for electroluminescence in semiconducting polymers," *Nature*, vol. 397, pp. 414–417, 1999.

[37] P. K. H. Ho, J. Kim, J. H. Burroughes, H. Becker, S. F. Y. Li, T. M. Brown, F. Cacialli, and R. H. Friend, "Molecular-scale interface engineering for polymer light-emitting diodes," *Nature*, vol. 404, pp. 481–484, 2000.

[38] N. S. Sariciftci, ed., *Primary Photoexcitations in Conjugated Polymers; Molecular Excitons vs. Semiconductor Band Model*. World Scientific, Singapore, 1997.

[39] M. Segal, M. A. Baldo, R. J. Holmes, S. R. Forrest, and Z. G. Soos, "Excitonic singlet-triplet ratios in molecular and polymeric organic materials," *Phys. Rev. B*, vol. 68, p. 075211, 2003.

[40] M. Wohlgenannt, K. Tandon, S. Mazumdar, S. Ramasesha, and Z. V. Vardeny, "Formation cross-sections of singlet and triplet excitons in pi-conjugated polymers," *Nature*, vol. 409, pp. 494–497, 2001.

[41] J. S. Wilson, A. S. Dhoot, A. J. A. B. Seeley, M. S. Khan, A. Köhler, and R. H. Friend, "Spin-dependent exciton formation in pi-conjugated compounds," *Nature*, vol. 413, pp. 828–831, 2001.

[42] M. Wohlgenannt, X. M. Jiang, Z. V. Vardeny, and R. A. J. Janssen, "Conjugation-length dependence of spin-dependent exciton formation rates in pi-conjugated oligomers and polymers," *Phys. Rev. Lett.*, vol. 88, pp. 197401–197404, 2002.

[43] A. S. Dhoot, D. S. Ginger, D. Beljonne, Z. Shuai, and N. C. Greenham, "Triplet formation and decay in conjugated polymer devices," *Chem. Phys. Lett.*, vol. 360, pp. 195–201, 2002.

[44] Z. V. Vardeny and X. Wei, *Handbook of Conducting Polymers II*, ch. 22. Marcel Dekker: New York, 1997.

[45] T. Bally, K. Roth, W. Tang, R. R. Schrock, K. Knoll, and L. Y. Park, "Stable polarons in polyactelyne oligomers: optical spectra of

long polyene radical cations," *J. Am. Chem. Soc.*, vol. 114, pp. 2440–2446, 1992.

[46] R. Österbacka, M. Wohlgenannt, D. Chinn, and Z. V. Vardeny, "Optical studies of triplet excitations in poly(p-phenylene vinylene)," *Phys. Rev. B*, vol. 60, pp. R11253–R11256, 1999.

[47] OP-lengths are based on quaterphenyl [72]: intra-ring bond = 1.4 Å; inter-ring bond = 1.45 Å. OPV-lengths are based on trans-stilbene [73]: intra-ring bond = 1.39 Å; inter-ring single (double) bond is 1.45 Å (1.33 Å). OT-lengths are based on quaterthiophene [74]: intra-ring C-C bond = 1.411 Å; inter-ring bond = 1.45 Å. OTV length is based on thiophene ring and the vinyl bond length.

[48] R. K. Khanna, Y. M. Jiang, B. Srinivas, C. B. Smithhart, and D. L. Wertz, "Electronic-transitions in polarons and bipolarons of poly(p-phenylene) oligomers," *Chem. Mat.*, vol. 5, pp. 1792–1798, 1993.

[49] R. Schenk, H. Gregorius, and K. Müllen, "Absorption-spectra of charged oligo(phenylenevinylene)s – on the detection of polaronic and bipolaronic states," *Adv. Mat.*, vol. 3, pp. 492–493, 1991.

[50] J. Guay, P. Kasai, A. Diaz, R. L. Wu, J. M. Tour, and L. H. Dao, "Chain-length dependence of electrochemical and electronic-properties of neutral and oxidized soluble alpha-alpha-coupled thiophene oligomers," *Chem. Mat.*, vol. 4, pp. 1097–1105, 1992.

[51] J. J. Apperloo, J.-M. Raimundo, P. Frere, J. Roncali, and R. A. J. Janssen, "Redox states and associated interchain processes in thienelynevinylene oligomers," *Chem. Eur. J.*, vol. 6, pp. 1698–1707, 2000.

[52] J. Seixas de Melo, L. M. Silva, L. G. Arnaut, and R. S. Becker, "Singlet and triplet energies of α-oligothiophenes: S spectroscopic, theoretical, and photoacoustic study: extrapolation to polythiophene," *J. Chem. Phys.*, vol. 111, pp. 5427–5433, 1999.

[53] A. P. Monkman, H. D. Burrows, I. Hamblett, S. Navarathnam, M. Svensson, and M. R. Andersson, "The effect of conjugation length on triplet energies, electron delocalization and electron-electron correlation in soluble polythiophenes," *J. Chem. Phys.*, vol. 115, pp. 9046–9049, 2001.

[54] M. Wohlgenannt, X. M. Jiang, and Z. V. Vardeny, "Confined and delocalized

polarons in pi-conjugated oligomers and polymers; a study of the effective conjugation-length," *Phys. Rev. B*, vol. 69, p. R241204, 2004.

[55] M. Wohlgenannt, C. Yang, and Z. V. Vardeny, "Spin-dependent delayed luminescence from nongeminate pairs of polarons in pi-conjugated polymers," *Phys. Rev. B*, vol. 66, p. 241201, 2002.

[56] M. Wohlgenannt and Z. V. Vardeny, "Spin-dependent exciton formation rates in -conjugated materials," *J. Phys.: Condens. Matter*, vol. 15, pp. R83–R107, 2003.

[57] X. Wei, B. C. Hess, Z. V. Vardeny, and F. Wudl, "Studies of photoexcited states in polyacetylene and poly(paraphenyleneviny-lene) by absorption detected magnetic resonance: The case of neutral photoexcitations," *Phys. Rev. Lett.*, vol. 68, pp. 666–669, 1992.

[58] P. A. Lane, X. Wei, and Z. V. Vardeny, "Spin and spectral signatures of polaron pairs in pi -conjugated polymers," *Phys. Rev. B*, vol. 56, pp. 4626–4637, 1997.

[59] J. M. Leng, X. Wei, Z. V. Vardeny, K. C. Khemani, D. Moses, and F. Wudl, "Magneto-optical studies of photoexcitations in c61," *Phys. Rev. B*, vol. 48, pp. 18250–18253, 1993.

[60] X. Wei, Z. V. Vardeny, N. S. Sariciftci, and A. J. Heeger, "Absorption-detected magnetic-resonance studies of photoexcitations in conjugated-polymer/c60 composites," *Phys. Rev. B*, vol. 53, pp. 2187–2190, 1996.

[61] P. A. Lane, X. Wei, and Z. V. Vardeny, "Studies of charged excitations in pi-conjugated oligomers and polymers by optical modulation," *Phys. Rev. Lett.*, vol. 77, pp. 1544–1547, 1996.

[62] X. Wei. PhD thesis, University of Utah, 1996.

[63] S. Depinna, B. C. Cavenett, I. G. Austin, T. M. Searle, M. J. Thompson, J. Allison, and P. G. Lecomber, "Characterization of radiative recombination in amorphous-silicon by optically detected magnetic-resonance .1.," *Phil. Mag. B*, vol. 46, pp. 473–500, 1982.

[64] L. S. Swanson, J. Shinar, and K. Yoshino, "Optically detected magnetic resonance study of polaron and triplet-exciton dynamics in poly(3-hexylthiophene) and poly(3-dodecylthiophene) films and solu-

tions," *Phys. Rev. Lett.*, vol. 65, pp. 1140–1143, 1990.

[65] F. Boulitrop, "Recombination processes in a-si: H. a study by optically detected magnetic resonance," *Phys. Rev. B*, vol. 28, pp. 6192–6208, 1983.

[66] P. A. Lane, L. S. Swanson, Q.-X. Ni, J. Shinar, J. P. Engel, T. J. Barton, and L. Jones, "Dynamics of photoexcited states in c60: An optically detected magnetic resonance, esr, and light-induced esr study," *Phys. Rev. Lett.*, vol. 68, pp. 887–890, 1992.

[67] E. J. W. List, C.-H. Kim, A. K. Naik, U. Scherf, G. Leising, W. Graupner, and J. Shinar, "Interaction of singlet excitons with polarons in wide band-gap organic semiconductors: A quantitative study," *Phys. Rev. B*, vol. 64, pp. 155204–155204/11, 2001.

[68] E. L. Frankevich, "On mechanisms of population of spin substates of polaron pairs." Chem. Phys., in press.

[69] M. Wohlgenannt, W. Graupner, G. Leising, , and Z. V. Vardeny, "Photogeneration and recombination processes of neutral and charged excitations in films of a ladder-type poly(para-phenylene)," *Phys. Rev. B*, vol. 60, pp. 5321–5330, 1999.

[70] O. Epshtein, Y. Eichen, E. Ehrenfreund, M. Wohlgenannt, and Z. V. Vardeny, "Linear and nonlinear photoexcitation dynamics in pi-conjugated polymers," *Phys. Rev. Lett.*, vol. 90, p. 046804, 2003.

[71] N. C. Greenham, J. Shinar, J. Partee, P. A. Lane, O. Amir, F. Lu, and R. H. Friend, "Optically detected magnetic resonance study of efficient two-layer conjugated polymer light-emitting diodes," *Phys. Rev. B*, vol. 53, pp. 13528–13533, 1996.

[72] Y. Delugeard, J. Desuche, and J. L. Baudour, "Structural transition in polyphenyls. ii. the crystal structure of the high-temperature phase in quaterphenyl," *Acta Cryst. B*, vol. 32, pp. 702–705, 1976.

[73] C. J. Finder, M. G. Newton, and N. L. Allinger, "An improved structure of trans-stilbene," *Acta Cryst. B*, vol. 30, pp. 411–415, 1974.

[74] S. Wang, F. Brisse, F. Belanger-Gariepy, A. Donat-Bouillud, and M. Leclerc, "3',4"-didecyl-2,2': 5',2": 5",2'''-quater-thiophene," *Acta Cryst. B*, vol. 54, pp. 553–555, 1998.

9

Phosphorescence as a Probe of Exciton Formation and Energy Transfer in Organic Light Emitting Diodes

M. Baldo and M. Segal

9.1
Introduction

Semiconducting organic materials have found an important application in organic light emitting devices (OLEDs). Brightly emissive across the visible spectrum, OLEDs may be fabricated on glass or flexible plastic substrates, and they are Lambertian sources, producing color independent of viewing angle. Based on these advantages, it is envisaged that OLEDs will provide a platform for the next generation of video displays. But if they are to compete with alternative display technologies, OLEDs must be efficient, converting as many injected charges into photons as possible.

OLEDs are typically constructed of thin films of a molecular or polymeric amorphous organic semiconductor. Electronic states are localized in OLEDs because the energetic and positional disorder in the constituent amorphous organic semiconductors overwhelms relatively weak intermolecular interactions. Thus, when an electron and hole combine within the emissive layer of an OLED, a localized excited state with a binding energy as high as 1eV is formed. This excited state, or 'exciton', possesses a spin. In fluorescent materials, only excitons with total spin $S=0$ may emit light, while in phosphorescent materials, excitons with total spin $S=1$ also luminesce, allowing these typically dark states to be harnessed and studied. [1]

The chemistry and detailed performance of phosphorescent materials in OLEDs has been reviewed in detail elsewhere. [2] This review is concerned with phosphorescent studies of the crucial exciton processes that start with the formation of an exciton from injected charge. An overview of exciton formation and triplet energy transfer in guest-host systems is described in part II. Experimental studies that have employed phosphors to directly study the spin dependence of exciton formation are discussed in part III. In part IV, phosphorescent studies of exciton diffusion and triplet energy transfer in phosphorescent blue OLEDs are described. We summarize the results of phosphorescent studies of triplet excitons in part V.

Physics of Organic Semiconductors. Edited by W. Brütting
Copyright © 2005 WILEY-VCH Verlag GmbH & Co. KGaA, Weinheim
ISBN 3-527-40550-X

9.2
Phosphorescence, the Spin Dependence of Exciton Formation and Triplet Energy Transfer

In an OLED, the conversion of electrical energy into light is mediated by excitons, whose properties are the primary determinant of the overall luminescent efficiency. Excitons may be thought of as two-electron systems: one electron is excited into an unfilled orbital of a given molecule or polymer, while the second remains in a partially filled ground state. The total spin of a two-electron system is either $S = 0$, or $S = 1$. The $S = 0$ spin wavefunction is antisymmetric under particle exchange:

$$\sigma_- = \frac{1}{\sqrt{2}} \{\uparrow (1) \downarrow (2) - \downarrow (1) \uparrow (2)\} \tag{1}$$

where $\uparrow$ and $\downarrow$ represent the possible spin states of each electron. The electrons are signified by (1) and (2), and σ is the spin wavefunction. There are three possible spin wavefunctions with $S = 1$, all symmetric under particle exchange:

$$\sigma_+ = \frac{1}{\sqrt{2}} \{\uparrow (1) \downarrow (2) + \downarrow (1) \uparrow (2)\}$$
$$\sigma_+ = \uparrow (1) \uparrow (2) \tag{2}$$
$$\sigma_+ = \downarrow (1) \downarrow (2)$$

The degeneracy of each state is reflected in its title: the $S = 0$ state is known as a singlet, and the $S = 1$ is a triplet. If the excited state is formed from the combination of two uncorrelated electrons, then in a completely random formation process the relative degeneracies of the singlet and triplet states result in a 1:3 singlet:triplet ratio, *i.e.* the fraction of singlet excitons is $\chi_S = 0.25$.

Exciton spin is therefore crucially important because it determines whether an exciton can efficiently emit light in a fluorescent material. The ground state of most molecules is a singlet state, and because the emission of a photon conserves the symmetry of the spin wavefunction, typically only singlet excited states can efficiently decay to the ground state and emit light. Radiative singlet decay is termed 'fluorescence'. Radiative triplet decay is termed 'phosphorescence'. But the probability of luminescence from triplet states is generally so low that almost all their energy is lost to non-radiative processes. Thus, the excitonic singlet-triplet ratio imposes a fundamental limit, χ_S, on the efficiency of fluorescent organic materials. It follows that $1/\chi_S$ expresses the gain to be achieved if luminescence from triplets can be harnessed as well.

Although radiation from triplet states is rare, the process can be quite efficient in some materials. [2] For example, the decay of the triplet state is partially allowed if the excited singlet and triplet states are mixed such that the triplet gains some singlet character. [2] Typically, singlet-triplet mixing and efficient phosphorescence is achieved in molecules with large spin-orbit coupling due to the presence of heavy metal atoms such as Pt or Ir. [2] To exploit an efficiently phosphorescent material in an OLED, we require the transfer of both singlet and triplet excitons from the charge transport layer (henceforth called the host) to the phosphorescent guest. [2]

Since the initial studies of Tang, *et al.* [3] OLED designers have found it advantageous to separate the functions of charge transport and luminescence within the emissive layer of an OLED. One method for achieving this is to mix a small concentration of a highly luminescent phosphorescent guest into a host material with suitable charge transport abilities. This technique confines excitons onto phosphorescent molecules and therefore has the ancillary advantage of minimizing exciton quenching [4] by other excitons in the emissive material, by charges in the emissive material, and by metallic contacts. Four processes determine the overall efficiency of energy transfer between a host and a guest molecule as shown in Fig. 2: the rates of exciton relaxation on the guest and host, k_G and k_H, respectively; and the forward and reverse triplet transfer rates between guest and host, k_F and k_R, respectively.

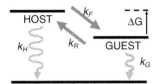

Figure 1 Triplet dynamics in a guest-host system: the rates of forward and back transfer, k_F and k_B, respectively, are determined by the free energy change (ΔG) and the molecular overlap; also significant are the rates of decay from the guest and host triplet states, labeled k_G and k_H, respectively. Adapted from Ref. [24].

Host-to-guest triplet energy transfer is exothermic when $\Delta G < 0$, and endothermic when $\Delta G > 0$. In fluorescent materials, endothermic energy transfer is very inefficient and results in a large population of excitons remaining confined on the host, where they rapidly decay via fluorescent or non-radiative processes. But, as described below, endothermic energy transfer may be successfully employed in phosphorescent devices since the decay of excitons in the host is retarded by spin conservation, i.e. $k_G \gg k_F \gg k_H$.

The triplet energy difference ΔG can be estimated by measuring the relaxed triplet state energies of both the donor and acceptor molecules from the highest-energy transition observed in their phosphorescent spectra, taken at low temperature to minimize non-radiative transitions. [5] Alternately, the triplet energy levels of a molecule can be determined using pulse radiolysis [6] by mediating triplet-triplet energy transfer from a solvent to the molecule with an energy acceptor of a known energy. Triplet populations in the molecule of interest are monitored using photoabsorption, and will be significant only when the energy level of the acceptor is greater than the molecular triplet energy level. [6]

Triplet energy transfer is a hopping process known as Dexter transfer and follows

$$^3D^* + {}^1A \rightarrow {}^1D + {}^3A^*,$$ (3)

where D is the donor and A, the acceptor. Triplet and singlet states are represented by superscript 3 and 1, respectively, and the asterisk signifies an excited state. Singlet energy transfer follows

$$^1D^* + {}^1A \rightarrow {}^1D + {}^1A^*.$$ (4)

Host singlet states may transfer to guest singlet states via Dexter transfer, except when guest-host singlet spectral overlap is strong, in which case long-range dipole-dipole or Förster energy transfer predominates. [7] The triplet state of the donor may also Förster transfer to the singlet state of the acceptor, following:

$$^3D^* + {}^1A \rightarrow {}^1D + {}^1A^*.$$ (5)

This process is known as sensitized fluorescence when it results in the transfer of triplet excitons to the singlet state of a fluorescent dye. It may be very efficient if the donor is phosphorescent. [7,8]

9.3
Phosphorescent Studies of Exciton Formation

Quantifying the fraction of singlet excitons formed in OLEDs is an active area of research. A simple model for interpreting experimental studies of χ_S is shown in Fig. 2. It assumes that the precursors to excitons, charged encounter complexes, relax into singlet and triplet excitons at different rates, k_S and k_T, respectively. In the absence of a process that mixes the spin of the encounter complex, a quarter of excitons form as singlets, regardless of the relative formation rates of singlets and triplets. But if a mixing process is present, then either singlet or triplet formation will dominate, depending on which is fastest.

It has been suggested both that $k_S > k_T$ [9-11] and $k_S < k_T$ [12] Tandon *et al.* [13] have critically reviewed theoretical estimates of k_S and k_T. They model these rates in polymers using parallel interacting chains and argue that electron correlations between delocalized states can account for $\chi_S > 0.25$ in polymers. In contrast, analyses of the magnetic field dependence of photoconductivity in polymers suggest that $k_S < k_T$. [12] But it is important to note that differences in exciton formation rates only influence the ultimate singlet fraction, χ_S, if there is significant mixing of singlet and triplet precursors to excitons. Quantification of this mixing rate must remain a major objective of ongoing studies of exciton formation.

By acting as a probe for the normally dark triplet states, phosphorescence provides an important experimental tool for the determination of χ_S. Here we discuss three techniques used in phosphorescent investigations of the spin dependence of exciton formation:

(i) *Comparison of OLED efficiencies with fluorescent or phosphorescent guest molecules*

Perhaps the first measurement [14] of χ_S in a small molecular weight organic semiconductor used fluorescent and phosphorescent guest molecules to capture sin-

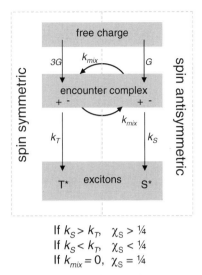

$$\text{If } k_S > k_T, \quad \chi_S > \tfrac{1}{4}$$
$$\text{If } k_S < k_T, \quad \chi_S < \tfrac{1}{4}$$
$$\text{If } k_{mix} = 0, \quad \chi_S = \tfrac{1}{4}$$

Figure 2 A model of exciton formation. The formation rates of triplet and singlet excitons are k_S and k_T, respectively. The mixing rate at room temperature is k_{mix}. G is a constant determined by the formation rate of encounter complexes. In the absence of mixing the singlet fraction $\chi_S = 0.25$ irrespective of the spin-dependent exciton formation rates.

glet and triplet excitons formed in a host film of tris(8-hydroxyquinoline) aluminum (Alq₃). Captured singlets were identified by fluorescence and triplets by phosphorescence. By comparing the ratio of phosphorescence to fluorescence, the singlet-to-triplet ratio was estimated [14] to be (22±3)%; see Fig. 3. Although it is apparently straightforward, this technique has several disadvantages. First, it is necessary to correct for the differing photoluminescent efficiencies of the fluorescent and phosphorescent guest molecules. Secondly, the efficiency of energy transfer to both guest materials must be determined to ensure that radiative emission accurately reflects the number of excitons formed within the device. Thirdly, the impact of quenching phenomena must be calculated for both singlet and triplet excitons. Finally, when applied to polymeric systems, it is especially important to ensure that excitons are formed on the polymeric host and not on the small molecular weight phosphorescent guest.

(ii) *Sensitized fluorescence*

As discussed in Part III, phosphor-sensitized fluorescence is a technique that uses a phosphorescent material to transfer energy from triplet excitons to the singlet state of a guest molecule; see Fig. 4. [8] To determine χ_S using this method two OLEDs must be built, one with and one without the phosphorescent sensitizer. In the absence of the sensitizer, only singlet excitons luminesce, but singlets *and* triplets are detected when the sensitizer is included. In both cases, radiation is emitted from the same fluorescent guest molecule, thereby eliminating errors in comparisons of fluorescence and phosphorescence due to differing radiative efficiencies, or quenching phenomena. [8,15] This technique yielded a singlet fraction of (22±2)%

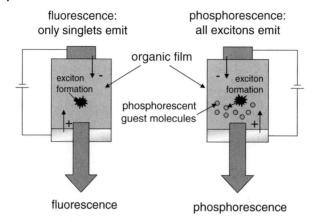

fluorescence:
only singlets emit

phosphorescence:
all excitons emit

$\chi_S \sim$ fluorescence/phosphorescence

Figure 3 Phosphorescent guest molecules can be used to sense triplets. After correcting for photoluminescent efficiency, comparing fluorescence and phosphorescence then gives the singlet fraction. Note that the intersystem crossing rate in phosphorescent molecules is very high, converting nearly all singlet excitons to triplets prior to luminescence. Adapted from Ref. [14].

in the material 4,4′-N,N′-dicarbazole-biphenyl (CBP). The technique may, however, overestimate the singlet fraction if triplets form directly on the fluorescent guest rather than on the host or sensitizer. Thus, the most accurate measurements [15] employ very low concentrations (~0.2%) of the fluorescent dye and high concentrations (~8%) of the phosphorescent sensitizer.

(iii) *Simultaneous measurement of fluorescence and phosphorescence*

Wilson and co-workers [16] proposed that χ_S can be readily measured in materials that simultaneously fluoresce from their singlet states and phosphoresce from their triplet states. If the intersystem crossing rate in the material is large, the singlet fraction is obtained from the ratio of fluorescence to phosphorescence in both electroluminescence (EL), which is luminescence resulting from an electrical pump, and photoluminescence (PL), which is luminescence resulting from an optical pump:

$$\chi_s = \left(\Phi_f / \Phi_p\right)_{EL} \bigg/ \left(\Phi_f / \Phi_p\right)_{PL}. \tag{6}$$

Here the fluorescent and phosphorescent radiation intensity is given by Φ_f and Φ_p, respectively. The singlet exciton formation fraction in a Pt-containing monomer was determined to be $\chi_S = (0.22 \pm 0.01)$. This technique has the advantage that it does not require any absolute measurements of fluorescent or phosphorescent quantum efficiencies. It may be necessary, however, to determine whether quenching phenomena affect triplets and singlets identically in electroluminescence and or photoluminescence. It is also necessary to ensure that significant numbers of photoexcited excitons do not dissociate into charges, as this can lead to erroneously large measured values of χ_S.

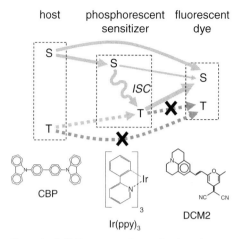

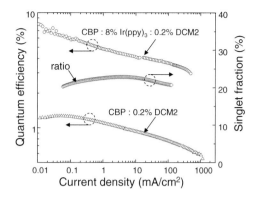

Figure 4 **(left)** Energy transfer mechanisms in phosphor sensitized fluorescence and the materials used to demonstrate the principle. Non-radiative dipole-dipole transfers are represented by solid lines and hopping transfers by dotted lines. The presence of the phosphorescent sensitizer redirects triplets to the singlet state of the fluorescent dye, where they may emit. **(right)** The external quantum efficiencies of the fluorescent dye DCM2 with and without phosphorescent sensitization by tris-(2-phenyl-pyridine) iridium (Ir(ppy)$_3$). The host material is 4,4'-N,N'-dicarbazole-biphenyl (CBP). The ratio between sensitized and unsensitized emission gives $\chi_S = (22\pm2)\%$. From Refs [8,34] and [18].

The leading phosphorescent study of exciton formation in polymers was also reported by Wilson, *et al.* [16] From a simultaneous measurement of fluorescence and phosphorescence in a Pt-containing polymer, these workers measured $\chi_S = (0.57\pm0.04)$, consistent with predictions [9] that the singlet fraction should be higher in polymeric materials. But overall, experimental results concerning exciton formation in polymeric materials remain contradictory. In a study similar to that of Wilson *et al.*, [16] measurements of the EL and PL spectra of polymers containing phosphorescent iridium complexes grafted onto a polyfluorene backbone do not show $\chi_S > 0.25$. [17] Furthermore, a comparison [18] of EL and PL efficiencies in MEH-PPV yielded only $\chi_S = (0.20\pm0.04)$, and photo-absorption detected magnetic resonance (PADMR) studies [10,11] that suggest $\chi_S > 0.25$ in polymers appear to be contradicted by electroluminescent-detected magnetic resonance (ELDMR) studies that show only small resonant changes in EL. [19,20]

Conclusive evidence for $\chi_S > 0.25$ in polymers may come from measurements of external EL quantum efficiencies. At present the external quantum efficiency of fluorescent polymer devices appears to be limited to 4-6%. [21,22] Given the high PL efficiencies of some polymers, it might be expected that measurements of $\chi_S > 25\%$ should be reflected in polymeric EL fluorescent quantum efficiencies that approach the $\approx 20\%$ observed in phosphorescent small molecular weight materials. [18] Thus, pending significant advances in the quantum efficiency of fluorescent polymers, achieving high efficiency electroluminescence will continue to require the harnessing of triplet excitons using phosphorescence. Phosphorescence will also continue to be an attractive experimental probe of χ_S in new materials.

9.4
Energy Transfer in Blue Phosphorescent OLEDs

Perhaps the best example of the importance of triplet energy transfer is the development of highly efficient blue phosphorescent OLEDs, one of the most challenging applications for phosphorescent materials. In most small molecular weight phosphors, exchange interactions typically lower the triplet excited state by approximately 1 eV relative to the singlet state. Thus relative to blue fluorescent devices, blue phosphorescent OLEDs must be capable of generating charge encounter complexes that are higher in energy by approximately 1 eV. [23] This complicates energy transfer in these devices. Here we examine three techniques, endothermic energy transfer, exothermic energy transfer and charge trapping, as they are applied to blue phosphorescent devices.

(i) *Endothermic energy transfer*

Endothermic energy transfer to a phosphorescent guest was demonstrated [24] and quantified [25] in transient studies of the guest:host system of $Ir(ppy)_3$:TPD, where $Ir(ppy)_3$ is *fac*-tris(2-phenyl-pyridine)iridium and TPD is N,N'-diphenyl-N,N'-bis(3-methylphenyl)-[1,1'-biphenyl]-4,4'-diamine. As noted above, triplet excitons are difficult to observe because in most materials, triplet radiative decay rates are much slower than competing non-radiative losses. Transient phosphorescence of molecules with high phosphorescent efficiencies, however, can be used to observe the diffusion of triplets within a host material, or it may give evidence for charge trapping and direct exciton formation on guest molecules within the host. Thus, it provides a convenient tool for examination of energy transfer, either from one host molecule to another, or from the host to the guest. [24,26]

Low temperature phosphorescent spectra give peak triplet energies of (2.3 ± 0.1)eV and (2.4 ± 0.1)eV for TPD and $Ir(ppy)_3$, respectively, [24] and as shown in Fig. 5, the transient response of $Ir(ppy)_3$:TPD exhibits a phosphorescent lifetime that is much longer than that typical of $Ir(ppy)_3$ ($<1\,\mu s$). [24] Since after the initial excitation, the phosphorescent emission is entirely due to $Ir(ppy)_3$, the likely explanation for the delayed rate of phosphorescence from $Ir(ppy)_3$ is slow endothermic energy transfer from TPD. In this system, the energy barrier between host and guest forces excitons to reside primarily in the host. Thus, TPD acts as a reservoir for triplet excitons. [25] But because triplet decay to the singlet ground state in the host TPD is retarded by spin conservation, triplets decay instead by endothermic energy transfer to the guest, $Ir(ppy)_3$. The energy transfer is slow, and as shown in Fig. 5, thermally activated.

The first demonstration of blue phosphorescence also employed endothermic energy transfer from the host N,N'-dicarbazolyl-4-4'-biphenyl (CBP) to the blue phosphorescent guest iridium(III)bis[(4,6-difluorophenyl)-pyridinato-$N,C^{2'}$]picolinate (FIrpic). [27] Low temperature phosphorescent spectra exhibit peak triplet energies of (2.56 ± 0.1)eV and (2.65 ± 0.1)eV for CBP and FIrpic, respectively, suggesting endothermic energy transfer. [27] This is confirmed by the transient response of FIrpic:CBP, which exhibits reduced energy transfer at lower temperatures, consistent with the archetypal endothermic system $Ir(ppy)_3$:TPD.

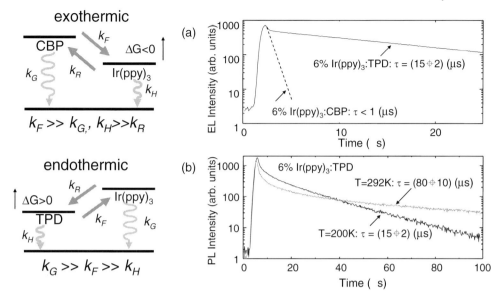

Figure 5 **(a)** The electroluminescent response of the endothermic system Ir(ppy)$_3$:TPD, as compared to the exothermic system Ir(ppy)$_3$:CBP. The lifetime of Ir(ppy)$_3$ in a TPD host is significantly longer (15 μs) than the natural radiative lifetime of Ir(ppy)$_3$ (<1 μs). The initial peak in the response is principally due to fluorescence from TPD. **(b)** The photoluminescent response of 8% Ir(ppy)$_3$ in TPD at T = 92 K and T = 200 K. The lifetime increases at low temperatures, consistent with a thermally activated process such as endothermic energy transfer. However, unlike the EL response, the initial transient in the photoluminescent response is comprised entirely of emission from photo-excited Ir(ppy)$_3$. Adapted from Ref. [24].

(ii) Exothermic energy transfer

Cleave, *et al.* [26] introduced transient phosphorescence as a technique for studying triplet energy transfer. Their experiments conclusively demonstrated exothermic triplet energy transfer from the polymer poly[4-(*N*-4-vinylbenzyloxyethyl, *N*-methylamino)-*N*-(2,5-di-*tert*-butylphenylnaphthalimide)] (PNP) to the phosphorescent guest platinum octaethylporphyrin (PtOEP).

Similar results were subsequently obtained in studies of triplet energy transfer from the small molecular weight host Alq$_3$ to the phosphorescent guest PtOEP. [24] In Fig. 6, the delay in the PtOEP phosphorescence after an electrical excitation pulse is due to triplet diffusion through the Alq$_3$ host. The delay is observed to decrease to zero as the exciton formation zone is moved closer to the PtOEP layer, confirming that excitons are formed initially in the Alq$_3$ host. These experiments were used to estimate the triplet exciton lifetime in Alq$_3$ to be (25±15)μs. [24] In both PtOEP:PNP and PtOEP:Alq$_3$, energy transfer is exothermic from the guest to PtOEP.

Exothermic energy transfer has also been recently demonstrated in blue phosphorescent OLEDs using the phosphorescent guest iridium(III)bis[[(4,6-difluorophenyl)-pyridinato-*N,C*$^{2'}$]picolinate (FIrpic) and the host *N,N'*-dicarbazolyl-3,5-benzene (mCP). [28] Low temperature phosphorescent spectra confirm that the triplet energy

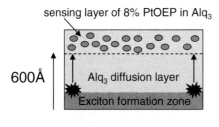

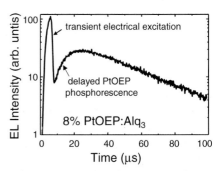

Figure 6 Delayed phosphorescence is a signature of triplet diffusion and subsequent capture by exothermic energy transfer from an Alq_3 host matrix to the phosphorescent guest molecule PtOEP. The transient response was measured at $\lambda = (650\pm10)$ nm. The initial electrical pulse excites some host fluorescence and marks the formation of excitons. The triplet excitons then diffuse through a 600Å-thick Alq_3 diffusion layer before being transferred to PtOEP. Adapted from Ref. [24].

of mCP is higher than that of FIrpic. [28] Exothermic energy transfer is expected to result in higher EL phosphorescent efficiency than endothermic energy transfer. In endothermic transfer, a large population of relatively mobile triplet excitons is expected to develop in the host, increasing losses due to triplet quenching. But in exothermic transfer, triplet excitons are confined on guest molecules and the triplet density and associated quenching losses in the host are small. Consequently, the peak quantum efficiency of FIrpic:mCP is higher, (7.5±0.8)% than that of the comparable endothermic system FIrpic:CBP, (6.1±0.6)%. [28]

(iii) *Charge trapping*

Perhaps the optimum energy transfer scheme for blue phosphorescent OLEDs is direct charge trapping on the phosphorescent guest. This mechanism was demonstrated by Lane, *et al.* [29] in mixtures of the polymeric host poly(9,9-dioctylfluorene) (PFO) and the phosphorescent guest PtOEP. Lane, *et al.* [29] studied triplet populations in PFO OLEDs using photoinduced absorption. No significant difference in the PFO triplet lifetime was observed with the addition of up to 8% PtOEP, demonstrating that triplet energy transfer from PFO to PtOEP is weak. Thus, the observation of significant phosphorescence from PtOEP in PFO must be due to direct exciton formation on the phosphorescent guest.

Eliminating the host material altogether provides the most energetically efficient means for harvesting triplet excitons. [30] Several methods for eliminating the host have been demonstrated. Lupton, *et al.* [30] exploited residual concentrations of a Pd catalyst employed in the synthesis of a diphenyl-substituted ladder-type poly(para-phenylene) PhLPPP to demonstrate enhanced phosphorescence from PhLPPP. [30] Wang, *et al.* [31] employed a fluorinated variant of the archetypal [32] green phosphor $Ir(ppy)_3$, fac-tris[5-fluoro-2-(5-trifluoromethyl-2-pyridinyl)phenyl-C,N]iridium (Ir-2h) as a neat emissive layer in an OLED. Ir-2h exhibits much lower self-quenching than $Ir(ppy)_3$ allowing it to be employed successfully without dilution into a host matrix. The external quantum efficiency of devices employing Ir-2h is approximately 6%. [31]

Recently, Holmes, *et al.* [28] demonstrated charge trapping in blue phosphorescent devices, yielding among the most efficient blue phosphorescent devices to date. The blue phosphor in this work, bis(4′,6′-difluorophenylpyridinato)tetrakis(1-pyrazolyl)borate (FIr6), was mixed into an inert matrix comprised of either diphenyldi (o-tolyl)silane (UGH1) or *p*-bis(triphenylsilyl)benzene (UGH2). The energy gaps of UGH1 and UGH2 are sufficiently large that they do not participate in electron or energy transfer within the device. For 10% FIr6 in UGH2, the peak quantum efficiency is (8.8±0.6)%. [28] Because the phosphor must perform both the roles of charge transport and luminescence, the major disadvantage of this technique is the increase in voltage necessitated by reducing the concentration of the phosphor to avoid self-quenching. Indeed, for devices containing neat films of FIr6, the operating voltage falls by nearly 50%, but the peak external quantum efficiency also falls to (5.0±0.5)%. [28] Although FIr6 shows much promise, synthesizing blue phosphorescent molecules that luminesce efficiently in neat films remains a key goal for the development of high efficiency blue phosphorescent OLEDs.

9.5
Conclusions

Phosphorescence has provided an important probe of exciton physics and has made possible dramatic gains in the electroluminescent efficiency achievable in OLEDs. In small molecular weight materials, the improvement is approximately a factor of four, while in polymers the benefit is debated, but is probably between a factor of two to four. To fully exploit the potential benefit of phosphorescence, energy transfer in the luminescent layer must be carefully controlled to maximize the efficiency of triplet and singlet exciton transfer to the phosphorescent guest molecules, and minimize quenching of triplet excitons prior to luminescence. With a careful choice of materials, phosphorescent studies have shown that it is possible to control the lifetime of triplet excitons from $\tau > 1$s to $\tau < 500$ ns. This unique property of triplet excitons can be exploited using a variety of energy transfer schemes to enhance the performance of electroluminescent devices. For example, sensitized fluorescence reduces triplet lifetimes, minimizing triplet-triplet annihilation, and endothermic triplet transfer exploits long-lived triplets to pump a blue phosphor guest without a blue host material. Thus, the development of efficient phosphorescence has significantly enhanced our ability to observe and control the behavior of excitons in organic electroluminescent devices.

References

[1] Baldo, M. A., O'Brien, D. F., You, Y., Shoustikov, A., Sibley, S., Thompson, M. E. & Forrest, S. R. High efficiency phosphorescent emission from organic electroluminescent devices. *Nature* **395**, 151–154 (1998).

[2] Baldo, M. A., Thompson, M. E. & Forrest, S. R. in *Organic Light Emitting Devices* (ed. Kafafi, Z.) (2004).

[3] Tang, C. W., VanSlyke, S. A. & Chen, C. H. Electroluminescence of doped organic thin films. *Journal of Applied Physics* **65**, 3610–3616 (1989).

[4] Baldo, M. A., Adachi, C. & Forrest, S. R. Transient analysis of organic electrophosphorescence. II. Transient analysis of triplet-triplet annihilation. *Physical Review B* **62**, 10967–10977 (2000).

[5] Sinha, S., Rothe, C., Güntner, R., Scherf, U. & Monkman, A. P. Electrophosphorescence and Delayed Electroluminescence from Pristine Polyfluorene Thin-Film Devices at Low Temperature. *Physical Review Letters* **90**, 127402 (2003).

[6] Monkman, A. P., Burrows, H. D., Hartwell, L. J., Horsburgh, L. E., Hamblett, I. & Navaratnam, S. Triplet Energies of pi-Conjugated Polymers. *Physical Review Letters* **86**, 1358–1361 (2001).

[7] Förster, T. Transfer mechanisms of electronic excitation. *Discussions of the Faraday Society* **27**, 7–17 (1959).

[8] Baldo, M. A., Thompson, M. E. & Forrest, S. R. High efficiency fluorescent organic light-emitting devices using a phosphorescent sensitizer. *Nature* **403**, 750–753 (2000).

[9] Shuai, Z., Beljonne, D., Silbey, R. J. & Bredas, J. L. Singlet and Triplet Exciton Formation Rates in Conjugated Polymer Light-Emitting Diodes. *Physical Review Letters* **84**, 131–134 (2000).

[10] Wohlgenannt, M., Tandon, K., Mazumdar, S., Ramasesha, S. & Vardeny, Z. V. Formation cross-sections of singlet and triplet excitons in pi-conjugated polymers. *Nature* **409**, 494–497 (2001).

[11] Wohlgenannt, M., Jiang, X. M., Vardeny, Z. V. & Janssen, R. A. J. Conjugation-Length Dependence of Spin-Dependent Exciton Formation Rates in pi-Conjugated Oligomers and Polymers. *Physical Review Letters* **88**, 197401 (2002).

[12] Frankevich, E. L., Lymarev, A. A., Sokolik, I., Karasz, F. E., Blumstengel, S., Baughman, R. H. & Hörhold, H. H. Polaron-pair generation in poly(phenylene vinylene). *Physical Review B* **46**, 9320–9324 (1992).

[13] Tandon, K., Ramasesha, S. & Mazumdar, S. Electron correlation effects in electron-hole recombination in organic light-emitting diodes. *Physical Review B* **67**, 045109 (2003).

[14] Baldo, M. A., O'Brien, D. F., Thompson, M. E. & Forrest, S. R. The excitonic singlet-triplet ratio in a semiconducting organic thin film. *Physical Review B* **60**, 14422–14428 (1999).

[15] D'Andrade, B. W., Brooks, J., Adamovich, V., Thompson, M. E. & Forrest, S. R. White light emission using triplet excimers in electrophosphorescent organic light-emitting devices. *Advanced Materials* **14**, 1032–1036 (2002).

[16] Wilson, J. S., Dhoot, A. S., Seeley, A. J. A. B., Khan, M. S., Köhler, A. & Friend, R. H. Spin-dependent exciton formation in pi-conjugated compounds. *Nature* **413**, 828–831 (2001).

[17] Chen, X., Liao, J.-L., Liang, Y., Ahmed, M. O., Tseng, H.-E. & Chen, S.-A. High-Efficiency Red-Light Emission from Polyfluorenes Grafted with Cyclometalated Iridium Complexes and Charge Transport Moiety. *Journal of the American Chemical Society* **125**, 636–637 (2003).

[18] Segal, M., Baldo, M. A., Holmes, R. J., Forrest, S. R. & Soos, Z. G. Excitonic singlet-triplet ratios in molecular and polymeric organic materials. *Physical Review B* **68**, 075211 (2003).

[19] Segal, M. & Baldo, M. A. Reverse bias measurements of the photoluminescent efficiency of semiconducting organic thin films. *Organic Electronics* **4**, 191–197 (2003).

[20] Greenham, N. C., Shinar, J., Partee, J., Lane, P. A., Amir, O., Lu, F. & Friend, R. H. Optically detected magnetic resonance study of efficient two-layer conjugated polymer light-emitting devices. *Physical Review B* **53**, 13528–13533 (1996).

[21] Kim, J.-S., Ho, P. K. H., Greenham, N. C. & Friend, R. H. Electroluminescence emission pattern of organic light-emitting diodes:

Implications for device efficiency calculations. *Journal of Applied Physics* **88**, 1073–1081 (2000).

[22] Friend, R. H. et al. Electroluminescence in conjugated polymers. *Nature* **397**, 121–128 (1999).

[23] Kohler, A. & Beljonne, D. The singlet-triplet exchange energy in conjugated polymers. *Advanced Functional Materials* **14**, 11–18 (2004).

[24] Baldo, M. A. & Forrest, S. R. Transient analysis of organic electrophosphorescence. I. Transient analysis of triplet energy transfer. *Physical Review B* **62**, 10958–10966 (2000).

[25] Kalinowski, J., Stampor, W., Cocchi, M., Virgili, D., Fattori, V. & Marco, P. D. Triplet energy exchange between fluorescent and phosphorescent organic molecules in a solid state matrix. *Chemical Physics* **297**, 39–48 (2004).

[26] Cleave, V., Yahioglu, G., Le Barny, P., Friend, R. & Tessler, N. Harvesting Singlet and Triplet Energy in Polymer LEDs. *Advanced Materials* **11**, 285-288 (1999).

[27] Adachi, C., Kwong, R. C., Djurovich, P., Adamovich, V., Baldo, M. A., Thompson, M. E. & Forrest, S. R. Endothermic energy transfer: A mechanism for generating very efficient high-energy phosphorescent emission in organic materials. *Applied Physics Letters* **79**, 2082–2084 (2001).

[28] Holmes, R. J., Forrest, S. R., Tung, Y.-J., Kwong, R. C., Brown, J. J., Garon, S. & Thompson, M. E. Blue organic electrophosphorescence using exothermic host-guest energy transfer. *Applied Physics Letters* **82**, 2422–2424 (2003).

[29] Lane, P. A., Palilis, L. C., O'Brien, D. F., Giebeler, C., Cadby, A. J., Lidzey, D. G., Campbell, A. J., Blau, W. & Bradley, D. D. C. Origin of electrophosphorescence from a doped polymer light emitting diode. *Physical Review B* **63**, 235206 (2001).

[30] Lupton, J. M., Pogantsch, A., Piok, T., List, E. J. W., Patil, S. & U, S. Intrinsic Room-Temperature Electrophosphorescence from a pi-Conjugated Polymer. *Physical Review Letters* **89**, 167401 (2002).

[31] Wang, Y., Herron, N., Grushin, V. V., LeCloux, D. & Petrov, V. Highly efficient electroluminescent materials based on fluorinated organometallic iridium compounds. *Applied Physics Letters* **79**, 449–451 (2001).

[32] Baldo, M. A., Lamansky, S., Burrows, P. E., Thompson, M. E. & Forrest, S. R. Very high-efficiency green organic light-emitting devices based on electrophosphorescence. *Applied Physics Letters* **75**, 4–6 (1999).

[33] Holmes, R. J., D'Andrade, B. W., Forrest, S. R., Ren, X., Li, J. & Thompson, M. E. Efficient, deep-blue organic electrophosphorescence by guest charge trapping. *Applied Physics Letters* **83**, 3818-3820 (2003).

[34] D'Andrade, B. W., Baldo, M. A., Adachi, C., Brooks, J., Thompson, M. E. & Forrest, S. R. High-efficiency yellow double-doped organic light-emitting devices based on phosphor sensitized fluorescence. *Applied Physics Letters* **79**, 1045–1047 (2001).

Part III
Transport and Devices

10

Electronic Traps in Organic Transport Layers

R. Schmechel and H. von Seggern

10.1
Introduction

Electrical transport is always accompanied by more or less frequent capture of the involved charge carriers in localized states. Such trapped carriers may be released after a specific retention period or may recombine with carriers of opposite charge sign. In case the release rate is higher than the recombination rate, the localized state is called a *trap* while for a dominant recombination rate the localized state forms a *recombination centre*. The same state may act as recombination centre or as trap depending on conditions like temperature or the ratio of the minority to majority carrier concentration. In the following only the term *trap* will be used but it should be kept in mind that the same state may act as recombination centre under certain conditions.

Traps affect strongly the charge transport properties since trapped charge carriers do no longer take part in the charge transport. However, their columbic charge will influence the electric field distribution in a device and therewith the transport. Further, if the release rate for trapped carriers is sufficiently low, there will be a significant time necessary to reach quasi-thermal equilibrium conditions. This causes delay and hysteresis effects in alternately operated devices. For all those reasons, it is very important for basic understanding and technical application to know the origin of such trap states and to have means to control them.

Since trap states are favourite energy sites they are situated in the energy gap of the semiconductor. In terms of classical semiconductor physics each localized state below the conduction band edge, which is able to capture an electron, is an electron trap and each localized state above the valance band edge, which is able to capture a hole is a hole trap. But in organic semiconductors the width of the bands can be very narrow and extended states are rarely observed. The conduction band and the valance band are usually replaced by the lowest unoccupied molecular orbital (LUMO) and the highest occupied molecular orbital (HOMO), respectively. Especially in amorphous layers of organic thin films the density of states (DOS) is quite well represented by a Gaussian-like distribution of localized molecular orbitals of individual molecules as presented in Fig. 1.

Physics of Organic Semiconductors. Edited by W. Brütting
Copyright © 2005 WILEY-VCH Verlag GmbH & Co. KGaA, Weinheim
ISBN 3-527-40550-X

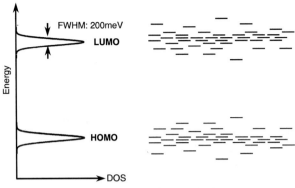

Figure 1 Distribution of HOMO and LUMO levels in amorphous organic semiconductors.

The charge transport in such amorphous layers is mainly determined by hopping processes between strongly localized molecular states. Therefore, it is not obvious how to distinguish between a trap state and a regular transport state. A way out is given by the *transport energy concept* first introduced by Monro [1] for amorphous inorganic semiconductors and later extended to amorphous organic semiconductors by several authors [2–6]. This concept is based on a statistical rule, namely, that a carrier in a deep tail state will most probably escape to a state of energy E_t independent on its initial energy in the tail (Fig. 2). This makes the energy level E_t to a protruding quantity. E_t is called *transport energy* or *escape energy* since it describes the level from which a trapped carrier is most probably released to move to a neighbouring site. The transport energy has only a statistical meaning, but it role is similar to that of the band edge or mobility edge in inorganic semiconductors. Consequently, each state below the transport energy is a trap state while states above the transport energy are regular transport states, despite all states are localized. However, the transport energy is a function of temperature. A state acting as trap state at room temperature may become a transport state at lower temperatures. Besides the trap

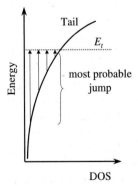

Figure 2 Most probable jump of a carrier from a tail state to the transport energy E_t.

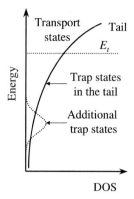

Figure 3 Trap states in the tail, additional trap states and regular transport states.

states formed by the tail of the regular HOMO/LUMO level distribution there may exist additional trap states at a discrete energy level or with any arbitrary energy distribution in the gap below the transport energy (see Fig. 3).

10.2
Origin of trap states

There are several sources for trap states:

1. *Impurities*: Since the interaction between molecules in an organic solid is only weak, a specific molecule keeps its HOMO/LUMO position independent of the surrounding matrix. If the HOMO or LUMO of an incorporated molecule is positioned in the gap of the host molecules it will form a trap state [7, 8]

2. *Structural Defects*: Even if there are only molecules of the same species, the HOMO/LUMO levels may vary from molecule to molecule. The exact energy position of the HOMO/LUMO level is not only determined by the chemical structure of the molecule itself but also by the electronic polarisation of its surrounding. In case of polymers the effective conjugation length also affects the position of the HOMO/LUMO level. Structural imperfections will lead to a fluctuating surrounding and in case of polymers to a fluctuating conjugation length. In consequence a distribution of HOMO/LUMO levels has to be expect as presented in Fig. 1. The few states in the tail of this distribution below the transport energy will form the trap states. But structural defects are not necessarily restricted to the formation of tail states alone. If a specific kind of structural defect occurs with enhanced probability, for example, on grain boundaries in polycrystalline layers, structural defects may result in more or less discrete trap states deep in the gap.

3. *Geminate pairs:* Due to the low dielectric function of organic solids Coulomb interactions are strong in these materials. A hole and an electron will underlie a Coulomb interaction, even if several molecules separate them. Such a pair of Coulomb bond charge carriers is called *geminate pair*. If the recombination probability between the hole and the electron is suppressed by selection rules, a geminate pair forms a Coulomb trap. However, such traps can only occur if both types of charge carriers are present.

4. *Self-trapping:* An excess charge carrier residing on an organic molecule leads usually to a molecular deformation, which causes a lowering in energy for the excess charge carrier. Such a carrier together with its produced molecular deformation is a new quasi-particle called *polaron*. If two charge carriers sharing the same molecular deformation a *bipolaron* is formed. Polarons or bipolarons are not really traps in the original meaning, since they are mobile. However, the mobility of a polaron or bipolaron is at least one or two orders of magnitude lower than the mobility of a "free" carrier. As long as the lowering in energy is only several tenth of meV the effect can be neglected compared to other trap states. But in some polymers like polythiophene or polyacetylene the polaron or bipolaron formation causes a lowering in energy of several hundreds of meV [9, 10]. In such cases the charge carrier forms its own trap state on the polymer chain. Such a trapping mechanism is called *self-trapping*.

10.3
Trap detection techniques

The information on the energy distribution of trap states is important for the disclosure of their origin. However, usually there is no direct access to this information. There are several experimental methods for the detection of trap states, but most of them allowing only for an indirect conclusion on the energy distribution. In most techniques the trap states are first filled, usually at low temperatures, to prevent a fast escape. The filling process can be done by photogeneration of charge carriers, which, however, produces simultaneously both types of charge carriers, or by electrical injection, which allows control over its polarity. In the second experimental step the trapped charge carriers are released in a controlled way. In the case of the *optically stimulated current* (OSC) technique, trapped charge carriers are detrapped by interaction with light and the resulting current is recorded as a function of the wavelength of the light. In principle, such an OSC-spectrum would directly yield the energy distribution of the trap states if an optical transition from a trap state to the transport states is possible. Unfortunately, in many organic semiconductors a direct transition from the trap state to the transport states is not allowed. Usually, the incident light excites the carrier into an excited state of the same molecule, from where a free carrier is generated by autoionisation. Thus, the required optical transition energy is not related to the energy difference between trap and transport state [11]. Only in some rare cases a direct optical transition to the transport states has been reported [12]. To overcome the restriction due to optical selection rules, the release process can be stimulated by thermal energy. This leads to the method of *thermally stimulated currents* (TSC), which is applied by many researchers to obtain information on trap states [11, 13–18]. In this technique, the trapped charge carriers are released by heating up the sample with a linear temperature ramp, while the stimulated current is recorded as function of temperature. This directly yields the required activation energies for the charge transport independent of any selection rules. However, if one wants to relate the obtained activation energies to the energy position of the trap state, one has to take into account that both the thermal release of trapped charge carriers as well as the temperature dependence of the mobility determines the recorded current. Usually, the temperature dependence of the mobility is neglected, which is reasonable if retrapping can be neglected such as in sufficiently thin layers [19, 20]. To overcome the problem of the unknown temperature dependence of the mobility the *thermally stimulated luminescence* (TSL) can be recorded instead. An experiment to detect the TSL is performed in a similar way as the TSC experiment but instead of recording the current due to the released charge carriers, the luminescence due to radiative recombination is recorded. Such luminescence is most probably related to recombination of geminate pairs. Such pairs do not recombine at lower temperatures, since both carriers are trapped on independent but close-by molecules. However, if one of the charge carriers of the geminate pair is released from its trap state it will recombine with its counterpart. Since both carriers are separated only over short distance, an intermediate retrapping is very unlikely and consequently the release process is the rate-limiting step. Due to this attribute, the TSL technique is preferred by several authors [21–26]. However, TSL has also some essential drawbacks: (1) TSL can be only applied on materials, which are luminescent. However, many important transport materials such as pentacene, metal-phtalocyanines and C_{60} are not luminescent. (2) TSL requires both types of charge carriers but it yields information only on the shallower trap, since the participating carrier in the deeper trap is annihilated by the recombination event. Further, it is

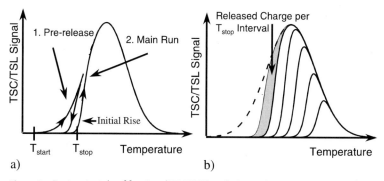

Figure 4 Basic principle of fractional TSC/ TSL technique: a) Basic cycle consisting of pre-release and main run, (b) different main runs for different T_{stop} temperatures of the pre-run; spectra are only schematically.

not possible to distinguish in a direct way, if the released charge carriers are originated from electron or hole traps. This has to be done in subsequent experiments. TSL and TSC are therefore complementary techniques. Each carrier, which recombines and produces a TSL signal, will lower the TSC signal and vice versa each not recombining carrier enhances the TSC signal. Thus, if possible TSL and TSC should be recorded simultaneously.

However, in all experiments based on thermal release of trapped charges it is not possible to focus the release process on a specific trap energy as it would be possible by optical excitation. This becomes a problem, if an energy distribution of the trap states exists. In such a case, TSC and TSL spectra are a complicated convolution of contributions from different traps at different energies. To demerge the information, fractional techniques have to be applied. The basic principle of such techniques is presented in Fig. 4. The main idea is the following: the shallowest occupied traps mainly determine the initial rise of a TSC/TSL signal. If, however, a previously fractional heating process up to a specific temperature T_{stop} already has emptied the shallowest traps, the next deeper lying occupied traps determine the initial rise (Fig. 4a). The whole trap spectrum can be scanned by a stepwise increase of the preheating temperature T_{stop}. (Fig. 4b). The area between different main runs is a measure for the released charge in the respective T_{stop} interval. Finally, each released charge per T_{stop} interval can be related to the activation energy obtained from the initial rise. This yields an image of the density of occupied states. Up to now, the thermally stimulated current and luminescence techniques seem to be the best methods to obtain information on the energy distribution of traps.

However, there are other methods to further obtain information on trap states: Similar to the OSC technique the *photo- induced absorption* (PIA) can be measured [10, 27]. This absorption process is related to an optical transition within a charged molecule. The related absorption spectra contain information on the charged states or can be used to probe the degree of occupation of the trap states [11]. Unfortunately, the absorption energy itself is usually not related to the trap depth. Additional information on traps can be obtained by electroabsorption experiments, which yield knowledge about the internal field strength produced by trapped charge carriers [28], by time-of-flight (TOF) techniques [29] and *I-V* characteristics in the space charge limited current (SCLC) regime [30], which both yield information on the charge carrier mobility, and by impedance spectroscopy [31], which yield information on trap depth and trap energy distributions.

In Chapter 4, some selected TSC/TSL experiments are presented, which reveal essential properties of traps in organic semiconductors. The electronic structure and influences of neutral doping and morphology thereon and on transport properties will be presented for the most prominent small molecule systems such as Alq$_3$, α-NPD and 1-NaphDATA. To be able to understand the basic concepts of neutral doping, model systems will be prepared and investigated. Chapter 5 summarizes first results on trap states in polymer layers.

10.4

Traps in non-doped and doped small molecule semiconductors

10.4.1

Pristine traps and their electronic structure

To be able to judge any changes induced upon doping, pristine organic semiconductors have to be investigated. Therefore 200nm thick single layer structures of the most prominent organic transport materials Alq$_3$, α-NPD and 1-NaphDATA were vacuum-deposited between ITO and Al contacts (for details see [18]). If not stated otherwise, traps were optically filled at 80 K for 5 minutes followed by a thermalization time of another 5 minutes at 80 K without optical illumination before starting the TSC temperature scan. The charging time of 5 minutes is necessary for the

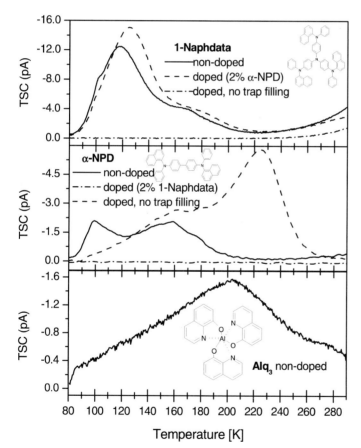

Figure 5 Typical TSC results of single layer devices of the small molecule electron transport material Alq$_3$ and hole transport materials α-NPD and 1-NaphDATA. In case of the hole transport materials TSCs of neutrally doped samples are also displayed as indicated. Results will be discussed in detail in Section 10.4.1 and 10.4.2.

photocurrent to reach a steady state. Subsequently single-scan TSC spectra were taken and are displayed as solid lines in Fig. 5. It can be seen that Alq_3 displays a much broader TSC spectrum than α-NPD and 1-NaphDATA. Such a broad featureless spectrum like in the case of Alq_3 indicates a broad energetic trap distribution whereas α-NPD and 1-NaphDATA exhibit two relatively narrow current peaks indicating two more discrete trap levels. For all TSC spectra it can be noticed that the low temperature TSC peak is truncated at temperatures below 100 K, which is due to trap emptying during the above introduced thermalization time. Further all TSC spectra even if not shown in Fig.5 exhibit an exponentially increasing current at high temperatures. This current is due to injection of charge carriers resulting from a small applied bias during the TSC scan.

In order to determine the density of occupied states, fractional spectra were taken in accordance with the procedure described in Chapter 3. In Fig. 6 the obtained activation energies as a function of the T_{stop} values are displayed indicating the presence of traps at the probed temperature. One realizes the principally different characteristics for Alq_3 and 1-NaphDATA. Alq_3 exhibits a more continuous increase of the activation energies whereas 1-NaphDATA exhibits a more step-like increase with T_{stop}. The latter is also the case for α-NPD, not displayed here. The correlation of the difference of the released charge of the individual fractional scans with the activation energies results in the density of occupied states (DOOS) displayed in Fig. 7. The results support the conclusion concerning the width of the distributions made above from the single TSC scans of Fig. 5 namely a single broad distribution in case of Alq_3 and two narrower distributions in case of 1-NaphDATA. The Gaussian shapes represented in Fig. 7 only describe the general features of the observed trap distributions and not the actually determined distributions as can be seen by comparing the spectra of Fig. 7 with Figs. 9 and 19 displaying examples of actually determined distributions. The integral of the TSC spectra allows one also to calculate a lower limit

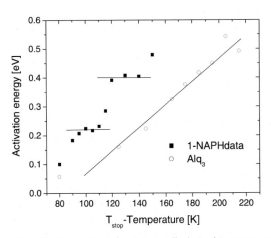

Figure 6 Comparison of experimentally derived activation energies for 1-NaphDATA and Alq_3 as a function of T_{stop} temperature derived from the T_{start}-T_{stop} method.

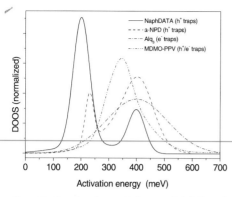

Figure 7 Density of occupied states (DOOS) as determined by a fractional TSC method for the most prominent small molecule transport materials as indicated. All curves were obtained from optical trap filling at 80K under a small bias of –0.3V. The curves are normalized to an equal number of charges.

for the trap density since in TSC measurements not all released charge contributes to the TSC spectra and not all traps are necessarily filled. For more details see reference [18].

The trap distribution of Alq$_3$ displays traps at activation energies ranging from 100meV to 700meV. On the other hand the hole transport materials α-NPD and 1-NaphDATA indicate rather discrete trap structures with narrow energy distributions. It is remarkable that the trap distributions of both hole-transporters are very similar. The first question that arises is, whether these observed trap states are belonging to the tail states of the regular transport states or if they have a different origin (see Fig. 3 in Chapter 1). Tail states as origin for the trap states were favored from temperature-dependent I-V characteristics of Alq$_3$ devices, since they could be well fitted under the assumption of an exponential decreasing trap distribution [30]. However, a more recent investigation by Arkhipov et al. [32] on TSC modeling assumed a Gaussian trap distribution with similar shapes as those determined in the present study. Even more recently, Steiger et al. [18] proved that for principal reasons a distinction between an exponential and a Gaussian DOS shape is not possible from I-V-characteristics in the investigated temperature regime. Therefore the question whether the observed trap distributions are just occupied tail states of the generally assumed Gaussian distribution of the HOMO and LUMO levels in disordered systems or traps from different origin has to be answered by taking a closer look at the TSC spectra in Fig. 5. From the generally assumed Gaussian distribution with a total width of about 200 meV [33] and amplitude of 10^{21} J^{-1}cm^{-3} one would expect a single TSC peak rising right after the filling temperature of T_{Start}. In the case of 1-NaphDATA the TSC peaks are composed of two isolated peaks, one at 115K, the other at 170K. In the case of α-NPD also two peaks are observed, whereby the first peak is truncated and its maximum is located below 100K as additional TSC measurements with trap filling at liquid helium temperature have proven. The second peak is located at 160K. The TSC spectrum of Alq$_3$ exhibits the above-mentioned truncation below 100K but its peak maximum is located at a much higher temperature and therefore resembles a DOOS, which cannot be appointed to the DOS of the HOMOs or LUMOs. Hence, all observed density of state distributions can be considered as independent from Gaussian tail states and therewith from a random distribution of frontier orbital tail states.

The nature of such isolated TSC peaks is related either to impurity defects or structural defects. As mention before, a significant difference in the trap spectrum of the hole transport materials (α-NPD or 1-NaphDATA) and Alq$_3$ is that the hole transport materials show a more discrete trap spectrum while Alq$_3$ reveals a broad distribution of trap states. An essential difference between the hole transport molecules (α-NPD and 1-NaphDATA) and Alq$_3$ molecules is the anisotropic molecular structure of the hole transporters compared to a more globular appearance of Alq$_3$ [34]. This may lead to preferred orientations during layer growth allowing specific orbital overlap to dominate. Recent investigations indicate that the second peak in the α-NPD DOOS is related to structural ordering effects as can be seen below (see Fig. 20). In contrast, the more globular appearance of Alq$_3$ may allow a variety of energetically and structurally different neighborhoods leading to the observed broad trap distribution. This explanation favors the molecular structure and, consequently, structural defects as origin for trap states. However, different batches of Alq$_3$ from different suppliers yield different TSC peaks [35] (Fig. 8). This, on the other hand, can be interpreted by the presence of specific impurities, however, different Alq$_3$ conformations, reported recently, could also sign responsible for such behavior [36]. Unfortunately, chemical analysis performed on these substances did not allow to appoint chemical impurities as the nature of such different traps.

To investigate the effect of impurities and structural order on the electronic structure, model experiments were performed to distinguish between different sources

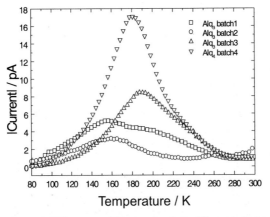

Figure 8 TSC spectra of Alq$_3$ batches from different industrial suppliers obtained for identical deposition and TSC conditions. These results were obtained within confidential industrial cooperations and therefore the suppliers are not mentioned explicitly ($V_{OL} = -2$ V for 5 min; $V_{TSC} = -2$ V; $\beta = 10$ K/min).

of trap origin. Therefore doping experiment were carried out with purposely incorporated impurities chosen such that the energetic position of the molecular frontier orbitals compared to the involved "transport states" of the matrix appears energetically favorable. In order to explore traps related to structural defects, samples were treated at different deposition rates and different substrate temperatures in order to allow for local ordering processes, which should lead to differently positioned LUMO and HOMO levels compared to the statistically ordered regimes [33], e.g., partially crystalline regimes in an otherwise amorphous solid.

10.4.2
Effect of doping on electronic traps

The influence of impurities on the TSC spectrum is demonstrated by purposely doping 2% of 1-NaphDATA into an otherwise α-NPD single layer device and 2% of α-NPD into a 1-NaphDATA single layer device. The single layer devices consist of ITO/α-NPD:1-NaphDATA/Al and ITO/1-NaphDATA:α-NPD/Al structures, respectively. The model systems α-NPD and 1-NaphDATA were selected since they exhibit strongly different HOMO positions at 5.2 to 5.5eV and 5.0eV, respectively, determined by UPS and almost equal LUMO positions at about 2.3eV, estimated from the optical gap and the HOMO position. The single layer devices were optically charged and the resulting TSC spectra are shown in Fig. 5. A comparison with the non-doped layers clearly exhibits the appearance of a new TSC peak at 230K in case of the incorporation of 1-NaphDATA into the α-NPD matrix [8, 37] (see Fig. 5b). Only minor changes are observed in the case of α-NPD doped into the 1-NaphDATA matrix (see Fig. 5a). The new TSC peak rises at the expense of the 100K peak of the pristine α-NPD, which is strongly reduced. A fractional T_{Start}-T_{Stop} experiment performed on such a device supports the appearance of a discrete new trap state as shown in Fig. 9.

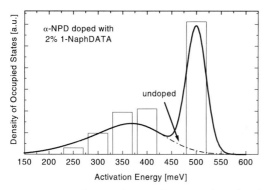

Figure 9 Density of occupied states of an α-NPD layer doped with 2% of 1-NaphDATA. The traps were optically filled at 80 K for 5 min. The bars presents the released charge at specific activation energies (measured as described in chapter 3), the line is a guideline for the eye.

The resulting activation energy is in agreement with published differences of the HOMO levels of the two involved materials. The HOMO of the 1-NaphDATA is reported to be 0.2 eV to 0.5 eV higher than the HOMO level of the α-NPD [38, 39]. Therefore it seems reasonable that the dopant resumes the same HOMO level in the solid state than in its molecular state. One other fact that has to be mentioned is that the low temperature TSC peak of the non-doped α-NPD at 100 K is strongly reduced in the doped sample (see Fig. 5b). The TSC peak at 160 K, on the other hand, seems to be almost unchanged. Such a reduction can generally be understood as a competing process between the dopant induced trap level and the matrix related trap. A high number of deep traps at 500 meV capture carriers released from the matrix states at low temperature with a very high efficiency. Therefore the mean free path of the released charges is very small compared to the mean free path of carriers in the non-doped sample. This leads to a strongly reduced TSC current for the carriers liberated at low temperatures and explains the strongly reduced TSC peak. The reason why the 160 K TSC peak is not influenced is not clear at present. The doping of 1-NaphDATA with 2% of α-NPD, on the other hand, has no drastic effect on the TSC spectrum as can be seen from the TSC spectra of Fig. 5a. There is only a small difference in the position of the low temperature TSC peak but the general feature of the matrix related traps is still preserved. Different from the dopant 1-NaphDATA in α-NPD the dopant α-NPD in 1-NaphDATA acts not as a trap. This can be seen from the resulting HOMO/LUMO level diagram of Fig. 10. In Fig. 10a 1-NaphDATA acts a trap level, which is retarding hole motion through the device. Generalizing the good experimental agreement between the molecularly determined HOMO and LUMO positions and those in the solid state one can propose that a molecular solid can be understood just as a spatial combination of the molecules of the matrix and the dopant without changing their individual HOMO-LUMO levels. This, on the other hand, is plausible from the small interaction of neighboring molecules due the weak van-der-Waals binding. Taking this principle of energetic independence of neighboring molecules with different HOMO-LUMO levels as granted, one can propose that due to the position of the HOMO level of α-NPD with respect to 1-NaphDATA, α-NPD acts more like a scattering center than like a trap. Since holes can easily circumvent such "potential hills" by scattering around the obstacle

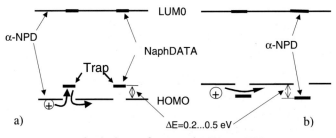

Figure 10 Energy level scheme of (a) 1-NaphDATA in α-NPD acting as a trap state and (b) α-NPD in 1-NaphDATA acting as a scattering center.

one would not expect that such additional scattering centers will strongly change the trap properties of the matrix since scattering is anyhow a dominant process in such disordered systems.

Very similar results were obtained by doping α-NPD with traces of 4,4′,4″-tris-(N-(3-methylphenyl)-N-phenylamino)-triphenylamine (MTDATA) as can be seen in detail in ref. 37. MTDATA exhibits like 1-NaphDATA a HOMO level at 5.0 eV but a LUMO level at 1.85 eV, which is 0.45 eV higher than the respective level of 1-Naph-DATA. Again only the influence of the hole trap on the TSC spectrum is seen whereas the scattering center in the LUMO has no impact on the TSC spectra. The obtained energetic position of the dopant is found at 0.6 eV above the HOMO of the α-NPD matrix again in agreement with the difference in the involved HOMO levels.

Up to now the influences of neutral doping on the electronic structure of hole transport materials were considered. In the following the effect of neutral doping on electron transport material will be investigated. Therefore a material combination of the classical electron conductor Alq_3 with the fluorescent dye Rubrene will be studied as a model system. Rubrene has been reported to be incorporable both in electron and hole conductors [40–44] and to increase the device lifetime considerably [40]. From the electronic structure of Rubrene [41] with its LUMO at 3.15 eV and its HOMO at 5.36 eV and the one of Alq_3 [44] with the HOMO at 3.1 eV and the LUMO at 5.7 eV one would expect from the above said a hole trap of a depth of about 0.35 eV. On the other hand, the observed trap states in pure Alq_3 are related to electrons [18]. The question that has to be answered is whether such an incorporated hole trap has an influence on the occupation of the electron traps. Therefore single layers of Alq_3 were prepared with different amounts of Rubrene, ranging from 0.3% to 5.0%. The resulting TSC and TSL spectra due to optical trap filling at 80 K are displayed in Fig. 11. It should be mentioned that the optical and electrical measurements

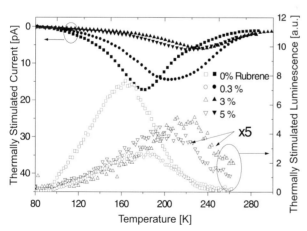

Figure 11 Effect of Rubrene doping on the TSC and TSL spectra of Alq_3. The Rubrene concentration was varied between 0 and 5%. ($V_{OL} = -1$ V, optically charged for 10 min, $V_{TSC} = -1$ V, heating rate 10 K/min).

were performed simultaneously. In case of TSC one observes a pronounced shift in the current maximum from 180K to 235K with increasing dopant concentration. One observes further that the released total charge decreases from 7.7 nC for 0% Rubrene to 3.1 nC in case of 5% doping. For the thermally stimulated luminescence (TSL) one observes a temperature shift of similar strength. However, the integral luminescence intensity drops even stronger than the released charge in the TSC. Since the utilized photomultiplier has almost constant sensitivity up to 800 nm this intensity drop cannot be accounted for by the shift of the emission wavelength to longer wavelength as can be seen from the PL in Fig. 12. Obviously, the doping of Alq_3 with 3–5 % Rubrene generates a new trap level with a peak maximum at 235 K. The 0.3% doped sample shows, however, a superposition of currents released from the Alq_3 traps and from the Rubrene related trapping level, which can be shown by curve fitting techniques.

To understand the effect of Rubrene on the electronic structure of Alq_3, photoluminescence (PL) spectra of the doped materials have been taken (see Fig. 12). One realizes the well-known wavelength shift in the emission spectra from Alq_3 to Rubrene indicating an effective energy transfer or charge transfer from the host to the dopant. One also realizes that in the case of the dopant concentration of 0.3 % a mixed emission spectrum is observed as already deduced from the TSC spectra. A small fraction of Alq_3 emission is still visible, which indicates an incomplete energy or charge transfer to the Rubrene molecule. For higher concentrations the Rubrene emission spectra dominates completely.

The TSC and TSL spectra then become explainable in the following way: Under zero doping the electrons and holes generated during optical trap filling at LN_2 are captured in Alq_3 related trap states with a density of states as displayed in Fig. 7. Due to the built-in potential and the electric potential applied during trap filling, part of the carriers will spatially separate from each other in the sense that there will be a space charge regime of excess holes near the cathode and of excess electrons

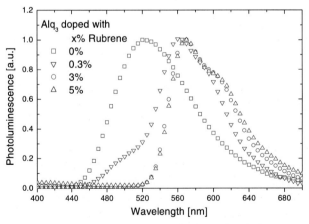

Figure 12 Photoluminescence spectra of Rubrene doped Alq_3 for different Rubrene concentrations as indicated.

near the anode. In addition, electrons and holes will be distributed throughout the volume of the layer compensating each other so that they do not contribute to the excess charge in the sample. Upon heating two basic processes are taking place: the one is a radiative recombination of neighboring electrons and holes and the other an electron motion due to the externally applied electric field or the excess-charge-induced internal electric field during the TSC/ TSL cycle. Both processes are not independent of each other, since the release of electrons is also the rate limiting step for the TSL as can be seen from the similar peak temperature of TSC and TSL. The slight temperature shift between TSC and TSL is caused by the fact, that the TSC signal is also influenced by the temperature dependence of the mobility, while for the TSL the release process is the only rate-limiting step. In summary, the TSL describes the radiative annihilation of the geminate pairs right after their thermal liberation whereas the TSC describes the annihilation of non-compensated excess carriers through charge motion in the sample. However, in pure Alq_3 these currents are caused by the release of trapped electrons only.

For weak doping (0.3%) the TSC results in Fig. 11 indicate the coexistence of Alq_3 (T_{max} = 181K) and Rubrene (T_{max} = 232K) related traps. This is in accordance with the room-temperature photoluminescence results of Fig. 12 that exhibit emission from Alq_3 and Rubrene simultaneously indicating that at room temperature and therefore even more at low temperatures no complete transfer of the optically generated charge carriers to the Rubrene molecules has to be expected. The TSC consists now at lower temperatures of an electron current caused by released electrons from traps in Alq_3 and at higher temperatures of a current of released holes originally trapped on Rubrene molecules. It should be noticed that the integral TSL signal for this weak doping of 0.3% has decreased already by a factor of almost three, which indicates that geminate pairs are much more inefficiently generated if Rubrene exists in the matrix. Obviously, Rubrene acts as strong recombination center, which prevents the formation of geminate pairs. On the other hand, the totally flown charge during TSC has not decreased for the small doping, but the amount of trapped electrons in Alq_3 has decreased.

At higher degrees of doping (3% and 5%) the integral TSL decreases by a factor of almost 7 compared to the non-doped sample. Obviously, the generation of geminate pairs becomes even more unlikely due to the high recombination probability on Rubrene during trap filling. The reason is that each Alq_3 has a very high probability to find one Rubrene molecule in its immediate vicinity. The average distance of two Rubrene molecules for the high doping concentrations is of the order of 1.3 nm, which means that they are separated by 3 to 4 Alq_3 molecules. In such situation energy transfer from Alq_3 molecules to Rubrene molecules is very likely and the generation of geminate pairs becomes less probable. The total charge flown during the TSC experiment drops also, but only by a factor of almost two compared to the above non-doped and weakly doped sample. The much weaker reduction in the total charge observed in the TSC experiments clearly indicates that charge storage of spatially macroscopically separated charges is still possible and not affected that strongly by the high Rubrene concentration. However, the TSC shows now mainly contributions from (excess) holes most probably trapped on Rubrene, while the original TSC contribution from electrons on Alq_3 sites has vanished or at least reduced to some very deep trapped electrons. A similar change of occupation probability of electron traps by incorporation of additional hole traps has also been reported for polymers [45].

10.3.3
Doping Induced Electrical and Optical properties

The question that now has to be addressed is whether these observed traps are actively contributing to charge transport at room temperature. In order to answer this question the effect of the additional trap states on the transport properties of the HTL materials was studied. Therefore three different experiments were performed to determine: (1) the temperature dependence of the I-V characteristics of a single-layer ITO/HTL/Al device, (2) the I-V characteristics of a two layer ITO/HTL/Alq_3/Al OLED and (3) the charge carrier mobility of a single layer device like in (1), all comparing a doped and a non-doped hole transport layer. The first two experiments are steady state investigations, whereas the last one is a dynamic experiment.

The temperature dependent I-V characteristics of a single layer device for doped and non-doped α-NPD at different temperatures are displayed in Fig. 13. One realizes that in case of the non-doped α-NPD the I-V current increases by about one order of magnitude at driving voltages $V_0 < 5$ V when the temperature is raised from 80 K to 160 K. At this temperature the intrinsic traps (see Fig. 5) are almost completely empty. Therefore, a temperature rise to 275 K does not change the current anymore. At higher voltages $V_0 > 5$ V the increase in current with temperature amounts to about one order of magnitude for a temperature change from 80 K to 160 K and three orders of magnitude for a temperature change from 80 K to 275 K. The reason could be seen in an increasing influence of the temperature on the charge carrier injection with higher electric fields.

The doped α-NPD with 1-NaphDATA shows a different behavior. At low temperatures of 80 K and 160 K the diode current is by at least three orders of magnitude smaller compared to the non-doped sample indicating the strong influence of the dopant induced traps (see Fig. 5, peak at 225 K). Only at 275 K the I-V current of the doped layer is comparable in amplitude to the corresponding values of the non-doped sample. This agreement is almost perfect for driving voltages below 5 V but deviates at higher applied voltages in the sense that in the non-doped sample the current supercedes the doped one by two orders of magnitude. Concerning the trap states assigned to the 1-NaphDATA molecules it can be recognized that these traps are not or very slowly emptying at temperatures below 180 K (see Fig. 5). Thus, those trapped charge carriers generate an internal electric field that has an opposite polarity with respect to the external field and therefore results in a reduced effective field leading to the observed small current density below 180 K.

A comparison between the room temperature I-V characteristics of a single layer device containing the MTDATA-doped α-NPD and the non-doped α-NPD (not shown here) reveals that at higher voltages the I-V current of the non-doped single layer device is more than three orders of magnitude higher. The influence of the

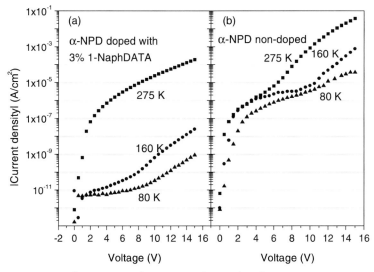

Figure 13 I-V characteristics of (a) 3 % 1-NaphDATA-doped α-NPD and (b) non-doped α-NPD single layer device at different temperatures

dopant-induced traps in this case is even stronger than for the 1-NaphDATA-doped α-NPD device (see Fig. 13). The stronger drop in the current is understandable in respect to the slightly larger activation energy of the MTDATA introduced traps as can be seen in [37].

In order to get a deeper inside into the modifications induced by the neutral traps in the HTL the hole mobility was measured by means of the time-of-flight technique. A representative selection of current responses to a 760ps UV pulse is displayed in Fig. 14. From the indicated transit times a mobility of $\mu_H = (5.56 \pm 0.1) \cdot 10^{-4}$ cm^2 V^{-1}s^{-1} was derived.

The time-of-flight signals of the 2% 1-NaphDATA doped α-NPD sample are displayed in Fig. 14. One realizes that all current transients quickly vanish in the noise indicating a rapid trapping of the generated charge carriers during the first microsecond. The difference to the non-doped α-NPD can be seen in the additional traps introduced by the dopant 1-NaphDATA, which quickly remove any free charge carrier from the transport levels. An extrapolation of the expected mobility in case of the doped sample can be obtained from the mobility of the non-doped sample of Fig. 14 and the reduction of the mobility by a factor of roughly 100, obtained from the doping induced change of the I-V characteristics in Fig. 16e. A mobility of about $\mu_H(\text{doped}) = 5 \cdot 10^{-6}$ cm^{2}V^{-1}s^{-1} can be estimated resulting on the other hand in a transit time of about 400µs for the 10V experiment. However, the relatively large noise level of the TOF-currents in case of the doped α-NPD indicates that a direct detection of such a time would be problematic (Fig. 15). An analysis according to Sher-Montrol in a double logarithmic current-time plot yields unphysical results

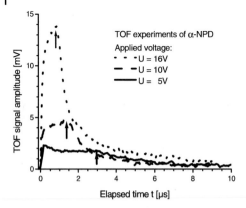

Figure 14 Exemplary time of flight results of a 1μm thick α-NPD sample, exited by a 760ps UV light pulse. The arrows indicate the obvious transit times for the utilized voltages. (obtained in collaboration with J. Pflaum, Univ. of Stuttgart, Germany).

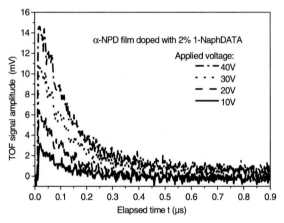

Figure 15 Exemplary time of flight results of a 1μm thick 1-NaphDATA doped α-NPD sample, exited by a 760ps UV light pulse. No indication for possible transit time (obtained in collaboration with J. Pflaum, Univ. of Stuttgart, Germany).

in the sense of an increasing transient time with increasing voltage and was therefore not further pursued.

In order to demonstrate the influence of the additionally introduced traps in the HTL on a complete OLED triple layer devices of [ITO/ pristine α-NPD(5nm)/ doped and non-doped α-NPD (55nm)/Alq$_3$ (60nm)/Al] were investigated whereby a pristine α-NPD injection layer is added to guarantee equal injection properties for the doped and non-doped devices. The room temperature results of the obtained I-V and L-V characteristics are displayed in Fig. 16 and compared to a single layer device of doped and non-doped α-NPD. It can be observed that the I-V and L-V characteristics

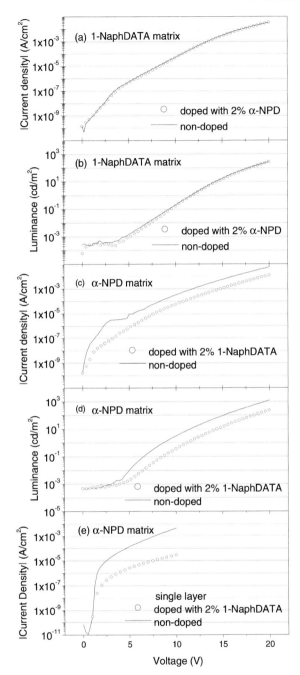

Figure 16   I-V and L-V characteristics of doped (circles) and
non-doped (lines) α-NPD and 1-NaphDATA double layer devices
(a-d) and α-NPD single layer device (e) measured in ambient air
at room temperature.

differ considerably for the doped and non-doped OLEDs. Current as well as luminance of the non-doped device are nearly one order of magnitude higher than those containing the 1-NaphDATA doped α-NPD layer. This indicates that charge transport is strongly hindered by the additionally incorporated trap states characterized by the higher lying HOMO level of 1-NaphDATA compared to that of the matrix. It should be noted that for the respective single layer device the current difference between the doped and the non-doped sample comes out to two orders of magnitude, which is even higher than for the double layer devices. The drop in current of the single layer device (see Fig. 16e) would mean by otherwise unchanged I-V characteristics that the mobility has dropped by two orders of magnitude due to doping to a value of $\mu_{hole}(doped) = 5 \cdot 10^{-6}$ cm^{2}V^{-1}s^{-1}. The reason for the difference in the current characteristics of the single and double layer devices is that in the case of the OLED with a non-doped hole transport layer the transport is determined by the electron transport layer due to the smaller electron mobility. In the case of the OLED with the doped hole transport layer the dopant reduced hole mobility is now determining the device performance. From taking the value of the reduced hole mobility one obtains for the electron mobility in Alq$_3$ a value of $\mu_{Alq} = 10 \cdot \mu_{hole}(doped) = 5 \cdot 10^{-5}$ cm^{2}V^{-1}s^{-1}, which is in agreement with published data [46,47]. This demonstrates the self-consistency of the results.

In case of the 1-NaphDATA as the hole transport matrix the the I-V as well as L-V characteristics are almost identical for the doped and the non-doped samples. This is in agreement with TSC spectra, which indicate minor changes to the DOOS as can be seen from Fig. 5a.

In order to judge the performance changes of both OLED devices of Fig. 16 the luminance vs. current densities are plotted in Fig.17 for both sets of samples, namely the 1-NaphDATA doped into α-NPD and the α-NPD doped into 1-Naph-DATA. In case of the α-NPD:1-NaphDATA a reduction of the diode current of one order of magnitude is obtained. The luminance-voltage (L-V) characteristics, however, also drops by one order of magnitude. The resulting quantum efficiencies, which are the slopes of the upper curves in Fig. 17 are therefore not changing. In case of α-NPD doped into 1- NaphDATA, no significant change is observed neither in the I-V nor in the L-V characteristics. The slope of the luminance vs. current density representing the efficiency in Cd/A is also not changing for the doped or non-doped HTL in the OLED. This observation indicates that the presence of additional trap states in the HTL influences the hole mobility but not the recombination probability in the device. The reason is that the doping does not introduce any additional charge annihilation channels in the HTL. In the case of the 1-NaphDATA matrix no changes are observed in the I-V and L-V characteristics due to doping also indicating no loss in quantum efficiency. The resulting efficiencies of the devices are 2.0 Cd/A in case of α-NPD:1-NaphDATA and 0.7 Cd/A in case of α-NPD doped into 1- Naph-DATA.

Coming back to the question raised at the beginning of this section concerning the influence of the additionally induced neutral traps on the I-V and L-V characteristics of multilayer OLED devices one can state that the influence of the purposely induced traps is definitely decreasing the overall performance of the device in terms

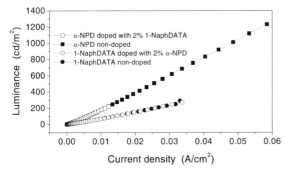

Figure 17 Luminance vs. current density for 1-NaphDATA doped in α-NPD and α-NPD doped in 1-NaphDATA.

of lower light output. However, the quantum efficiency stays the same. For practical considerations, the last observation allows for an optimisation of electron and hole currents in an OLED, which may lead on the other hand to longer lifetimes.

10.3.4
Structure related traps and their influence on device performance

The above results are by definition related to dopants purposely introduced into the transport materials. So they can be considered as model systems for an impurity related trap. The question that still remains is whether the determined DOSs of the pristine materials, as can be seen in Fig. 7, are also related to impurities or whether structural order is responsible for the observed density of states. Since the most probable parameters, which could influence any local molecular order in the transport layers are the deposition rate and substrate temperature, TSC and fractional TSC experiments of differently prepared samples were performed. To test the hypothesis of different deposition rates three $Al/Alq_3/ITO$ where fabricated with strongly different deposition rates of 0.2 Å/s, 2 Å/s and 10 Å/s [48]. The corresponding TSC signals are displayed in Fig. 18. One observes small but characteristic variations in the in the TSC current shapes. All spectra exhibit a maximum between 180 K and 190 K. The slight differences in the peak values are most probably related to differences in the thickness resulting in slightly different bias fields and thereby in different TSC currents. The samples at the high evaporation rates of 2 Å/s and 10 Å/s exhibit in addition a small but clearly visible shoulder in the TSC current around 125 K which is not visible for the sample prepared with the evaporation rate of 0.2 Å/s. This latter sample, however, exhibits contrary to the other samples a strong high-temperature shoulder at temperatures greater than 230 K. A first interpretation would assign a common trap to all samples related to the currents peaking around 180 K to 190 K; however, the samples prepared at the higher evaporation rates show in addition traps with smaller activation energies (125 K). Traps with larger activation energy are found only in case of the smallest evaporation rate. This implies that for the smallest evaporation rate even at room temperature a large

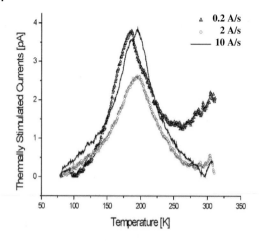

Figure 18 TSC spectra of Al/Alq$_3$/ITO samples fabricated utilizing different evaporation rates of 0.2 Å/s, 2 Å/s and 10 Å/s. Trap filling was performed optically at 80K at a wavelength of 400 nm.

amount of charge is trapped in the sample. In order to support this interpretation of trap assignment, fractional TSC measurements (T_{Start}–T_{Stop}) were preformed. The results are displayed in Fig. 19. In all cases the trap distributions are rather broad. It can be seen that the trap states are located at larger activation energies for smaller evaporation rates and become shallower at larger evaporation rates. For an evaporation rate of 0.2 Å/s the maximum of the distribution is located at about 500 meV, for 2 Å/s at about 250 meV and for 10 Å/s at about 200 meV. These results indicate that the obtained trap structure is strongly influenced by the preparation technique and in the present case by the evaporation rate.

A possible origin for such different trap contributions can be seen in a different molecular structure or packing induced by the different evaporation rates. For larger evaporation rates one would expect a more amorphous structure, and for smaller evaporation rates a more ordered one. In a first glance this interpretation would lead to the conclusion that the more ordered structures should exhibit shallower traps, which is contrary to the experimental observation. Taking, however, a closer look at the origin of the electronic levels the experimental observation may become understandable. It is widely assumed that the distributions of the HOMO and LUMO levels in organic small molecules are caused by an induced electronic polarization of those molecules surrounding the molecule carrying the excess charge. This polarization leads to a strong local binding of the charge and an energetic relaxation of up to 1.5 eV [49]. However, this energy relaxation is strongly dependent on the local surrounding of each transport molecule leading in case of a statistically ordered neighborhood to the well-accepted and frequently used Gaussian HOMO and LUMO distributions as already mentioned in chapter 1. Presumed that the different evaporation rates have an affect on the molecular order of the sublimed film, one would expect for the large evaporation rates a more amorphous molecular surrounding

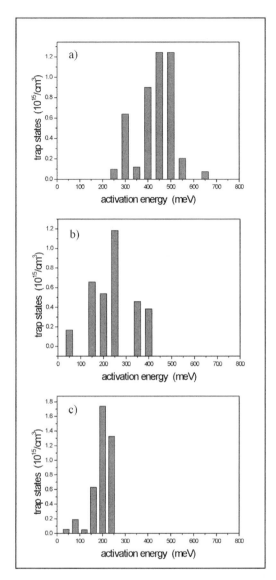

Figure 19 Trap distributions derived from T_{Start}-T_{Stop} spectra:
(a) evaporation rate: 0,2 Å/s, (b) 2 Å/s and (c) 10 Å/s.

and in the case of the small evaporation rates a more crystalline one. Since no complete ordering of the whole organic layer as one single-crystal can be expected, a mixed state of partially amorphous and partially crystalline phases is more likely. The presence of the more crystalline phase would thereby mean the introduction of deeper traps due to larger exchange forces and hence smaller HOMO-LUMO distances in the higher ordered phases.

The energetically broad distribution in Alq$_3$ indicates that there is no specific structural defect responsible for the trap distribution. At high evaporation rates the film seems to be more amorphous, which means that the relative arrangement of the molecules to each other is irregular and therefore their distribution of states is more Gaussian-like distributed. In this case shallow traps should be formed. At smaller evaporation rates the system has more time to form structural order and microcrystalline areas or areas with a specifically extended short-range order may be formed. Such areas can be made responsible to cause deeper trap states. Unfortunately, an attempt to confirm the different degree of order by x-ray investigations failed, which can be attributed to the very low absolute concentration of such ordered areas.

The different trap depths also resulted in different I-V and L-V-characteristics of the investigated Alq$_3$ devices [48], not displayed here. It was observed that the electroluminescence as well as the diode efficiency with voltage is much higher in the case of the samples evaporated at the highest rate of 10 Å/s. This can again be explained in the framework of the above picture by assuming that the evaporation-rate induced deeper traps are more likely to be occupied at room temperature and thereby form an internal counter field opposing the applied electric field strength. Whether these traps are all purely electron traps as was found in a previous study [50] is not clear at present.

In order to examine whether pristine trap levels of hole transport materials also originate from structural defects, α-NPD was investigated in a single-layer device deposited at different substrate temperatures. The TSC results for substrate temperatures of 25 °C and 50 °C and optical trap filling are displayed in Fig. 20. The spectrum of the device deposited on the heated substrate (50 °C) differs strongly from that of the device fabricated on a substrate kept at room temperature during evaporation. The sample deposited on the non-heated substrate exhibits two TSC peaks,

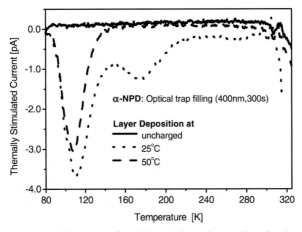

Figure 20 TSC spectra of α-NPD single layer devices, the substrate of one sample was heated during evaporation as indicated. Traps were filled by optical exposure under a small electrical bias of –0.3V.

one at 120 K and the other at 170 K. In the case of the heated substrate, however, the TSC peak at 170 K disappears completely. The relative narrow shape of the low temperature TSC peak confirms again a narrower energetic trap distribution for α-NPD than observed for Alq_3 (see Fig. 7).

The TSC peak at 170 K is, however, broader and its energetic distribution is published in ref. [51]. The strong difference of the shape of the low temperature TSC peak of α-NPD compared to the Alq_3 peak can be related to the different molecular symmetry of the molecules: Alq_3 is a spherical molecule whereas α-NPD is rather anisotropic and extended in one dimension. This suggests that in a film of α-NPD more structural ordering can be expected due to a certain affinity between the molecules. This could result in specific structural defects, which, on the other hand, could be responsible for the occurrence of specific trap states. Therefore in a sample deposited on a hot substrate only one ordered phase or preferred symmetry exists, whereas the samples deposited on a room temperature substrate exhibit an additional TSC peak, which indicates that a different molecular arrangement exists. The experimental observation that the 170 K TSC peak is energetically broader than the low temperature trap may point to a partially crystalline situation as in the case of Alq_3.

10.5
Traps in polymeric semiconductors

Traps in polymers were investigated in two different materials: DMO-PPV also known as MDMO-PPV or OC_1C_{10}-PPV of chemical composition poly(2-methoxy-5-(3′,7′-dimethyloctyloxy)-1,4-phenylenevinylene and a blockpolymer DMO-PPV (BP) chemically composed of copoly [(2-methoxy-5-(3′,7′-dimethyloctyloxy)-1,4-phenylene-vinylene]-[(4′-(3″,7″-dimethyloctyl-oxy)-phenyl-5)-1,4-phenylenevinylene]. The polymers are fabricated as ITO/polymer/Al devices. Traps were optically filled at 80 K. The resulting TSC spectra are illustrated in Fig. 21. Both spectra exhibit one dominant TSC peak located at 173 K in case of P1 and at a slightly higher temperature (178 K) in the case of the blockpolymer. The current truncation at 95 K marked by an arrow can be attributed again to a thermal release of trapped charge carriers during the thermalization period between the optical filling and the start of the TSC scan. This observation strongly suggests that the polymers possess further traps below 80 K most likely being related to smaller activation energies. Since the observed TSC peak at 175 K is clearly isolated from those shallower traps it cannot be considered just as tail states of the HOMO or LUMO level distribution. The TSC spectrum of DMO-PPV exhibits an additional weak TSC peak, which can be realized as a small shoulder peaking at about 220 K. Such a current peak is not visible in the TSC spectrum of the blockpolymer. For even higher temperatures the currents of both samples first decrease to a minimum followed by an exponential increase with an activation energy of about 270meV. This increase can be understood by a thermally induced charge carrier injection from one of the electrodes driven by the applied bias voltage of –0.3V during the TSC sweep.

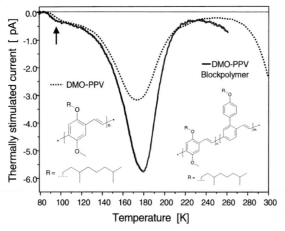

Figure 21 Conventional TSC spectra of the investigated polymers DMO-PPV and DMO-PPV (BP) in an ITO/polymer/Al device.

Apart from small differences, the two TSC spectra have a quite similar appearance and the broad peaks indicate a rather wide trap distribution. In order to support this assumption the energetic trap level distribution was determined by the above-introduced fractional T_{start}–T_{stop} method. The resulting density of occupied states (DOOS) for both polymers is shown in Fig. 22. Both diagrams indicate a broad trap distribution with trap depth between 200 and 600 meV. However, some features are different for both polymers. In contrast to the blockpolymer the distribution of DMO-PPV seems to consist of two density maxima at 350 and 450 meV, respectively. The additional deep trap in DMO-PPV is in agreement with the appearance of the second TSC peak at 220 K in the conventional TSC spectrum of Fig. 21. The shift of the main trap density from 350 meV in DMO-PPV to 400 meV in the blockpolymer could be related to the chemical and electronic difference of the two polymers, which probably leads to less ordered polymer chains in films made of the blockpolymer. Townsend and Friend found activation energy for charge carrier transport in polyacetylene in the range of 300 meV and interpreted their results in terms of a hopping mechanism of charged-soliton pairs [10]. Applying their viewpoint a difference in the TSC peak maximum would represent different binding energies for polarons in the two polymers.

After the confirmation of a broad trap distribution the question still has to be answered whether those described traps after optical filling are mainly electron traps or hole traps. Therefore, TSC measurements with electrical trap filling were performed for both polymers. For a positive bias at the ITO electrode of the ITO/polymer/Al device holes are quite easily injected in the polymer due to a small injection barrier of about 0.2 eV between the Fermi-level of ITO and the HOMO level of the organic semiconductor. In contrast, the electron injection barrier between aluminum and the LUMO level of the polymer is in the range of 1.5 eV. Therefore, at least for lower voltages, currents through the device are only transported by holes and only hole

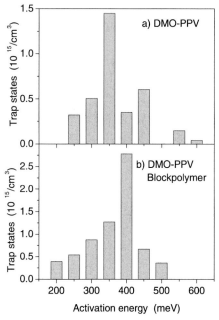

Figure 22 Density of occupied states of DMO-PPV (a) and the DMO-PPV blockpolymer (b).

traps will be filled. In the opposite case where only electron traps are present the injected holes cannot be trapped in the polymer and consequently no TSC signal can be measured.

In Fig. 23 TSC spectra of the electrically trap filled DMO-PPV polymer are depicted. The results are compared to the optically filled polymer added as a line graph. For each applied voltage the loading current I_L through the device is given in brackets and for reference the data for the optical loading are also shown. The current peaks below 110 K are always directed opposite to the loading field, and are identified as a polarization effect of molecular segments of the polymer. They will not be discussed in this paper. Considering the loading currents one realizes that the optical one with I = −2.5pA exhibits by far the smallest loading current of all measurements. When the ITO/DMO-PPV/Al device is loaded with +12 V for 10 minutes at a temperature of 80 K, the resulting TSC sweep delivers a TSC signal similar to the TSC spectrum measured for optical charging. Although the charging mode for a loading bias of +12 V suggests strongly the presence of hole traps only, the final existence of electron cannot be completely excluded by this experiment. The possibility of electron injection from the Al electrode into the polymer at such high voltages has to be taken into account. Therefore, an "electron-only" device was fabricated where the DMO-PPV polymer is positioned between two calcium electrodes. Since from energy consideration the Ca/polymer interface is expected to represent an Ohmic contact, any current through the device can be considered as a pure electron current. The injection of holes from the anode seems to be rather unlikely due to a potential barrier of 2.1 eV. The TSC result of such an experiment is also displayed in Fig. 23. One recognizes two small peaks at about 160 and 220 K, which are definitely located at different temperatures than the dominant TSC peak for positive applied voltages. This observation clearly indicates the presents of a different set of traps, which due to the most probable electron injection have to be attributed to the existence of electron traps in DMO-PPV. Whether an additional deeper electron trap exists around room temperature cannot be answered from the present experiments. The increasing current for temperatures above 240 K can also be due to injection under the utilized readout bias during TSC. The small shoulder at 220 K observed for the optical trap filling under positive voltage could be a sign of that electron trap superimposed on the TSC peak of the hole traps.

There is one peculiarity that should be mentioned here. It is amazing to notice that for the smaller trap-filling voltages of +6 V no TSC peak is observable especially since the loading currents exceed the optical ones by a factor of 160. This indicates that despite of large electrical loading currents compared to the optical trap filling, the trapping probability of holes in the DMO-PPV polymer is much smaller than in the optical case. Fig. 23 reveals that traps obviously can only be filled at electric field strengths exceeding 5×10^5 V/cm. For smaller field strength the TSC signal is not measurable ($< 10^{-13}$A). One possible explanation for such an effect is the existence of geminate pairs meaning that a trap only exists in the presence of a counter charge in its immediate surrounding. In this interpretation trapping will not occur until both holes and electrons are injected simultaneously into the device only occurring at high field strengths. Another explanation is the existence of percolation pathways through the polymer. In this picture the holes are transported via one-dimensional energetically favourable pathways from one electrode to the other at lower electric field strengths. This means that only a few molecules in such a device are carrying the complete current. This implies that such pathways have only a limited access to

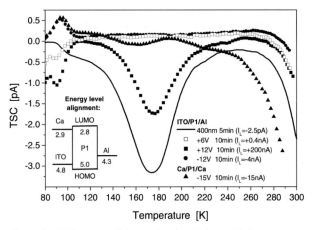

Figure 23 TSC spectra of electrical and optical trap filled DMO-PPV. The trap-filling conditions are indicated in the figure. TSC is always performed under a bias voltage of –0.3 V.

deep traps surrounding such pathways, which could explain the missing TSC-peaks for small voltages. At higher electric field strengths the bending of the transport levels in the polymer can be considered as so strong indication that a global injection may take place from the electrode into the polymer. Thereby the injected charge carriers are able to reach all of the existing trapping sites similar to the optical trap filling. In the latter case the whole volume of the device is excited optically and thereby a large number of traps can be accessed and filled even by a small photocurrent. Which of the two models is finally valid is too early to answer, but the existence of traps of completely different signature for a hole-only device and an electron only device to our understanding favours the percolation model.

10.6
Conclusions

Despite the fact that organic electronics is bearing their first commercial products, the basic understanding of the underlying physical processes such as the charge transport mechanisms is still not finally understood. In the present article the density-of-states of the most prominent organic transport layers has been investigated. Further the origin of traps has been addressed by selecting specific model systems. In this context impurity type defects as well as structural defects are probable candidates for traps in these materials. Finally the influence of the detected traps on the charge transport properties and device characteristics were demonstrated. Future research is still necessary to get an even deeper insight into organic materials and establish a sound basis for the commercial success of organic based products.

Acknowledgement

The authors like to express their gratitude to Norwin von Malm and Jürgen Steiger who performed most of the experiments, to Dr. Jens Pflaum from University of Stuttgart for performing the TOF measurements and extensive discussions thereon and to financial support by the BMBF and DFG (Project No: KA 1207/5).

References

[1] D. Monro, Phys. Rev. Lett. **54**, 146 (1985).

[2] S. D. Baranovskii, T. Faber, F. Hensel, and G. Leising, J. Phys.: Condens. Matter **9**, 2699 (1997).

[3] S. D. Baranovskii, H. Cordes, F. Hensel, and G. Leising, Phys. Rev. B **62**, 7934 (2001).

[4] V. I. Arkipov, E. V. Emelianova, G. J. Adriaenssens, and H. Bassler, J. Non-Cryst. Solids **299–302**, 1047 (2001).

[5] V. I. Arkipov, E. V. Emelianova, and G. J. Adriaenssens, Phys. Rev. B **64**, 125125 (2001).

[6] R. Schmechel, *J. Appl. Phys.* **93**, 4653 (2003).

[7] F. Gutman and L.E. Lyons, *Organic Semiconductors* Wiley, New York, (1967); N. Karl, *Organic Semiconductors* in Festkörperprobleme, Vol. XIV (Vieweg, Braunschweig, 1974).

[8] N. von Malm, R. Schmechel and H. von Seggern, "Trap engineering in organic hole transport materials", *J. Appl. Phys.* **89**, pp. 5559–5563, (2001).

[9] A. J. Heeger, S. Kivelson, J. Schrieffer, and W.-P. Su, Rev. Mod. Phys. **60**, 781 (1988).

[10] P.D. Townsend, R.H. Friend *Phys. Rev. B* **40**, 3112 (1989).

[11] N. Karl: *Getting beyond impurity-limited transport in organic photocundocturs* in *Defect Control in Semiconductors* , Elsevier Science Publishers 1990 p. 1725.

[12] R. Stehle, Diplomarbeit, University Stuttgard (1980).

[13] W. Graupner, G. Leditzky, G. Leising, *Phys. Rev. B.* 54, 7610 (1996).

[14] M. Meier, K. Zuleeg, S. Karg, et al., *J. Appl. Phys.* 84, 87 (1998). (and references therein).

[15] S. Karg, J. Steiger, H. von Seggern, *Synth. Met.* 111–112, 277 (2000).

[16] E. J. W. List, C. H. Kim, J. Shinar, W. Graupner, *Appl. Phys. Lett.* 76, 2083 (2000).

[17] A. G. Werner, J. Blochwitz, M. Pfeiffer, and K. Leo, *J. Appl. Phys.* **90**, 123 (2001).

[18] J. Steiger, R. Schmechel, H. von Seggern, *Synt. Met.* **129**, 1–7 (2002).

[19] J.G. Simmons, M.C. Tam, *Phys. Rev. B* **7** 3706 (1973).

[20] J.G. Simmons, G.W. Taylor, M.C. Tam, *Phys. Rev. B* **7** 3714 (1973).

[21] E.W. Forsythe, D.C. Morton, C.W. Tang, Y. Gao, *Appl. Phys. Lett.* 73, 1457 (1998).

[22] A. Kadashchuk, Yu. Skryshevski, A. Vaknin, E.V. Emelianova, V.I. Arkhipov, H. Bässler, *Phys.Rev.B* 63, 115205 (2001).

[23] A. Kadashchuk, Yu. Skryshevski, Yu. Piryatinski, A. Vakhnin, E. V. Emelianova, V. I. Arkhipov, H. Bässler, J. Shinar, *J. Appl. Phys.*, **91**, 5016 (2002).

[24] V. I. Arkhipov, E. V. Emelianova, A. Kadashchuk, H. Bässler, *Chem. Phys.* 266, 97 (2001).

[25] V. I. Arkhipov, E. V. Emelianova, A. Kadashchuk, I. Blonsky, S. Nešpurek, D. S. Weiss, H. Bässler, *Phys. Rev. B.* **65**, 165218 (2002).

[26] A. Kadashchuk, D. S. Weiss, P. M. Borsenberger, S. Nešpurek, N. Ostapenko, and V. Zaika, *Chem. Phys.*, **247**, 307 (1999).

[27] W. Graupner, G. Leditzky, G. Leising, U. Scherf *Phys. Rev. B* **54** 7610 (1996).

[28] C. Giebeler, S.A. Whitelegg, A.G. Campbell, M. Liess, et al., *Appl. Phys. Lett.* 74, 3714 (1999).

[29] H. Antoniadis, M. A. Abkowitz, B. R. Hsieh, *Appl. Phys. Lett.* 65, 2030 (1994).

[30] S. Berleb, A. G. Mückl, W. Brütting, M. Schwoerer, *Synth. Met.* **111–112**, 341 (2000).

[31] A. J. Campbell, D. D. C. Bradley, D. G. Lidzey, *J. Appl. Phys.* 82, 6326 (1997).

[32] V. I. Arkhipov, E. V. Emelianova, A. Kadashchuk., H.Bässler, "Hopping model of thermally stimulated photoluminescence in disordered organic materials", *Chem. Phys.* **266**, pp. 97–108, 2001.

[33] H. Bässler, "Charge transport in disordered organic photoconductors", *Phys. Stat. Sol. B* **175**, pp.15-56, 1993.

[34] M. D. Halls, C. P. Tripp and H. B. Schlegel, "Structure and infrared (IR) assignments for the OLED material NPB", *Phys. Chem. Chem. Phys.* 3, pp. 2131–2136, 2001.

[35] Juergen Steiger, S. Karg, Heinz von Seggern, "Electronic Traps in OLED Transport Layers: Influence of Doping and Accelerated Aging", Proc. SPIE Vol. 4105, p. 256–264 (2001)

[36] Colle M., Forero-Lenger S., Gmeiner J., Brütting W., *Physical Chemistry Chemical Physics* **5** 2958 (2003).

[37] N. von Malm, R. Schmechel and H. von Seggern, "Distribution of occupied states in doped organic hole transport materials", *Synt. Met.* **126**, pp. 87–95, 2002.

[38] Y. Sato, S. Ichinosawa and H. Kanai, "Operation characteristics and degradation of organic electroluminescent devices", *IEEE J. Sel. Top. Quantum Electronics* **4**, pp. 40–47, 1998.

[39] I. G. Hill and A. Kahn, "Energy level alignment at interfaces of organic semiconductor heterostructures", *J. Appl. Phys.* **84**, pp. 5583–5586, 1998.

[40] H. Fujh, T. Sano, Y. Nishio, Y. Hamada, K. Shibata, "Improved stability of molecular organic EL devices", Macomol. Symp. 125, 77–82 (1997)

[41] T. Sano, Y. Hamada, K. Shibata, "Energy band scheme of highly stable organic electroluminescent devices", *IEEE J. Selected Topics in Quantum Electronics* **4(1)**, 34–39 (1998)

[42] H. Murata, C.D. Merrit, Z.H. Kafafi, "Emission mechanism in Rubrene-doped molecular organic light emitting diodes: direct carrier recombination at luminescent centers", *IEEE J. Selected Topics in Quantum Electronics* **4(1)**, 119–124 (1998)

[43] F. Steuber, "Untersuchung zur Effizienz, Farbabstimmung und Degradation von organischen Lumineszenz-Anzeigen", Dissertation, Technische Fakultät Erlangen-Nürnberg (1999)

[44] Z. Zang, X. Jiang, S. Xu, T. Nagatomo, O. Omoto, "Stability enhancement of organic electroluminescent diode through buffer layer or Rubrene doping in hole transporting layer", *Synth. Met.* **91**, 131–132 (1997)

[45] Colle M., Forero-Lenger S., Gmeiner J., Brutting W., *Physical Chemistry Chemical Physics* **5** 2958 (2003).

[46] Shigeka Naka, Hiroyuki Okada, Hiroyoshi Omnagawa, Yoshihisa Yamaguchi and Tetsuo Tsutsui, *Synth. Met.* **111–112** 331 (2000).

[47] Deng Zhenbo, S. T. Lee, D. P. Webb, Y. C Chan, W. A. Cambling *Synth. Met.* **107–109** 107 (1999).

[48] A. Hepp, N. von Malm, R. Schmechel, H. von Seggern, Synth. Metals 138 201–207 (2003)

[49] H. Bässler, Phys. Stat. Sol. (B) **107**, 9–53 (1981)

[50] J. Steiger; PhD thesis; Darmstadt University of Technology; 2001

[51] N. von Malm, J. Steiger, T. Finnberg, R. Schmechel, H. von Seggern, Proc. SPIE Vol. **4800**, pp. 164–171, (2003).

11

Charge Carrier Density Dependence of the Hole Mobility in poly(p-phenylene vinylene)

C. Tanase, P. W. M. Blom, D. M. de Leeuw, E. J. Meijer

11.1
Introduction

Solution-processable conjugated polymers have attracted attention because of their potential advantages in developing low-cost microelectronic devices such as light-emitting diodes (LEDs) [1,2] and field-effect transistors (FETs) [3,4]. An important factor in understanding and developing the devices based on organic semiconductors is the mechanism of charge carrier transport. It has become clear that the charge transport in these devices is dominated by structural and energetic disorder both in LEDs [5,6] and FETs [7,8].

Poly(p-phenylene vinylene) (PPV) is considered a suitable semiconductor for LEDs, being extensively studied in the last decade due to its high photoluminescence yields and high values of the hole mobility [9]. The charge transport in PPV is described by a theoretical model based on thermally assisted intermolecular hopping of charges in a correlated Gaussian disordered system [6]. The Gaussian density of states (DOS) reflects the energetic spread of the charge transport sites. This energetic spread is the result of the fluctuation in the local conjugation length and structural disorder.

Field-effect transistors have shown great improvements during the last years. Reference 10 gives an overview of the scientific and technological knowledge of organic thin-films transistors. Recently, flexible active-matrix monochrome electrophoretic displays based on solution processed organic transistors on polyimide substrates have been demonstrated [11]. The charge transport in p-type disordered organic FETs is described by variable range hopping of localised charges in an exponential DOS [12]. It has been recently demonstrated that the charge carrier mobility and the charge transport properties are dependent on the charge carrier density [12–14].

The focus in this review is on the charge transport in the conjugated polymer PPV and its derivatives studied as the active layer in LEDs and FETs. Chemical modification of PPV provides a tool to control the backbone and side chains of the polymer and influences both the electronic and morphologic properties. Different side chains attached to the polymer result in different structural configurations and hence the conductive properties of the polymer can be modified. Therefore, the molecular structure of the polymer chains can strongly affect the charge carrier mobil-

Physics of Organic Semiconductors. Edited by W. Brütting
Copyright © 2005 WILEY-VCH Verlag GmbH & Co. KGaA, Weinheim
ISBN 3-527-40550-X

ity. Theoretically, the hole transport in LEDs is described differently than in FETs. We study the dependence of the hole mobility in PPV derivatives with different side chains and correlate the hole mobility obtained from diodes with the hole mobilities obtained from field-effect transistors.

For a single PPV, the hole mobility determined from LEDs shows big differences with the hole mobility determined from FETs, up to 3 orders of magnitude [13]. The large mobility differences obtained from the two types of devices originate from the strong charge carrier dependence of the mobility [13]. The mobility description at low charge carrier density (in a LED) using a Gaussian DOS is correlated to the mobility description at high charge carrier density (in a FET) using an exponential DOS. The only difference in this case is the energy position and energy range over which the Fermi level moves in the DOS, in the operational regime of the two devices. The temperature and charge carrier dependence of the hole mobility are unified in a single charge transport model for disordered PPV derivatives. In more ordered PPVs the charge transport is observed to be anisotropic with respect to the in-plane (FET) and out-of-plane (LED) transport and cannot be unified.

11.2
Experimental results and Discussion

In the present study we focus on three PPV-derivatives, which are used as active layers in LEDs and FETs. The materials are poly(2-methoxy-5-(3′,7′-dimethyloctyloxy)-p-phenylene vinylene) (OC$_1$C$_{10}$-PPV), poly[2,5-bis(3′,7′-dimethyloctyloxy)-p-phenylene vinylene] (OC$_{10}$C$_{10}$-PPV) and a random copolymer of poly(2-methoxy-5-(3′,7′-dimethyloctyloxy)-p-phenylene vinylene) and poly[4′-(3,7-dimethyloctyloxy)-1,1′-biphenylene-2,5-vinylene] (NRS-PPV). The chemical structures corresponding to these PPV-derivatives are presented in Fig. 1. In all devices the polymer has been spin-coated from a toluene solution.

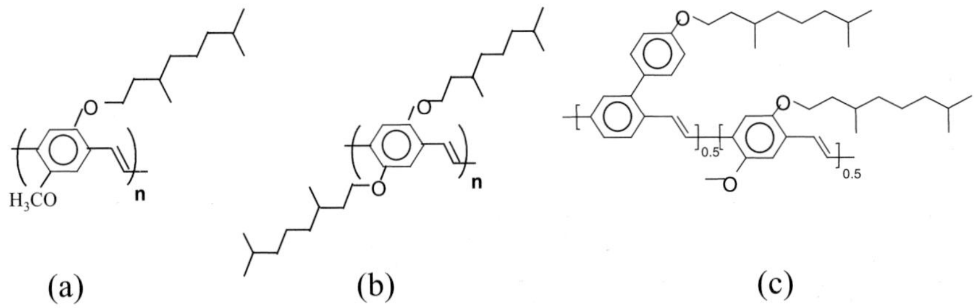

(a)　　　　　　　　(b)　　　　　　　　(c)

Figure 1 Chemical structures of OC$_1$C$_{10}$-PPV (a), OC$_{10}$C$_{10}$-PPV (b) and NRS-PPV (c).

11.2.1

Charge carrier mobility in hole-only diodes

The schematic structure of the hole-only diode used in the experiments is presented in the inset of Fig. 2. Fig. 2 displays a typical example of the temperature dependent current density-voltage characteristics (*J-V*) of a PPV hole only device (NRS-PPV). The current density through PPVs contacted with ITO is space-charge limited (SCL) and shows a strong dependence on both the temperature, *T*, and the applied electric field, *E* [15]. At low bias voltages the hole mobility is constant, whereas at high bias voltages the dependence of the mobility on the voltage has to be taken into account. The increase of μ_{LED} with electric field reflects the lowering of the hopping barriers in the direction of the applied electric field, which exponentially enhances the hopping probability. The charge transport in disordered organic semiconductors proceeds by means of hopping in a Gaussian site-energy ε distribution:

$$DOS_{Gauss} = \frac{N_t}{\sqrt{2\pi}\sigma}\exp\left(-\frac{\varepsilon}{2\sigma^2}\right), \tag{1}$$

where N_t is the number of states per unit volume and σ is the width of the Gaussian [5]. It has been demonstrated that the hole mobility in PPV derivatives is well described by a 3-D transport model based on hopping in a correlated Gaussian disordered model [6]:

$$\mu_{LED} = \mu_0\exp\left[-\left(\frac{3\sigma}{5k_BT}\right)^2 + 0.78\left(\left(\frac{\sigma}{k_BT}\right)^{1.5} - \Gamma\right)\sqrt{\frac{eEa}{\sigma}}\right], \tag{2}$$

where μ_0 is the zero-field mobility in the limit $T\to\infty$, σ the width of the Gaussian density of states (DOS), Γ gives the geometrical disorder, and *a* the average intersite spacing. The lines in Fig. 2 represent the prediction of SCL model with field depen-

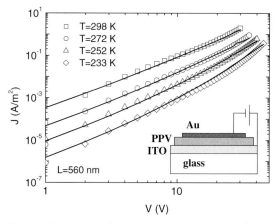

Figure 2 Temperature dependent current-density vs. voltage characteristics of NRS-PPV hole-only diode. The solid lines represent the prediction from the SCL model including the field dependent mobility.

Table 1 Parameters $\mu_{LED}(E=0)$ (zero-field mobility) at room temperature and σ (the width of the Gaussian density of states) for the PPV-derivatives as determined from hole-only diode.

Polymer	$\mu_{LED}(E=0)$ (m^2/Vs)	σ (meV)
OC$_{10}$C$_{10}$-PPV	9.0×10^{-10}	93
OC$_1$C$_{10}$-PPV	5.0×10^{-11}	110
NRS-PPV	1.5×10^{-12}	125

dent mobility given by Eq. [2] using $\Gamma=6$ and a=1.1 nm and are in excellent agreement with the experimental data. Similar study has been done for OC$_1$C$_{10}$-PPV and OC$_{10}$C$_{10}$-PPV [16]. The hole mobility for low electric fields at room temperature for the polymers studied in this paper are presented in Table 1. The zero-field mobility for all three polymers is plotted in Fig. 3 as function of T^{-2}. Modelling the mobility data with $\ln(\mu_{LED}(E=0)) \approx T^{-2}$ the width of the Gaussian DOS is determined and the values of σ for the PPV-derivatives are presented in Table 1. Within the temperature range studied, the zero-field mobility of the samples examined is in good agreement with the predictions of the correlated Gaussian disorder model (solid lines in Fig. 3). From Fig. 3 we find that chemical modification of PPV can result in a change in the zero-field mobility by orders of magnitude. Furthermore, the magnitude of the mobility is correlated to the amount of energetic disorder. For stronger disorder the energy barriers which the charge carriers have to jump become higher, leading to a lower mobility. The lowest value for σ is obtained for OC$_{10}$C$_{10}$-PPV. Compared to OC$_1$C$_{10}$-PPV, which has two asymmetric side chains, OC$_{10}$C$_{10}$-PPV has two long symmetric OC$_{10}$ side chains which cancel any effect of the interaction between side chains on the conformation. From studies of the morphology of these two polymer

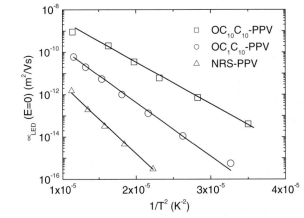

Figure 3 Temperature dependence of the zero-field mobility $\mu_{LED}(E=0)$ for NRS-PPV, OC$_1$C$_{10}$-PPV and OC$_{10}$C$_{10}$-PPV hole-only diode.

films with phase-imaging scanning force microscopy it has been demonstrated that the symmetry of substitution is related to surface morphology and aggregation behaviour [17]. $OC_{10}C_{10}$-PPV shows straight, aligned individuals chains and strong aggregation, while spiralling chains and no aggregation were observed for the asymmetrical OC_1C_{10}-PPV. In comparison with $OC_{10}C_{10}$-PPV, the configurationally freedom of OC_1C_{10}-PPV will result in a larger energetic spread between the electronic levels of individual chain segments and therefore a larger σ for the Gaussian DOS. NRS-PPV is a random copolymer, which means from the aggregation point of view a much stronger structural disorder than the other two polymers and as a consequence it has a larger σ.

11.2.2
Local mobility versus field-effect mobility in FETs

The schematic structure of the p-type field-effect transistor used in the experiments is presented in the inset of Fig. 4. A negative voltage applied at the gate electrode (G) of the p-type FET forces the top of the valence band to bend upwards closer to the Fermi level and induces an accumulation of holes at the semiconductor/insulator interface [18]. A small voltage V_d applied between the source (S) and the drain (D) electrodes gives rise to a source-drain current I_{ds}.

Figs. 4 and 5 display typical examples for the temperature-dependent transfer characteristics of FETs based on NRS-PPV and $OC_{10}C_{10}$-PPV, respectively. The transfer characteristics have been measured in dark and in vacuum, in the linear operating regime of the transistor, by using a drain voltage $V_d = -0.1$ V, which is much smaller than the applied gate voltage ($V_g = -1$ to -20 V). The conductive channel has

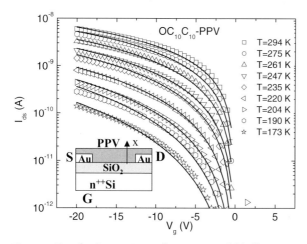

Figure 4 Transfer characteristics of $OC_{10}C_{10}$-PPV field-effect transistor. The solid lines indicate the calculated source-drain currents. The inset shows a schematic view of an organic field-effect transistor.

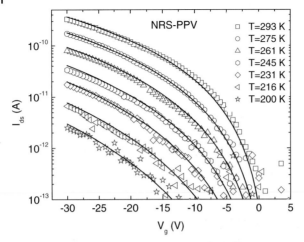

Figure 5 Transfer characteristics of NRS-PPV field-effect transistor. The solid lines indicate the calculated source-drain currents.

a width, W, of 2500 µm and a length, L, of 10 µm. The electric field in the active channel is small, such that any field dependence of the field-effect mobility in the source-drain direction can be ignored. The typical thickness of the polymer spin-coated on top of the gold source and drain contacts is 200 nm. The insulator capacitance per unit area, C_i, is 17 nF/cm². From the transfer characteristics the experimental field-effect mobility is determined using the following equation [19]:

$$\mu_{FET}(V_g) = \frac{\partial I_{ds}}{\partial V_g} \frac{L}{WC_i V_d}. \tag{3}$$

In Fig. 6 the experimental field-effect mobility, μ_{FET}, is presented as determined from Eq. [3] for OC$_1$C$_{10}$-PPV for different temperatures as function of gate voltage, V_g. By using Eq. [3] we assume that all the charge carriers in the accumulation layer have the same mobility.

In disordered organic semiconductors, where the charge carriers are strongly localised, the charge transport is described by a variable range hopping model [12,18], in contrast to conventional monocrystalline silicon. By increasing the gate voltage the induced charge carrier density in the accumulation channel increases, the lower states of the organic semiconductor are filled and any additional charges in the system will need less activation energy for the jumps to neighbouring sites. As a result the charge carrier mobility will be enhanced. In this way the dependence of the mobility on the gate bias is due to the dependence of the mobility on the charge carrier density. For the understanding of the transfer characteristics of organic semiconductors it is important to realise that in a FET the charge carrier density is not uniformly distributed in the accumulation channel in the direction perpendicular to the semiconductor/insulator interface, but depends on the distance from the interface. Therefore, the mobility that is charge carrier dependent also

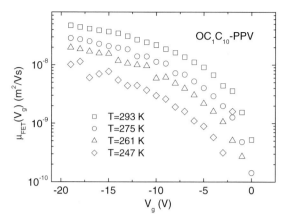

Figure 6 Temperature dependence of the OC$_1$C$_{10}$-PPV filed-effect mobility as a function of gate voltage as determined from Eq. [3] (symbols).

depends on the position in the accumulation layer. The consequence is that for a given V_g a distribution of charge carrier mobilities is present in the organic FET.

The question that arises is how the distribution of the local mobility of the charge carriers in the accumulation channel compares to the field-effect mobility as calculated from Eq. [3] for a certain gate voltage. In order to answer this question the distribution of the charge carrier density in the accumulation channel has to be calculated. An unintentionally doped system is considered. The energy levels, which are responsible for charge transport, are characterized by an exponential density of states:

$$DOS_{expon} = \frac{N_t}{k_B T_0} \exp\left(\frac{\varepsilon}{k_B T_0}\right) \tag{4}$$

where k_B is Boltzmann's constant, T_0 indicates the width of the exponential distribution. The total concentrations of charges in the semiconductor layer are given by the formula:

$$p = \int_\varepsilon DOS_p(\varepsilon) \cdot f_p(\varepsilon)d\varepsilon \tag{5}$$

$$n = \int_\varepsilon DOS_n(\varepsilon) \cdot f_n(\varepsilon)d\varepsilon \tag{6}$$

where p, n represent the density of majority and minority charge carriers, $f_p(\varepsilon), f_n(\varepsilon)$ represent the Fermi-Dirac distribution function for holes and electrons. The position of the Fermi level can be found from the preservation of the electrical charges $n + N_A^- = p$, where N_A^- is the number of ionized acceptors. By applying a negative voltage the equilibrium between the charges in the semiconductor is changed due to the local internal potential or band bending $V(x)$, where the x direction is perpendicular to the semiconductor/insulator interface. In this situation the Fermi level becomes $E_f(x)=E_f-qV(x)$. Combining this with Eqs. [5-6] then provides the relation between the charge concentration (p,n) and the potential (V). From the Poisson equation and the relation between the electric field and the potential in the channel, the electric field distribution in the accumulation channel is:

$$F_x = \left[\left(\frac{2}{\varepsilon_0 \varepsilon_r} \right) \left| \int_0^V e\rho(V')dV' \right| \right]^{1/2}, \tag{7}$$

where V' is the local potential, which varies from zero far away in the semiconductor bulk to V in the accumulation channel, ε_r is the relative dielectric constant of the semiconductor, and ρ is the density of charge carriers. The potential distribution if found from $x = \int_V^{V_0} \frac{dV'}{F_x(V')}$, where V_0 is the surface potential at the semiconductor/insulator interface. The boundary conditions are given by: $F_x(0) = F_x(V=V_0)$, which is the electrical field at the S/I interface and $\rho_{ind} = \varepsilon F_x(V_0)$, which is the induced charge per unis area. The gate voltage is related to ρ_{ind} as follows: $V_g = \rho_{ind}/C_i + V_{fb}$, where V_{fb} is the flat-band voltage and is neglected in these calculations. Now the distribution of the charge carrier density in the accumulation layer can be calculated as function of distance x for every gate voltage [20].

In order to model the experimental transfer characteristics obtained on the three polymer based FETs, we use the variable range hopping model developed by Vissenberg and Matters [12]. Using a percolation model of variable range hopping, the conductivity has been determined as a function of the density of charge carriers and temperature [12]. From the conductivity an expression for the local mobility as a function of charge carrier density can be derived:

$$\mu_{FET}(p) = \frac{\sigma_0}{e} \left[\frac{\left(\frac{T_0}{T} \right)^4 \sin\left(\pi \frac{T}{T_0} \right)}{(2\alpha)^3 B_c} \right]^{T_0/T} p^{\frac{T_0}{T} - 1}, \tag{8}$$

where σ_0 is a prefactor for the conductivity, α^{-1} is the effective overlap parameter between localised states and $B_c \cong 2.8$ is the critical number for the onset of percolation. Taking into account the distribution of the charge carrier density perpendicular to the channel the field-effect current is calculated using the integration over the accumulation channel $I_{ds} = WV_d/L \int_0^t ep(x)\mu(p(x))dx$, where t represents the thickness of the accumulation channel. Using this formalism the dc transfer characteristics of the PPV-derivatives can be modelled as a function of T and V_g. The fit parameters T_0, σ_0, α^{-1} are given in Table 2. The calculated transfer characteristics are shown in Figs. 4 and 5 as solid lines. Good agreement has been obtained for all three semiconductors [13]. It should be noted that at high gate voltages the charge carriers are strongly confined to the interface, approaching a 2-D transport system. Therefore, one could expect that the percolation threshold B_c is modified as compared to the 3-D model used by Vissenberg and Matters. However, the model based on the 3-D percolation threshold consistently describes the charge transport both a low gate bias, where the system is more 3-D like, and high gate voltages. This indicates that the effect of an eventual 3-D to 2-D change in the type of percolation will not be very large. A description about a possible gate voltage dependence of B_c is beyond the scope of this paper.

Table 2 Parameters T_0 (the width of the exponential density of states), σ_0 (the conductivity prefactor), α^{-1} (the effective overlap parameter), $\mu_{FET}(V_g)$ (the field-effect mobility determined from Eq. [3] at $V_g = -19$ V and room temperature).

Polymer	T_0 (K)	σ_0 (10^6 S/m)	α^{-1} (Å)	$\mu_{FET}(V_g)$ (m²/Vs)
$OC_{10}C_{10}$-PPV	340	0.13	2.6	8.7×10^{-8}
OC_1C_{10}-PPV	540	31	1.4	4.7×10^{-8}
NRS-PPV	560	3.5	1.36	4.0×10^{-9}

Subsequently, by combining Eq. [8], using the parameters as determined from the experimental transfer characteristics, with the distribution of the charge carrier density as a function of x, the local mobility in the accumulation channel can be calculated. In Fig. 7 the distribution of the local mobility versus distance x from the interface is presented together with the experimental field-effect mobility from Eq. [3] for OC_1C_{10}-PPV at a gate voltage $V_g = -19$ V. We find that the local mobility decreases about one order of magnitude in the first 2 nm from the semiconductor/insulator interface into the bulk. The local mobility of the charge carriers directly at the semiconductor/insulator interface is about 15% larger as compared with the experimental field-effect mobility determined from Eq. [3], which represents an average over all the induced carriers. As a consequence, the error due to the approximation in Eq. [3] that all the charge carriers in the accumulation layer have the same

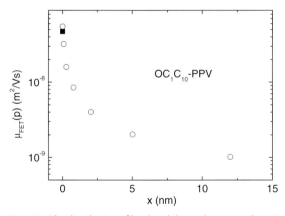

Figure 7 The distribution of local mobility in the accumulation channel as a function of the distance from the semiconductor/insulator interface of an undoped semiconductor for $V_g = -19$ V.

mobility is relatively small. The reason for this relatively small difference is that not only a major part of the charge carriers is located close to the interface, but also that these charge carriers have the highest mobility. As a result the field-effect current is mainly determined by the charge carriers directly at the interface.

11.2.3
Comparison of the hole mobility in LEDs and FETs

The charge carrier mobility as function of charge carrier density as obtained from the field-effect transistors at room temperature is shown in Fig. 8, together with the mobility of the hole-only diodes. At low bias voltages the SCL current of the hole-only diode is exactly quadratic, indicative of a constant mobility. Consequently, at low bias voltage the hole mobility is independent of both electric field and charge carrier density. The lowest charge carrier density p_L in a space-charge limited diode is found at the non-injecting contact and is given by:

$$p_L = \frac{3}{4}\left(\frac{\varepsilon_0 \varepsilon_r V}{eL^2}\right), \tag{9}$$

where L represents the thickness of the polymer layer. For the diode structures studied here the low bias range corresponds to hole densities of typically 2.5×10^{20} to 2.5×10^{21} m^{-3}. For high bias voltage the dependence of the mobility on the charge carrier density and on the electric field cannot be disentangled due to the fact that in a space charge limited diode both these parameters increase simultaneously. The values of the experimental field-effect mobility have been calculated using Eq. [3]. Combination of the results from the diode and field-effect measurements shows that typically the hole mobility is constant for charge carrier densities $<10^{22}$ m^{-3} and increases with a power law for charge carrier densities $>10^{22}$ m^{-3}. As demonstrated

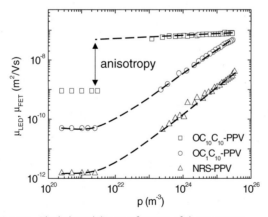

Figure 8 The hole mobility as a function of charge carrier density in diode and field-effect transistor for NRS-PPV, OC$_1$C$_{10}$-PPVand OC$_{10}$C$_{10}$-PPV. The dashed line is a guide for the eye.

in Fig. 8 the mobility differences of up to three orders of magnitude obtained from diodes and FETs, based on a single disordered polymer (OC$_1$C$_{10}$-PPV or NRS-PPV), originates from the different charge density regimes in these devices. Although it has been demonstrated that in OC$_1$C$_{10}$-PPV the optical properties exhibit a significant anisotropy [21], a possible anisotropy in the charge transport properties would obscure a direct comparison between diodes and FETs. As shown from Eq. [2] the amount of energetic disorder is directly reflected in the thermal activation of the low-field mobility. For the (logarithm of) low-field mobility obtained from PPV-based LEDs it is difficult to discriminate whether its temperature dependence scales with $1/T$ or $1/T^2$ (Eq. [2]), due to the limited temperature range accessible in the experiments (150-300 K). When plotted against $1/T$ the mobility is also well described by $\mu \sim \exp(-E_a/k_B T)$, and an activation energy E_a of typically 0.48 eV has been reported [15]. Also for transistors the field-effect mobility is thermally activated by a gate-voltage dependent activation energy [22]. In Fig. 9 the activation energy E_a of the field-effect mobility is plotted for OC$_1$C$_{10}$-PPV as a function of gate voltage from −1 to −19 V. Extrapolation towards V_g=0 V yields an E_a of 0.46 eV, exactly equal to the activation energy as obtained from the diode measurements on this polymer. This is a strong indication for the absence of anisotropy in the charge transport properties of this polymer. The absence of anisotropy has also been observed for NRS-PPV. In contrast, for OC$_{10}$C$_{10}$-PPV the mobility behaviour in Fig. 8 clearly shows a lack of correlation in mobility between diode and field-effect transistor measurements. The explanation is that OC$_1$C$_{10}$-PPV and NRS-PPV are highly disordered systems in which the charge transport takes place in 3-D. OC$_{10}$C$_{10}$-PPV is on the other hand more ordered due to the two symmetric side-chains and the charge transport is different in the directions parallel and perpendicular to the polymeric film.

In order to compare the two theoretical models used to explain the charge carrier transport in LEDs and FETs, we plotted the Gaussian DOS (Eq. [1]) and the exponential DOS (Eq. [4]) as function of energy for our devices. In Fig. 10 and 11 the Gaussian

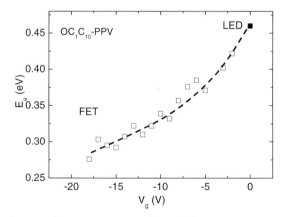

Figure 9 The activation energy of the mobility in OC$_1$C$_{10}$-PPV diode (circle) and FET (squares) as a function of voltage.

DOS, as obtained from the temperature dependent diode measurements, is plotted as a function of energy for NRS-PPV and OC_1C_{10}-PPV, respectively. For the total number of states per unit volume N_t we have used a value of 3×10^{26} m^{-3} for both OC_1C_{10}-PPV and NRS-PPV. Additionally, the exponential DOS, described by T_0 (See Table 2), of OC_1C_{10}-PPV and NRS-PPV as obtained from the FET characteristics are shown. With increasing gate voltage up to –19 V the Fermi-level in the Gaussian DOS ranges from 0.4 eV to 0.16 eV with respect to the centre of the Gaussian DOS for the OC_1C_{10}-PPV based FET, and from 0.42 eV to 0.17 eV for the NRS-PPV based FET. From Figs. 10 and 11 we find that in this energy range the exponential distribution is a good approximation of the Gaussian DOS. In this way the two models

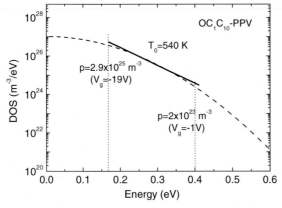

Figure 10 The Gaussian density of states (DOS) (dashed line) and the exponential DOS (solid line) as a function of energy for NRS-PPV.

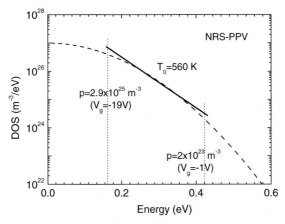

Figure 11 The Gaussian density of states (DOS) (dashed line) and the exponential DOS (solid line) as a function of energy for OC_1C_{10}-PPV.

are unified, in the sense that the exponential DOS accurately describes the Gaussian DOS in the energy range in which the field-effect transistors operate. Consequently, the temperature- and charge carrier density dependencies of the hole mobility in these disordered conjugated polymers are unified in one single charge transport model. An effect that we ignored in this comparison is that the density of states could be modified due to the induced charges. Recent model calculations demonstrated that an increase of the doping level also increases the energetic disorder due to potential fluctuations caused by the Coulomb field of randomly distributed dopant ions [23]. In a FET the charge carrier density is increased by a gate field. The fact that the models for LED and FET can be unified indicates that the effect of the induced charge alone, without ions, is less pronounced in the used gate voltage range.

11.3
Conclusions

In conclusion, we investigated the hole mobility for three PPV derivatives in hole-only diodes and field-effect transistors as function of bias and temperature. It has been show that the experimental hole mobility for a single polymer based diode and FET can differ by 3 orders of magnitude. This discrepancy originates from the strong dependence of the hole mobility on the charge carrier density. For highly disordered PPVs it is demonstrated that the exponential density of states (DOS), which theoretically describes the charge transport in FETs, is a good approximation of the tail states of the Gaussian DOS, which describes the charge transport in LEDs. Increase of the directional order in the polymeric PPV film leads to an increase of the hole mobility in both diodes and FETs, but also to a strong anisotropy in the charge transport between the two types of devices.

Acknowledgements

This work is part of the research programme of the Dutch Polymer Institute (project no. 276).

References

[1] J. H. Burroughes, D. D. C. Bradley, A. R. Brown, R. N. Marks, K. Mackay, R. H. Friend, P. L. Burn, A. B. Holmes, Nature (London) **347**, 539 (1990).

[2] P. W. M. Blom, M. C. J. M. Vissenberg, J. N. Huiberts, H. C. F. Martens, H. F. M. Schoo, Appl. Phys. Lett. **77**, 2057 (2000).

[3] H. Sirringhaus, N. Tessler, R. H. Friend, Science **280**, 1741 (1998).

[4] H. E. A. Huitema, G. H. Gelinck, J. B. P. H. van der Putten, K. E. Kuijk, C. M. Hart, E. Cantatore, P. T. Herwig, A. J. J. M. van Breemen, D. M. de Leeuw, Nature **414**, 599 (2001).

[5] H. Bässler, Phys. Status Solidi B **175**, 15 (1993).

[6] S. V. Novikov, D. H. Dunlap, V. M. Kenkre, P. E. Parris, and A. V. Vannikov, Phys. Rev. Lett. **81**, 4472 (1998).

[7] H. Sirringhaus, P. J. Brown, R. H. Friend, M. M. Nielsen, K. Bechgaard, B. M. W. Langeveld-Voss, A. J. H. Spiering, R. A. J. Janssen, E. W. Meijer, P. T. Herwig, D. M. de Leeuw, Nature (London) **401**, 685 (1999).

[8] S. F. Nelson, Y. -Y. Lin, D. J. Gundlach, T. N. Jackson, Appl. Phys. Lett. **77**, 1854 (1998).

[9] R. H. Friend, Y. -Y. Lin, R. W. Gymer, A. B. Holmes, J. H. Burroughes, R. N. Marks, C. Taliani, D. D. C. Bradley, D. A. Dos Santos, J. L. Brédas, M. Lögdlung, W. R. Salaneck, Nature (London) **397**, 121 (1999).

[10] C. D. Dimitrakopoulos, P. R. L. Malenfant, Advanced Materials **14**, 99 (2002).

[11] G. H Gelinck, H. E. Huitema, E. van Veenendaal, E. Cantatore, L. Schrijnemakers, J. B. P. H. van der Putten, T. C. T. Geuns, M. Beenhakkers, J. B. Giesbers, B-H. Huisman, E. J. Meijer, E. Mena Benito, F. J. Touwslager, A. W. Marsman, B. J. E. van Rens, and D. M. de Leeuw, Nature Materials **3**, 106 (2004).

[12] M. C. J. M. Vissenberg and M. Matters, Phys. Rev. B **57**, 12964 (1998).

[13] C. Tanase, E. J. Meijer, P. W. M. Blom, D. M. de Leeuw, Phys. Rev. Lett. **91**, 216601 (2003).

[14] Y. Roichman, N. Tessler, Appl. Phys. Lett. **80**, 1948 (2002).

[15] P. W. M. Blom and M. C. J. M. Vissenberg, Mat. Sc. and Eng. **27**, 53 (2000).

[16] H. C. F. Martens, P. W. M. Blom, H. F. M. Schoo, Phys. Rev. B **61**, 7489 (2000).

[17] M. Kemerink, J. K. J. van Duren, P. Jonkheijm, W. F. Pasveer, P. M. Koenraad, R. A. J. Janssen, H. W. M. Salemink, J. H. Wolter, Nano Lett. **3**, 1191 (2003).

[18] A. R. Brown, C. P. Jarrett, D. M. de Leeuw, M. Matters, Synt. Met. **88**, 37 (1997).

[19] S. M. Sze, Physics of Semiconductor Devices (Wiley, New York, 1981).

[20] C. Tanase, E. J. Meijer, P. W. M. Blom, D. M. de Leeuw, Org. Electronics **4**, 33 (2003).

[21] C. M. Ramsdale and N. C. Greenham, Adv. Mater. **14**, 212 (2002).

[22] E. J. Meijer, M. Matters, P. T. Herwig, D. M. de Leeuw, T. M. Klapwijk, Appl. Phys. Lett. **76**, 3433 (2000).

12
Analysis and Modeling of Organic Devices

Y. Roichman, Y. Preezant, N. Rappaport, and N. Tessler

12.1
Introduction

Charge injection and transport phenomena have been studied for many years and in many material systems [1, 2] including that of organic semiconductors [3–13]. However, the analysis and modeling of polymer or small molecule based devices are often performed using different models and sometimes even different physical pictures despite the fact that similar (or even identical) materials are being used. In this paper we consider the physical picture, promoted by Bassler et. al. [5], of a disordered material having a Gaussian density of localized states as the one that can be used to shed light on the performance of field-effect transistors (FETs) light-emitting diodes (LEDs) or photo-cells (PCs). To do so we have expanded the theoretical treatment of localized density of states to include the notion of charge density and its effect on the transport parameters [10, 14, 15]. This was done within the framework of the mean medium approximation (MMA) that will be briefly described below. In this paper we will show that one can use as a basic model the well known device model framework where the unique properties of organic semiconductors are accounted for using appropriate expressions for the mobility (μ) and/or diffusion coefficient (D) that are derived from a single physical picture. The general formalism we have developed [10, 14, 15] can be applied to any shape of DOS (Gaussian, exponential, etc.). However, our philosophy is that given a material and a processing procedure one should not change the physical picture (DOS shape) between LEDs, FETs, or solar cells. Namely, one should be able to reproduce not only the shape of the experimental curves but also the absolute values when applying a physical picture to different devices. As we show below, the extended Gaussian disorder model can self consistently and qualitatively explain the operation of a range of device structures. It also shows that predictions made for the low density limit are not necessarily correct at the high density regime found in LEDs or FETs. Finally, we point out that for a complete quantitative fit we are still missing an experimentally derived density of charge transport states (both the shape and the total density) as well as a hopping rate that will account for the polaronic effect.

Physics of Organic Semiconductors. Edited by W. Brütting
Copyright © 2005 WILEY-VCH Verlag GmbH & Co. KGaA, Weinheim
ISBN 3-527-40550-X

12.1.1
Motivation

Before embarking on the formalism we set the general concepts we will use to describe the mechanism underlying the concept of diffusion and mobility. Let us consider a uniform media where part of the electronic sites is occupied by charge. A carrier in such a site has a finite probability to jump to a neighboring site which is independent of any spatial direction. The value of probability may depend on many variables, as energy or distance, depending on the transfer mechanism but due to symmetry it is independent of the jump direction. This direction-independent motion is the diffusion.

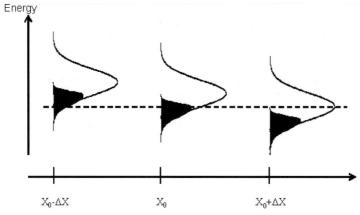

Figure 1 Schematic description of a uniform media having a Gaussian density of states and under applied bias.

When an external field is applied it creates a gradient in the relative position of the energy states thus creating a preferred direction and breaking the symmetry (see Figure 1). As typically a jump downwards, in energy, is favorable to a jump upwards a net flow in the slope direction is created. The proportionality constant between this flow and the electric field is the mobility. Since diffusion and mobility are closely related, through the medium and motion mechanism, one can define a ratio between the two which is called the Einstein relation.

An important question would be the extent of the density dependence of the transport properties. This may tell us the importance of this issue and the level of treatment required (or what kind of approximation one can/can't use). To illustrate this issue we plot in Figure 2 the normalized total jump rate of carriers as a function of the initial energy assuming a Gaussian DOS with a width (σ) of 5kT (see eq (12)). The full line was calculated for a charge density of $3\times10^{18} \text{cm}^{-3}$. The dashed line is for an extremely low density corresponding to Fermi level at -35 kT. The shape of the dashed line is almost constant for any charge density bellow few time 10^{16} cm^{-3} and only its magnitude would drop (almost linearly) with the reduced density.

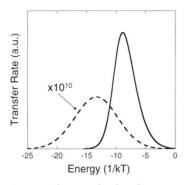

Figure 2   The normalized total jump rate of carriers as a function of the initial energy assuming a Gaussian DOS with a width (σ) of 5 kT. The full line was calculated for a density that corresponds to a Fermi level at –10 kT relative to the Gaussian DOS centre ($\sim 10^{18}$cm^{-3}) and the dashed line for a very low density with Fermi level at –35 kT.

Figure 2 actually shows the energy through which most of the transport takes place. We note that it is rather broad and that it is charge density dependent. Namely, on one hand the mobility would be density dependent and on the other approximating the physical picture by introducing traps relative to an effective "band" may be too simplistic.

12.2
Charge transport

As the properties of organic devices are highly dependent on the transport properties and are often used to extract such properties it is important to define the framework used. In this contribution we limit ourselves to the regime where one can use the concepts of charge-mobility and charge-diffusion to describe the electronic characteristics of the material and the device made of. This implies that we touch only the part of the physical picture that can be described through the following equations:

$$\frac{d}{dy}E = -\frac{q}{\varepsilon_\pi \varepsilon_0}n(y) \tag{1}$$

$$J_n = qn\mu_n E + qD_n \frac{\partial}{\partial y}n \tag{2}$$

$$\frac{\partial}{\partial t}n(y,t) = \frac{1}{q}\frac{\partial}{\partial y}J_n = \frac{\partial}{\partial y}\left[D_n\frac{\partial}{\partial y}n(y,t) + \mu_n n(y,t)E(y,t)\right] \tag{3}$$

where equation (1) is the Poisson equation, equation (2) is the current equation and equation (3) is the current continuity equation. All the symbols have their conventional meaning. To expand the regime for which these equations hold it is common to allow D and μ to be dependent on device/material parameters and specifically on the temperature, electric field, and charge density. The most straight forward manner of using these equations for device analysis is to solve them numerically. These equations have been shown to describe LEDs and Solar cells and the 2D (or 3D) [16–19] version of them have been used for FETs. The other approach would be to choose a small subset of material/device parameters and analytically derive approximate expressions that are many times more powerful then a full numerical model. As mentioned above, the unique device/material properties are represented through ε_π, D, and μ and hence, choosing a self consistent representation is crucial.

12.2.1
The Gaussian DOS

In order to analyze the transport in a system of localized states, one needs to know the energy position of each site and the topology of the sites so that the distance to neighboring sites can be determined. In cases where the transport involves a large enough number of sites the exact topology of the system is not required and one is only concerned with the statistical functions describing it. As in this chapter we are interested in the notion of mobility and diffusion we a priory limit ourselves to the macroscopic scale where one can apply statistics to the charge motion.

The most common function used to describe the probability-distribution associated with a large number of events is the Gaussian distribution. This physical framework was developed by Baessler and co-workers [5] for the low charge density limit and later extended towards the higher density regime by Arkhipov and Baessler [20, 21] and Roichman and Tessler [10, 14]. The DOS function is hence described by:

$$DOS(E) = \frac{N_V}{\sqrt{2\pi \cdot \sigma}} \exp\left[-\left(\frac{E-E_0}{\sqrt{2\pi \cdot \sigma}}\right)^2\right] \tag{4}$$

Where E is the energy, E_0 is the Gaussian centre, N_V is the effective DOS, and σ is the Gaussian variance (σ is sometimes called the width of the DOS). The spatial distribution function could be either homogenous (which describes uniform spatial distribution) or slightly different (to account for a specific morphology). For the sake of the present discussion we are only interested in the typical distance between sites defined as $r_{typ} = \sqrt[3]{N_V^{-1}}$. Issues relating to long range energetic correlation [22–24] (off diagonal disorder) and morphological effects are not discussed here.

Our use of equations (1) to (3) implies that we are only dealing with cases where the system can be said to be at equilibrium. Practically, one often states that a system is at equilibrium if the charge energy-distribution can be described as

$$p(E) = DOS(E)f(E) \tag{5}$$

or the total density (P) as

$$P = \int_{-\infty}^{\infty} p(E)dE = \int_{-\infty}^{\infty} DOS(E)f(E)dE \tag{6}$$

where $f(E)$ is the Fermi-Dirac distribution function and $DOS(E)$ is the DOS that we "see" when we map the entire sample (or sometimes even for a hypothetically large system).

12.2.2
The Equilibrium Charge Density

As we are interested in the equilibrium state where the charge density and the Fermi-level are directly related through the DOS we first examine this relation. For inorganic non-degenerate semiconductors it is common to describe the effect of a change in a Fermi level energy as: $P = N_V \exp \frac{-\Delta E}{kT}$. However, we have already shown that organic semiconductors are degenerate [25] at all practical densities and hence such an expression is wrong. To illustrate this point we plot in Figure 3 the charge density as a function of the position of the Fermi-level relative to the centre of the Gaussian DOS.

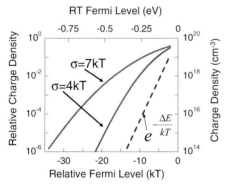

Figure 3 Relative filling of the DOS as a function of the position of the Fermi-level relative to the Gaussian centre. The calculation was done for σ=7kt and 4kT. The dashed line shows the commonly used expression based on Boltzman statistics.

The top solid line shows the charge density dependence for a DOS width of 7kT and the low solid line for DOS width of 4kT. The dashed line shows the use of exponential, Boltzman, dependence. We note that the slope is not constant and at reasonable densities is much smaller then the prediction made by $\left(\frac{\Delta E}{kT}\right)$. The implication is, for example, that the effect of an injection barrier is significantly smaller for a Gaussian DOS and hence conventional methods for extracting injection barrier will underestimate the energy offset at the contact interface. In fact such methods will effectively seek to linearize the function of P (Figure 3) in a small range around a given working-point $P_{(T_0,\Delta E_0)}$ as:

$$P_{(T,\Delta E)} = P_{(T_0,\Delta E_0)} \exp\left(-\frac{\delta E}{kT}\right) \tag{7}$$

However, the results are that $P_{(T_0,\Delta E_0)} \neq N_V$ and $\delta E \neq \Delta E_0$.

12.2.3
The diffusion coefficient

The notion of diffusion, mobility and the ratio between the two is strictly valid only when the charge density population can be considered to be at equilibrium (i.e. equations (5) and (6) are valid). The relation between the diffusion coefficient (D) and the mobility (μ) in the low density limit is given by $D/\mu = kT/q$ [2] (Einstein relation) where k is Boltzmann coefficient, q is the charge of the particle and T is the characteristic temperature. A generalized relation between the diffusion coefficient and the mobility (i.e. generalized Einstein relation) can be derived for a general charge-carrier energy-distribution function, and a general DOS function [2]:

$$\frac{D}{\mu} = \frac{P}{q\frac{\partial P}{\partial E_F}} \tag{8}$$

Where p is the particle concentration and E_F is the chemical potential (or quasi Fermi-level). Applying the Gaussian DOS, equation (4), to equation (8) results in [10]:

$$\frac{D}{\mu} = \frac{kT}{q} \frac{\int\limits_{-\infty}^{\infty} \exp\left[-\left(\frac{E-E_0}{\sqrt{2}\cdot\sigma}\right)^2\right]\cdot\frac{1}{1+\exp\left(\frac{E-E_F}{kT}\right)}dE}{\int\limits_{-\infty}^{\infty} \exp\left[-\left(\frac{E-E_0}{\sqrt{2}\cdot\sigma}\right)^2\right]\cdot\frac{\exp\left(\frac{E-E_F}{kT}\right)}{\left[1+\exp\left(\frac{E-E_F}{kT}\right)\right]^2}dE}\frac{kT}{q}\cdot\eta$$

As has been discussed in [10] η assume values larger then one for any practical charge density and the effect is more pronounced for larger σ values (higher disorder). The dependence of η on the normalized charge density $\bar{p} = \frac{p}{N_V}$ can be fitted by the following expression:

$$\eta^{-1} = a + b\cdot\log(\bar{p}) + c\cdot\exp\left(-\frac{\bar{p}}{c_1}\right) + d\cdot\exp\left(-\frac{\bar{p}}{d_1}\right) \tag{9}$$

the results of such a fit for T=300 k, 200 k and several DOS widths are shown in the tables below. The fit was made for the range of $\bar{p}\in\left[10^{-7}to0.5\right]$.

12.2.4
The mobility coefficient

It has been shown, within the context of the electric field dependence of the mobility, by Baessler and co-workers that the addition of traps has a similar effect to that of making the Gaussian distribution wider as in the two cases lower energy states are being added. Since the most known effect of traps is that of making the effective-mobility density-dependent one would also expect that the Gaussian DOS will also give rise to density dependence of the charge mobility [14, 15, 20].

Such a calculation can be done in a straight forward calculation similarly to that of the Einstein relation discussed above [14, 15]. To calculate the mobility one needs

Table 1 Coefficients for equation (9) which is used to calculate the generalised Einstein relation.

					T=300k			
σ (eV)	σ/kT	a	b	c	c_1	d	d_1	
σ=0.05	1.92	1450.852786	−0.004220806	−1450.141488	−1719.759355	0.260996005	0.03394920872	
σ=0.10	3.85	0.1869241014	−0.0354867371	0.3647272841	0.1595124618	0.182902459	0.00205054279	
σ=0.15	5.77	0.1100758809	−0.0631063584	0.2032429204	0.1918673121	0.084043262	0.00252621548	
					T=200k			
σ (eV)	σ/kT	a	b	c	c_1	d	d_1	
σ=0.05	2.88	13349.82696	−0.0207680376	−13349.25429	−17839.39656	0.2886843652	0.01450610423	
σ=0.10	5.77	0.1100758809	−0.0631063584	0.2032429204	0.1918673121	0.084043262	0.00252621548	
σ=0.15	8.65	0.0663089357	−0.0517996542	0.1288063362	0.2274425917	0.0455107742	0.003701326288	

also to adopt a hopping mechanism. Here we neglect the effect of polarization of the volume adjacent to the charge carrier (the polaronic effect), and hence we assume that the charge carriers hop between localized states according to "Miller-Abrahams" rate [26]:

$$v_{ij} = v_0 \exp\left(-2\gamma r_{ij}\right) \begin{cases} \exp\dfrac{E_j - E_i}{kT} & E_j \geq E_i \\ 1 & E_j < E_i \end{cases}. \tag{10}$$

The potential drop between the sites is calculated as the distance between sites times the electric field $(\vec{r}_{ij} \cdot \vec{E})$. The occupation probability of the state at energy ε_i at a certain site is determined by Fermi-Dirac distribution $f(\varepsilon_i)$. The total particle current is the integration on the contribution from the hops from state i to state j, according to the following expression:

$$J_{ij} = [DOS(E_i)f(E_i, E_F) \times \left[DOS(E_j)\left(1 - f(E_j, E_F)\right)\right] v_{ij}\, \vec{r}_{ij} \cdot \hat{E} \tag{11}$$

The two square brackets define the probability that a hopping event will occur as the probability that the initial state is occupied times the probability that the final state is empty. $\vec{r}_{ij}\hat{E}$ is the propagation distance at the electric-field direction and v_{ij} is the hopping rate (equation (10)) from site i to site j. Integrating (11) over space and energy of final sites results in the total jump rate per initial energy (see also Figure 2).

$$J_i = \int_{-\infty}^{\infty} dE_j \int_{\Re} \vec{dr}_{ij} [DOS(E_i)f(E_i, E_F) \times \left[DOS(E_j)\left(1 - f(E_j, E_F)\right)\right] v_{ij}\, \vec{r}_{ij}\, \hat{E} \tag{12}$$

and finally integrating over the energy of initial sites we find the total current:

$$J_i = \int_{-\infty}^{\infty} dE_j \int_{-\infty}^{\infty} dE_j \int_{\Re} \vec{dr}_{ij} [DOS(E_i)f(E_i, E_F) \times \left[DOS(E_j)\left(1 - f(E_j, E_F)\right)\right] v_{ij}\, \vec{r}_{ij}\, \hat{E}$$

$$\tag{13}$$

From the total particle current (J) one can then deduce the mobility as $\mu = \frac{J}{PE}$ with P being the charge density (equation (6)). Equations (12) and (13) actually describe the averaging procedure over space and energy to arrive at an effective mobility.

12.3
The operation regime of organic devices

After establishing the physical framework we base our study on it is important to derive the implications for device performance. Examining the equations involved in the derivation of D and μ it is evident that the knowledge of the shape of the DOS is essential to derive the "correct" values. Unfortunately, we are not aware of a method for extracting the charge DOS without invoking too many assumptions. Therefore, we will only demonstrate the effects for Gaussian DOS and show the trends expected once its width is varied. Figure 4 shows the low-field mobility (Figure 4B) and the Einstein relation (Figure 4A) as a function of the filling factor of the DOS. The top x-axis shows the corresponding charge density assuming a packing that leads to a total DOS of 10^{20}cm^{-3}.

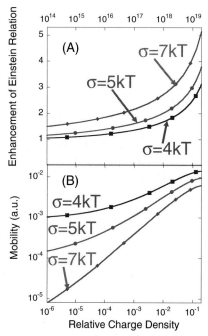

Figure 4 (A) enhancement of the Einstein relation, relative to the low density limit (kT/q), as a function of the filling of the density of states. (B) The low field mobility as a function of the filling of the density of states. The calculation is shown for a DOS width of 7kT (◆), 5kT (●), and 4kT (■).

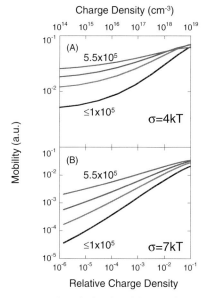

Figure 5 The calculated mobility as a function of charge density at T=300k. The lowest curve is for low electric fields.

We note that both D and μ are density dependent above a critical density and that the critical density for the diffusion coefficient (Einstein relation) is somewhat lower then the critical density for the mobility. Figure 4 also shows that the density dependence is significantly enhanced as a function of disorder (DOS width). To complete the picture we examine next the electric field dependence of the mobility and the effect of the charge density on it (Figure 5).

Note that the density dependence at the relatively high range (above 10^{17}cm^{-3}) resembles a power law ($\mu \propto P^{\kappa}$), but the coefficient is electric field dependent. For example, for σ=4 kT the low field and high field (in Figure 5) exhibit a power law of κ=0.33 and κ=0.11, respectively. For σ=7 kT the low field and high field (in Figure 5) exhibit a power law of κ =0.62 and κ=0.28, respectively. We have calculated κ for a

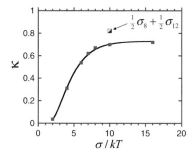

Figure 6 The dependence of the power law coefficient (κ) on the DOS width σ/kT for the low electric field regime at T=300 k.

range of DOS widths (see Figure 6) and at the low field regime. It was found that it could be numerically fitted using $\kappa = 0.73 - 1.17 \cdot \frac{\sigma}{kT} \cdot \exp\left(-\frac{\sigma}{1.65 \cdot kT}\right)$. Namely, for a system described by a single Gaussian DOS the maximum slope is $\kappa = 0.73$ and it is close to this maximum at $\sigma = 8kT$. It has been suggested that organic films are composed of domains that would make the effective density of states a bit more complicated then a single Gaussian [15, 27-30]. Following the notion in [15, 27] we also show in Figure 6 the slope found for a DOS that is a linear combination of two Gaussians having $\sigma = 8$ kT and $\sigma = 12$ kT (1:1 composite).

Figure 7 shows the mobility calculated as a function of the electric field and for a range of charge densities (noted next to each curve assuming $N_V = 10^{20}$cm^{-3}). The top of Figure 7 was calculated for $\sigma = 4$ kT and the bottom of Figure 7 for $\sigma = 7$ kT. From the shape of the curves it seems that this simple model does not necessarily capture the entire physical picture since the field dependence is sometimes believed to extend to very low electric fields. This has been shown to be "corrected" via the introduction of site-correlation effects that are beyond our scope. Also, at the very high fields, where the voltage-drop across $r_{typ} = \sqrt[3]{N_V^{-1}}$ exceeds kT/q, new effects make come into play [31]. However, this picture captures a significant part of the physics and it illustrates the importance of the charge concentration. It is instructive to compare our results with the predictions previously made for the Gaussian model at the low density limit [32]:

$$\mu = \mu_{E=0} \exp\left\{ c\left[\left(\frac{\sigma}{kT}\right)^2 - 2.25\right]\sqrt{E} \right\} \tag{14}$$

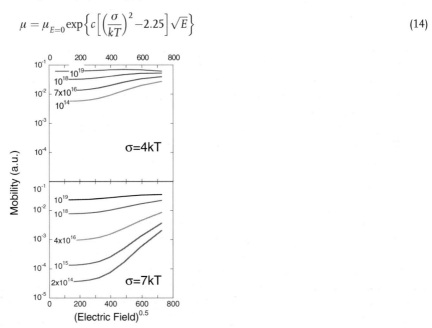

Figure 7 The calculated mobility as a function of normalized applied electric field for several charge densities at T=300k. We note that at most device operation conditions (above 10^{16}cm^{-3}) the field dependence is almost independent of σ. The scale of the electric field was set assuming inter-site distance of 1nm.

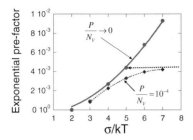

Figure 8 The exponential factor as a function of the Gaussian width at T=300k. The circle marks were calculated for the low density limit and the diamond marks were calculated for a charge density typical to space-charge limited diodes (10^{16}–10^{17}cm^{-3}). The solid line is a fit based on equation (14) with c=2x10^{-4} (cm/V)$^{0.5}$. The dashed line connecting the diamond marks is only a guide to the eye.

Figure 8 shows the exponential factor as a function of the Gaussian width. The circle marks were calculated for the low density limit and the diamond marks were calculated for a charge density typical to space-charge limited diodes (10^{16}–10^{17} cm^{-3}). The solid line is a fit based on equation (14) with c=2x10^{-4} (cm/V)$^{0.5}$. The dashed line connecting the diamond marks is only a guide to the eye. Projecting the curve found at LED charge density onto the low density limit-prediction (see horizontal arrow) tells us that the low density prediction would yield values between σ=100 meV to σ=120 meV almost independent of the material properties (actual σ value). This conclusion is consistent with reported comprehensive experimental analysis of LEDs reported in [33]. Namely, the contribution of disorder to the exponential factor at standard device charge-densities saturates at about σ=100 meV and the difference between samples would mainly reflect the total state density (N_V) or the inter-site. We also not that we do not consider here temperature effects since the simple hopping rate we use (eq. (10)) does not include the polaronic nature which will add effects as enhanced activation energy or more.

12.3.1
Contact workfunction

Figure 9 shows the charge density for a fixed Fermi-level energy (or metal workfunction) as a function of temperature. The offsets between the center of the Gaussian DOS and the Fermi-level were chosen to result in a set of densities that may be relevant to devices (for the corresponding energy shift see Figure 3). The solid lines were calculated for DOS of 182 meV and the dashed line for DOS of 104 meV. We note that the temperature dependence of the density, at the contact interface, is small to negligible.

To summarize the effects of a Gaussian DOS:

1. The charge density dependence on Fermi-level (or contact workfunction) is small at practical charge densities.
   a. The temperature dependence is extremely weak.

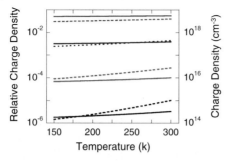

Figure 9 Charge density as a function of temperature for constant Fermi-levels. For each curve the relative position of the Fermi-level was held fixed (see Figure 3) and the temperature was scanned. The solid lines are for $\sigma=7*0.026$ eV and the dashed are for $\sigma=4*0.026$ eV.

2. The Einstein relation is higher at high charge concentration.
   a. The density dependence increases with the density.
3. The mobility is higher at high charge density.
   a. The density dependence is most pronounced at charge density that fills about 1% of the total DOS ($\sim 10^{18}$cm^{-3}).
   b. The density dependence is reduced once a high enough electric field is applied.
   c. It is dependant on the exact DOS shape.
4. The electric field dependence of the mobility is smaller at higher densities.
   a. At densities above 10^{16}cm^{-3} the electric field dependence is almost independent of the width of the DOS (σ).

12.4
Organic Light-Emitting Diodes

12.4.1
Space charge regime

In the space charge limited regime one can approximate equations (1) to (3) using the assumption that the diffusion current is negligible[1]:

$$V = \sqrt{\frac{8j}{9\varepsilon\mu}\left(d+\frac{j\varepsilon}{2P^2q^2\mu}\right)^{\frac{3}{2}}-\left(\frac{j\varepsilon}{2P^2q^2\mu}\right)^{\frac{3}{2}}}$$

If we further assume that the charge density near the injection contact is high enough to screen the electric field (E=0 , near the contact) then:

$$d >> \frac{j\varepsilon}{2P^2q^2\mu}$$

Resulting in the following two expressions:

$$J_{SCL} = \frac{9}{8}\varepsilon\mu\frac{V^2}{d^3} \tag{15}$$

$$\overline{P} = \frac{1}{d}\int_0^d P(x)dx = \frac{3}{2}\left(\frac{\varepsilon V}{ed^2}\right) and\, P_{(d)} = \frac{3}{4}\left(\frac{\varepsilon V}{ed^2}\right) \tag{16}$$

Here V is the applied minus built in voltage and using common values for ε_r=3 and d=100nm we find $\overline{P} = 2.5 \cdot 10^{16} \cdot V(cm^{-3})$. We note that the average electric-field (V/d) as well as the average density scale with the applied voltage and hence $\mu_V = \mu_{(P,E)}$ (see Figure 5 and Figure 7). Examining Figure 5 we note that at about E=5x10^{5}Vcm^{-1} the density dependence of μ is weak and one may be able to extract a "pure" field dependence. On the other hand, examining Figure 7 we note that above 10^{17}cm^{-3} there is almost no electric field dependence. Since $\overline{P} \propto d^{-2}$ and $E \propto d^{-1}$ the use of a relatively thick LED (above 100nm) will allow to access a small range where one can measure an E dependence with only a small dependence on P. On the other hand, very thin LEDs (bellow 100nm) are operating at a high density regime where μ is mainly dependent on P.

12.4.2
Contact limited regime

The analysis of injection limited current can be roughly divided into two categories as shown in Table 2.

Table 2 Emission and Equilibrium driven models developed for inorganic semiconductors and for Gaussian DOS. (1) reference [34] (2) reference [35] (3) this work following reference [36]

	Emission Rate limited	Equilibrium Driven
Boltzman regime	$J = A * T^2 \exp\left(-\dfrac{d\Delta E}{kT}\right)^{(1)}$	$J = q\mu_{(T)} EN_V \exp\left(-\dfrac{\Delta E}{kT}\right)^{(1)}$
Gaussian DOS	$J = ev_0 \int_{+\infty}^a dx_0 \exp(-2\gamma x_0)w_{esc}(x_0)$ $\times \int_{-\infty}^{+\infty} dEBol(E)g(U_0(x_0) - E)^{(2)}$	$J = q\mu_{(p,E,T)} EP_{(\Delta E,T)}^{(3)}$

Historically speaking, the emission models were developed first [34, 35, 37] and captured most of the physics involved where the emission-backflow is a later modification that is mainly important for low mobility semiconductors. In fact one can consider the two mechanisms as possibly limiting the injection with the final value being dictated by the slow process. In the context of organic semiconductors it has been argued [38] that one should adopt the equilibrium driven (emission+backflow or mobility limited) formalism. It was shown [34, 38] that the emission model may reproduce the shape of the curves but the absolute values are orders of magnitude too high for low mobility semiconductors. Moreover, examining the two relevant equations, for Gaussian DOS, it is obvious that the difference between the two models is temperature dependent (largely due to $\mu(T)$). In the context of Gaussian DOS the equilibrium driven model has been implemented before using a numerical semiconductor device model modified to account for Gaussian DOS [36] that treated the contact and the bulk in a self consistent manner allowing to account for both space charge limited and contact limited regime in a self consistent manner. Alternatively, for the injection limited regime only, one can use a simpler form which follows the same formalism developed for inorganic semiconductors but taking the effect of the DOS into account by calculating $P(\Delta E, T)$ as in Figure 9 (see Table 2):

$$J = e\mu_{(P,E,T)} EP_{(\Delta E,T)} \tag{17}$$

There may be arguments against equilibrium being achieved within the time scale associated with transport across a device length of 100 nm let alone within the effective contact region. However, the situation close to the contact is very different to that found in the bulk of the device since the metal provides a huge bath that is at equilibrium. Due to localization of the organic sites, the direct jump from the metal to the organic film is to the first mono-layer mainly (nearest neighbour) resulting in:

1. These initial states see a dilute DOS at the polymer side and on the other a very high DOS belonging to the metal. Namely, these sites will have significantly stronger interaction with the metal then with transport states in the polymer and hence equilibrium with the metal will be achieved before transport commence.

2. The transport from the initial sites into the polymer is via diffusion and against the electric field set up by the image force. Again, the interaction with the metal states is favored and should lead to equilibrium with it.

3. The presence of a thin (nm) insulating layer between the metal and the organic layer will decouple the metal and make the injection emission rate limited.

12.4.2.1 Image force
To add the effect of the image force to the expression in Table 2 (eq. (17)) we make use of the linearized expression in equation (7) and add the image force lowering by the electric field (E):

$$P = P_{(T_0,\Delta E_0)} \exp\left(-\frac{\delta E}{kT}\right) \exp\left(\frac{1}{kT}\sqrt{q\frac{E}{4\pi\varepsilon_0\varepsilon}}\right) \tag{18}$$

We note that due to the linearization procedure that led to equation (7) this expression is adequate only for not too high fields ($<10^6 \mathrm{V\,cm^{-1}}$) such that the change in N will be well below an order of magnitude.

12.4.3

I-V Analysis of LEDs

We have constructed a hole only LED having a structure of ITO|PEDOT|MEH-PPV| gold and measured it in forward and reverse bias to compare the injection from PE-DOT to that from gold. We varied the thickness of the MEH-PPV layer and found that the current injected from PEDOT scales almost with d^3 and the current injected from gold almost scales with the applied field. Namely, we consider the injection from PEDOT as space charge limited and from the gold as injection limited. To get better understanding of the device we measured the characteristics between 160 k and 300 k at 20 k intervals. Figure 10 shows the results achieved at the two extremes for both PEDOT and gold injection. The striking [39, 40] "strange" feature is that while the SCLC drops by more then 2 orders of magnitude the injection-limited current hardly changes. Below, we examine an explanation for this "strange" behaviour.

We start by analysing PEDOT injection as space charge limited current and fit the current to $I = CV^2 \exp\left(\gamma \left[V/120 \cdot 10^{-7}\right]^{\frac{1}{2}}\right)$. As in the discussion following equations (15) and (16) the fit is done for a small voltage range. We find that at 300 k $C = 1.06 \times 10^{-4}$ $((\mu_{SCL}|_{V=1v} = 1.3 \cdot 10^{-6} cm^2 V^{-1} s^{-1}))$, $\gamma = 1.09 \times 10^{-3}$, $\bar{P}|_{V=1v} = \frac{3}{2}\left(\frac{\varepsilon V}{ed^2}\right) = 1.7 \cdot 10^{16} cm^{-3}$ and that the zero field activation energy is 0.4eV. If one uses the formalism developed for the low charge density limit (equation (14)) then the γ value at 300 k suggests that the DOS width is of about 100 meV which is 4 kT at 300 k. Unfortunately, the space charge limited diode operate at the degenerate regime (see $\bar{P}$) where the DOS width has a very small effect on γ (see discussion of Figure 7). Hence, the determination of σ from SCL currents is not so reliable.

Moving to the gold injection: at 1.5 V and 300 k the current is about 2.5 orders of magnitude lower then the PEDOT injection curve suggesting that the device is at the injection limited regime. If we use the mobility derived from the SCLC in equation (17) we find that at 1 V (J=5 × 10^{-6}A cm^{-2}) P = 5 × 10^{14} cm^{-3}. If we assume a

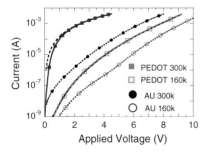

Figure 10 Measured I-V curves for 120 nm thick MEH-PPV device sandwiched between PEDOT and gold (pixel area 2 × 2 mm). The squares denote injection from the PEDOT and the circles from the gold (reverse bias). The full markers are for 300 k and the empty ones for 160 k. The dashed lines superimposed on the PEDOT curves are the fits discussed in the text.

commonly used value of $\sigma \sim 110$ meV and state density of 10^{20} cm^{-3} such charge density (P) would correspond to a metal situated $\Delta E = 0.54$ eV below the Gaussian centre. Using these values one can now test the values predicted by the emission model [35, 41]. Inserting σ, ΔE, and total DOS (10^{20}) (and assuming for this model a distance to first site of a=1 nm, a wavefunction decay of $\gamma = 0.5$ nm, and attempt frequency $\nu = 10^{12}$ s^{-1}) into the emission model we get a current of $\sim 7 \times 10^{-6}$A cm^{-2}. Namely, using this set of parameters (which is in the range used for organic semiconductors) it is difficult to decide which model is limiting as both give a similar result. Next, we examine the properties at 160k. If we use the same parameters to predict the current at 160 k and at 1 V ($\mu_{SCLC}|_{T=160} = 6 \cdot 10^{-11}$ cm^{2}V^{-1}s^{-1}) we find that J = 4×10^{-10}A cm^{-2} for the equilibrium driven model and J = 6×10^{-7}A cm^{-2} for the emission model. Namely, in this case the injection must be equilibrium-driven (mobility limited). We argue that hence the more appropriate model to be used between 160 k and 300 k, for our results, is the equilibrium model (see also discussion of contact limited regime).

Looking critically at the model prediction we note that it predicts a current that is too low. Unfortunately, the experimentally measured value (2×10^{-8}A cm^{-2}) is about two orders of magnitude higher. As it seems that we can not 100% account for the temperature dependence of the contact we sought the aid of temperature dependent UPS measurements but so far the results [42] do not seem to suggest that there is a change regarding energy level alignment. However, it seems to be known to spectroscopists [42] that the metal workfunction is not as well defined as a device-engineer would like it to be. For example, it has been shown that the workfunction of gold is very sensitive to surface interactions [43] resulting in a possible shift of anywhere between zero and 1 eV. In the context of Si Schottky diodes it has been suggested [44, 45] that the barrier height is fluctuating across the device giving rise to an effective barrier height that is reducing at low temperature and suggested an expression suitable for the injection limited regime in Si Schottky diodes: $\Delta E_{eff} = \Delta E - \frac{\sigma_m^2}{2kT}$ where σ_m is the width of the barrier fluctuation distribution. We suggest that such effects may be playing an important role in organic devices too especially since it is known that organic materials tend to form domains [28-30] which has large consequences on transport [15, 27].

12.5
Field Effect Transistors

In the context of field effect transistors (FETs) which are constructed using conductor, insulator, and a π-conjugated layer (CIπ technology) equations (1) to (3) are reduced under several assumption [46] into:

$$I_{DS} = \frac{W}{L}\mu C_{ins}\left[(V_{GS} - V_T)V_{DS} - \frac{V_{DS}^2}{2}\right] \tag{19}$$

for V_{GS} in accumulation regime and $|V_{DS}| \leq |V_{GS}|$

$$I_{DS\_SAT} = \frac{W}{L_{EFF}} \mu C_{ins} [V_G - V_T]^2 \tag{20}$$

for V_{GS} in accumulation regime and $|V_{DS}| \geq |V_{GS}|$

This device structure is most suitable to study charge density effect on the mobility at low electric fields. If V_{DS} is small then the charge density across the channel is quasi constant and the extracted mobility is well suited to compare with theoretical predictions. The most straight forward way to extract the mobility would be to solve the above equation for μ and arrive at:

$$\mu = \begin{cases} \dfrac{I_{DS}}{\frac{W}{L} C_{ins} \left[(V_{GS} - V_T) V_{DS} - \frac{V_{DS}^2}{2} \right]} & \text{Linear regime} \\[3ex] \dfrac{I_{DS\_SAT}}{\frac{W}{L_{EFF}} C_{ins} [Vg - V_T]^2} & \text{Saturation regime} \end{cases} \tag{21}$$

As it is important to keep V_{DS} small the relevant regime is the linear one. It is common to use the derivative of the above equation in order to extract the mobility [47, 48] as a function of gate voltage. However, if the mobility is density (gate voltage) dependent the method is not useful and the common formula may lead to large errors. To demonstrate this we write the derivative of the equation for linear regime:

$$\frac{\partial}{\partial V_{GS}} I_{DS} = \frac{\partial}{\partial V_{GS}} \left(\frac{W}{L} \mu C_{ins} \left[(V_{GS} - V_T) V_{DS} - \frac{V_{DS}^2}{2} \right] \right) \tag{22}$$

Applying the derivation:

$$\frac{\partial}{\partial V_{GS}} I_{DS} = \frac{\partial \mu}{\partial V_{GS}} \left(\frac{W}{L} C_{ins} \left[(V_{GS} - V_T) V_{DS} - \frac{V_{DS}^2}{2} \right] \right) + \mu \left(\frac{W}{L} C_{ins} V_{DS} \right) \tag{23}$$

Only in cases where the mobility is independent of the gate bias (or charge density) one can neglect the first term and equation (23) reduces to:

$$\mu = \frac{\frac{\partial}{\partial V_{GS}} I_{DS}}{\left(\frac{W}{L} C_{ins} V_{DS} \right)} \tag{24}$$

Figure 11 shows the current measured in an MEH-PPV CIπ-FET and the mobility as derived using equation (21). The mobility shows about an order of magnitude increase between 10^{-6} and $10^{-5}\,\text{cm}^2\,\text{V}^{-1}\text{s}^{-1}$ which is typical to organic materials [15, 47]. More details regarding these FETs can be found in [15].

To demonstrate the problem with the use of equation (24) we show in Figure 12 the first and second terms in equation (23) (using the data of Figure 11). As discussed above, equation (24) is derived by neglecting the term that is proportional to $\frac{\partial \mu}{\partial V}$. However, according to Figure 12 the two terms are very similar and hence the error in equation (24) is of the order of 100% and it also exhibits an incorrect voltage dependence.

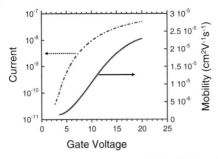

Figure 11 Drain source current (dashed line) and the mobility (full line) deduced using equation (21) as a function of gate voltage.

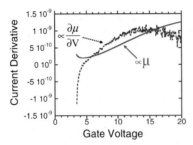

Figure 12 The first (dotted line) and second (full line) terms of equation (23) applied to the data of Figure 11.

Assuming the voltage dependence was correctly found one still needs to translate the gate voltage into a charge density. If one assumes that D/μ is approximately constant across the π-conjugated layer then it is possible to derive [46] the following approximate expressions:

$$E = \sqrt{\frac{2CD_h}{\mu_h}} \tan\left[\sqrt{\frac{C\mu_h}{2D_h}}(x - L)\right] \tag{25}$$

Where E is the electric field perpendicular to the substrate or a cross the π-conjugated layer that has a thickness L. This electric field is associated with band bending across the π-conjugated layer:

$$\Delta V_{channel} = -\frac{2D_h}{\mu_h} \log\left(\cos\left[\sqrt{\frac{C\mu_h}{2D_h}}L\right]\right) \tag{26}$$

And the charge density across the π-conjugated layer is given by:

$$p = \frac{\varepsilon_\pi\varepsilon_0}{q}\frac{\partial E}{\partial x} = \frac{C\varepsilon_\pi\varepsilon_0}{q}\left(\tan\left[\sqrt{\frac{C\mu_h}{2D_h}}(x - L)\right]^2 + 1\right) \tag{27}$$

where C is a constant to be determined by:

$$\frac{\varepsilon_{ins}}{\varepsilon_{\pi}} E_{ins} = E\Big|_{x=0} = \sqrt{\frac{2CD_h}{\mu_h} \tan\left[\sqrt{\frac{C\mu_h}{2D_h}}(0-L)\right]} \tag{28}$$

Where E_{ins} is the electric field inside the insulator and ε_{ins} is the insulator permittivity.

Using a different approach, one can also derive [46] the effective channel width (depth) to be:

$$X_{Channel} \approx \frac{d_{ins}}{V_G - V(y)} \frac{\varepsilon_{\pi}}{\varepsilon_{ins}} \frac{kT}{q} \eta \tag{29}$$

Where d_{ins} is the thickness of the insulator and $V(y)$ is the potential at the channel/insulator boundary. In the present context it is important to mention that the channel thickness is proportional to the temperature and hence the charge density is inversely proportional to the temperature. Also, the channel thickness is enhanced due to the η factor that enhances the Einstein relation. And finally, it is directly proportional to the insulator thickness.

In Figure 13 we plot the charge density dependence on the gate voltage using the following formula: $P = q^{-1} \cdot (V_G - V_T) \cdot C_{INS}/X_{Channel}$. The dott-dashed line was calculated by forcing $\eta = 1$ and the full line was calculated using $\eta(P)$ and assuming $\sigma = 7\,kT$ as well as $N_V = 10^{20}$ cm^{-3}. Also shown in the figure is the channel depth ($X_{Channel}$) for the $\sigma = 7\,kT$ case. We note that at high densities the enhancement of η stops the channel from shrinking and effectively prevents from the density to exceed the total DOS density. The dott-dashed line is clearly unphysical as it shows a density that can easily exceed any total DOS value.

Using the data in Figure 13 and in Figure 11 we can now plot the charge density dependence of the mobility as shown in Figure 14. Since we can only use data that satisfy $V_{GS} - V_T > V_{DS}$ and the V_{DS} used was 2 V we do not have access to the very low density regime but nevertheless one can still extract the slope (in loglog scale) of μ versus P. The power law found for this data is $\mu \propto P^1$. If we assume $N_V = 5 \times 10^{20}$ cm^{-3}

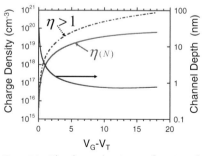

Figure 13 The charge density as a function of gate voltage (left axis). The calculation was done (dott-dashed) neglecting any change in the Einstein relation as well as for a density dependent Einstein relation assuming $\sigma = 7\,kT$. The right axis shows the channel depth as a function of gate voltage for the case of $\sigma = 7\,kT$.

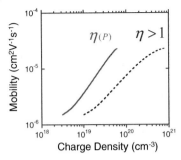

Figure 14 The mobility as a function of charge density. The solid line was derived accounting for the density dependent Einstein relation (solid line in Figure 13). The dashed line was calculated assuming the DOS is infinitely high such that η=1.

instead of 10^{20} the power law would be $\mu \propto P^{0.86}$ and if we do not correcting for $\eta>1$ at all (i.e. for $N_v \rightarrow \infty$) then the power law is $\mu \propto P^{0.67}$. We note that κ=1 (or κ=0.86) is larger then the maximum value (0.72) predicted for a single Gaussian DOS (see Figure 6). As was discussed in [15] this high slope is due to the presence of domains in the MEH-PPV film giving rise to effective DOS of a double Gaussian shape (see also discussions in [27-30]).

Finally, if the mobility is density dependent then one of the assumption underlying the derivation of equations (19) and (20) break down and one has to recalculate the current equation. The charge 2D-density at the channel can be written as:

$$n_s(x) = (C_{g/q})[V_{GS} - V_T - v(x)] \tag{30}$$

Where $v(x \cdot \Delta \cdot$ is the potential in the channel at point x. The current at each point can be written then as:

$$I_{DS} = Wqn_s(x)\mu(x)\frac{-dv(x)}{dx} \tag{31}$$

Integrating equation (31) from 0 to L:

$$\int_0^L I_{DS}dx = \int_0^L Wqn_s(x)\mu(x)\frac{-dv(x)}{ds}dx$$

$$I_{DS} = \frac{W}{L}\int_0^L C_g[V_{GS} - V_T - v(x)]\mu(x)\frac{dv(x)}{dx}dx \tag{32}$$

And finally replacing the integration over x with that of the channel potential:

$$\begin{cases} |I_{DS\_Lin}| = \frac{W}{L}\int_0^{V_{DS_{DS}}} C_g[V_{GS} - V_T - v]\mu(V_{GS} - V_T - v)dv \\[4mm] |I_{DS\_Sat}| = \frac{W}{L}\int_0^{V_{GS}-V_T} C_g[V_{GS} - V_T - v]\mu(V_{GS} - V_T - v)dv \end{cases} \tag{33}$$

For example, the simplest way of fitting the mobility is through a simple polynomial [49] function:

$$\mu(V_{GS} - V_T - v) = \sum_{n=0}^{N} \mu_n [V_{GS} - V_T - v]^n$$

Where μ_n are coefficients to be fitted. The equations describing the current take the form (assuming V_{GS}-V_T>0):

$$I_{DS\_Lin} = C_g \frac{W}{L} \sum_{n=0}^{N} \frac{\mu_n}{n+2} \left[(V_{GS} - V_T)^{n+2} - (V_{GD} - V_T)^{n+2} \right]$$

for V_{GS} in accumulation regime and $V_{DS} \leq V_{GS}$
And

$$I_{DS\_sat} = C_g \frac{W}{L} \sum_{n=0}^{N} \frac{\mu_n}{n+2} (V_{GS} - V_T)^{n+2}$$

for V_{GS} in accumulation regime and $V_{DS} \geq V_{GS}$. Using this formalism it is possible to fit the current and find the coefficients ($\mu_0..\mu_N$) and thus derive the mobility even for the low density regime which is inevitably associated with non-uniform charge distribution across the channel.

12.6
Photo-Cells

We have numerically analyzed the photocell using equations (1) to (3) modified to account for a position dependent optical pump rate [50]. It was shown that at the low excitation regime where the quantum efficiency is independent of the pump power the device can be considered as (optical) injection-limited. At higher excitation density where the efficiency starts to drop the device becomes limited by space charge and charge-recombination effects. Within the scope of Langevin recombination the two effects occur almost simultaneously and we have shown that as long as the efficiency hasn't dropped below half of its low density value it can be described following:

$$Eff_{SCL} = \frac{A}{\left(1 + 0.2373 \frac{A \cdot L}{J_{SCL}}\right)} \tag{34}$$

$$Eff_{REC} = A \left\{ 1 - \frac{\sqrt{1 + \frac{AL\, 9(\mu_h + \mu_e)}{J_{SCL}\, 8\ \ \mu_h}}}{\frac{AL\, 9(\mu_h + \mu_e)}{J_{SCL}\, 8\ \ \mu_h}} \right\} \tag{35}$$

Here L is the photon flux, J_{SCL} is the space charge limited current (eq. (15)) at the given voltage ($V=V_{appl}$-V_{bi}), and A is the efficiency (electron/photon) at the low excitation density regime. Experimental results made with high quality photocells [51] show that at sun brightness most organic cells would be at the high excitation regime hence, the space charge limited regime [50] or charge density higher then 10^{16} cm^{-3}.

Acknowledgements

This research (No. 56/00-11.6) was supported by the Israel science foundation. Y. R. thanks Israel science Ministry and Israel science foundation for the generous scholarships. We acknowledge fruitful discussions with V. I. Arkhipov

References

[1] N. F. Mott and R. W. Gurney, (Oxford university press, London, 1940).

[2] N. W. Ashcroft and N. D. Mermin, (HOLT, RINEHART AND WINSTON, New York, 1988).

[3] M. Pope and C. E. Swenberg, (Clarendon Press, Oxford, 1982).

[4] R. H. Friend, R. W. Gymer, A. B. Holmes, J. H. Burroughes, R. N. Marks, C. Taliani, D. D. C. Bradley, D. A. Dossantos, J. L. Bredas, M. Logdlund and W. R. Salaneck, Nature **397**, 121-128 (1999).

[5] M. Van der Auweraer, F. C. Deschryver, P. M. Borsenberger and H. Bassler, Advanced Materials **6**, 199–213 (1994).

[6] H. Scher, M. F. Shlesinger and J. T. Bendler, Physics Today **44**, 26–34 (1991).

[7] E. M. Conwell and M. W. Wu, Applied Physics Letters **70**, 1867–1869 (1997).

[8] G. G. Malliaras and J. C. Scott, J. Appl. Phys. **85**, 7426-7432 (1999).

[9] P. S. Davids, I. H. Campbell and D. L. Smith, J. Appl. Phys. **82**, 6319–6325 (1997).

[10] Y. Roichman and N. Tessler, Applied Physics Letters **80**, 1948–1950 (2002).

[11] P. W. M. Blom, J. M. J. deJong and C. Liedenbaum, Polymers For Advanced Technologies **9**, 390–401 (1998).

[12] D. J. Pinner, R. H. Friend and N. Tessler, J. Appl. Phys. **86**, 5116–5130 (1999).

[13] A. J. Campbell, D. D. C. Bradley, H. Antoniadis, M. Inbasekaran, W. S. W. Wu and E. P. Woo, Appl. Phys. Lett. **76**, 1734–1736 (2000).

[14] Y. Roichman and N. Tessler, Synthetic Metals **135-136**, 443-444 (2003).

[15] S. Shaked, S. Tal, Y. Roichman, A. Razin, S. Xiao, Y. Eichen and N. Tessler, Advanced Materials **15**, 913-+ (2003).

[16] A. A. Muhammad, A. Dodabalapur and M. R. Pinto, IEEE trans. elect. dev. **44**, 1332–1337 (1997).

[17] N. Tessler and Y. Roichman, Appl. Phys. Lett. **79**, 2987–2989 (2001).

[18] T. Li, J. W. Balk, P. P. Ruden, I. H. Campbell and D. L. Smith, J. Appl. Phys. **91**, 4312–4318 (2002).

[19] P. V. Necliudov, M. S. Shur, D. J. Gundlach and T. N. Jackson, J. Appl. Phys. **88**, 6594-6597 (2000).

[20] V. I. Arkhipov, P. Heremans, E. V. Emelianova, G. J. Adriaenssens and H. Bassler, Journal of Physics-Condensed Matter **14**, 9899–9911 (2002).

[21] V. I. Arkhipov, P. Heremans, E. V. Emelianova, G. J. Adriaenssens and H. Bassler, Applied Physics Letters **82**, 3245–3247 (2003).

[22] Y. N. Gartstein and E. M. Conwell, Chemical Physics Letters **245**, 351–358 (1995).

[23] Z. G. Yu, D. L. Smith, A. Saxena, R. L. Martin and A. R. Bishop, Phys. Rev. B **63**, 085202 (2001).

[24] L. Zuppiroli, M. N. Bussac, S. Paschen, O. Chauvet and L. Forro, Physical Review B **50**, 5196–5203 (1994).

[25] Y. Preezant, Y. Roichman and N. Tessler, J. Phys. Cond. Matt. **14**, 9913–9924 (2002).

[26] A. Miller and E. Abrahams, Phys. Rev. **120**, 745–755 (1960).

[27] S. V. Rakhmanova and E. M. Conwell, Applied Physics Letters **76**, 3822–3824 (2000).

[28] A. Kadashchuk, Y. Skryshevski, Y. Piryatinski, A. Vakhnin, E. V. Emelianova, V. I. Arkhipov, H. Bassler and J. Shinar, Journal of Applied Physics **91**, 5016–5023 (2002).

[29] M. A. Baldo, Z. G. Soos and S. R. Forrest, Chemical Physics Letters **347**, 297–303 (2001).

[30] T. Q. Nguyen, I. B. Martini, J. Liu and B. J. Schwartz, J. Phys. Chem. B **104**, 237–255 (2000).

[31] D. M. Basko and E. M. Conwell, Physical Review Letters **88** (2002).

[32] H. Bassler, Phys. Stat. Sol. (b) **175**, 15–56 (1993).

[33] H. C. F. Martens, P. W. M. Blom and H. F. M. Schoo, Physical Review B **61**, 7489–7493 (2000).

[34] J. G. Simmons, Physical Review Letters **15**, 967–& (1965).

[35] V. I. Arkhipov, U. Wolf and H. Bassler, Phys. Rev. B **59**, 7514–7520 (1999).

[36] Y. Preezant and N. Tessler, Journal of Applied Physics **93**, 2059–2064 (2003).

[37] V. I. Arkhipov, H. von Seggern and E. V. Emelianova, Applied Physics Letters **83**, 5074–5076 (2003).

[38] P. S. Davids, S. M. Kogan, I. D. Parker and D. L. Smith, Appl. Phys. Lett. **69**, 2270–2272 (1996).

[39] M. A. Baldo and S. R. Forrest, Physical Review B **64**, 85201 (2001).

[40] T. van Woudenbergh, P. W. M. Blom, M. Vissenberg and J. N. Huiberts, Applied Physics Letters **79**, 1697–1699 (2001).

[41] V. I. Arkhipov, E. V. Emelianova, Y. H. Tak and H. Bassler, Journal of Applied Physics **84**, 848–856 (1998).

[42] W. Osikowicz and W. R. Salaneck, private communication).

[43] J. M. Gottfried, K. J. Schmidt, S. L. M. Schroeder and K. Christmann, Surface Science **536**, 206–224 (2003).

[44] J. H. Werner and H. H. Guttler, Journal of Applied Physics **69**, 1522–1533 (1991).

[45] K. Maeda, Surface Science **493**, 644–652 (2001).

[46] N. Tessler and Y. Roichman, edited by J. C. deMello and J. J. M. Halls (Wiley and Son.

[47] C. Tanase, E. J. Meijer, P. W. M. Blom and D. M. de Leeuw, Physical Review Letters **91**, 216601 (2003).

[48] H. Sirringhaus, N. Tessler and R. H. Friend, Synth. Met. **102**, 857–860 (1999).

[49] O. Katz, Y. Roichman, G. Bahir, N. Tessler, and J. Salzman, Semicond. Sci. Technol. **20**, 90–94 (2005).

[50] N. Tessler and N. Rappaport, J. Appl. Phys. **96** (2), 1083–1087 (2004).

[51] W. U. Huynh, J. J. Dittmer, N. Teclemariam, D. J. Milliron, A. P. Alivisatos and K. W. J. Barnham, Physical Review B **67** (2003).

13
Fabrication and Analysis of Polymer Field-effect Transistors

S. Scheinert and G. Paasch

13.1
Introduction

Recent developments on organic field-effect transistors (OFET) are aimed at different applications as organic displays [1–3], complementary circuits [4, 5] and all-polymer integrated circuits [6, 7]. Fast organic circuits [8] have been prepared with vacuum-deposited films of pentacene with a comparatively high mobility. Solution processing seems to be better suited for low-cost fabrication, but such films show usually a low mobility [9]. However, in the last years mobilities up to 0.1 cm^{2}V^{-1}s^{-1} have been demonstrated for soluble regioregular poly(3-alkylthiophene) [10, 11, 7], poly-(9,9-dioctylfluorene-bithiophene) [12] and pentacene from soluble precursors [13]. Nevertheless, application-relevant performance requires rather short channels. Different patterning techniques such as screen printing [14], soft lithographic stamping [15], or inkjet printing [16] have so far not demonstrated both resolution and alignment accuracy desired. Photolithography [6] is expected to be too costly for the submicrometer regime. Recently, Stutzmann et al. [17] used embossing to fabricate vertical-channel field-effect transistors with submicrometer channel lengths, but could not observe saturation in the measured output characteristics. We developed [18] an underetching technique to define submicrometer channel length polymer field-effect transistors with high performance. By using source and drain as opaque optical mask for the gate definition in Ref. [19] a self-aligned gate has been fabricated recently for long-channel (20 μm) devices.

In spite of the achieved progress one is confronted with several problems as the observed hysteresis effects [20–22], a high inverse subthreshold slope [23–26] (which can be caused by rechargeable traps), or short-channel effects even at larger channel lengths up to 12 μm [27]. We reduced the inverse subthreshold slope in a poly(3-octylthiophene)/SiO$_2$ OFET almost to the lowest possible value of $\approx$ 200 mV/dec [28, 29] and also for the new submicrometer devices [18] low values of $\approx$ 500 mV/dec have been reached.

Modelling is required in order to analyze the measured data, to achieve an adequate understanding of the device operation and to develop optimized new device designs compatible with low-cost fabrication requirement. Of course, such modelling is based on the general physics of semiconductor devices [30]. Several analytical

Physics of Organic Semiconductors. Edited by W. Brütting
Copyright © 2005 WILEY-VCH Verlag GmbH & Co. KGaA, Weinheim
ISBN 3-527-40550-X

models have been developed [27, 31–34]. In Ref. [31] bulk traps are included in the current equations and their impact on the charge transport is investigated. Short channel effects such as channel length shortening, apparent field-effect mobility and series parasitic resistance have been considered in [27] and a possible explanation of the drain current dependence has been proposed. In Ref. [34] an analytical model for the specifics of the thin-film transistor has been developed and also the possible influence of bipolaron formation has been described. It has been argued also [7, 35–38] that the so-called amorphous-silicon model should be appropriate for OFET's. Consequences of this assumption have been clarified by a simulation study [39].

Refinements may concern a possible Gaussian distribution of the transport states (more likely in the case of evaporated low molecular weight materials than for polymers) and the dependence of the mobility on the applied field and on doping, the influence of the source/drain contacts, or internal series resistances due to the chosen transistor design. Evidently, the whole complexity of influences can be taken into account only in numerical two-dimensional (2D) simulations. Successful simulations of organic light emitting diodes [40–44] demonstrated the applicability of simulation tools from microelectronics [45, 46] based on the drift-diffusion model, and possible limitations have been clarified. Also, simulations of field-effect devices [21, 28, 25, 26, 29] have been used to analyze our prepared transistors and capacitors. Important design parameters of our submicrometer transistors [18] have been determined in advance by simulation of short-channel effects. Earlier, in Ref. [47] an organic thin film transistor simulation on the same basis has been reported including the field dependence of the mobility. Further numerical simulations have been reported in Refs. [48–51].

In this article, subsuming our experiences of the last years, we focus on how to make analytical estimates and numerical simulations as integral parts of the research on OFET's from design and fabrication up to the analysis of the devices, comprising both the transistors and the corresponding capacitors. Starting with the Shockley current characteristics, material models are discussed which are needed in the drift-diffusion-model enabling both numerical 2D simulations and analytical estimations connecting material, design and performance, which are helpful in both designing and analyzing devices. The next section deals with the preparation of devices and details of measuring techniques. Then we outline two examples of extensive analysis of OFET's and capacitors fabricated with the same materials in the same technological run. Results on other two transistor structures are presented including our new submicrometer devices. Next, also subthreshold currents and hysteresis effects are analyzed. Finally, three simulation studies are discussed concerning, inversion, the influence of a trap distribution and of different types of source/drain contacts for top and bottom contact OFET's, and short-channel effects in submicrometer devices.

13.2
Models

13.2.1
The Shockley current characteristics

Let us consider a thin-film transistor as shown below in Fig. 5 (b) of channel length and width L and w, respectively, and gate insulator thickness d_{ox}. The active layer material is assumed to be unintentionally p-doped and in the on state one has a p-accumulation channel with the hole mobility μ. In the most simple model [30], hereafter referred to as the Shockley model[1], the (above threshold) drain current of the transistor depends on drain-source and gate-source voltages V_{DS} and V_{GS}, respectively, as

$$I_D = \begin{cases} \mu \dfrac{w}{L} C''_{ox} \left[(V_{GS} - V_{th}) V_{DS} - \dfrac{1}{2} V_{DS}^2 \right] & V_{DS} < V_{GS} - V_{th} \\[2ex] \mu \dfrac{w}{L} C''_{ox} \dfrac{1}{2} (V_{GS} - V_{th})^2 & V_{DS} > V_{GS} - V_{th} \end{cases} \tag{1}$$

Here $C''_{ox} = \varepsilon_0 \varepsilon_{ox}/d_{ox}$ is the insulator capacitance per unit area and V_{th} is the threshold voltage. Evidently, all the material properties in these equations are the mobility and the threshold voltage (to be discussed later), whereby the latter can be influenced also by the interface to the gate insulator. It is amazing that nevertheless these simple expressions are often rather good for first estimates when analyzing measured characteristics. Although many refinements are necessary being just the subject of this article, the main voltage dependencies arise solely from the Coulomb law in the given transistor design with an active semiconducting (non-metallic) layer.

For illustration of the power of this approximation, we compare the maximum transconductance $g_m = (\partial I_D/\partial V_{GS})^{max}|_{V_{DS}}$ of two rather different devices, an ultra-short inversion-channel silicon transistor ('70nmSi') [52] and an organic micrometer-channel thin-film transistor ('μmP3HT') [18]. The ratio of the experimental values (10^3 mS/mm for '70nmSi' and 0.5×10^{-5} mS/mm for 'μmP3HT') for comparable low drain voltage (about 2 V) is

$$\left(\frac{g_m}{w} \right)^{70nmSi} \approx 2 \times 10^8 \left(\frac{g_m}{w} \right)^{\mu mP3HT}. \tag{2}$$

According to Eq. (1), one should expect a scaling as $g_m/w \propto \mu \varepsilon_{ox}/(L d_{ox})$. For the 'μmP3HT' device with channel length $L = 1\mu$m the mobility estimated from the measured current characteristics is $\mu \approx 3 \times 10^{-5}$ cm^2 V^{-1}s^{-1}, the gate insulator is SiO$_2$ with $d_{ox} = 30$ nm. For the '70nmSi' device with $L = 0.07\mu$m the mobility

---

[1] Textbook derivations are usually given for the inversion channel transistor. This model is often characterized as graduate channel approximation. Thereby in addition the diffusion current has been neglected.

(n-channel) due to high doping and field will be of the order of $\mu \approx 100 \, \text{cm}^2 \, \text{V}^{-1} \text{s}^{-1}$ and the silicon dioxide gate insulator is 3 nm thick. Thus, simple scaling gives

$$\left(\frac{\mu}{Ld_{ox}}\right)^{70\text{nmSi}} \approx 5 \times 10^8 \left(\frac{\mu}{Ld_{ox}}\right)^{\mu\text{mP3HT}}. \tag{3}$$

In view of the enormous magnitude and the different types of transistors, the astonishing agreement of this scaling with the experimental ratio (2) allows for the following conclusions: (i) The main operation mode is actually reflected by the simple Shockley equation and (ii) the main material property is the averaged mobility independent from the nature of the transport process (band conduction in crystalline silicon or hopping conduction in the disordered polymer).

Modifications of the current characteristics can arise from specifics of the material properties as distributions of transport and/or trap states, doping, of the nature of contacts, or they can be caused by a specific design. Thereby sometimes the functional dependence on the voltages might be apparently the same as in Eq. (1), or qualitative differences may occur, or only the subthreshold current is modified.

13.2.2
Material models

In the past, for the polymers and low molecular weight materials being of interest for field-effect devices, many properties as electrical, optical, structural, electrochemical, have been investigated both experimentally and theoretically. However, only few of these properties determine the operation of field-effect devices. In general, one can describe the conjugated polymers as organic wide-gap semiconductors, in most cases disordered ones. Depending on the preparation, in the case of the low molecular weight materials, but also e.g. for regioregular poly(3-alkylthiophene), some crystallinity can occur. In such cases, considering typical device dimensions, averaged material properties must be considered.

For the relative dielectric constant one has typically $\varepsilon = 2.5...3$. The energy gap in these devices is usually larger than $\varepsilon_g \approx 2 \, \text{eV}$ which is connected with a low intrinsic density. In unipolar devices one has accumulation or inversion of one kind of carriers. Correspondingly, the gap is the energetic difference between a state with one additional electron and one with an additional hole. Strictly speaking, these are polaronic states, connected with a distortion of the polymer chain or the molecule around the additional charge. However, for simplicity these states will be denoted either as the conduction band, CB (valence band, VB) edge or lowest unoccupied molecular orbital, LUMO (highest occupied molecular orbital, HOMO). The gap is in this sense a transport gap which might be different from the optically detected gap. The electron affinity, according to its definition (typically $\chi = 2....3 \, \text{eV}$) gives just the LUMO relative to the vacuum level. In addition to these states, due to disorder, tail states and traps will be present usually. Their influence will be discussed below.

In the organics suitable for field-effect devices the mobility (of holes or electrons) is clearly lower than $1\,\mathrm{cm}^2\,\mathrm{V}^{-1}\mathrm{s}^{-1}$, typically[2] $10^{-5}....10^{-2}\,\mathrm{cm}^2\mathrm{V}^{-1}\mathrm{s}^{-1}$. That means one has in any case some kind of hopping transport. It should be mentioned that the mobilities determining the transport in the device are *defined* only as averages, e.g. for the holes by the ratio of conductivity σ_p and density p as $\mu_p = \sigma_p/ep$. The exact values can be regarded as to be determined from fitting e.g. current characteristics, however, this mobility is not directly accessible from experiments such as time-of-flight measurements. Further, a possible field dependence [53, 54–56] of the mobilities for higher fields can be taken into account. But it is worth to check whether the electrical field in the transistor becomes really large enough this effect to become important. At present there is only little information on the possible dependence of the mobilities on the carrier concentration for a given doping, which in principle depends on the type of the hopping transport.

For the transport, besides the mobility one needs the carrier concentrations. It is supposed that one can divide the whole energy spectrum into current carrying states and trap states. In some cases this assumption introduces an ambiguity. With the densities of states (DOS) of the mobile electrons and holes ($\mathcal{D}_c$, $\mathcal{D}_v$) the electron and hole concentrations are given by

$$n = \int d\varepsilon \mathcal{D}_c(\varepsilon - \varepsilon_c) f(\varepsilon - \varepsilon_F) \quad , \quad p = \int d\varepsilon \mathcal{D}_v(\varepsilon_v - \varepsilon)[1 - f(\varepsilon - \varepsilon_F)] \tag{4}$$

and contain the Fermi distribution function $f(\varepsilon - \varepsilon_F)$ with the Fermi energy ε_F. In the non-degenerate limit one obtains e.g. for the electron concentration

$$n = N_C \exp\{(\varepsilon_F - \varepsilon_c)/k_B T\} \tag{5}$$

$$N_C = \int d\varepsilon \mathcal{D}_c(\varepsilon) \exp\{-(\varepsilon - \varepsilon_c)/k_B T\} \tag{6}$$

where N_C is to be evaluated for the given DOS. Especially with the parabolic DOS of band states it is the so called effective density of states (actually a density, unit cm^{-3}) given by the well known textbook expression [30].

However, the functional dependence given by Eq. (5) is valid in many other cases. For a molecular material, Eq. (5) (apart from the notation[3]) is just the Nernst equation describing reduction for low concentrations. As the gap one has the difference between reduction and oxidation potentials (in the solid state). They contain contributions from the deformation of the molecules and possibly entropy contributions for the case of multiple redox sites. The situation is similar in polymers with polarons (and bipolarons) as charged states of the chains, where the gap is twice the polaron formation energy [57], containing the same influences as just mentioned. In both cases, instead of 'effective densities of states' one has to use the molecular

---

[2] Higher mobilities are possible for ordered layers or single crystals. But they seem not to be suited for low-cost fabrication.

[3] In electrochemistry electrical potentials are used instead of energies as in Eq. (5), $-\varepsilon_F/e$ is then the electrochemical potential and $-\varepsilon_c/e$ the standard potential for reduction.

or monomer density which is of the order of $N_C = N_V = 10^{21}\,\text{cm}^{-3}$ (alternatively the degeneration factor due to spin and possible multiple redox sites can be included here). In addition, a distribution of the redox potentials or the polaron formation energies is characterized by a density of states formally similar to the DOS of the band states (Electrochemical measurements on conducting polymer layers indicate at least in some cases a rather narrow distribution which is reflected by the small width of the oxidation peak [58]). Then N_C is again given by Eq. (6) where ε_c is the position of the maximum of the distribution. For a width of the distribution small compared to $k_B T$ one obtains for N_C and N_V again the above mentioned values, and smaller ones for broader distributions. The non-degenerate approximation (5)-(6) can be applied usually if the Fermi energy lies in a region where the DOS is zero and its distance to the nonzero DOS is large compared to $k_B T$. For both an exponential and a Gaussian DOS [56, 59, 60] the non-degenerate approximation Eqs. (5) and (6) remains valid if their exponential tails decrease faster than the Fermi-distribution tail raises. This case is disfavored for lower temperatures used in analyzing transport properties. Thus according to Ref. [61] the degenerate case should apply to the organics. For OFETs (accumulation, operation at about or above room temperature) we show in Appendix A the actual limits for the applicability of the non-degenerate approximation in the case of the Gaussian distribution. It should be mentioned that until now for a broad distribution when the non-degenerate approximation fails, implications and comparisons with experiment are unknown for both the subthreshold and saturation regions of OFETs. For polymers with polarons as mobile charges as the main subject of the present article we assume that the concentrations are sufficiently well described by Eq. (5).

For a long period, the interest in doped conjugated polymers has been concentrated on the 'synthetic metals' aspect with doping levels above 1 Mol%. But for electronic devices one needs semiconductors with a level of electrically active dopants clearly below $N_A = 10^{18}\,\text{cm}^{-3}$, i.e. roughly below 0.01 Mol%. Otherwise the screening lengths determining the device operation become too small. In this range purity of the polymers becomes a problem, there are remains of catalysts and unintentional oxidation is possible. Indeed, unintentional doping usually leads to either p- or n-type material. In the materials used for transistors one has usually p-doping. There is only little information both on the concentration and the chemical/structural nature of dopants in these practically used materials. By analyzing capacitors and transistors we found usually (for P3AT and modified PPV) doping levels of the order $N_A = 10^{16}\,\text{cm}^{-3}\ldots10^{17}\,\text{cm}^{-3}$. Even in high quality organic single crystals acceptor concentrations of $10^{13}\,\text{cm}^{-3}$ and trap concentrations of 10^{15} to $10^{16}\,\text{cm}^{-3}$ have been found [62]. What one needs is a controlled low doping at the above mentioned level. But, for many disordered organics of interest for electronics, this is contradictory to the need of a high mobility. Actually, an empirical relation between conductivity and mobility has been established [63] ($\mu \propto \sigma^{0.76}$, Fig. 1) in addition to the doping dependence of the conductivity $\sigma \propto N^{4.5}$. Both can arise for variable range hopping as transport mechanism [64] as demonstrated in Fig. 1. According to these dependencies, high doping level would favor the required high mobility in contrast to the desired controlled low doping. Oligothiophenes belong to these materials and

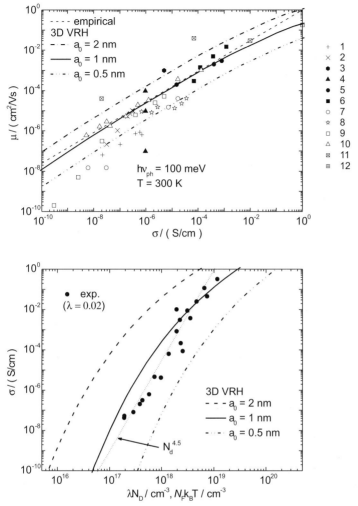

Figure 1   (a) The $\mu - \sigma$ relation: Experimental values from Ref. [63] (for the references see also [64]) with the empirical fit (dashed) and the 3D VRH dependence (Parameters: phonon energy of 100 meV, $T = 300$ K and three different Bohr radii a_0). (b) Conductivity in dependence on the carrier concentration expressed for the experimental values (taken from Ref. [63], symbols, 300 K) by λN_D and for 3D VRH on the effective concentration by $N_F k_B T$ (Parameters: phonon energy of 100 meV, $T = 300$ K and three different Bohr radii a_0). The empirical dependence is shown also.

hence probably also polythiophenes. Their unintentional oxidation might be partly responsible for the obtained high mobility. We compared recently transistors with P3HT and the same material after chemical purification showing clearly that purification reduces the mobility. However, the mobility of the polymer layers can be increased significantly by ordering. The challenge is to combine low-cost solution based layer deposition with methods to achieve some degree of order.

13.2.3
Drift-diffusion-model and numerical simulation

Estimations which are usually employed are based on approximations of the basic system of semiconductor equations. Full solution of the same system of equations is achieved by numerical simulation, which in the case of most field-effect devices must be done two-dimensionally. Variables of this system are the electrical potential φ and the hole and electron quasi-Fermi potentials φ_{Fp} and φ_{Fn}, all depending on position and on time. For the concentrations given in the equilibrium by Eq. (5) and the corresponding expression for holes, in the local equilibrium approximation the three potentials are connected with the concentrations as

$$p = n_i \exp\left\{\left(\varphi_{Fp} - \varphi\right)/V_T\right\} \quad , \quad n = n_i \exp\left\{(\varphi - \varphi_{Fn})/V_T\right\}$$

$$n_i = \sqrt{N_V N_C} \exp\left\{-\varepsilon_g/2eV_T\right\}. \tag{7}$$

The intrinsic density n_i is connected with the gap energy ε_g and N_V and N_C. $V_T = k_B T/e$ is the thermal voltage.

The system of partial differential equations consists of the Poisson equation for the electrical potential φ (position dependent intrinsic potential) and the continuity equations for the hole and electron densities. After introducing the current density equations

$$\vec{j}_p = e p \mu_p \left(-\nabla \varphi_{Fp}\right) \quad , \quad \vec{j}_n = \nabla e n \mu_n \left(-\nabla \varphi_{Fn}\right). \tag{8}$$

into the continuity equations, the system of equations reads as

$$\nabla \varepsilon \varepsilon_0 (-\nabla \varphi) = e(p - n - N^-)$$

$$\nabla e p \mu_p \left(-\nabla \varphi_{Fp}\right) = -e \frac{\partial}{\partial t} p - e(U_{SRH} + U_{direct}) \tag{9}$$

$$\nabla e n \mu_n \left(-\nabla \varphi_{Fn}\right) = +e \frac{\partial}{\partial t} n + e(U_{SRH} + U_{direct}).$$

Here a negative net concentration N^- of ionized dopants is taken into account, traps will be included below. U_{SRH} is the Shockley-Read-Hall recombination rate [30]. This mechanism is usually negligible in the unipolar devices. Exceptions are the creation of an inversion layer in a MIS capacitor and the normally unwanted case of Schottky contacts at source/drain. The direct radiative recombination rate becomes important only in the ambipolar [65, 66] or double-injection field-effect transistor. It is given by

$$U_{direct} = r_L \left(np - n_i^2\right) \quad , \quad r_L = \frac{e}{\varepsilon \varepsilon_0}(\mu_p + \mu_n) \tag{10}$$

where r_L is the (so-called bimolecular) Langevin recombination constant. For the organics and for the device dimensions realized until now other recombination/generation processes (as Auger recombination, avalanche generation) are negligible.

The system of equations (9) is the drift-diffusion-model (DDM): The hole and electron currents (8) are driven by the gradients of the two quasi-Fermi potentials, but they can be expressed as drift and diffusion currents since the mobilities are connected with the diffusion coefficients D_p and D_n by the (non-degenerate) Einstein relations $\mu_p = D_p/V_T$ and $\mu_n = D_n/V_T$, respectively. These Eqs. (8) do not depend on the type of the transport mechanism and they are especially also valid in the case of hopping transport[4]. It is only assumed that the transport is linear in the gradient of the quasi-Fermi potential. Essentially the same model has been applied to OFET's also by other groups [47–51].

Dependencies of the mobilities on concentration and field can be taken into account directly. The dependence of the mobility on the lateral field due to increased hopping mobility by barrier lowering becomes important for fields which are in most cases larger than in the present OFET's. A model for the dependence on the perpendicular field, or the gate voltage, has been discussed in Ref. [56]. The corresponding experimental information has been extracted from the linear region of the current characteristics assuming the validity of Eq. (1). However, other influences as e.g. an exponential trap distribution, a Schottky-type contact or an Ohmic series resistance in a top-contact structure will lead also to significant deviations from this equation (see Section 13.6), which can be misinterpreted into the mobility. Thus we prefer usually to start with a constant mobility. The extension for the inclusion of traps or even trap distributions are described in Appendix B.

Apart from initial conditions, boundary conditions have to be used at source and drain. In the simulations either thermal equilibrium alone (diffusion approximation) or thermal equilibrium in addition with thermionic emission can be used and also the image charge correction can been added. Simulations have shown that for the typical parameters of the organics both thermionic emission and the image charge correction lead to almost negligible modifications when taking into account the barrier lowering. This is in contrast to a larger influence of back diffusion due to the spatial dependence of the image charge potential which has been found for injection into empty states of a broad Gaussian distribution. For favorable Ohmic or accumulation contacts in OFET's this effect will be less important. Also, tunneling is not of interest for the required ohmic source/drain contacts. As (hole) barrier at the source/drain contacts we choose the differences between the metal work function and the energy of the transport states ($-\chi$ and $-(\chi + \varepsilon_g)$). Possible dipole layers or Fermi level pinning can be modelled by choosing a corrected work function. The nature of the contact, ohmic/accumulation or Schottky, is determined by the positions (relative to the vacuum level) of the metal Fermi energy and the semiconductor Fermi energy. The latter is determined by bulk neutrality and depends therefore on the doping level and, may be on the traps.

Numerical two-dimensional simulations with the DDM are carried out with the device simulation programs ATLAS [45] and ISE TCAD [46]. The material parameters as specified to the organics are summarized in Table 1.

---

[4] Only in the case of variable range hopping there is the peculiarity that for increasing band bending the type of the hopping mechanism can be changed.

Table 1 Standard material parameters used in the simulations.

	P3AT	PPV	SiO$_2$	P4VP
ε	3.24	3.0	3.9	2.56
χ (eV)	3.0	2.6		
ε_g (eV)	2.0	2.4		
N_C, N_V (cm^{-3})	10^{21}	10^{21}		
N_A^- (cm^{-3})	$10^{16}...10^{17}$	$10^{16}...10^{17}$		
μ_p, μ_n (cm^{2}V^{-1}s^{-1})	10^{-3}	10^{-3}		
σ_p, σ_n (cm^2)	10^{-15}	10^{-15}		
$v_{th,p}, v_{th,n}$ (cm s^{-1})	10	10		

13.3
Analytical estimates

13.3.1
Basic dependencies

In a MIS structure with a thick semiconductor, the electric field vanishes in the bulk and its value at the interface to the semiconductor, denoted as surface electric field, according to Gauss law is connected with the areal charge of the space charge layer $Q'' = -\varepsilon\varepsilon_0 E_s$. For the concentrations (7), due to the Poisson equation, the electric field and the potential at the same position are connected with each other, e.g. at the interface by (11)

$$E_s = \text{sgn}(\varphi_s - \varphi_b)\sqrt{2}\frac{V_T}{L_D}\left(\cosh\frac{\varphi_F - \varphi_s}{V_T} - \cosh\frac{\varphi_F - \varphi_b}{V_T} + \frac{\varphi_s - \varphi_b}{V_T}\frac{N_A^-}{2n_i}\right)^{1/2} \quad (11)$$

$$E_s = \frac{\varepsilon_{ox}}{\varepsilon}\frac{(V_{GS} - V_{FB}) - (\varphi_s - \varphi_b)}{d_{ox}} \quad (12)$$

whereas Eq. (12) follows from the gate voltage drop over the whole MIS structure. The (intrinsic) Debye screening length is $L_D = \sqrt{\varepsilon\varepsilon_0 V_T/2en_i}$. V_{FB} is the flat-band voltage (i.e. the difference $\Phi_{M,S}$ between the gate work function and that one of the semiconductor, both expressed in Volt). The actual values of the surface potential and the surface electric field are obtained from the solution of the two coupled equations. The semiconductor capacitance follows from $C'' = \varepsilon\varepsilon_0 |dE_s/d\varphi_s|$.

The current (8) is driven by the gradient of the quasi-Fermi potential, hence one has in direction of the channel $j = -e\mu_p p(d\varphi_{Fp}/dx)$. Now, from the two equations (11)-(12) only the second one is used. The total areal charge in the Gauss law consists of ionized acceptors $(-eN_A^- d)$, the amount of holes (Q_p'') which is controlled by the gate voltage within the depletion length

$$l_{dep} = \sqrt{\frac{2\varepsilon\varepsilon_0 |2\varphi_b|}{eN_A^-}} \quad , \quad \varphi_b = -V_T\ln\frac{N_A^-}{n_i} \quad (13)$$

and out of the depletion layer there is the constant bulk hole concentration $p_0 = N_A^-$ resulting in an ohmic current. Thus, the total current $I_{total} = I + I_\Omega$ consists of the gate voltage controlled transistor current I and the ohmic current I_Ω outside the depletion layer according to

$$I = -\mu_p w Q_p'' \frac{d\varphi_{Fp}}{dx} \quad , \quad I_\Omega = e\mu_p p_0 \frac{w(d - l_{dep})}{L} V_{DS},$$

$$Q_p'' \approx C_{ox}'' \left\{ (V_{GS} - V_{th}) - \varphi_{Fp} \right\}, \tag{14}$$

$$V_{th} = V_{FB} + \frac{eN_A^- d}{C_{ox}''} - 2\varphi_b,$$

Here in the second line it is assumed that the surface potential depends only on the voltage drop at a given position (graduate channel approximation) and that at this position the surface potential is shifted, compared to source, by the quasi-Fermi potential (neglect of diffusion current). Integration of Idx leads to the Shockley equation (1) and one has the additional ohmic current I_Ω with the consequence that there is no off-state below the threshold voltage and one has also no saturation of the current.

Therefore the layer thickness must be less than the depletion length, $d < l_{dep}$. Then the ohmic current contribution vanishes, but the potential at the outer surface of the layer is floating, in Eqs. (11)-(12) one has to replace the bulk potential $\varphi_b \to \varphi_d$ where φ_d must be determined by numerical integration. An approximation [34] leads to a modified threshold voltage

$$V_{th} = V_{FB} + \frac{eN_A^- d}{C_{ox}''} - 2\varphi_b \left(\frac{d}{l_{dep}} \right)^2 \quad , \quad d < l_{dep}. \tag{15}$$

An extended analytical model (beyond Eq. (1)) has been developed by us in Ref. ([34]). However, for a more detailed analysis full numerical two-dimensional simulation seems to be preferable. Another extension of the Shockley model is the Pao-Sah model which is obtained by integrating just the first of the two equations (11)-(12) and using Eq. (12) in the limit of integration. The graduate channel approximation is still used in this model, but in addition to the field current also the diffusion current is taken into account. The needed numerical evaluation has not been carried out for the thin film transistor.

13.3.2
Interrelation between material and device properties

Here several estimations will be discussed which are helpful both in designing devices and in the first analysis of measurements, leading to input parameters for simulations. At first, operation of the thin film transistor with off-state and saturation requires a layer thickness less than the depletion length (13),

$$d < l_{dep}. \tag{16}$$

For the parameters of PPV one has $l_{dep} \approx 240, 76, 24$ nm for doping levels of $10^{16}, 10^{17}, 10^{18}$ cm^{-3} and only slightly lower values for P3AT. For high unintentional doping or high intended doping (in order to have a high mobility) it becomes then important to prepare layers in the 50 nm range with roughness small compared to the thickness.

The next important quantity is the threshold voltage V_{th} (15). Generally, low operation voltage requires at first a low threshold voltage. For a *p*-channel device operating at negative voltages, a negative threshold voltage would be preferable, since a positive one must be shifted by additional circuitry. In Eq. (15) only the first term can be negative by choosing a low-work function gate. Especially the second contribution acts with high positive values for high doping into the unwanted direction. In Fig. 2 several dependencies are visualized (for $\varepsilon_g = 2$ eV, $\varepsilon = 3.24$, $\chi = 3$ eV, gate work function 5 eV, organic insulator with $\varepsilon_{ox} = 2.56$, $V_{FB} = 1$ V $+ \varphi_b$). Only for low and moderate doping ($< 10^{16}...10^{17}$ cm^{-3}) the threshold voltage is sufficiently low for the considered layer thicknesses (Fig. 2 (a)). For higher doping (which might be favorable to achieve a high mobility) a strong increase of the threshold voltage can be prevented only by choosing both a thin insulator and a thin active layer. For high doping of 10^{18} cm^{-3} (Fig. 2 (b)) and layer thickness less than the depletion length, the extreme increase of the threshold voltage with the gate insulator thickness is visible. Thus for all-polymer circuits the most stringent requirement concerns the rather thin organic gate insulator.

In addition, a shift of the threshold voltage can result from a flat band voltage shift caused by interface charges (areal charge Q_{if}'') at the interface between the active layer and the gate insulator according to

$$V_{FB} = \Phi_{MS} - \frac{Q_{if}''}{C_{ox}''} = \Phi_{MS} - \frac{e N_{if}'' d_{ox}}{\varepsilon_0 \varepsilon_{ox}}. \tag{17}$$

Such interface charges can occur unintentionally and evidence for their existence has been reported [34]. For $N_{if}'' = 10^{12}$ cm^{-2} and $\varepsilon_{ox} = 3$ the flat band voltage is shifted by 3 V (18 V) for an gate insulator thickness of 50 nm (300 nm) showing again the importance of a thin organic insulator. A procedure to intentionally introduce such a charged layer in order to shift the threshold voltage into the needed direction has not yet been reported.

The inverse subthreshold slope $S = \partial V_{GS}/\partial \lg(I_D/A)$ gives the gate voltage change needed in order to change the subthreshold current by one order of magnitude. For a low-voltage operation of the transistor, S must be as low as possible. Its minimum value is at room temperature $S \geq V_T \ln 10 \approx 60$ mV/dec and is determined by the Boltzmann statistics for the carriers. By contrast, experimental values for the organic transistors are usually with several V/dec rather large [23–26][5]. Actually, recharging of capacitances leads to higher values according to [30]

$$S = V_T \ln 10 \left(1 + \frac{C_d'' + C_{it}''}{C_{ox}''} \right). \tag{18}$$

---

[5] Such large values cannot be explained by the assumption of an exponential or Gaussian distribution of transport states.

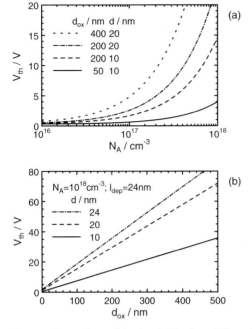

Figure 2 Dependence of the threshold voltage (15) on doping (a) and gate insulator thickness (b) for parameters as indicated.

Here $C''_d = \varepsilon\varepsilon_0/l_{dep}$ is the depletion-layer capacitance. Bulk traps contribute to the depletion capacitance. Interface charges Q''_{it} due to rechargeable traps with energy states in the band gap of the semiconductor are associated with a capacitance C''_{it} which is in parallel with the depletion-layer capacitance. For the trap free case in a thin layer, according to Eq. (16), there is no further recharging since the layer is anyway depleted and one would have the ideal minimum value. Thus, in the thin layer bulk and interface traps cause the deviation from the ideal minimum value.

In order to analyze fabricated transistors one should in any case prepare also MIS capacitors in the same technological run. The maximum value of the total MIS capacitance measured in accumulation is the oxide capacitance

$$C^{MIS}_{max} = C_{ox} = \frac{\varepsilon_0\varepsilon_{ox}A}{d_{ox}} \tag{19}$$

(A is the area of the capacitor). The minimum capacitance

$$C^{MIS}_{min} = \frac{C_{ox}C_{min}}{C_{ox}+C_{min}} \quad , \quad C_{min} = \frac{\varepsilon_0\varepsilon_r A}{l_{max}} \tag{20}$$

is determined by the achievable minimum of the semiconductor capacitance C_{min} which in turn is determined by the maximum extension l_{max} of the depletion region. For a thick layer, $d > l_{dep}$, this is the depletion length $l_{max} = l_{dep}$ (13). In this case one can determine the doping concentration N_A^- from the minimum capacitance (20).

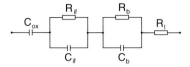

Figure 3 Equivalent circuit of an organic MIS capacitor.

However, for a thin layer ($d < l_{dep}$) one has $l_{max} = d$ and one can estimate only a maximum value of the doping level.

For the preceding consideration the MIS capacitor has been approximated as a series connection of the insulator and semiconductor capacitances. For the frequency dependence the common model for the MIS structure consists of the oxide capacitance and a parallel RC-term for the semiconductor. Actually, in dynamic measurements one determines the reciprocal of the complex impedance Z, the admittance $Z^{-1} = Y = G_p + j\omega C_p$. The modulus is defined as $M = j\omega C_{geo} Z = j\omega C_{geo} Y^{-1}$ with C_{geo} as the geometrical capacitance of the whole MIS structure. The real part of $M^{-1} = \varepsilon$ is proportional to the capacitance discussed above and the second quantity of interest is its imaginary part which shows more structure for this system. It turns out that for the organic MIS structures it might be that one has to include all elements of the equivalent circuit depicted in Fig. 3: the lead resistance R_l, the oxide capacitance C_{ox}, the resistance and capacitance of the bulk (R_b, C_b) layer and the resistance (R_{if}) in parallel with a capacitance (C_{if}) accounting for the accumulation layer (and/or interface states). The imaginary part of the modulus is then given by

$$\text{Im}(M) = \omega C_{geo} \left(\frac{R_{if}}{1+(\omega C_{if} R_{if})^2} + \frac{R_b}{1+(\omega C_b R_b)^2} + R_l \right). \tag{21}$$

Fitting this expression to the experimental data yields the various capacitances and resistances determined by layer thicknesses and conductivities. In combination with transistor data on can then extract input parameters for more detailed simulation studies.

In general, in the presence of traps the equivalent circuit must be extended further. The corresponding analysis can be complicated if the different relaxation times do not differ from each other.

13.3.3
Cut-off frequency

As a matter of course, goals of the OFET developments were higher mobility of the active layer material and shorter channel (or gate) length, both leading to higher drain currents according to Eq. (1). Actually, for any applicability minimum requirements on the speed of operation must be fulfilled. The relevant quantity is the cut-off frequency, $f_0 = g_m/(2\pi C_{GS})$, at which the voltage gain (ratio of output and input voltages) comes down to unity. It is determined by the maximum transconductance and the gate-source capacitance, which due to overlap capacitances is larger than the gate insulator capacitance, $C_{GS} \geq C_{ox}'' wL$. Hence as an upper limit we obtain with Eq. (1)

$$f_0 \leq \frac{1}{2\pi} \frac{g_m}{C_{ox}'' wL} = \frac{\mu}{2\pi L^2} V_{GS,eff} \qquad , \qquad V_{DS} = V_{GS,eff} = V_{GS} - V_{th}. \qquad (22)$$

where the second equation gives the voltage (where pinch-off occurs) at which the transconductance has its maximum. The channel width cancels out in Eq. (22). Consequences of this relation are seen at best by a graphics (Fig. 4) showing the channel length as function of the mobility with the cut-off frequency and $V_{GS,eff}$ as parameters. Applicable circuits should operate at sufficiently low voltage. For $V_{GS,eff} = 10\,V$ in Fig. 4 three cut-off frequencies (100 kHz, 1 MHz, 10 MHz) have been chosen, where the lowest one is clearly the limit for applications. Only for a high $\mu = 1\,cm^2V^{-1}s^{-1}$ (unrealistic for solution processing) for this frequency a long channel ($\approx 30\,\mu m$) is sufficient which can be fabricated with present printing technique. Even for the still high value $\mu = 0.01\,cm^2V^{-1}s^{-1}$ a short channel length of $\approx 3\,\mu m$ is required. Considering that overlap capacitances are still neglected, one can expect that only transistors with submicrometer channel length lead to the required speed of plastic circuits. In Fig. 4 in addition an experimental value for our fabricated submicrometer OFET is depicted together with the corresponding dependence (22). Both will be discussed below in Section 13.5.3.2. It should be mentioned that for such devices one must take care to avoid short-channel effects. Corresponding conditions are analyzed below by means of two-dimensional simulation in Section 12.6.3.

The same limitation is obtained from the average transit time $\tau_t = L/v_d$ with the average drift velocity $v_d = \mu V_{DS}/L$. At the transition from active to saturation regime one has $V_{DS} = V_{GS,eff}$. Then $(2\pi\tau_t)^{-1}$ coincides with the right hand side of Eq. (22). This is of interest from the following reason. Since the transistor can operate only for frequencies clearly below the cut-off frequency, dispersive transport will not influence the device since it becomes important for shorter times than the average transit time.

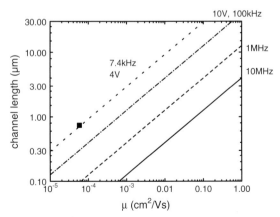

Figure 4 Channel length as function of the mobility leading for a given effective voltage of 10 V to three different cut-off frequencies 100 kHz, 10 MHz, 1 MHz. Also one experimental value for our fabricated submicrometer OFET is depicted together with the dependence (22) for the corresponding voltage and maximum cut-off frequency.

13.4
Experimental

In this section, preparation and measuring of organic devices are described. Low-cost fabrication requires the use of solution-processible polymers instead of vacuum deposited low-molecular weight organics. We use poly(3-alcylthiophene) (P3AT) and different types of poly-(phenylene-vinylene) (PPV) such as arylamino PPV and poly(2-methoxy, 5 ethyl (2'hexyloxy)paraphenylenevinylene) (MEH) PPV for the active layer. To impede the doping by oxygen, the preparation of the polymer solution, the spin coating process and the subsequent annealing step are realized in a glove-box with a nitrogen atmosphere. The devices are prepared either on top of a highly n-doped silicon substrate or on a plastic foil. In the latter case, an organic insulator is used and consequently only the source, drain and gate contacts are made from gold. For the devices on the silicon wafer a thermally grown oxide is used for the gate insulator. In this case, at any technological run MOS capacitors and OFET's are prepared on the same wafer to get more information with regard to material parameters.

13.4.1
Preparation of transistors and capacitors

Cross sections of the prepared devices are shown in Fig. 5. Highly doped silicon substrate is used as the gate contact. The thickness of the silicon dioxide is as low as 30 or 50 nm in order to realize low voltage operation and to reduce short channel effects in the case of submicrometer devices. For the transistor design of Fig. 5 (a), the active layer is spin coated on top of the silicon dioxide followed by an annealing step. The necessary layer thickness in the range of 10 to 50 nm is adjustable by controlling the viscosity of the solution and the spin coating velocity. For PPV layers a 0.5 wt.% solution in toluene is used and the P3AT layer is made from a 0.3 ... 0.4 wt.% solution in $CHCl_3$. The spin coating velocity is 3000 rpm. Finally, circular contacts for the MOS capacitors and source/drain contacts for the transistors are evaporated through a shadow mask. The channel lengths of the transistors are 25 and 50 μm, respectively. Because of the last evaporation process the maximum channel width is as low as 1 mm. Consequently, the realized w/L ratios of the transistors are very low resulting in low currents. Higher w/L ratios are possible applying a finger structure for the source/drain contacts used in the design of Fig. 5 (b). For these transistors channel length of 40 μm down to 2 μm have been realized with the following fabrication steps. First, a 50 nm thick gold layer is large-area sputtered on top of the oxidized silicon wafer. Subsequently, a photoresist is spin-coated onto the gold layer followed by a low-resolution photolithographic step to fabricate the finger-structured photoresist layer. Then, a wet chemical etching is applied to remove the gold. After removing the photoresist, the active organic layer is spin-coated. The realized w/L ratios are in the range of 1000 up to 16000 resulting in higher transistor currents. Another transistor with finger-structured source/drain contacts is described by us in [25]. For this device (Fig. 5 (c)), poly(ethylenetherephthalate)

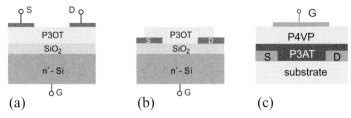

Figure 5 Schematic cross sections of prepared thin film transistors.

(PET) substrates carry the source and drain electrodes (Au) which were patterned by conventional lithographic processes. Then the active layer made from poly(3-dodecylthiophene) and subsequently the organic insulator poly(4-vinylphenol) (P4VP) are spin coated. The thickness of the insulator is 500 nm. Finally, the gates were made from gold by vacuum evaporation using a hard mask.

As described above, because of the low mobility of polymer materials, channel lengths in the submicrometer range are necessary. For such transistors the same design as used for the finger structures (Fig. 5 (b)) has been realized. We used underetching technique, in combination with low-resolution photolithographic steps and further standard microelectronic processes, for the inexpensive and manufacturable definition of the submicrometer channel length devices. The fabrication steps are described in [18]. Applying this technology, channel lengths $<1\,\mu$m with sufficiently high w/L ratios are possible.

13.4.2
Measuring techniques

Characterization of the devices comprise both the investigation of the layer properties and the measurement of the electrical device characteristics. Spectral ellipsometry is used to measure the index of refraction, the band gap energy and the layer thickness. The roughness of the prepared layers is investigated by means of atomic force microscopy (AFM).

Electrical characterizations have been carried out for both the MOS capacitor and the OFET. From the capacitance-voltage curves of the capacitor one gets the doping concentration and in the case of thin layers the layer thickness. Furthermore, the flat band voltage shift gives an information regarding the interface properties. Quasi-static capacitance-voltage (CV) curves have been measured using the HP 4140B pA-Meter/dc V source which allows for a variation of the linear gate voltage ramp from $1\,$mV s^{-1} to $1\,$V s^{-1}. The gate voltage is swept from positive voltages (depletion) to negative values (accumulation) and subsequently in the inverse direction. In this manner, hysteresis effects can be investigated. Dynamic measurements of the CV-curves have been carried out with the Solartron impedance/gain-phase analyzer 1260A in combination with the dielectric interface 1296. Consequently, measurements in the frequency range from 10^{-3} Hz to 10^5 Hz are possible. The gate voltage is again swept in different directions. The measurement of the frequency response of the impedance in accumulation allows for the estimate of the mobility vertical to the interface. Besides the frequency, the magnitude of the dc-voltage step and range, the value of the ac-voltage, the delay time t_D and hold time t_h are variable parameters in the dynamic measurements.

Quantification of the transistor performance requires both the output and transfer characteristics. Furthermore, the mobility in the lateral direction to the interface can be estimated from these curves. The current-voltage (IV) characteristics of the transistors have been measured using the Keithley 4200 which is provided with low current pre-amplifiers with a resolution of 0.1 fA. This configuration allows for an exact measurement of both the gate leakage currents and the drain currents in the subthreshold region. The choice of the measuring parameters depends on the characteristics to be measured. For the OFET output characteristics short delay times ($t_D = 1s$) are possible. The time before the measurement starts after the variation of the gate voltage (hold time) is $t_H = 60s$. To investigate hysteresis effects the gate voltage is varied from negative to positive voltages and in the inverse direction. In a further measurement, at each gate voltage the drain voltage is swept from 0 V to a negative voltage and back to 0 V. The transfer characteristics have been measured with a hold time of $t_H = 180s$. Since for these characteristics the gate voltage is up and down swept, higher delay times of $t_D = 10s$ are required.

13.5
Device characterization and analyzes

13.5.1
Basic procedure: P3OT devices as an example

13.5.1.1 Layer characterization
For the P3AT layers, the static dielectric constant $\varepsilon_r = n^2 \approx 3.24$ and the gap energy of about $E_g \approx 2\,eV$ have been measured by means of spectral ellipsometry [29]. Both values are used below in the numerical simulations. The prepared layer thickness has been ellipsometrically measured for each sample with a constant wave length of 632.8 nm. The results for the different layers are in the range of 10 to 50 nm. Another important parameter is the roughness of the layer because according to Eq. (15) the layer thickness determines the threshold voltage of the thin film transistor. Fig. 6 shows a result of an AFM measurement for 12 nm thick P3OT. The roughness of the layer is with $\approx 3.5\,nm$ relatively small. Further information regarding the material properties has been obtained from measured characteristics of the MOS capacitor and the polymer transistor.

13.5.1.2 MOS capacitor
Measured characteristics of a MOS capacitor with an active layer made from P3OT are shown in Fig. 7. Noticeable is the large hysteresis of the dynamic CV curves (Fig. 7 (a)) reported also e.g. for arylamino-poly-(p-phenylenevinylene) in [29, 21] and poly(3-hexylthiophene) (P3HT) [67]. Until now, the reasons for this effect are not

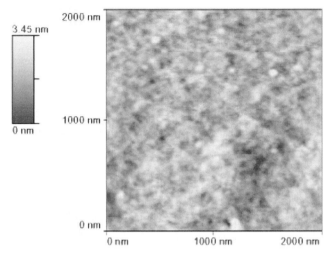

Figure 6 Atomic force micrograph of a P3OT layer with $d_{P3OT} = 12$ nm.

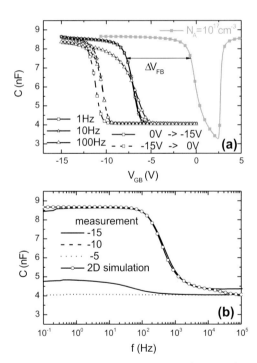

Figure 7 Measured dynamic CV curves of a P3OT capacitor at different frequencies and sweep directions and analytical CV curve (grey line, $N_A = 10^{17}$ cm^{-3}) (a) and the frequency response of the capacitance at different voltages and one simulated curve (-15 V, $\mu = 5 \times 10^{-8}$ cm^2/Vs, $N_A = 10^{17}$ cm^{-3}).

clear. Besides, the dynamic CV curves show the accumulation (maximum capacitance) at high negative voltages as expected for P3OT which is easily (unintentionally) p-doped by oxidation. Reduction of the absolute value of negative voltage decreases the capacitance down to a constant minimal value indicating that the layer is fully depleted. Consequently, with $C^{MIS} = 4\,\text{nF}$ the layer thickness determined by Eq. (19) is $d_{P3OT} = 48\,\text{nm}$. The maximum capacitance results from Eq. (19). With the silicon dioxide thickness of $d_{ox} = 50\,\text{nm}$ and the contact area of $A = 0.126\,\text{cm}^2$ we obtain $C^{MIS} = 8.7\,\text{nF}$ as measured. However, this capacitance is only achieved at low frequencies. Since for the formation of the accumulation layer the frequency must be lower than the corresponding reciprocal relaxation time, this indicates a low conductivity caused either by a low doping or a low mobility. The doping concentration can be determined from the depletion region of the dynamic CV curve measured at a frequency lower than the cut-off frequency visible in Fig. 7 (b). For these purposes, we solved the analytical equations Eq. (11) and (12). The capacitance of the MOS capacitor results from the series connection of the semiconductor capacitance and the oxide one. Parameters in the calculations are the gap and dielectric constant as given above, the affinity $\chi = 3\,\text{eV}$ (we use the value for the similar material P3HT [68], the effective densities of states $10^{21}\,\text{cm}^{-3}$ [57, 41] and the work function of the gold contact $\Phi_{Au} = 5.0\,\text{eV}$. The analytical model does not account for the layer thickness and for generation of inversion charges, therefore the inversion region is calculated which cannot be measured due to the large generation time of the wide gap organic material. Apart from a shift of the flat band voltage the slope in the depletion region is described well by the analytical curve (also shown in Fig. 7 (a)) with a doping of $1 \times 10^{17}\,\text{cm}^{-3}$. For this value the work function difference is $\Phi_{MS} = -0.592\,\text{V}$. The slope of the CV curves in depletion is independent from the sweep direction. Consequently, for the sweep from depletion to accumulation the flat band voltage shift is $\Delta V_{FB} = -7.138\,\text{V}$ and for the down sweep is $\Delta V_{FB} = -10.468\,\text{V}$. According to Eq. (17), the shift can be described by variation of fixed interface charges, in this case in the range of $2.56 \times 10^{12} \ldots 3.75 \times 10^{12}\,\text{cm}^{-2}$. These measurements give numbers but neither an information on the physico-chemical nature of these charges nor on the mechanism which leads to the switching between the two values of the interface charge within seconds when the sweep *direction* is reversed.

Fig. 7 (b) shows the frequency response of the capacitance at different gate voltages. The transition frequency is rather low, about $f_c \approx 10^2\,\text{Hz}$ at a voltage of $-15\,\text{V}$. Using this transition frequency of the dielectric relaxation and the doping concentration determined above, we obtain from $\tau_R = 1/(2\pi f_c) = \varepsilon_0 \varepsilon_R/(eN_A\mu)$ the very low mobility (perpendicular to the layer) of $\mu \approx 10^{-8}\,\text{cm}^2\,\text{V}^{-1}\,\text{s}^{-1}$. A more exact value is obtained by 2D simulations because in this case the accumulation layer of the holes is included. Such simulations were carried out with the same parameters used as for the simulation of the CV curve. As visible in the figure, the simulated curve with a mobility of $\mu = 5 \times 10^{-8}\,\text{cm}^2\,\text{V}^{-1}\,\text{s}^{-1}$ describes well the measured one. More information about characteristic frequencies is obtained from the imaginary part of the modulus (Eq. 21) depicted in Fig. 8. There is only one maximum at a frequency of $f_c = 797\,\text{Hz}$. Nevertheless, there are two RC elements necessary to describe well the curve in the measured frequency range. From the given fitting parameters we

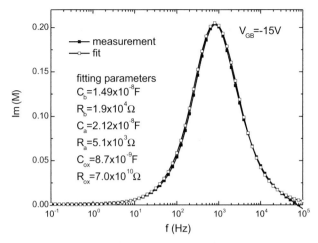

Figure 8 Measured and fitted imaginary part of the modulus function of the device of Fig. 7 at $V_{GB} = -15V$.

calculated the time constant $\tau_b = 2.81 \times 10^{-4}$s of the bulk element resulting in a mobility of $\mu = 6.3 \times 10^{-8}\,\text{cm}^2\,\text{V}^{-1}\text{s}^{-1}$, a value very close to the result of the 2D simulation. The time constant $\tau_{if} = 1.07 \times 10^{-4}$s of the accumulation layer differs only marginally from the bulk one so that only one maximum of the imaginary part of the modulus is visible. In addition, as a result of the analysis of the admittance at low frequencies a resistance parallel to the oxide capacitance has to be included to account for the oxide leakage currents. However, the estimated value of $R_{ox} = 7 \times 10^{10}\,\Omega$ is very high indicating the good quality of the oxid.

13.5.1.3 Thin film transistors

Organic transistors with the design of Fig. 5. (a) are prepared on the same wafer as the capacitors. The output characteristics of a transistor with a channel length $L = 50\,\mu m$ is shown in Fig. 9. Although the layer is relatively thick for the estimated doping concentration, there is a pronounced saturation region at least at low gate voltages. The measured characteristics show a small hysteresis for different sweep directions of the gate voltage. Using the simple Shockley equation Eq. (1) a mobility of $\mu = 6.6 \times 10^{-5}\,\text{cm}^2\,\text{V}^{-1}\text{s}^{-1}$ has been estimated from the saturation region. Fig. 10 shows the transfer characteristics of the transistor. The influence of series resistance becomes clear from the depiction in a linear scale (Fig. 10 (a)) because there is a non-linear increase of the current in the active region at higher gate voltages. Nevertheless, the mobility in the active region can be estimated from the linear part of the curve. The calculated value of $\mu = 5.9 \times 10^{-5}\,\text{cm}^2\,\text{V}^{-1}\text{s}^{-1}$ is close to that one obtained from the output characteristics. However, the comparison with the mobility estimated above from the frequency response of the MOS capacitor makes clear the (lateral) mobility in the transistor is about three orders of magnitude higher than that one in the capacitor (vertical). This indicates an extreme anisotropy of the charge transport vertical and lateral to the interface and we conclude that there

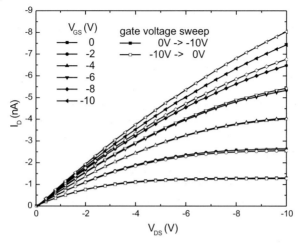

Figure 9 Output characteristics of a P3OT transistor with
$L = 50\,\mu m$ and $w = 1$ mm at different gate voltage sweep directions. The hold and delay times are $t_h = 180$ s and $t_D = 1$ s,
respectively.

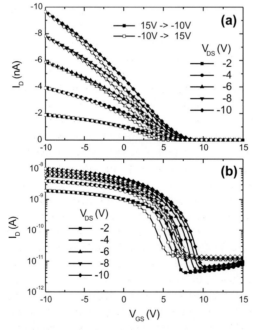

Figure 10 Transfer characteristics of the transistor of Fig. 9
for different drain voltages and sweep directions (a) - linear
scale, (b) - logarithmic scale. The hold and delay times are
$t_h = 180$ s and $t_D = 10$ s, respectively.

is a preferential orientation of the polymer chains parallel to the interface. The same conclusion has been drawn from the measurements of the optical properties of the P3OT layers.

More information regarding the turn-off behavior of the transistor is obtained from the logarithmic characteristics (Fig. 10 (b)). The inverse subthreshold slope is with $S \approx 1\,\text{V/dec}$ a good value for organic transistors. The theoretical value of the threshold voltage is given by Eq. (15). With the material parameters described above and the doping concentration determined from the CV curves we obtain $V_{th} = 1.5\,\text{V}$. The comparison with the measured curves shows higher positive threshold voltages of $V_{th} \approx 3.5\,\text{V}$ for the sweep from negative to positive gate voltages and $V_{th} \approx 5\,\text{V}$ for the inverse sweep direction. Contrary to the CV curves, negative interface charges of $8.6 \times 10^{11}\,\text{cm}^{-2}$ to $1.5 \times 10^{12}\,\text{cm}^{-2}$ are necessary for such threshold voltages. In addition, one can see in Fig. 10 (b) the satisfying on-off ratio of $\approx 10^3$ for larger drain voltage.

13.5.2
Devices made from MEH-PPV

The devices described above have been also prepared using poly(2-methoxy, 5 ethyl (2'hexyloxy) paraphenylenevinylene) (MEH-PPV) for the active layer. Here a short summary regarding the differences and similarities ought to be considered. In Fig. 11 (a) the dynamic CV curves are shown. Contrary to the P3OT device, the curves were measured at a very low frequency of $f = 0.01\,\text{Hz}$. The reason is visible in Fig. 11 (b) where the frequency response of the capacitance and the imaginary part of the modulus are depicted. The oxide capacitance of $C^{\text{MIS}} = 8.7\,\text{nF}$ is only achieved at a frequency as low as 0.01 Hz indicating a lower conductivity in the vertical direction as estimated for the P3OT capacitor. Again, the constant minimum capacitance at low negative voltages shows that the layer is completely depleted. In the same manner as described above a layer thickness of $d_{\text{PPV}} \approx 63\,\text{nm}$ can be deduced. To determine the doping concentration the analytical CV curve has been calculated with the following parameters, the affinity $\chi = 2.6\,\text{eV}$, the gap $E_g = 2.4\,\text{eV}$, the relative dielectric constant $\varepsilon = 3$ and the other values as used for P3OT. The best fit with the measured curve was obtained for a concentration of $N_A = 10^{17}\,\text{cm}^{-3}$, the same value as for the P3OT capacitor. The work function difference for this system is $\phi_{MS} = -0.592\,\text{V}$. The curve in Fig. 11 (a) is shown for a flat band voltage shift of $\Delta V_{FB} = -6.7\,\text{V}$ to realize a good fit to the measured curve for the sweep direction from depletion to accumulation. The inverse sweep can be described with $\Delta V_{FB} = -7.4\,\text{V}$. Consequently, positive interface charges of $N_{it} = 2.9 \ldots 3.2 \times 10^{12}\,\text{cm}^{-2}$ are necessary, that means the same order of magnitude as for the P3OT/SiO$_2$ interface. However there is a deviation of the analytical curve from the measured one for voltages $V_{GB} < -6\,\text{V}$ because the measuring frequency is very close to the limit frequency. With the estimated doping concentration and the limit frequency of $f_c = 0.154\,\text{Hz}$ obtained from the maximum of the modulus a value of $1.6 \times 10^{-11}\,\text{cm}^2\,\text{V}^{-1}\text{s}^{-1}$ for the vertical component of the mobility has been calculated. This value is conspicuously smaller as estimated for P3OT.

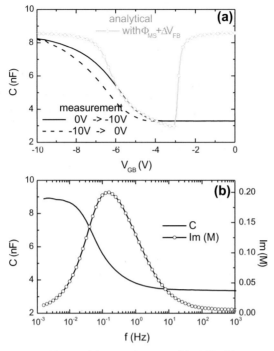

Figure 11 Measured dynamic CV curve of the MEH-PPV capacitor ($f = 0.01$ Hz) and analytical curve (grey line, $N_A = 10^{17}$ cm^{-3}, $\Delta V_{FB} = -6.7$ V) (a) and the frequency response of the capacitance and imaginary part of the modulus for $V_{GB} = -10$ V.

The organic transistors have been prepared with channel length of 25 μm and 50 μm. Fig. 12 shows the output characteristics at different gate sweep directions for the 50 μm transistor and so not only the doping concentration is the same as for the P3OT device but also the geometrical dimensions. Comparison of the two Figs. 9 and 12 makes clear the drain currents are of the same order of magnitude. However, the MEH-PPV device shows a better saturation behavior even thought the layer is thicker. In addition, the estimated mobility in the saturation region is with 1.6×10^{-4} cm^2 V^{-1}s^{-1} by a factor of 10 higher as for P3OT. The reason for approximately equal currents despite of the shown differences is visible in the transfer characteristics of Fig. 13. From the linear depiction the threshold voltage of $V_{th} \approx -4.5$ V is obtained. In contrast to the P3OT transistor, the value is negative resulting in a lower effective gate voltage causing lower currents at a given V_{GS}. The value deviates from the theoretical threshold voltage of $V_{th} = 2.4$ V calculated with Eq. (15) and the above mentioned material parameters for MEH-PPV. However, the necessary flat band voltage shift of $\Delta V_{FB} \approx -6.9$ V to obtain the measured V_{th} is in the same direction and has approximately the value as determined from the CV curve. The mobility of 2×10^{-4} cm^2 V^{-1}s^{-1} estimated from the active region of the transfer curves is

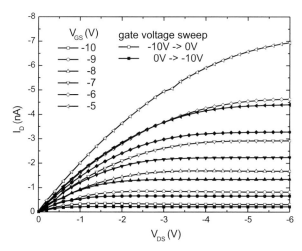

Figure 12 Output characteristics of a MEH-PPV transistor with $L = 50\,\mu m$ and $w = 1$ mm at different gate voltage sweep directions. The hold and delay times are $t_h = 180\,s$ and $t_D = 1\,s$, respectively.

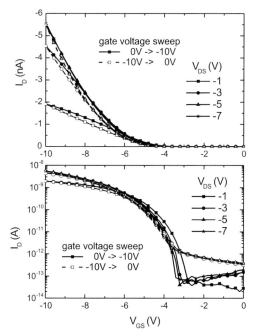

Figure 13 Transfer characteristics of the transistor of Fig. 12 for different drain voltages and sweep directions (a) - linear scale, (b) - logarithmic scale. The hold and delay times are $t_h = 180\,s$ and $t_D = 10\,s$, respectively.

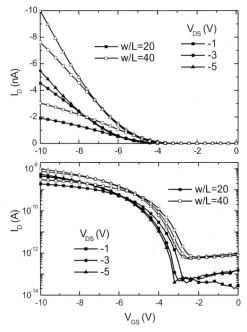

Figure 14 Comparison of the transfer characteristics of two transistors with different w/L ratios and different drain voltages (gate voltage sweep from 0 V to -10 V) (a) - linear scale, (b) - logarithmic scale. The hold and delay times are $t_h = 180$ s and $t_D = 10$ s, respectively.

close to the value obtained from the output characteristics. The linearity of the transfer characteristics in the active region indicates that series resistances at the source and drain contacts are negligible. In addition, the very good inverse subthreshold slope of $S \approx 0.5$ V/dec and the small hysteresis of $\Delta V \approx 0.3$ V have to be emphasized. Finally, the comparison of the vertical and lateral component of the mobility shows a high anisotropy of 10^7 indicating a very good order of the polymer chains in parallel to the interface. Until now, it is not clear wether this order is the reason for such small hysteresis.

In addition, in Fig. 14 the transfer characteristics of the device are compared with those ones of a transistor with the same width and a channel length of $L = 25\,\mu m$. The mobility of the shorter transistor is with $1.8 \times 10^{-4}\,cm^2\,V^{-1}\,s^{-1}$ close to the value of the 50 μm transistor. The threshold voltage of $V_{th} = -4$ V deviates marginally from the devices with the longer channel. Consequently, as expected the current for the shorter transistor is higher by a factor of approximately two.

Summarizing the comparison of both materials we conclude that in this case the MEH-PPV transistor shows a better performance than the devices made from P3OT.

13.5.3
Further transistor designs

13.5.3.1 Finger structure

Shorter channels (down to few micrometer) are producible employing low resolution lithography as also demonstrated in [69–71]. With this technology also interdigitate electrodes are possible [72] allowing for higher w/L ratios for high drain currents despite of the low mobilities. We have prepared such finger structures with the design of Fig. 5 (b) using P3HT as the active material. For these devices the thicknesses of the silicon dioxide insulator and of the active layer are 30 nm and 28 nm, respectively. The output characteristics of these transistors (not shown here) show a good saturation region and drain currents in the μA-range. In Fig. 15 the transfer characteristics are shown for a transistor with a channel length of $L = 10\,\mu$m and a total w/L ratio of 4000. From the curves on the linear scale we conclude that there are no series resistances influencing the drain current. Furthermore, a mobility of $\mu = 4 \times 10^{-5}\,\mathrm{cm^2\,V^{-1}\,s^{-1}}$ has been estimated, i.e. a value as low as calculated for the P3OT device. The threshold voltage is $V_{\mathrm{th}} \approx 1.7\,$V for the sweep from positive to negative gate voltages and $V_{\mathrm{th}} \approx 1\,$V for the inverse sweep direction resulting in a small hysteresis. In addition, the inverse subthreshold slope of $S \approx 0.7\,$V/dec is good and a high on/off current ratio of $> 10^4$ in a small voltage range is achieved.

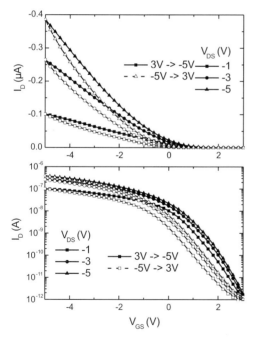

Figure 15 Transfer characteristics of a transistor with $L = 10\,\mu$m and $w/L = 4000$ for different drain voltages and sweep directions (a) - linear scale, (b) - logarithmic scale. The hold and delay times are $t_{\mathrm{h}} = 180\,$s and $t_{\mathrm{D}} = 10\,$s, respectively.

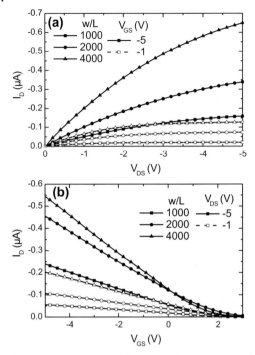

Figure 16 Comparison of the output characteristics at different gate voltages (a) and transfer characteristics at different drain voltages (b) for three transistors with different w/L ratios. The channel length are $L = 40\,\mu m$ for $w/L = 1000$, $L = 10\,\mu m$ for $w/L = 2000$ and $L = 20\,\mu m$ for $w/L = 4000$.

A comparison of three transistors with $w = 2000\,\mu m$ but different w/L-ratios is shown in Fig. 16, realized by both different channel length and different number of fingers. Channel length of $40\,\mu m$, $10\,\mu m$ and $20\,\mu m$ have been prepared for the ratios of 1000, 2000 and 4000, respectively. The estimated mobility is $\mu = 6 \times 10^{-5}\,cm^2\,V^{-1}s^{-1}$ for these different transistors. There are again only small differences in the threshold voltage and therefore the currents differ from each other essentially according to the w/L-ratio.

Comprising the demonstrated properties one can conclude that the finger structure is a suitable design to achieve high currents. However, because of the low mobility of the polymer materials the channel length has to be smaller than $L = 2\,\mu m$ for higher frequency applications. Consequently, a short channel transistor is described in the following section.

13.5.3.2 Short channel design

Organic short channel pentacene transistors with $L = 30\,nm$ have been described in [73]. However, the used electron beam lithography for the structuring is to costly for cheap organic electronics. In [17] a simple embossing technique is used to realize

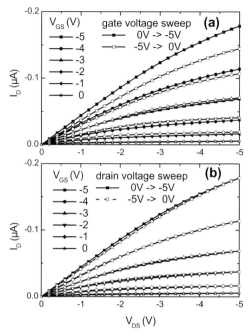

Figure 17 Output characteristics of a P3HT transistor with $L = 740$ nm and $w = 1000\,\mu$m at different gate voltages and different sweep directions of the gate voltage (a) and the drain voltage (b). The hold and delay times are $t_h = 180$ s and $t_D = 1$ s, respectively.

channel length in the submicrometer range, but there is no saturation for a transistor with a channel length of 900 nm. We applied also a simple technique for the preparation of short channel transistors. The realized design and the gate insulator thicknesses are the same as used for the finger structures. With an underechting technique in combination with low resolution lithography channel length $< 1\,\mu$m have been realized. In Fig. 17 the output characteristics of a short channel transistor are depicted. The channel length and width of $L = 740$ nm and $w = 1000\,\mu$m have been measured by means of scanning electron microscopy. In spite of such a short channel there is a pronounced saturation region since the gate insulator is thin enough as demonstrated later in Section 13.6. Comparing the two diagrams of the figure it becomes clear that there is only a hysteresis for different gate voltage sweep directions (Fig. 17 (a)) but not for a variation of the drain voltage sweep direction (Fig. 17 (b)). Therefore, the variation of the gate voltage is more critical.

The transfer characteristics of the same transistor are shown in Fig. 18. From the linear dependence in the active region (Fig. 18 (a)) we conclude there are no problems caused by series resistance. Therefore, despite of the short channel the polymer should fill in the space between source and drain without cavities near the contacts. Furthermore, a mobility of $5.8 \times 10^{-5}\,\mathrm{cm}^2\,\mathrm{V}^{-1}\,\mathrm{s}^{-1}$ is estimated from the active

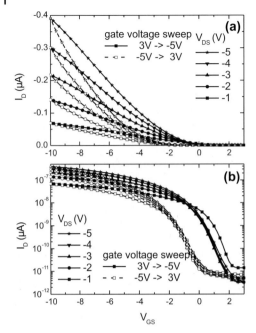

Figure 18 Transfer characteristics of the transistor of Fig. 17. for different drain voltages and sweep directions (a) - linear scale, (b) - logarithmic scale. The hold and delay times are $t_h = 180$ s and $t_D = 10$ s, respectively.

region. The small inverse subthreshold slope of 0.73 V/dec estimated from the curves of Fig. 18 (b) indicates again that short channel effects are negligible for this device. The hysteresis is approximately 1.7 V.

Another important point for the application of the devices in circuits is the possible cut-off frequency. As described above, the value can be calculated with the equation $f_0 = g_m / (2\pi C_{GS})$. We read off the maximum transconductance from the derivative of the curve for $V_{DS} = -4$ V measured from depletion to accumulation. With the obtained value of $g_m = 3.96 \times 10^{-8}$ S and the assumption $C_{GS} = C_{OX} = C''_{OX} w L$ a theoretical cut-off frequency of 7.4 kHz is calculated. On this account, in Fig. 4 the curve for cut-off limit frequency at an effective voltage of 4 V is also depicted. Furthermore, the channel length and the estimated mobility give a point in the graphics which is very close to the line for the obtainable limit frequency. From this result we can conclude, that the estimate of the necessary channel length for a given mobility of the material is possible based on the simple Shockley equation as demonstrated in Section 13.3.3. However, the measurable limit frequency is determined not only by the oxide capacitance. High cut-off frequencies are only possible for transistors with low parasitic capacitances. In our first short channel transistors the silicon wafer serves as the gate and this leads to high overlap capacitances between the gate and the source/drain electrodes. Consequently, the challenge for the future is the development of a design with reduced parasitic capacitances.

13.5.4
Subthreshold currents and trap recharging

In Fig. 5 (c) the design of a transistor with an organic insulator and active layer is shown. The source/drain contacts are again prepared as finger structures. The thicknesses of the active layer made from P3DDT and the P4VP gate insulator are 30 nm and 500 nm, respectively. For this device the peculiarities in the subthreshold region have been investigated. In [25, 26] the results are described detailed. Hence, in this section we give only a short summary of the essential results.

In Fig. 19 the transfer characteristics of a device with a channel length of $L = 2\,\mu m$ and a width of $w = 10\,mm$ is shown. These curves have been obtained from the measured output data. The transistor turns on close to a threshold voltage of $V_{th} = 0\,V$ indicating that the layer is fully depleted at positive gate voltages. Consequently, the maximum possible value of the doping concentration is $N_A \approx 6 \times 10^{17}\,cm^{-3}$. The mobility is again estimated from the linear region of the transfer characteristics of Fig. 19 (a). One obtains $\mu_p = 5 \times 10^{-3}\,cm^2\,V^{-1}\,s^{-1}$ at drain voltages of $-5\,V$ and $-10\,V$ and $\mu_p = 2 \times 10^{-3}\,cm^2\,V^{-1}\,s^{-1}$ at $V_{DS} = -1\,V$. The inverse subthreshold slope estimated from the transfer curves in a logarithmic scale is very high. In the case of $V_{DS} = -1\,V$ one has $S = 7.7\,V/dec$. Furthermore, the subthreshold current depends on the drain voltage, which is not expected for such a long channel device.

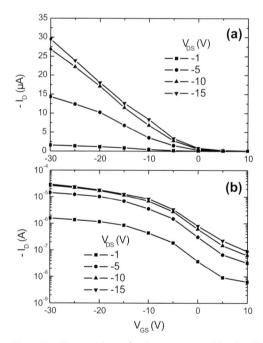

Figure 19 Measured transfer characteristics of the thin film transistor with $w/L = 5000$, $L = 2\,\mu m$, $d_{P3DDT} = 30\,nm$ and $d_{P4VP} = 500\,nm$, linear (a) and logarithmic (b) scale.

These peculiarities have been investigated by means of two-dimensional simulations using the numerical program ATLAS [45]. Besides the above determined mobilities the following parameters have been used: the static dielectric constants of both materials, $\varepsilon_r = 2.56$ for P4VP and 3.24 for P3DDT, the monomer density of $10^{21}\,\mathrm{cm}^{-3}$ and the work function of the gate, drain and source electrodes of 5.0 eV.

In a first step, the doping concentration has been varied to fit the simulated transfer curves to the measured ones. According to Eq. (15) with the flat band voltage of Eq. (17) the threshold voltage and the associated run of the curves are not only determined by the doping but also by fixed interface states. Simulations show indeed [25] that the measured curves above the threshold voltage are described well by simulations either with $N_A = 10^{16}\,\mathrm{cm}^{-3}$ and without fixed interface states or with a higher doping concentration and a positive interface charge. However, there is a large difference in the subthreshold region between the measured and the simulated characteristics. Whereas the measured curve at $V_{DS} = -1\,\mathrm{V}$ shows an inverse subthreshold slope of $S = 7.7\,\mathrm{V/dec}$ one obtains a value of $S \approx 200\,\mathrm{mV/dec}$ for the simulated one. Furthermore, the measured subthreshold current strongly depends on the drain voltage whereas the simulated transfer characteristics approach a common subthreshold dependence. According to Eq. (18) the inverse subthreshold slope can be influenced either by interface or by bulk traps. The basics of their influence ought to be explained at the example of acceptor-like traps N_{ia} (neutral if empty and negative if occupied by an electron). In this case a doping concentration of $10^{16}\,\mathrm{cm}^{-3}$ is considered because acceptor-like traps change the threshold voltage towards a positive value. The simulations have been carried out for different interface trap densities (N_{ia}) and different trap levels (E_{it}). As an example, the simulated influence of the trap density with an energy of 0.4 eV above the valence band is shown in Fig. 20.

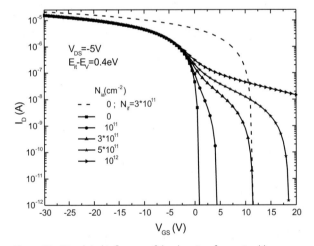

Figure 20 Simulated influence of the density of acceptor-like interface traps at a constant energy $E_t - E_V = 0.4\,\mathrm{eV}$ on the transfer characteristic for $V_{DS} = -5\,\mathrm{V}$. In this figure also the dependence with negative fixed interface states of $N_{if} = 3 \times 10^{11}\,\mathrm{cm}^{-2}$ is shown.

There is no influence of the trapped interface charges on the drain current above the threshold voltage. However, in the off-state the inverse subthreshold slope depends strongly on the occupancy of the interface traps. As an example the curves for a density of $3 \times 10^{11}\,\mathrm{cm^{-2}}$ of either fixed charges (also depicted in the figure) or interface trap states shall be discussed. At a positive gate voltage larger than 11 V the curves coincide and the low value of $S \approx 200\,\mathrm{mV/dec}$ is obtained by simulation. That means the trapped interface states are completely occupied by an electron and the trap level is below the Fermi level position. Recharging of the interface traps takes place by reducing the positive gate voltage up to approximately 0.5 V. Consequently, in this voltage region the inverse subthreshold slope degrades drastically. In addition, the higher the trap density the higher the drain current since with increasing negative interface charge a higher hole concentration is caused in the layer. This is also the reason for higher drain currents for shallow traps compared to deeper ones. At even lower voltages all of the interface traps are neutral and do not influence the drain current.

A further peculiarity was the drain voltage dependency of the subthreshold current. To investigate whether traps can cause such a behavior, in Fig. 21 the comparison of curves simulated with and without interface traps is shown for different drain voltages. In the case with traps we assumed $N_{ia} = 10^{12}\,\mathrm{cm^{-2}}$ at an energy level of 0.4 eV above the valence band. In the linear and saturation regions there is no difference between these two cases and the drain current depends only on the drain voltage. However, in the subthreshold region the situation is completely different. As expected for a long channel device the drain current does not depend on the drain voltage if trapped interface charges are not present. Inclusion of the traps at the interface does not only increase the subthreshold slope but, in addition, the drain current is higher and depends on the drain voltage. The drain voltage dependency can be explained by investigating the hole concentration, the occupancy of traps and the quasi-Fermi level of the holes (not shown here). The occupancy of the interface

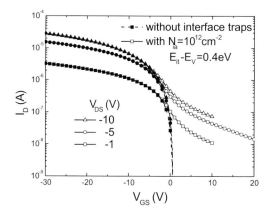

Figure 21 Simulated logarithmic transfer characteristics with acceptor-like and without interface traps for different drain voltages.

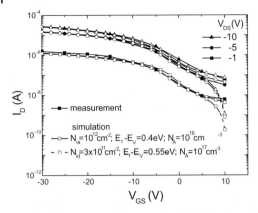

Figure 22 Comparison of measured and simulated transfer characteristics for different drain voltages. The acceptor-like interface density is $N_{ia} = 10^{12}$ cm^{-2} at an energy of 0.4 eV from the valence band and the donor-like density is $N_{id} = 3 \times 10^{11}$ cm^{-2} at 0.55 eV from the same band.

traps give a value larger than 2×10^{10} cm^{-2} which moreover increases with increasing drain voltage. As a result, the hole concentration is higher, approximately by a factor of 10^6 even at the interface of source and drain. Consequently, at such a high hole concentration (which does almost not depend on the drain voltage at all) the increase of the gradient of the quasi-Fermi potential with increasing drain voltage causes the drain voltage dependence of the current in the subthreshold region.

As mentioned before, the linear and saturation regions of the measured transfer characteristics at $V_{DS} = -5$V can be described also by a doping concentration of 10^{17} cm^{-3} assuming fixed *positive* interface charges or donor-like traps (positive if empty and neutral if occupied by an electron). Therefore, in Fig. 22 the comparison of the measured curves with the simulated transfer characteristics is shown for both the acceptor-like traps and the donor-like ones. For the donor-like traps a density of $N_{id} = 3 \times 10^{11}$ cm^{-2} at the energy of 0.55 eV above the valence band has been assumed. There is a good agreement of the curves above the threshold voltage and also in the subthreshold regime up to a gate voltage of 5 V. At higher voltages the simulated curves with acceptor-like interface traps describe the experimental data better. Thus, the simulations reveal that the anomalous subthreshold characteristics with an extremely large inverse subthreshold slope can arise from the recharging of interface states.

Apart from interface traps, the inverse subthreshold slope of the transfer characteristics may be influenced also by variations of bulk charges as caused for example by bulk traps. In this case, the simulated transfer characteristics are similar to those obtained by supposing interface traps however with slightly different energy levels. In this way the simulations have shown that traps either at the interface or in the bulk are possible origins of the anomalous subthreshold behavior.

13.5.5
Hysteresis

One of the problems of organic devices is the occurrence of hysteresis effects. Even though they are very often ignored in the literature, Brown et al. [67] described hysteresis effects for MIS capacitors based on poly(3-hexylthiophene) and attributed them to carrier trapping or migration of dopants. Traps and defects also cause a non-optimal behavior of the transistors [74, 31]. In [19] the reduction of such effects have been realized by using of patterned gate structures. Residual hysteresis were attributed to border effects. Our investigations regarding these effects are described in Ref. [21, 29]. As a result, mobile ions could indeed be a possible origin. However, the increase of the hysteresis with increasing temperature indicates also other reasons for this effect. It has been suggested that transport of a low mobility species in connection with at least one reaction must be taken into account.

Here we describe hysteresis effects occurring in devices of Section 13.5. In the output characteristics (Fig. 17) of the short channel transistor a hysteresis does occur only for different gate voltage sweep direction but not for the variation of the drain voltage sweep direction. As a consequence, the hysteresis should occur in the MOS capacitor in the same manner as in the transistor. Therefore, in Fig. 23 the CV

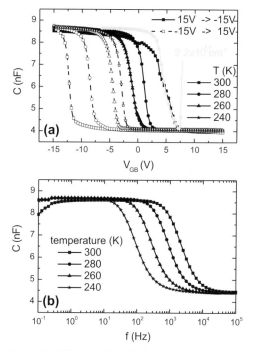

Figure 23 CV curve at 1 Hz for both sweep directions (a) and the frequency dependent capacitance at $V_{GB} = -15$ V (b) for different temperatures. In (a) also the analytical calculated CV curve at room temperature ($N_A = 2.2 \times 10^{17}$ cm^{-3}) is depicted.

curves and the frequency response at different temperatures are shown for the P3OT device of Section 13.5.1. At first, comparing the CV curves at $T = 300$ K with the first measurements (one week earlier) of Fig. 7 (a) it becomes clear, that there is a large shift towards positive gate voltages for the curve measured from depletion to accumulation. In addition, the slope in the depletion region is smaller and, as shown in the figure, can be described by a slightly higher doping concentration of $N_A = 2.2 \times 10^{17}$ cm^{-3}. However, the curve for the other sweep direction is only slightly shifted without variations in the slope. Consequently, as also described in [29] this sweep direction seems to be nearer to equilibrium than the other one. A further conclusion from this comparison is that the hysteresis effects can not definitely distinguished from the ageing process. Nevertheless, the influence of the temperature is clearly visible in Fig. 23, i.e. the hysteresis increases with increasing temperature as also described in [29], whereas the shift of the curves measured from negative to positive gate voltages is larger as for the other direction. Complemental, in Fig. 23 (b) the temperature dependence of the frequency response of the capacitance is shown. The cut-off frequency for the curve $T = 300$ K is higher (about by a factor of 10) as obtained for the first measurement of Fig. 7 (b). The reason for this is not clear until now. However, in contrast to the results in [29], this frequency becomes smaller for decreasing temperature because of the reduced mobility as expected for the hopping transport.

Finally, the hysteresis and their temperature dependence of the P3OT transistors will be considered. For this purpose, in Fig. 24 the transfer characteristics are shown for a transistor with channel length of $L = 50 \mu$m at a drain voltage of -10 V. As observed for the MOS capacitance also the hysteresis in the transistor current characteristics increases with increasing the temperature. Furthermore, the curves of both

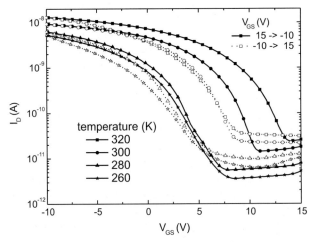

Figure 24 Transfer characteristics for of the transistor of Fig. 9 for $V_{DS} = -10$ V, different temperatures and sweep directions in a logarithmic scale. The hold and delay times are $t_h = 180$ s and $t_D = 10$ s, respectively.

sweep directions are shifted towards positive gate voltages for higher temperature caused for example by higher negative interface charges. The variation of the inverse subthreshold slope is hardly to explain. Whereas for $T = 300\,\text{K}$ the value of $S = 1\,\text{V/dec}$ is obtained (the same as obtained for the first measurement depicted in Fig. 10 (b)), the value increases up to $S = 2\,\text{V/dec}$ for $T = 260\,\text{K}$. According to Eq. (18) the inverse subthreshold slope is proportional to the temperature which is in contrast to the observation. However, there are additional dependencies on temperature for recharging processes leading to a temperature dependence of the capacitances. Therefore further investigations have to clarify the origin for the increased slope.

13.6
Selected simulation studies

13.6.1
Formation of inversion layers

One of the challenges for polymer electronics is the realization of complementary circuits (CMOS) requiring both n- and p-channel transistors on the chip. Depositing both an electron-conducting and a hole-conducting organic material is technologically cumbersome. Only recently both n- and p-channels have been achieved in the same material (pentacene) in a so-called ambipolar device [65] (It should be noted that a practical use of the inverter made with these ambipolar device can hardly be of interest due to its high power consumption). Apart from this result it was not possible to create both accumulation and inversion layers in the same material. This has been attributed [66] to the difficulty to engineer the metallic source/drain contacts for efficient injection of both electron and holes, and also to the fact that the electron and hole mobilities differ (often) by several orders of magnitude.

However, there is another problem, connected with the time scale for the formation of an inversion layer. Simulations have been done for both a MOS capacitor and a top-contact transistor [21] (50 nm silicon dioxide, 30 nm organic layer, parameters of the organic layer: $N_C = N_V = 10^{22}\,\text{cm}^{-3}$, band gap $\varepsilon_g = 2.4\,\text{eV}$, hence $n_i = 68.8\,\text{cm}^{-3}$, life time $\tau_n = \tau_p = 10^{-7}\,\text{s}$, mobility $\mu_n = \mu_p = 10^{-3}\,\text{cm}^2/\text{Vs}$, doping concentration $N_A = 10^{16}\,\text{cm}^{-3}$ and permittivity $\varepsilon = 3$). For the capacitor, even for a frequency as low as 0.01 Hz, the simulated dynamic capacitance-voltage (CV)-curve (Fig. 25) does not yield an increase of the capacitance due to inversion for positive voltages. Actually, in a MOS capacitor an inversion layer is formed by generation of minority carriers (and not by transport) and hence the minority carrier response time τ_{inv} is the relevant quantity. It is determined by [21]

$$\tau_{\text{inv}} = \frac{1}{\sqrt{2}} \frac{N_A}{n_i} \sqrt{\tau_n \tau_p} \tag{23}$$

and becomes extremely large due to the small intrinsic density of the organic semiconductor (large gap). With the numbers given above one gets $\tau_{\text{inv}}^{-1} \approx 10^{-7}\,\text{Hz}$. Under

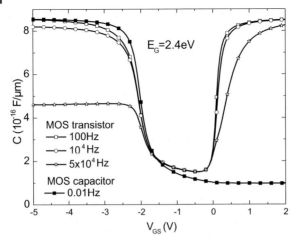

Figure 25 Simulated dynamic low-frequency CV-curve for an organic MOS capacitor with 40 nm oxide thickness and for the field-effect transistor with 1 μm channel length for different frequencies.

this condition inversion cannot be observed in a MOS capacitor. But for a lower band gap polymer (say $\approx 1.5\,\mathrm{eV}$) and eventually higher purity inversion should be observable for lower experimentally accessible frequencies.

The situation is different in the field-effect transistor. There is no need to generate minority carriers if the source/drain contacts with the channel are ohmic ones for the minority carriers allowing for their injection into the channel. The corresponding time constant is determined by the transmission line behavior of the minority carriers in the channel coupled with the oxide capacitance. It can be simply estimated from the resistance R_K of the inversion channel and the oxide capacitance C_{ox} as

$$\tau_{tl} \approx \frac{R_K}{2}\frac{C_{ox}}{2} = \frac{1}{4}\frac{L}{e\mu_n nd_K w}\frac{\varepsilon_0 \varepsilon wL}{d_{ox}} \tag{24}$$

where d_K is the thickness of the inversion channel. Thus one can estimate the limiting frequency as

$$f_{tl} = \tau_{tl}^{-1} = 4\frac{\sigma_n}{\varepsilon_0 \varepsilon}\frac{d_K d_{ox}}{L^2} = 4f_{d,n}\frac{d_K d_{ox}}{L^2} \tag{25}$$

where $f_{d,n}^{-1} = \varepsilon_0 \varepsilon/\sigma_n$ is the (electron) dielectric relaxation time. To achieve a larger f_{tl} enabling the formation of the inversion channel, smaller channel length L are needed. Simulations have been carried out with $L = 1\,\mu\mathrm{m}$ and supposing ohmic contacts for the minority carriers between the channel and both source and drain. The resulting dynamic CV-curves of this MOS transistor (Fig. 25) show indeed the increase of the capacitance at positive gate voltage due to inversion. It seems surprising at the first moment that now the decrease of the accumulation capacitance begins even at lower frequencies than on the inversion side. But this might becaused by the influence of different capacitances of the MOS structure which is not the same in inversion and accumulation. Thus from the general consideration and the simulation it is proposed that inversion should occur also in organic transistors if the channel length is sufficiently small and if ohmic contacts for the minority carriers between the channel and both source and drain can be prepared.

13.6.2
Trap distributions in top and bottom contact transistors with different source/drain work functions

Due to disorder, in the energy spectrum of organic materials exponentially distributed traps may occur in the gap as known for a-Si. The trap density of states is $N_0 \exp\{-|\varepsilon - \varepsilon_0|/\varepsilon_S\}$ with ε_0 as the valence or conduction band edge and ε_S is a decay constant. The traps are donor- (acceptor-) like near the valence (conduction) band. It has been supposed in the literature that such a model should be relevant for modelling OFETs [7, 35–38]. For such system simulations have been done for the bottom (source/drain) contact (BOC) and the top contact (TOP) transistors. The parameters of the gate insulator are $\varepsilon_{ox} = 3.9$, $d_{ox} = 40$ nm. For the active layer (thickness $d = 30$ nm) we used $\varepsilon = 3.24$, $\chi = 3.0$ eV, $\varepsilon_g = 2.0$ eV, $N_C = N_V = 10^{21}$ cm^{-3}, $N_A^- = 10^{17}$ cm^{-3}, $\mu_p = \mu_n = 10^{-3}$ cm^{2}V^{-1}s^{-1}. In the TOP structure, the layer thickness corresponds simultaneously to the vertical distance of the channel from source/drain, resulting in series resistance, whereas it is adjacent to them in the BOC structure. The channel length is $L = 5\,\mu$m. Here an example for a high trap distribution ($N_0 = 2.2 \times 10^{20}$ cm^{-3}(eV)$^{-1}$, $\varepsilon_S = 0.1$ eV) will be demonstrated. The bulk Fermi energy (4.451 eV below the vacuum level) is determined practically only by acceptor doping and traps, whereas the bulk hole concentration is many orders of magnitude lower. For a large source/drain work function ($\Phi_M = 5$ eV) one has accumulation (ohmic) contacts. They are neutral for $\Phi_M = 4.451$ eV due to the traps, but owing to the low hole concentration they act similarly as Schottky contacts.

Simulated output characteristics are depicted in Fig. 26. For the larger source/drain work function (Fig. 26. a) due to the trap distribution the currents are lower than for the trap-free case by about a factor ≈ 4 since immobile trapped charges are part of the accumulation layer. The difference between TOC and BOC structures is rather small and only caused by the mentioned series resistance. With increasing trap concentration there is a progressive decrease of the current. Furthermore, the threshold voltage is shifted to a higher negative voltage and the inverse subthreshold slope increases. Qualitatively, the characteristics resemble the Shockley model. One can extract an effective mobility from the current characteristics. It is approximately (error of 20%) given by the ratio of mobile to total areal charges in the accumulation layer according to the analytical approximation $\mu_{eff}/\mu_0 = Q''_{mobile}/Q''_{total}$. However, this mobility might be lower than the time-of-flight mobility also from other reasons.

For the lower source/drain work function the contacts become non-ohmic. This leads to some interesting peculiarities (Fig. 26. b) which do occur also in the trap-

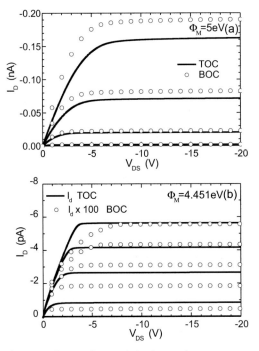

Figure 26 Output characteristics (gate voltage $-V_{GS} = 4(2)10\,V$) for the TOC and BOC transistors (with trap distribution) and the source/drain work function $\Phi_M = 5.0\,eV$ (a) and $\Phi_M = 4.451\,eV$ (b).

free case. The currents are much lower than for the ohmic contact. Now one has large differences between the two structures. The current in the BOC structure is by orders of magnitude smaller than in the TOC structure due to a Schottky-type contact for the mobile carriers in the BOC structure. In both cases, there is a non-quadratic (non-linear) gate voltage dependence of the current in the saturation (active) region. In the TOC structure, a low gate voltage dependence in the active region and a sharp transition into the saturation region result from the Ohmic resistance of the regions between source/drain and the channel.

The rather different modi of operation of these devices are clearly visible in the profiles of the hole quasi-Fermi level directly below the interface to the oxide (Fig. 27) for $-10\,V$ gate voltage and increasing drain voltage. In the BOC structure with higher work function on has the normal FET dependence, linear for lower drain voltages and above the transition into saturation the additional voltage drop occurs only near drain. For the same structure with the lower work function up to $-8\,V$ the whole voltage drop occurs immediately in a depletion layer at source and only the further voltage drop essentially at drain. For the same low work function in the TOC structure, one has up to $-4\,V$ essentially a voltage drop perpendicular to the layer in a low-concentration ohmic region, up to $-6\,V$ there is the additional

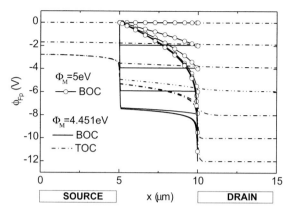

Figure 27 Quasi-Fermi potential 1 nm below the interface to the insulator for the BOC structure with $\Phi_M = 5.0$ eV, for the BOC structure with $\Phi_M = 4.451$ eV, and for the TOC structure with $\Phi_M = 4.451$ eV (all with trap distribution). Gate voltage $V_{GS} = -10$ V and varying drain voltage $-V_{DS} = 0(2)12$ V from top downwards.

drop at source and the further voltage increase occurs at drain. In any case, low source/drain work function should be avoided du to the accompanied strong lowering of the current.

13.6.3
Short-channel effects

In order to reach the off-state and saturation in an OFET, the active layer thickness must be less than the depletion length [29] as discussed in Section 13.3.2. In addition, sufficiently high speed of operation requires short channel length in the submicrometer region (Section 13.3.3). Channel length down-scaling comprises also the gate insulator thickness since the OFET operation requires the transverse field to be large compared to the longitudinal field. Otherwise short-channel effects will occur with missing saturation of the output characteristics and drain voltage dependence of the subthreshold current. To impede such short-channel effects, conditions for the appropriate device design have been analyzed by two-dimensional simulations. A bottom-contact transistor (Fig. 5 (b)) is considered with 30 nm P3HT as active layer and the organic insulator poly-4-vinylphenol (P4VP). In the current characteristics of Fig. 28 the channel length is varied and an insulator thickness of 400 nm has been chosen as practically needed in order to prevent leakage currents. The output characteristics (Fig. 28 (a)) show saturation only for channel length larger than 1 μm and a large supralinear current for shorter channels. In the transfer characteristics (Fig. 28 (b)) there occurs a shift of the threshold voltage with decreasing the channel length below 1 μm. While the threshold voltage of the long channel devices is close to 9 V this value becomes 13 V for the shortest channel length. The origin of these effects can be understood by inspecting the potential profile between source and drain near the interface to the insulator for a positive gate voltage of 8 V

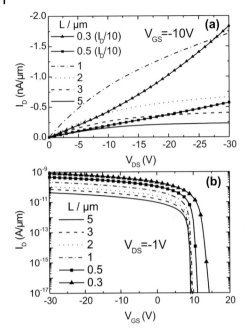

Figure 28 Simulated current characteristics (current per unit channel width) for $-10\,V$ gate-source voltage in the output characteristics (a) and $-1\,V$ drain-source voltage in the transfer characteristics (b) for a doping concentration of $N_A = 10^{17}\,cm^{-3}$. For a 400 nm thick P4VP gate insulator the channel length is varied as indicated.

close to the off-state of the $5\,\mu m$ device (Fig. 29). For zero drain voltage the potential in the channel has a constant plateau and due to the boundary condition, decreases near source and drain causing there accumulation of the holes. But for the shorter channels the plateau value is no longer reached, the maximum of the potential is lower and hence the hole concentration is enhanced. Applying now a drain voltage $(-1\,V)$, one has a further lowering of the maximum for the shorter channels, leading to a strong increase of the hole concentration and hence of the current. Consequently, a higher positive voltage is necessary to switch-off the transistor. For a further increase of the drain voltage the barrier near source is reduced more and more causing the supralinear rise of the current. This effect is well known in microelectronics as the drain induced barrier lowering. As described above, the reduction of this effect is possible reducing the insulator thickness. Therefore, in Fig. 30 the influence of the insulator thickness is demonstrated for $0.5\,\mu m$ and $0.3\,\mu m$ channel length. Indeed for 50 nm insulator thickness this short channel effect is clearly reduced. Furthermore, the threshold voltage is reduced down to $V_{th} \approx 1.5\,V$ and independent on both the channel length and the drain voltage. These results show clearly that for fast submicrometer devices all-polymer variants can only be realized with new materials or technologies allowing for gate insulator thickness as low as about 50 nm.

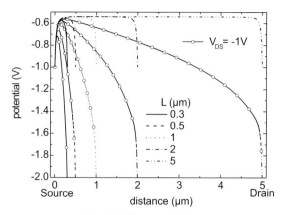

Figure 29 Potential profiles from source to drain 1 nm below the interface to the insulator for a gate voltage of $V_{GS} = 8\,V$ and two drain voltages $V_{DS} = 0\,V$ and $V_{DS} = -1\,V$ for the devices of Fig. 28. a.

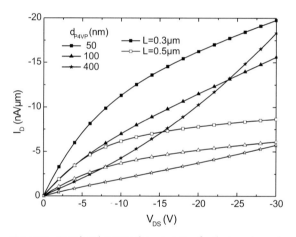

Figure 30 Simulated output characteristics for the two channel length $0.5\,\mu m$ (open symbols) and $0.3\,\mu m$ (filled symbols) the gate insulator thickness is varied. Further parameters: p-doping $10^{17}\,cm^{-3}$ of the 30 nm P3OT layer, mobility $\mu = 10^{-3}\,cm^2\,V^{-1}\,s^{-1}$.

13.7
Conclusions

Great efforts towards commercial polymer electronics are undertaken at present. Apart from many unresolved problems one is faced with one specific deficiency in the research methodology. It is the underdeveloped usage of quantitative theoretical device description by appropriate estimates and two-dimensional simulation as integral parts of both design/fabrication and analysis of experimental data.

Existing doubts on the applicability of the microelectronics simulation tools to polymer field-effect transistors are based on the unfounded assumption that these programs are valid only for crystalline semiconductors. In spite of such doubts, most simple approximations used for data analysis are based on the same basic equations as the simulation programs, the current density and continuity equations for electrons and holes and the Poisson equation. There is indeed one critical point, namely the dependence of the carrier density on the Fermi energy (or electrochemical potential). But the simple non-degenerate dependence following from Boltzmann statistics is valid in many cases also for the organics, and the effective density of states has to be replaced by an appropriate quantity close to the monomer density. On the other hand, a Gaussian or exponential density of transport states requires modifications which have not yet been implemented and therefore one can hardly discuss resulting influences on the OFET current characteristics. However this case might be important for evaporated small molecules rather than for solution processed polymers.

Analytical estimations, beyond the common ones, connecting device and material properties for both the transistor and the corresponding capacitor are indispensable for designing devices with targeted properties, and they yield at least some of the input parameters for simulating the devices. The approximate connection between cut-off frequency, mobility and channel length leads to rather restrictive requirements downsizing the window for material properties and device parameters to channel length $L < 1\,\mu m$, mobility $\mu > 0.01...0.1 cm^2 V^{-1} s^{-1}$ and organic gate insulator thickness $d_{ox} < 50\,nm$ for a low-voltage operation at frequency above 100 kHz. In addition overlap capacitances must be almost negligible. Further influences arise from doping, interface charges and rechargeable interface and bulk traps.

Preparations of OFET's and capacitors with P3OT, P3HT, Arylamino-PPV, MEH-TPD-PPV, and pentacene from a soluble precursor, with silicon dioxide or P4VP as gate insulator, and with rather different channel length, and the analysis of resulting experimental data demonstrate the advantage of combining all steps with analytical estimates and numerical simulation. This approach becomes of special importance in downscaling as demonstrated for the novel submicrometer field-effect transistors fabricated with an underetching technique. The decision to use a hybrid design with thin (30 nm) silicon dioxide as gate insulator was a clear consequence of the simulation. They have shown that short channel effects can be avoided only by using such thin gate insulator, which at present cannot be realized with an organic material. As expected from the estimates, as a positive side-effect we achieved the favorable low inverse subthreshold slope.

Furthermore, independently from actual experimental data, numerical simulations fulfill the self-contained task to clarify special effects as the possible occurrence of inversion, the influence of the type of the source/drain contacts (also in the case of double-injection), the influence of different geometries, or the role of trap distributions. The advantage of the simulations lies in the possibility to clarify the mode of operation by inspecting the internal distributions of concentrations and fields and by relating them to peculiarities of the current and capacitance characteristics.

A

Non-degenerate approximation for the Gaussian distribution

The Gaussian distribution

$$D^{\text{Gauss}} = \frac{N_0}{\varepsilon_0 \sqrt{2\pi}} \exp\left\{ -\frac{1}{2} \left(\frac{\varepsilon - V_0}{\varepsilon_0} \right)^2 \right\} \tag{26}$$

contains as parameter the width (variance) ε_0, the total concentration is N_0, and V_0 is the position of the maximum. The numerically calculated electron concentration (4) is depicted in Fig. 31 for the dimensionless parameter $\varepsilon_0/k_B T$ between 1 and 6 together with the (analytical) non-degenerate approximation (5), (6). In an accumulation layer in a field-effect transistor, due to the break-through voltage of the insulator the density cannot exceed about $10^{-3} N_0$. The figure demonstrates clearly, that up to this density the non-degenerate approximation is applicable up to about $\varepsilon_0/k_B T \approx 3...4$ whereas in Ref. [61] values up to 6 have been considered as typical. However, this concerns essentially lower temperatures typically for investigation of transport properties. OFET's, on the other hand will operate above about 300 K. Thus the approximation works well at least up to $\varepsilon_0 \approx 100$ meV. Practically the range of applicability is even larger from the following reason. In the MOS structure the surface electric field and hence the areal charge in the accumulation layer follow from the simultaneous solution of the Eqs. (11), (12) where the second one expresses the fact that most of the gate voltage drop occurs over the insulator. This is the reason (well known in MOS electronics) that in accumulation a change of the gate voltage leads to a comparatively small change in the areal charge. From the same reason, even for large differences in the surface electric field as function of the band bend-

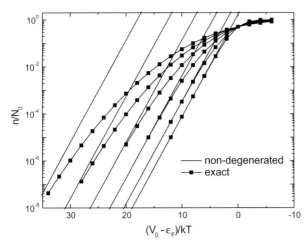

Figure 31 Electron density for the Gaussian DOS as function of the Fermi energy: exact (with symbols) and non-degenerate limit (without symbols) the parameter $\varepsilon_0/k_B T = 1, 2, 3, 4, 5, 6$ varies from right to left.

ing, the difference in the accumulation charge between non-degenerate and degenerate approximation is rather small. It becomes important only for an extremely thin insulator (see e.g. [75]) not yet used in OFETs.

It seems to be worthwhile to comment on the Einstein relation investigated in Ref. [61]. In a field-effect transistor, diffusion is rather unimportant in the linear region (homogeneous channel). Thus, it is important in the subthreshold region and in the transition into saturation with channel pinch-off. In the subthreshold region the density is anyway small enough to use the non-degenerate approximation. Although in the transition into saturation diffusion becomes more important, the influence on the current is rather small just since the current in saturation becomes almost independent from the drain voltage. Therefore, the operation of the transistor can be influenced only slightly by deviations from the non-degenerate Einstein relation even for a broader Gaussian DOS.

B
DDM with traps and trap distributions

Traps at different energies and rather high concentrations have been found in organic light emitting diodes (see [41] for references) their possible influence on organic transistors has been discussed also. Inclusion of traps at energies ε_t extends the three basic DDM equations (9) as follows

$$\nabla \varepsilon \varepsilon_0 (-\nabla \phi) = e \left(p - n - N^- + \sum_{\varepsilon_t} p_t^{(\varepsilon_t)} - \sum_{\varepsilon_t} n_t^{(\varepsilon_t)} \right),$$

$$\nabla e p \mu_p \left(-\nabla \phi_{F_p} \right) = -e \frac{\partial}{\partial t} p - e \left(U_{SRH} + U_{direct} + \sum_{\varepsilon_t} U_{pt}^{(\varepsilon_t)} \right), \tag{27}$$

$$\nabla e n \mu_n \left(-\nabla \phi_{F_n} \right) = +e \frac{\partial}{\partial t} n + e \left(U_{SRH} + U_{direct} + \sum_{\varepsilon_t} U_{nt}^{(\varepsilon_t)} \right).$$

Trapped hole and electron concentrations for one level are related to the occupation functions for holes f_p and electrons f_n according to

$$p_t^{(\varepsilon_t)} = N_{td}^{(\varepsilon_t)} f_p \quad , \quad n_t^{(\varepsilon_t)} = N_{ta}^{(\varepsilon_t)} f_n \tag{28}$$

where the concentrations of acceptor-like and donor-like traps are $N_{ta}^{(\varepsilon_t)}$ and $N_{td}^{(\varepsilon_t)}$, respectively. For the numerical treatment a continuous trap distribution is replaced by a number of discrete traps. $U_{pt}^{(\varepsilon_t)}$ is the net hole capture rate and $U_{nt}^{(\varepsilon_t)}$ the net electron capture rate by the traps. Now the system of equations must be complemented by as much kinetic equations as different traps are present. As an example, for one acceptor-like trap with an energy ε_t several $k_B T$ below the intrinsic energy level ε_i one has

$$-\frac{d}{dt} f_t = \frac{U_{pt}^{(\varepsilon_t)}}{N_{ta}} = K_p f_t - G_p (1 - f_t) \tag{29}$$

$$K_p = \sigma_p v_{th}^p p \quad , \quad G_p = \sigma_p v_{th}^p n_i \exp\{(\varepsilon_i - \varepsilon_t)/k_B T\}. \tag{30}$$

Here σ_p is the capture cross section and v_{th}^p is the thermal velocity of the holes. K_p is the rate of an occupied trap to capture one hole and G_p is the hole emission rate of an empty trap. This equation is general and applies also to organic semiconductors. There is only the peculiarity that for hopping transport process the thermal velocity is much smaller than for band conduction [41].

For the steady-state the derivatives with respect to time vanish. This means $dp_t/dt = -U_{pt}^{(\varepsilon_t)} = 0$, $dn_t/dt = -U_{nt}^{(\varepsilon_t)} = 0$. In this case, for one discrete level one obtains

$$f_n = \frac{\sigma_n v_{th}^n n + \sigma_p v_{th}^p p_1}{\sigma_n v_{th}^n (n+n_1) + \sigma_p v_{th}^p (p+p_1)} \quad , \quad f_p = 1 - f_n \tag{31}$$

$$p_1 = n_i \exp\left(\frac{\varepsilon_i - \varepsilon_t}{k_B T}\right) \quad , \quad n_1 = n_i \exp\left(\frac{\varepsilon_t - \varepsilon_i}{k_B T}\right). \tag{32}$$

σ_n and σ_p are the electron and hole capture cross-sections, v_{th}^n and v_{th}^p are the electron and hole thermal velocities. ε_i is the intrinsic level and ε_t the trap energy, both in the bulk, i.e. for zero potential $\phi = 0$. Of course, in this steady-state occupation function, p and n are the non-equilibrium steady-state densities which are determined by the potential and the respective quasi-Fermi potentials. Eq. (31) is simplified if both the trap level and the quasi-Fermi energies lie e.g. several $k_B T$ below the (bent) midgap energy. One obtains then

$$f_p \rightarrow \frac{1}{1 + \exp\left\{e(\phi + (\varepsilon_i - \varepsilon_t)/e - \phi_{Fp})/k_B T\right\}}. \tag{33}$$

This is the usual hole occupation function, where non-equilibriumis included via the potential and the hole quasi-Fermi potential.

References

[1] A. Dodabalapur, Z. Bao, A. Makhija, J. G. Laquindanum, V. R. Raju, Y. Feng, H. E. Katz, and J. Rogers, Appl. Phys. Lett. **73**, 142 (1998).

[2] Th. N. Jackson, Y-Y. Lin D. J. Gundlach, and H. Klauk, IEEE J. of selected Topics in Quantum Electronics **4**, 101 (1998).

[3] E. Y. Ma, S. D. Theiss, M. H. Lu, C. C. Wu, J. C. Sturm, and S. Wagner, IEDM (1997), Technical Digest, 535.

[4] D. M. de Leeuw, P. W. M. Blom, C. M. Hart, C. M. J. Mutsaers, C. J. Dury, M. Matters, and H. Termeer, IEDM (1997), Technical Digest, 331.

[5] Y.-Y. Lin, A. Dodabalapur, R. Sarpeshkar, Z. Bao, W. Li, K. Baldwin, V. R. Raju, and

H. E. Katz, Appl. Phys. Lett. **74**, 2714 (1999).

[6] C. J. Drury, C. M. J. Mutsaers, C. M. Hart, M. Matters, and D. M. Leeuw, Appl. Phys. Lett. **73**, 108 (1998).

[7] H. Sirringhaus, N. Tessler, and R. H. Friend, Science **280**, 1741 (1998).

[8] H. Klauk, D. J. Gundlach, and Th. N. Jackson, IEEE Electr. Dev. Lett. **20**, 289 (1999).

[9] A. R. Brown, C. P. Jarrett, D. M. de Leeuw, and M. Matters, Synthetic Metals **88**, 37 (1995).

[10] A. Assadi, C. Svensson, M. Willander, and O. Inganäs, Appl. Phys. Lett. **53**, 195 (1988).

[11] Z. Bao, A. Dodabalapur, and A. J. Lovinger, Appl. Phys. Lett. **69**, 4108 (1996).

[12] H. Sirringhaus, R. J. Wilson, R. H. Friend, M. Inbasekaran, W. Wu, E. P. Woo, M. Grell, D. D. C. Bradley, Appl. Phys. Lett. **77**, 406 (2000).

[13] A. Afzali, C. D. Dimitrakopoulos, and T. L. Breen, J. A. Chem. Soc. **12**, 8812 (2002).

[14] Z. Bao, Y. Feng, A. Dodabalapur, v. R. Raju, A.J. Loviger, Chem Mat. **9**, 1299 (1997).

[15] J. A. Rogers, Z. Bao, A. Makhija, P. Braun, Adv. Mater. **11**, 741 (1999).

[16] H. Sirringhaus, T. Kawase, R. H. Friend, T. Shimoda, M. Inbasekaran, W. Wu, and E. P. Woo, Science **290**, 2123 (2000).

[17] N. Stutzmann, R. H. Friend, H. Sirringhaus, Science **299**, 1881 (2003).

[18] S. Scheinert, A. Scherer, T. Doll, G. Paasch, I. Hörselmann, Appl. Phys. Lett. **84**, 4427 (2004).

[19] A. Bonfiglio, F. Mamelli, O. Sanna, Appl. Phys. Lett. **82**, 3550 (2003).

[20] G. Horowitz, R. Hajlaoui, D. Fichou, and A. El Kassmi, J. Appl. Phys. **85**, 3202 (1999).

[21] S. Scheinert, G. Paasch, S. Pohlmann, H.-H. Hörhold, R. Stockmann, Solid State Electronics, **44**, 845 (2000).

[22] A. R. Brown, A. Pomp, D. M. de Leeuw, D. B. M. Klaassen, E. E. Havinga, P. Herwig, and K. Müllen, J. Appl. Phys. **79**, 2136 (1996).

[23] Y.-Y. Lin, D. J. Gundlach, S. F. Nelson, and Th. N. Jackson, IEEE Trans. on Electr. Dev. **44**, 1325 (1997).

[24] H. Sirringhaus; N. Tessler, D. S. Thomas, P. J. Brown, R. H. Friend; Advances in Solid State Physics **39**, 101 (1999).

[25] S. Scheinert, G. Paasch, M. Schrödner, H.-K. Roth, S. Sensfuß, Th. Doll, J. Appl. Phys. **92**, 330 (2002).

[26] S. Scheinert, G. Paasch, T. Doll, Synthetic Metals **139**, 233 (2003).

[27] L. Torsi, A. Dodabalapur, and H. E. Katz, J. Appl. Phys. **78**, 1088 (1995).

[28] S. Scheinert, G. Paasch, Verhandl DPG (VI) **36**, 1 (2001), 207.

[29] S. Scheinert, W. Schliefke, Synthetic Metals **139**, 501 (2003).

[30] S. M. Sze, *Physics of Semiconductor Devices*, John Wiley & Sons, 2nd Edition (1981).

[31] G. Horowitz, Ph. Delannoy, J. Appl. Phys. **70**, 469 (1991).

[32] G. Horowitz, R. Hajlaoui, H. Bouchriha, R. Bourguiga, and M. Hajlaoui, Adv. Mat. **10**, 923 (1998).

[33] L. Aguilhon, J.-P. Bourgoin, A. Barraud, P. Hesto, Synthetic Metals **71**, 1971 (1995).

[34] R. Tecklenburg, G. Paasch, S. Scheinert, Adv. Mat. for Optics and Electronics **8**, 285 (1998).

[35] G. Horowitz, R. Hajlaoui, and P. Delannoy, J. Phys. *III* (France) **5**, 355 (1995).

[36] F. Schauer, J. Appl. Phys. **86**, 524 (1999).

[37] G. Horowitz, Adv. Mater. **10**, 365 (1998).

[38] G. Horowitz, M. E. Hajlaoui, and R. Hajlaoui, J. Appl. Phys. **87**, 4456 (2000).

[39] T. Lindner, G. Paasch, S. Scheinert, submitted, J. Materials Research.

[40] S. Scheinert, G. Paasch, P. H. Nguyen, S. Berleb, W. Brütting, Proc. ESS-DERC'2000, Frontier Group, Paris, 444 (2000).

[41] P. H. Nguyen, S. Scheinert, S. Berleb, W. Brütting, G. Paasch, Organic Electronics, **2**, 105 (2001).

[42] G. Paasch, S. Scheinert, Synthetic Metals **122**, 145 (2001).

[43] A. Nesterov, G. Paasch, S. Scheinert, T. Lindner, Synthetic Metals **130**, 165 (2002).

[44] G. Paasch, A. Nesterov, S. Scheinert, Synthetic Metals **139**, 425 (2003).

[45] ATLAS User's Manual: Device Simulation Software, SILVACO International, Santa Clara (1998).

[46] ISE TCAD, Integrated Systems Engineering AG, Zurich, Switzerland, 1995–1999.

[47] M. A. Alam, A. Dodabalapur, and M. R. Pint, IEEE Trans. on Electr. Dev. **44**, 332 (1997).

[48] N. Tessler and Y. Roichman, Appl. Phys. Lett. **79**, 2987 (2001).

[49] Y. Roichman and N. Tessler, Appl. Phys. Lett. **80**, 151 (2002).

[50] T. Li, J. W. Balk, P. P. Ruden, I. H. Campbell, and D. L. Smith, J. Appl. Phys. **91**, 4312 (2002).

[51] T. Li, P. P. Ruden, I. H. Campbell, and D. L. Smith, J. Appl. Phys. **93**, 4017 (2003).

[52] H. S. Momose, S. Nakamura, Y. Katsumata and H. Iwai, Proc. ESSDERC'1997, Frontier Group, Paris, 133 (1997).

[53] P. W. M. Blom, M. J. M. de Jong, and M. G. van Munster, Phys. Rev B **55**, R656 (1997).

[54] P. W. M. Blom and M. C. J. M. Vissenberg, Phys. Rev. Lett. **80**, 3819 (1998).

[55] A. Ioannidis, E. Forsythe, Y. Gao, M. W. Wu, E. M. Conwell, Appl. Phys. Lett. **72**, 3038 (1998).

[56] M. C. J. M. Vissenberg, M. Matters, Phys. Rev. **B 57**, 12964 (1998).

[57] G. Paasch, P. H. Nguyen, S.-L. Drechsler, Synth. Met. **97**, 255 (1998).

[58] G. Paasch, P. H. Nguyen, A. Fischer, Chemical Physics **227**, 219 (1998).

[59] H. Bässler, phys. stat. sol. (b) **175**, 15 (1993).

[60] S. D. Baranovskii, H. Cordes, F. Hensel, G. Leising, Phys. Rev **B 62**, 7934 (2000).

[61] Y. Roichman, N. Tessler, Appl. Phys. Lett. **80**, 1948 (2002).

[62] J. H. Schön, C. Kloc, R. A. Laudise, and B. Batlogg, Phys. Rev. **B 58**, 12952 (1998).

[63] A. R. Brown, D. M. de Leeuw, E. E. Havinga, A. Pomp, Synthetic Metals **68**, 65 (1994).

[64] G. Paasch, T. Lindner, S. Scheinert, Synthetic Metals **132**, 97 (2002).

[65] E. J. Meijer, D. M. De Leeuw, S. Setayesh, E. Van Veenendaal. B.-H. Huisman, P. W. M. Blom, J. C. Hummelen, U. Scherf, T. M. Klapwijk, Nature Materials **2**, 678 (2003).

[66] H. Sirringhaus, Nature Materials **2**, 641 (2003).

[67] P. J. Brown, H. Sirringhaus, M. Harrison, M. Shkunov, R. H. Friend, Phys. Rev. B **63**, 125204 (2001).

[68] P. Barta, J. Sanetra, P. Grybos, S. Niziol, M. Trznades, Synthetic Metals **94**, 115 (1998).

[69] M. Halik, H. Klauk, U. Zschieschang, G. Schmid, W. Radlik, S. Ponomarenko, St. Kirchmeyer, W. Weber, J. Appl. Phys. **93**, 2977 (2003).

[70] W. Fix, A. Ullmann, J. Ficker, W. Clemens, Appl. Phys. Lett. **81**, 1735 (2002).

[71] Ch.-K. Song, B.-W. Koo, S.-B. Lee, D.-H. Kim, Jpn. J. Appl. Phys. **41**, 2730 (2002).

[72] G. H. Gelinck, T. C. T. Geuns, D. M. de Leeuw, Appl. Phys. Lett. **77**, 1487 (2000).

[73] Y. Zhang, J. R. Petta, S. Ambily, Y. Shen, D. C. Ralph, G. G. Malliaras, Adv. Mater. **15**, 1632 2003.

[74] Y. S. Yang, S. H. Kim, J. L. Lee, H. Y. Chu, L. Do, H. Lee, J. Oh, T. Zyung, M. K. Ryu, and M. S. Jang, Appl. Phys. Lett. **80**, 1595 (2002).

[75] G. Paasch, S. Scheinert, K. Tarnay, phys.stat.sol.(a) **149**, 751(1995).

14
Organic Single-Crystal Field-Effect Transistors

R. W. I. de Boer, M. E. Gershenson, A. F. Morpurgo, and V. Podzorov

14.1
Introduction

The electronic properties of Van-der-Waals-bonded organic semiconductors are profoundly different from those of covalently/ionically-bonded inorganic semiconductors [1, 2]. In the highly-polarizable crystal lattices of organic semiconductors, the electron-phonon coupling is usually strong and the inter-molecular hopping amplitude small. This results in the formation of self-trapped states with a size comparable to the lattice constant, i.e., the small polarons. The electronic, molecular and lattice polarization plays a key role in determining transport in organic materials, as polaronic effects "shape" both the *dc* transport and optical properties of these materials. Because of a very complicated character of the many-particle interactions involved in polaron formation, this problem has been treated mainly at the phenomenological level (see Chapter 7 in Ref. [1]). Many basic aspects of this problem have not been addressed yet, and a well-developed microscopic description of the charge transport in organic materials is still lacking.

Until recently, the experimental study of the low-frequency *intrinsic* electronic properties of organic semiconductors have been performed only on *bulk* ultra-pure crystals [3, 4]. In the time-of-flight (TOF) experiments by the group of Norbert Karl at Stuttgart University [5, 6], it has been found that the mobility of non-equilibrium carriers generated by light absorption in ultra-high-purity oligomeric crystals can be as high as 400 cm^2/Vs at low temperatures (the latter μ value is comparable to the mobility of electrons in Si MOSFETs at room temperature). This behavior suggests that coherent, band-like polaronic transport is possible in crystal of small organic molecules.

To further investigate the electronic properties of organic materials, it is important to go beyond the TOF measurements. One of the alternative techniques to probe the charge transport on a semiconductor surface is based on the electric field effect [7]. Continuous tuning of the charge density induced by the transverse electric field enables the systematic study of charge transport, in particular the regime of large carrier density that cannot be accessed in the TOF experiments. The field effect forms the basis for operation of silicon field-effect transistors (FETs), the workhorses of modern inorganic electronics. The field-effect technique is also becoming increas-

Physics of Organic Semiconductors. Edited by W. Brütting

ingly popular in the fundamental studies as a convenient method to control the behavior of strongly correlated electron systems such as high-temperature superconductors (see, e.g., [8]) and colossal magnetoresistance manganites [9]. Other recent examples of applications of this remarkably simple and very successful principle are the electric-field tuning of the metal-insulator transition in cuprates [10] and vanadium oxides [11], and the electrostatic control of ferromagnetism in Mn-doped GaAs [12].

Organic semiconductors are, in principle, well suited for the field-effect experiments. Owing to the weak van-der-Waals bonding, the surface of organic semiconductors (e.g., polyacenes [13, 14] and conjugated polymers [15]) is characterized by an intrinsically low density of dangling bonds that can act as the charge traps, and, hence, by a low threshold for the field effect. This fact is at the origin of the rapid progress of organic field-effect transistors based on thin film technology, i.e., organic thin-film transistors (OTFTs) [16, 17].

Unfortunately, *thin-film* transistors are *not* suitable for the study of *intrinsic* electronic properties of organic conductors, because their characteristics are often strongly affected by imperfections of the film structure and by insufficient purity of organic materials (see, e.g., [17–19]). As a consequence, these devices commonly exhibit an exponential decrease of the mobility of field-induced charge carriers with lowering temperature [20]. This behavior contrasts sharply a rapid increase of μ with decreasing temperature, observed in the TOF experiments with bulk ultra-pure organic crystals [5, 6]. Because of a very strong dependence of the OTFT parameters on fabrication conditions, some researchers came to a pessimistic conclusion that even the best organic TFTs "may not be appropriate vehicles for illuminating basic transport mechanisms in organic materials" [21].

To explore the *intrinsic* electronic properties of organic materials and the physical limitations on the performance of organic FETs, devices based on *single-crystals* of organic semiconductors are needed, similar to the single-crystal structures of inorganic electronics. One of the major impediments to realization of the single-crystal OFETs is the lack of hetero-epitaxial growth technique for the Van-der-Waals-bonded organic films. In this situation, the only viable option to study the intrinsic charge transport on the surface of organic semiconductors is to fabricate the field-effect structures on the surface of free-standing organic molecular crystals (OMCs). However, fabrication of single-crystal OFETs poses a technological challenge. Because the surface of OMCs can be damaged much more easily than that of their inorganic counterparts, organic materials are by and large incompatible with conventional microelectronic processing techniques such as sputtering, photolithography, etc. This is why the systematic investigation of single-crystal OFETs has been carried out only very recently [22–27], after the successful development of a number of novel fabrication schemes (for earlier work see [28, 29]).

Realization of the single-crystal OFETs opens a new avenue for the study of charge transport in highly ordered molecular systems. The use of single-crystal OFETs as an experimental tool enables the investigation of aspects of charge transport in organic materials that could not be addressed in the TOF experiments. One of the important distinctions between these two types of experiments is the magni tude of carrier densities. Very low densities of charge carriers in the TOF experiments make interactions between them insignificant. At the same time, in the field-effect experiments with organic materials, where accumulation of ~1 carrier per molecule seems to be feasible with the use of high-k dielectrics, these interactions could play a major role. Indeed, it is well-known that at a sufficiently high density of chemically-induced carriers, the potassium-doped fullerene $K_x C_{60}$ exhibits superconductivity ($x = 3$) and a Mott-Hubbard insulating state ($x = 4$) [30]. This example illustrates a great potential of experiments with the single-crystal OFETs.

The first working FET on the surface of a free-standing organic molecular crystal has been fabricated a year ago [22]. Though this field is in its infancy, the progress has been remarkably rapid, with new record values of carrier mobility for OFETs being achieved, new promising organic materials being introduced, and new device processing techniques being developed. In this review, we discuss the new techniques responsible for the progress, the state-of-the-art characteristics of single-crystal OFETs, and the experiments that show that development of single-crystal OFETs enables investigation of intrinsic electronic properties of organic materials. In the future, the combined efforts of experimenters and theorists, physicists and chemists will be required to reveal the full potential of this research area. We hope that this paper will provide a timely source of information for the researchers who are interested in the progress of this new exciting field.

The structure of this review is as follows. In Section 14.2, we present an overview of the organic crystal growth and techniques for the single-crystal OFET fabrication, with a special attention paid to the crystal characterization. Discussion of the single-crystal OFET characteristics in Section 14.3 focuses on revealing the intrinsic transport properties of organic semiconductors. Finally, in Section 14.4, we summarize the main results, and attempt to predict the directions of rapid growth in this fascinating research field.

14.2
Fabrication of single-crystal organic FETs

The successful realization of FETs on the surface of organic molecular crystals (OMC) is an important milestone in the research of electronic transport in organic semiconductors. For the first time, it opens the opportunity to study the intrinsic behavior of charges at the organic surface, not limited by structural defects. Fabrication of single organic crystal FETs comprises two main steps: the growth of an organic crystal with atomically-flat surface and preparation of the field-effect structure on this surface. In this section we discuss both aspects, paying special attention to the crystal characterization and to the analysis of advantages and limitations of different fabrication methods.

14.2.1
Single-crystal growth

Most of the single crystals used so far for the fabrication of organic FETs have been grown from the vapor phase in a stream of transport gas, in horizontal reactors (glass or, better, quartz tube) [31, 32] (for a notable exception, in which the crystals have been grown from solution, see Ref. [33]). In the Physical Vapor Transport (PVT) method, the starting material is placed in the hottest region of the reactor, and the crystal growth occurs within a narrow temperature range near its cold end (see Fig. 1). For better separation of larger and, presumably, purer crystals from the rest of re-deposited material along the tube, the temperature gradient should be sufficiently small (typically, 2–5 °C/cm).

Several ultra-high-purity gases have been used as a carrier agent: in Ref. [24], the highest mobility of tetracene-based devices was realized with argon, whereas the best rubrene FETs fabricated so far have been grown in pure H_2 [22, 23]. In the latter case, hydrogen has been chosen after comparison of the field-effect characteristics of rubrene crystals grown in Ar, N_2, and H_2 atmospheres. It is unclear at present how exactly the transport gas affects the crystal quality; uncontrollable variations of the crystal quality might be caused by the residual water vapor and oxygen in the reactors. Photo-induced reactions with O_2 are known for most organic molecules [34] and the products of these reactions can act as traps for charge carriers. To minimize possible photo-activated oxidation of organic material, the reactors should be pumped down to a reduced pressure $P \simeq 10^{-2}$ mbar prior to the crystal growth, and the growth should be performed in the dark.

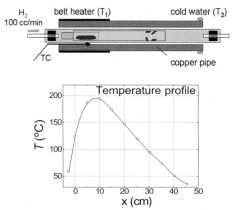

Figure 1 Schematic overview of crystal growth system. Organic material sublimes at temperature T_1, is transported through the system by the carrier gas and recrystallizes in the cooler end of the reactor. Heavy impurities (with a vapor pressure lower that that of the pure organic compound) remain at the position of the source material. Light impurities (with a vapor pressure higher than that of the pure organic compound) condense at a lower temperature, i.e. at a different position from where the crystals grow. Therefore, the crystal growth process also results in the purification of the material.

Several factors affect the growth process and the quality of the crystals. Important parameters are, for instance, the temperature in the sublimation zone, T_{sblm} and the gas flow rate. Many other factors can also play a role: e.g., acoustical vibrations of the reactor in the process of growth might affect the size, shape, and quality of the crystals. For each material and each reactor, the optimal parameters have to be determined empirically. At least in one case (the rubrene-based OFETs [23]), it has been verified that the slower the growth process, the higher the field-effect mobility. For this reason, the temperature of the sublimating organic material was chosen close to the sublimation threshold. The crystal growth in this regime proceeds by the flow of steps at a very low rate ($\leq 5 \times 10^{-7}$ cm/s in the direction perpendicular to the *a-b* plane), and results in a flat surface with a low density of growth steps [35]. As an example, sublimation of 300 mg of starting material in Ref. [23] took up to 48 hours at $T_{sblm} = 300\,^{\circ}\mathrm{C}$.

Another important parameter is the purity of the starting material. As the crystal growth process also results in the chemical purification of the material, several re-growth cycles may be required for improving the field-effect mobility, with the grown crystals used as the starting material for the subsequent re-growth. The number of required re-growth cycles depends strongly on the purity of starting material. Figure 2 illustrates the need for several re-growth cycles in the process of the growth of tetracene crystals. Despite the nominal 98% purity of the starting tetracene (Sigma-Aldrich), a large amount of residue left in the sublimation zone after the first growth cycle is clearly visible (Fig. 2a); this residue is not present at the end of the second growth cycle (fig. 2b). A word of caution is appropriate: in the authors' experience, different batches of as-purchased material, though being of the same nominal purity, might leave different amount of residue. Clearly, the better purity of the starting material, the fewer re-growth cycles are required for a high FET mobility: in Ref. [23] the rubrene OFETs with $\mu > 5$ cm^2/Vs have been fabricated from the "sublimed grade" material (Sigma-Aldrich) after 1–2 growth cycles.

It is likely that the purity of crystals for the OFET fabrication can be substantially improved if a zone refining process [3, 4] is used for pre-purification of the starting material. Indeed, in the time-of-flight studies of organic crystals, the best results and the highest mobilities have been obtained after multiple zone-refinement purification cycles [4, 5]. This process enabled reduction of the impurity concentration in the

(a)

(b)

Figure 2 (a) Result after first regrowth of as-purchased organic material. Purified crystals are visible in the middle; the dark residue present where the source material initially was and the light (yellow) material visible on the right are due to impurities. (b) At the end of the second regrowth no dark residue is present at the position of the source material, which demonstrates the purifying effect of the growth process.

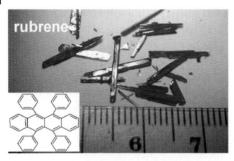

Figure 3 Result of rubrene crystal growth. Most of the organic crystals grown by the physical vapor transport are shaped as elongated "needles" or thin platelets.

bulk down to the part-per-billion level. It is unlikely that a comparable purity can be achieved simply by multiple vapor transport re-growth processes. It has to be noted that zone-refinement cannot be applied to all organic materials, since this technique requires the existence of a coherent liquid phase (i.e. the melting temperature of the substance has to be lower than the temperature of decomposition of its molecules) [3]. However, for several materials that have already been successfully used for fabrication of single crystal FETs (e.g., rubrene, perylene, anthracene), a coherent liquid phase does exist and zone refinement is possible. The zone-refining purification might be especially useful for realization of intrinsic polaronic transport at low temperatures, where trapping of polarons by defects becomes a serious problem.

Most of the organic crystals grown by the physical vapor transport are shaped as elongated "needles" or thin platelets (see Fig. 3). The crystal shape is controlled by the anisotropy of inter-molecular interactions: for many materials, a larger crystal dimension corresponds to the direction of the strongest interactions and, presumably, the strongest overlap between π-orbitals of adjacent molecules. For this reason, the direction of the fastest growth of needle-like rubrene crystals coincides with the direction of the highest mobility of field-induced carriers (see Sec. 14.3). For platelet-like crystals, the larger facets are parallel to the *a-b* plane. Typical in-plane dimensions range from a few square millimeters for rubrene to several square centimeters in the case of anthracene. The crystal thickness also varies over a wide range and, in most cases, can be controlled by stopping the growth process at an early stage. For example, the thickness of the tetracene crystals grown for 24 hours ranges between $\sim 10\ \mu$m and $\sim 200\ \mu$m [36], but it is possible to harvest several crystals of sub-micron thickness by stopping the growth process after ~ 30 minutes.

Because of a weak van der Waals bonding between the molecules, polymorphism is a common phenomenon in organic materials: the molecular packing and the shape of organic crystals can be easily affected by the growth conditions. For example, the thiophenes exhibit two different structures depending on the growth temperature [37]. In many cases, organic molecular crystals exhibit one or more structural phase transitions upon lowering the temperature. For the study of single-crystal FETs at low temperature, the occurrence of a structural phase transition can be

Figure 4 Two platelet-shaped anthracene single-crystals grown by physical vapor transport. The crystals are transparent and colorless. The left crystal is illuminated by UV light, and fluoresces in the blue.

Figure 5 A cm^2-sized platelet-shaped tetracene single-crystal grown by physical vapor transport.

detrimental. In tetracene crystals, for instance, a structural phase transition occurs below 200 K (see, e.g., [38]). Co-existence of two crystallographic phases at lower temperatures causes the formation of grain boundaries and stress, which are responsible for the trapping of charge carriers (see Fig. 10). In tetracene, in addition, occurrence of the structural phase transition often results in cracking of the crystals with cooling and a consequent device failure.

14.2.2
Crystal characterization

To understand better the effect of different factors on the crystal growth, a thorough characterization of the crystal properties is needed. Note, however, that many experiments provide information on the crystal properties that is only indirectly related to the performance of the single-crystal OFETs. For example, the x-ray analysis of organic crystals, though necessary for identification of the crystal structure and ori-

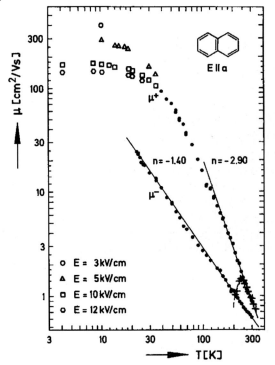

Figure 6 Electron and hole mobility μ versus temperature T in ultra-pure single crystals of naphthalene, as measured in Time-Of-Flight (TOF) experiments. The solid lines indicate a T^n power-law temperature dependence with exponents n as indicated in the figure. For holes, mobility values as high as 400 cm²/Vs are observed at low temperature. ([85], printed with permission of N. Karl, Crystal Laboratory, Univ. Stuttgart.)

entation of the crystallographic axes, is insufficiently sensitive for detection of a minute concentration of defects that might severely limit the field-effect mobility at low temperatures. Similarly, the TOF experiments, although useful in assessing the quality of the *bulk* of organic crystals, are not sensitive to the surface defects that limit the OFET performance. Below we briefly review several techniques that have been used for organic crystal characterization.

14.2.2.1
Polarized-light microscopy

Inspection of crystals under an optical microscope in the polarized light provides a fast and useful analysis of the crystalline domain structure. Visualization of domains is possible because crystals of most organic conjugated materials are birefringent [39]. Optical inspection also enables detection of the stress in crystals, which results in appearance of the interference fringes with orientation not related to any specific crystallographic direction. This technique simplifies the process of selection of single crystals for transport measurements.

14.2.2.2
The time-of-flight experiments

In the time-of-flight (TOF) experiments, a platelet-like crystal is flanked between two metal electrodes, one of which is semi-transparent [5]. A thin sheet of photo-excited charge carriers is generated near the semi-transparent electrode by a short laser pulse with the photon energy greater than the band gap. In the presence of a constant voltage bias between the two electrodes, the charge sheet propagates in the direction determined by the *dc* electric field and generates a displacement current, whose magnitude diminishes rapidly as soon as the sheet reaches the opposite electrode [3]. From the duration of the displacement current pulse and the known crystal thickness, the drift velocity and carrier mobility can be calculated. This method also provides indirect information on the concentration of (shallow) traps in the bulk: the decrease of mobility at low temperatures is caused by multiple trapping and release processes (for more details, see Ref. [40]). An important aspect of TOF measurements is that their results are not sensitive to contact effects, since the charge carriers are photo-generated (i.e., not injected from a metal electrode) and their motion is detected capacitively. This simplifies the contact preparation and improves the reproducibility of results for identically grown crystals.

For measuring the intrinsic mobility of charge carriers in the bulk, the lifetime of the carriers against charge trapping has to be greater than the time of flight between the electrodes. This requirement imposes severe limitation on the concentration of charge traps. As a result, only very pure and defect-free crystals can be characterized by the TOF method. For example, according to preliminary measurements by the Stuttgart group [41], the rubrene crystals used for fabrication of the high-mobility OFETs [23] are unsuitable for the TOF measurements. The crystals for the TOF measurements should also have sufficiently parallel opposite facets and be sufficiently thick for the displacement current pulse to be longer than the apparatus time resolution. Because of these limitations, the TOF measurements can be performed only on a small fraction of the crystals grown by the vapor phase deposition technique.

Despite the difficulties of application of the TOF method to the organic crystal grown by vapor transport, successful TOF measurements have been performed on vapor-grown tetracene crystals similar to those used in FET experiments [36]. The room-temperature mobility $\mu = 0.5 - 0.8$ cm^2/Vs, measured in the TOF experiment for three different crystals, is comparable to the highest mobility of the field-induced carriers in the OFET experiments; these quantities also exhibit similar temperature dependencies. Interestingly, the two types of measurements provided similar μ values for charge transport along different directions in anisotropic organic crystals: the FET measurements probe surface transport in the *a-b* plane, whereas TOF experiments on the platelet tetracene crystals probed motion along the *c*-axis. This observation seems to be in disagreement with what one would expect from the crystallographic structure of the material, namely a pronounced anisotropy of mobility along different crystallographic directions [6, 42, 43]. To better understand the origin and the implications of this result, further characterization of a larger variety of organic crystals by the TOF method and comparison with field-effect measurements are needed.

14.2.2.3
Space charge limited current spectroscopy

A rather common and, in principle, simple way to characterize the electrical properties of OMCs is the study of the *I–V* characteristics measured in the space charge limited current (SCLC) regime [44]. The value of the carrier mobility, as well as the density of deep traps can be inferred from these measurements. Similar to the TOF experiments, SCLC measurements require relatively large electric fields. Usually, these measurements are performed on thin platelet-like crystals; the metallic contacts are located on the opposite facets of a crystal of thickness *L*, and the current *I* in the direction perpendicular to the surface (along the *c*-axis) is measured as a function of the voltage *V* between the contacts. Measurements of *I–V* characteristics with contacts on the same surface of the crystal can also be performed, but in this geometry the extraction of the carrier mobility and the density of traps from the experimental data is more involved. Instrumentation-wise, the *dc* charge injection experiments are less challenging than the TOF measurements. However, a high quality of contacts is required for the former measurements, otherwise the results for nominally identical samples are not reproducible [36]. Because of a high sensitivity of the data to the contact quality, the SCLC technique typically requires acquisition of a large volume of data for many samples and, therefore, is not very efficient.

Recently, the trap-free space charge limited current (SCLC) regime has been observed for samples with both thermally-evaporated thin-film contacts [23] and silver epoxy contacts [36]. The *I–V* characteristics for a thin rubrene crystal with thin-film silver contacts demonstrate the crossovers from the Ohmic regime to the space-charge-limited-current (SCLC) regime, and, with a further voltage increase, to the trap-free (TF) regime (see Fig. 7) [23]. Observation of a linear Ohmic regime indicates that the non-linear contribution of Schottky barriers formed at the metal/rubrene interfaces is negligible in these measurements: the voltage drop across the Schottky barriers is smaller than the voltage drop across the highly resistive bulk of the crystal. The crossover to the SCLC regime at a low bias voltage $V_\Omega \approx 2.5$ V suggests that the charge carrier injection from the contacts is very efficient. From the threshold voltage of the TF regime, V_{TF}, the density of deep traps, N_t^d, can been estimated [44, 45]:

$$N_t^d = \frac{\varepsilon\varepsilon_0 V_{TF}}{eL^2} \qquad (1)$$

Here ε_0 is the permittivity of free space, ε is the relative dielectric constant of the material. Note that an assumption-free estimate of the trap density can be made only if a well-defined crossover between SCLC and TF regimes is observed. For this reason, the method can be applied only to sufficiently pure crystals. For the tetracene crystals studied in Ref. [36], $N_t \approx 5 \times 10^{13}$ cm^{-3} is significantly smaller than $N_t \approx 10^{15}$ cm^{-3} for rubrene crystals studied in Ref. [23]. The steep increase of current that signifies the transition to the trap-free regime, is also much more pronounced for tetracene crystals (see Fig. 7 and 8).

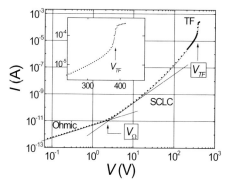

Figure 7 *I–V* characteristic of a ∼ 10 μm-thick rubrene crystal, measured along the *c*-axis. The inset is a blow-up of the cross-over to the trap free regime (also in a double-log scale). From the crossover to the trap-free regime, the density of deep traps $N_t^d \simeq 10^{15}$ cm^{-3} can be estimated [23].

The estimate of N_t^d is based on the assumption (not usually mentioned in the literature), that the deep traps are uniformly distributed throughout the entire crystal bulk. However, it is likely that the trap density is greater near the metal/organic interface because of the surface damage during the contact preparation. A small amount of traps located close to the surface can have a large effect on the current flow: the charges trapped near the surface strongly affect the electric field in the

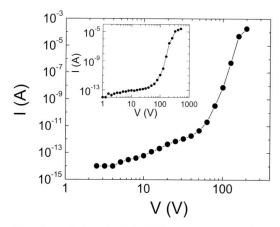

Figure 8 Typical result of a DC *I–V* measurement perpendicular to the *a-b* plane of a tetracene single-crystal, with a thickness $L = 30$ μm and a mobility $\mu_{min} = 0.59$ cm^2/Vs. The inset shows a similar measurement on a different crystal ($L = 25$ μm, $\mu_{min} = 0.014$ cm^2/Vs), in which a crossing over into an approximately quadratic dependence on voltage is visible at high voltage. In both cases, a very steep current increase occurs around or just above 100 V that we attribute to filling of deep traps. We observed a steep increase in current in most samples studied.

bulk of the crystal, which determines the current flow in the TF regime. For this reason, the value of N_t^d may be considered as an upper limit of the actual density of traps in the bulk (see Ref. [36] for a more detailed discussion).

In the TF regime, the mobility can be estimated from the Mott-Gurney law for the trap-free regime [44]:

$$J_{TF} = \frac{9\varepsilon\varepsilon_0\mu V^2}{8L^3} \tag{2}$$

where J_{TF} is the current density. Even when the TF limit is not experimentally accessible, the same formula can be used to extract a lower limit, μ_{min} for the intrinsic mobility, at least in materials in which one type of carriers (electrons or holes) is responsible for charge transport (see Ref. [36] for details). In tetracene crystals, the values of μ_{min} along the c-axis obtained from SCLC measurements performed on a large number of samples are shown in Fig. 9. The large spread in values for μ_{min} obtained from identically grown crystals is due to scattering in the contact parameters.

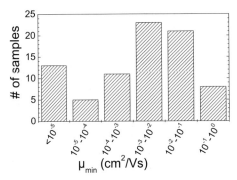

Figure 9 Histogram of values for μ_{min} calculated from dc I-V measurements performed on approximately 100 tetracene single-crystals. The large scattering in the observed values is due to the spread in contact quality.

For the samples with $\mu_{min} > 0.1$ cm^2/Vs, the mobility increases with cooling over the interval $T \simeq 180 - 300$ K (Fig. 10). Observation of the mobility increase with cooling in high-quality crystals is usually considered as a signature of the intrinsic (disorder-free) transport (for comparison, an increase of mobility with lowering temperature has never been observed in SCLC measurements performed on disordered thin organic films). With further cooling, however, the mobility decreases: Fig. 10 shows that for most of the tetracene crystals, a sharp drop of μ below 180 K is observed. This suggests that this drop might be related to a structural phase transition, which is known to occur in tetracene in this temperature range [38]. Both the observed effects of the structural phase transition on the carrier mobility, as well as the increase in μ_{min} with lowering temperature, indicate that these SCLC measurements reflect the intrinsic electronic properties of the material.

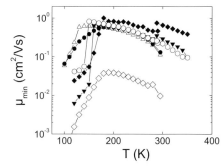

Figure 10 Temperature dependence of the lower limit to the mobility, μ_{min}, measured for tetracene single-crystals with a high μ_{min} value. Note the abrupt drop in mobility occurring at different temperatures below ≈ 180 K, originating from a known structural phase transition [38].

14.2.3
Fabrication of the field-effect structures

Fabrication of the field-effect structure on the surface of organic crystals poses a challenge, because many conventional fabrication processes irreversibly damage the surface of van-der-Waals-bonded organic crystals by disrupting molecular order, generating interfacial trapping sites, and creating barriers to charge injection. For example, sputtering of an insulator onto the crystal creates such a high density of defects on the organic surface that the field-effect is completely suppressed. Up to date, two techniques for single-crystal OFET fabrication have been successfully used: (a) electrostatic "bonding" of an organic crystal to a prefabricated source/drain/gate structure, and (b) direct deposition of the contacts and gate insulator onto the crystal surface. In this section, we address the technical aspects of these fabrication processes; effects of different fabrication methods on the electrical characteristics of the resulting single-crystal FETs will be discussed in Sec. 14.3.

14.2.3.1
Electrostatic bonding technique

In this approach, the transistor circuitry (both gate and source/drain electrodes) is fabricated on a separate substrate, which, at the final fabrication stage, is electrostatically bonded to the surface of organic crystal. The technological processes vary depending on the type of substrate for the transistor circuitry. Two kinds of substrates have been used: conventional silicon wafers [24, 25, 29], and flexible elastomer substrates (the so-called rubber stamps) [46].

Source/drain/gate structures on Si substrates In this method, the source/drain/gate structure is fabricated on the surface of a heavily doped (n-type or p-type) Si wafer, covered with a layer of thermally grown SiO_2 (typically, 0.2 μm thick). The conduct-

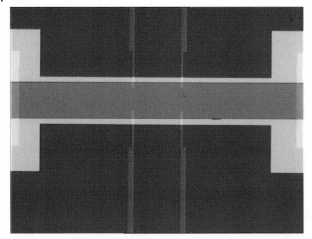

Figure 11 Optical microscope picture of a single-crystal rubrene FET, fabricated by electrostatic bonding. The crystal, which has a rectangular shape, overlaps with the source and drain contacts (at the left and right edge of the picture) and with four small contacts in the center, used to perform 4-probe electrical measurements. The purple area consists of a Ta_2O_5 layer sputtered on top of the substrate prior to the crystal bonding, which, for wider crystals, serves to confine the electrically active region of the FET.

ing Si wafer serves as the gate electrode, and the SiO_2 layer plays the role of the gate insulator. The source and drain gold contacts are deposited on top of the SiO_2 layer, and, as a final step, a sufficiently thin OMC crystal is electrostatically bonded to the source/drain/gate structure. It has been found in Ref. [24] that the reactive ion etching (RIE) of the contact/SiO_2 surface in an oxygen plasma prior to the OMC bonding improves significantly the characteristics of tetracene OFETs: the RIE cleaning reduces the spread of mobilities, the field-effect threshold voltage, and the hysteresis of transfer characteristics. The RIE cleaning also significantly improves adhesion of freshly grown tetracene crystals to SiO_2 surface. The technique works best for very thin crystals ($\leq 1\ \mu$m thick) that adhere spontaneously to the substrate, but it can also be applied (with a lower success yield) to much thicker crystals by gently pressing on the crystal to assist the adhesion process [47]. Fig. 11 shows a picture of a rubrene FET fabricated with the technique of electrostatic adhesion to SiO_2.

Source/drain/gate structures on flexible substrates In this approach, the FET circuitry is fabricated on top of a flexible elastomer substrate (polydimethylsiloxane, or PDMS), by sequential shadow-mask deposition of the gate and source/drain electrodes [46]. The fabrication process, illustrated in Fig. 12, begins with deposition of the gate electrode on top of a few-mm-thick PDMS substrate (1.5 nm of Ti as an adhesion promoter to the substrate, 20 nm of Au, and 3 nm of Ti as an adhesion promoter to the subsequent layers). A (2-4)μm-thick PDMS film, deposited by spin-coating on top of the structure, serves as a gate dielectric. Evaporation or transfer printing of source and drain electrodes (1.5 nm of Ti and 20 nm of Au) on top of the dielectric completes the stamps. Careful control of the fabrication processes results in electrode and dielectric surfaces with low surface roughness (the root-mean-squared value of ~0.6 nm, as measured by atomic force microscopy). The final assembly of the devices, similar to the Si-based technique, consists of positioning of the OMC crystal on the stamp surface, and applying a gentle pressure to one edge of the crystal. Van der Waals forces then spontaneously cause a "wetting" front to proceed

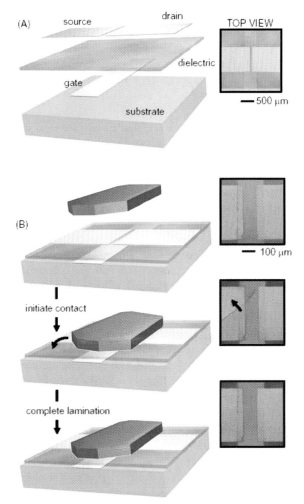

Figure 12 The PDMS-based stamp: source/drain/gate structures on a flexible substrate. (a) Schematic picture of the layout of the stamp. It consists of a few-mm-thick PDMS pad, a patterned gate electrode, a (2–4) μm-thick PDMS spin-coated film as a gate-dielectric, and gold source and drain electrodes. (b) The principle of lamination. The crystal is positioned on the stamp surface, and a gentle pressure is applied to one edge of the crystal. Van-der-Waals forces then spontaneously cause a "wetting" front to proceed across the crystal surface. The insets at the right show microscope pictures of the transparent crystal on top of the stamp at different stages of the lamination process.

across the crystal surface at a rate of a few tenths of a millimeter per second. This lamination process yields uniform contact, devoid of air gaps, bubbles or interference fringes.

The main advantage of both Si- and PDMS-based stamps technique is obvious: it eliminates the need for deposition of metals and dielectrics directly onto a very vulnerable organic surface. Since these techniques exploit the technologies well-developed in electronic industry, the dimension of the circuitry can be easily reduced, if desirable, well in the sub-micron range: specifically, a very small source/drain contact separation can be achieved by using electron-beam lithography. The PDMS-based masks work well not only for thin, but also for thick crystals: the flexible PDMS surface and the ductile Au contacts adjust easily to the crystal shape (i.e. no flexibility of the crystal is required). Remarkably, despite the fact that rubbers become rigid upon cooling, this technique has been shown to work well at low temperatures [48]. Another important advantage of the PDMS stamp technique is that it is non-destructive and *reversible:* it has been shown that the contact between the stamp and the rubrene crystal can be re-established many times without noticeable degradation of the OMC surface [46]. For this reason, PDMS-supported circuitry has been used for the first observation of the anisotropy of the field-effect mobility within the *a-b* plane of the single crystals of rubrene, as we discuss in Sec. 14.3.6. Interestingly, simple adhesion of organic crystals to metallic surface results in contacts with good electrical properties. This has been demonstrated for both Si- and PDMS-based stamps by comparing the results of two- and four-probe measurements. The contact resistance is similar to that observed in the devices for which the thin metal film contacts are directly deposited onto the organic surface (see below).

Although its simplicity makes electrostatic bonding particularly appealing for the fabrication of organic single crystal FETs, this technique also suffers from a number of limitations. For instance, the choice of metals for the source and drain contacts in the electrostatic "bonding" technique is limited by noble metals, since other materials are easily oxidized in air. The channel width is not well defined (unless it is limited by the crystal dimensions or by patterning the gate electrode), because the conduction channel is formed over the whole area of overlap between the OMC crystal and the Si wafer. In the current PDMS-stamp OFETs, the gate insulator is relatively thick and its dielectric constant is low – for this reason, the maximum density of induced charges is relatively small (typically, below 1×10^{11} carriers/cm^2). Finally, another potential problem of the Si- and PDMS stamps might be the mismatch between the coefficients of thermal expansion for the stamp and the organic crystal. Upon changing the temperature, this mismatch might cause a mechanical stress and formation of defects in the crystal. Since the surface defects can trap the field-induced charge, this might result in deterioration of the low-temperature OFET characteristics. This is an important issue that requires further studies.

OFETs with high-k dielectrics   The electrostatic bonding technique is compatible with the use of high-k dielectrics as gate insulators, which allow the accumulation of a large carrier density in OFETs. Particularly interesting is the possibility of reaching a surface charge density of the order of 1×10^{14} carriers/cm^2, which corresponds to approximately one charge carrier per molecule (this estimate assumes that all the charges are accumulated in a single molecular layer, as it is expected from calculations of the screening length). Indeed, the maximum surface charge density is $Q = \varepsilon\varepsilon_0 E_{bd}$, where ε is the relative dielectric constant of the gate dielectric and E_{bd} is its breakdown field. For typical high-k dielectrics, such as Ta_2O_5 or ZrO_2, $\varepsilon = 25$ and $E_{bd} > 6$ MV/cm, and the resulting charge density at the breakdown is 10^{14} carriers/cm^2. Many novel high-k materials hold the promise of even higher charge densities [49, 50].

The process of fabrication of OFETs with high-k dielectrics is similar to the aforementioned Si-based technique, with SiO_2 replaced by a high-k dielectric that is sputtered onto a heavily doped silicon substrate. In the experiments at TU Delft, Ta_2O_5 and ZrO_2 gate dielectrics have been sputtered onto the Si wafers at room temperature. Though the measured values of the dielectric constant and breakdown field for these layers are close to the best results reported in literature [51], the leakage currents were relatively large ($\sim 10^{-6}$ A/cm^2 for a Ta_2O_5-thickness of 350 nm). This is typical for deposition of dielectric films onto non-heated substrates, which results in a relatively high density of vacancies in the films. Substantially lower values of the leakage currents might be achieved in the future by sputtering on heated substrates, as already demonstrated in literature [51]. The characteristics of tetracene single-crystal FETs with high-k dielectrics will be discussed in Sec. 14.3.

14.2.3.2
"Direct" FET fabrication on the crystal surface

The "direct" fabrication of the single-crystal OFETs, in which a free-standing OMC is used as the substrate for subsequent deposition of the contacts and gate dielectric, is not trivial, because the organic crystals are incompatible with the standard processes of thin-film deposition/patterning. Fabrication of the field-effect structures based on single crystals of organic semiconductors became possible after several innovations have been introduced both for the source/drain fabrication and for the gate dielectric deposition [22, 23].

Source/drain contacts   The performance of the organic FETs is often limited by the injection barriers formed at the interface between the metal contacts and the semiconductor. The charge injection in such devices occurs by thermally assisted tunneling of the electrons or holes through the barrier, whose effective thickness depends on both gate and source-drain voltages. This is why reducing of the contact resistance is especially important for proper functioning of OFETs.

Different routes have been followed for the fabrication of source/drain contacts on the surface of organic crystals. The simplest (but also the crudest) is the "manual" application of a conducting paste. Among the tested materials, the water based solution of colloidal carbon provided the lowest contact resistance to organic crystals. A two-component, solvent-free silver epoxy (Epo-Tek E415G), which hardens at room temperature in a few hours, has been also used [36]. A disadvantage of this method is that it is difficult to prepare small and nicely-shaped contacts on hydrophobic OMC surfaces. In addition, this technique often results in formation of defects (traps) at the contact/organic interface, as shown, for instance, by the transport experiments in the space charge limited transport regime.

The thermal deposition of metals through a shadow mask is a more versatile method. However, the thermal load on the crystal surface in the deposition process (mostly because of radiation from the evaporation boat) has to be painstakingly minimized: deposition might generate traps at the metal/organic interface, or even result in the OMC sublimation. The effect of fabrication-induced traps has been regularly observed in both FET and SCLC measurements; presence of these traps is also a plausible cause for irreproducibility of the metal contacts evaporated on top of organic films used in applied devices [52–54]. This limits the choice of metals to the materials with a relatively low deposition temperature. As a first (and very crude) approximation, matching of the metal work-function to the HOMO(LUMO) levels of OMC for the p-type (n-type) conductivity can be used as the guideline in the metal selection.

Despite the technological difficulties, successful deposition of high-quality silver contacts by thermal evaporation has been recently performed, which demonstrates that the contact fabrication problems are not intrinsic. In order to minimize damage of the crystal surface, the authors of Ref. [23] fixed the deposition rate at a low level (~ 1 Å/s), increased the distance between the evaporation source and the sample up to 70 cm, maintained the crystal temperature during the deposition within the range $-20 - 0\,^\circ$C by using a Peltier cooler, and placed a diaphragm near the evaporation boat to shield IR radiation from the hottest parts of the boat. It has been also observed that, in order to achieve high mobilities, it is important to avoid contamination of the channel surface by metal atoms deposited at oblique angles under the shadow mask. Such contamination, which dramatically affects the device performance, presumably occurs because of scattering of silver atoms from residual gas molecules even at 5×10^{-7} Torr, the typical pressure in the chamber for contact deposition. In order to prevent oblique angle deposition in the shadowed regions, silver was deposited through a "collimator", a narrow (4 mm ID) and long (30 mm) tube, positioned close to the crystal surface. Following this process, high-quality OFETs have been fabricated on the surface of several organic crystals (rubrene, TCNQ, pentacene).

In the future, it would be useful to better understand the mechanism of damaging of organic crystals in the process of contact fabrication, in order to make the preparation of high-quality contacts routinely possible with many different metals. In particular, preparation of high-quality contacts will help to elucidate the role of the work function of the metallic electrodes, which seems to play a less prominent role than what was initially expected (see, e.g., [55]).

Parylene as a novel gate dielectric   After many unsuccessful attempts, it became clear that sputtering of Al_2O_3, as well as other dielectrics, onto the surface of organic molecular crystals unavoidably results in a very high density of traps and prohibitively high field-effect threshold: the field effect is completely suppressed even if the organic crystals were positioned in the shadow region of the vacuum chamber, where the deposition rate was zero. Presumably, the OMC surface is damaged by high-energy particles in the plasma. The attempts to shield the surface from high-energy charged particles by electrostatic deflection did not improve the situation. Thermal deposition of silicon monoxide was also unsuccessful, probably because of a too high temperature of the deposition source.

The breakthrough in the "direct" fabrication of free-standing single-crystal OFETs came with using thin polymer films of parylene as a gate-dielectric material [22]. Parylene coatings are widely used in the packaging applications; the equipment for parylene deposition is inexpensive and easy to build (see, e.g., *Parylene Conformal Coatings Specifications and Properties*, Technical notes, Specialties Coating Systems). This material with the dielectric constant $\varepsilon = 2.65$ forms transparent pinhole-free conformal coatings with excellent mechanical and dielectric properties: the breakdown electric field could be as high as ~ 10 MV/cm for a thickness of 0.1 μm.

In Ref. [22], parylene was deposited in a home-made reactor with three temperature zones (see Fig. 13). Prior to the deposition, the reactor (quartz tube) was evacuated to a pressure of ~ 1 mTorr. The dimer *para-xylylene* (generic name, parylene) is vaporized in the vaporization zone at $\sim 100\,°$C, cleaves in the pyrolysis zone at $\sim 700\,°$C, and polymerizes in the deposition zone (the sample location) at room temperature and pressure ~ 0.1 Torr. The precise value of these parameters during the parylene deposition is not critical. In Ref. [22], the parylene deposition rate was ~ 300 Å/min for the samples positioned ~ 35 cm away from the pyrolysis zone of the parylene reactor. Parylene was deposited onto the OMC crystals with pre-fabricated source and drain contacts with the attached wires (otherwise, contacting the contact pads might be difficult). The parylene thickness ~ 0.2 μm is sufficient to cover uniformly even the rough colloidal-graphite contacts. The capacitance of the gate electrode per unit area, C_i, was $C_i = 2 \pm 0.2$ nF/cm^2 for a $\sim 1-\mu$m-thick pary-

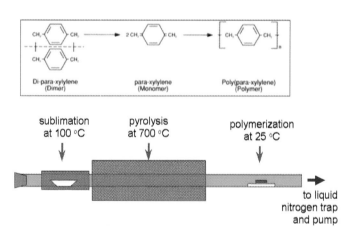

Figure 13   Parylene deposition. The top panel shows the reactions that occur during the deposition: the dimer of parylene sublimes at $\sim 100\,°$C; it splits up to monomers as it enters the pyrolysis zone at $\sim 700\,°$C; the monomers polymerize in the room temperature zone where the sample is placed. The bottom panel is a simple schematics of a homemade deposition chamber.

lene film. The output of working devices with the parylene gate insulator approached 100% and the parylene films deposited onto organic crystals withstand multiple thermal cycling between 300 K and 4.2 K.

There are several important advantages of using parylene as the gate dielectric: (a) it can be deposited while the crystal remain at room temperature, (b) being chemically inert, it does not react with OMCs, and (c) the parylene/OMC interface has a low density of surface states. Apart from that, parylene is a carbon-based polymer, and its thermal expansion coefficient is likely to be close to that of most organic crystals (but that remains to be tested). As it has already been emphasized above, different thermal expansion/contraction of the crystal and gate dielectric might result in the stress-induced carrier trapping. In this regard, the use of parylene is particularly promising for the operation of OFETs at low temperature. Parylene is also promising as the gate insulator for the future thin-film, flexible devices, where flexibility of the gate dielectric is required.

14.3
Characteristics of single-crystals OFETs

Fabrication of the single-crystal OFETs enables exploration of *the physical limits on the performance of organic thin-film FETs*. For the first time, one can study the characteristics of OFETs not limited by the disorder common for organic thin films. As the result, many important characteristics of OFETs, including the charge carrier mobility, the field-effect threshold, and the sub-threshold slope, have been significantly improved.

The organic semiconductors used in OFETs are undoped (or, at least, not intentionally doped). For this reason, OFETs belong to the class of injection, or Schottky-limited FETs, in which the charge carriers are injected into the conduction channel through the Schottky barriers at the metal/organic interface. For the same reason, the resistance of source and drain contacts is much higher than in Si MOSFETs, and depends strongly on the biasing regime. Since the contact resistance might be comparable or even greater than the channel resistance (especially at low temperatures), only the 4-probe measurements provide the intrinsic characteristics of the conduction channel, not affected by the contact resistance. However, in the limit of large V_G and V_{SD}, the contact resistance becomes small, and the results of 2-probe and 4-probe measurements typically converge, at least at room temperature (see Fig. 14).

It is worth mentioning that at this initial stage, when the research focuses mostly on the study of the intrinsic field-induced conductivity in organic semiconductors, the biasing regimes in the experiments with single-crystal OFETs often differ from the conventional FET biasing [13]. The difference is illustrated in Fig. 15: the polarity of the source-drain voltage is chosen to explore a wider range of the carrier densities. Note also that, because single-crystal devices have not been optimized for applications (e.g., the gate insulator is much thicker than in the commercial devices), the typical values of V_G and V_{SD} are an order of magnitude greater than that for the conventional Si MOSFETs. The main characteristics of the single-crystal FETs are summarized below.

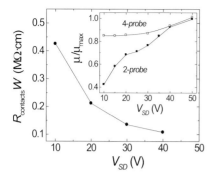

Figure 14 The contact resistance normalized by the channel width W measured using a 4-probe rubrene OFET as a function of the source-drain voltage ($V_G = -40$ V). The inset: the mobility of the same device measured in the 4-probe and 2-probe configurations, normalized by μ at $V_{SD} = 50$ V.

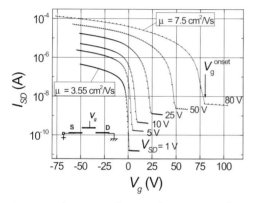

Figure 15 The trans-conductance characteristics of an OFET fabricated on the rubrene single crystal, measured at different values of the source-drain voltage V_{SD}. The in-plane dimensions of the conducting channel are $L \times W = 1 \times 1$ mm².

14.3.1
Unipolar operation

All the single-crystal devices fabricated up to date exhibited unipolar operation. Specifically, the p-type conductivity has been observed, for instance, in antracene, tetracene, pentacene, perylene, rubrene, whereas the n-type conductivity has been observed in TCNQ. Typical transistor characteristics for the rubrene and tetracene single crystal OFETs are shown in Fig. 16 and 17. In principle, the unipolar operation can be explained by the choice of metallic contacts that are efficient injectors of only one type of carriers. However, the presence of traps that selectively capture either electrons or holes cannot be excluded. For instance, the TOF experiments with pery-

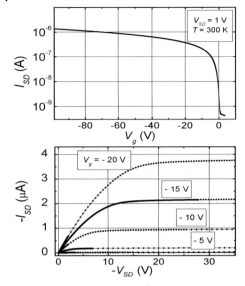

Figure 16 Two-probe characteristics of a single-crystal rubrene FET [22]. Upper panel: the dependence of the source-drain current, I_{SD}, on the gate voltage, V_G. Lower panel: I_{SD} versus the bias voltage V_{SD} at several fixed values of V_G. The source-drain distance is 0.5 mm, the width of the conduction channel is 1 mm, the parylene thickness is 0.2 μm.

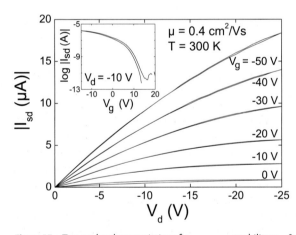

Figure 17 Two-probe characteristics of a single-crystal tetracene FET [24]. Source-drain current I_{SD} versus source-drain voltage V_{SD} measured at different values of V_G. The inset shows the dependence of log (I_{SD}) on V_G at fixed V_{SD}, for a different device, which has a mobility $\mu = 0.05$ cm^2/Vs and a threshold voltage $V_{th} \approx 0.3$ V. From this plot we calculate the subthreshold slope to be 1.6 V/decade. For both devices the source-drain distance is 25 μm, the width of the conduction channel is 225 μm, and the SiO$_2$ thickness is 0.2 μm.

lene [6, 42] have shown that in high quality crystals of this compound, both electrons and holes are sufficiently mobile at room temperature. However, in the single-crystal perylene FETs, only the hole conduction has been observed [56]. One of the reasons for that might be presence of oxygen in the crystal, which is known to act as a trap for electrons.

14.3.2
Field-effect threshold

The threshold voltage V_{th} is a measure of the amount of charge that it is necessary to induce electrostatically in order to switch-on electrical conduction in a FET. Using the equation that describes FET operation

$$I_{SD} = \frac{W}{L}\mu C(V_G - V_{th})V_{SD} \tag{3}$$

V_{th} can be obtained by extrapolating the quasi-linear (high-V_G) part of trans-conductance characteristics $I_{SD}(V_G)$ to zero current (here W and L are the width and the length of the conducting channel, respectively, and C is the specific capacitance between the channel and the gate electrode). The charge induced in the sub-threshold regime fills the traps that immobilize the charge carriers.

The magnitude of the field-effect threshold voltage depends on several factors, such as the density of charge traps on the interface between the organic crystal and the gate dielectric, the quality of the source/drain contacts (particularly important for Schottky transistors), and the absence/presence of a "built-in" conduction channel. Firstly, let's consider the situation when the built-in channel is absent; this is the case, for example, of the rubrene devices with parylene gate dielectric [22, 23]. The corresponding trans-conductance characteristics are shown in Fig. 15 on a semi-log scale. The field-effect onset is observed at a *positive* gate voltage, similar to the OFET based on well-ordered pentacene thin-films [21, 57, 58]. This behavior resembles the operation of a "normally-ON" p-type FET with a built-in channel. The resemblance, however, is superficial: in Ref. [23], the sharp onset was always observed at $V_G = V_{SD}$, which indicates that the channel was induced electrostatically. Indeed, an application of a positive voltage V_{SD} to the source electrode in the presence of the gate electrode ~1 μm away from the interface creates a strong electric field normal to the crystal surface. This field induces propagation of the conducting channel from the source electrode to the drain at any $V_G < V_{SD}$. Thus, the single-crystal rubrene OFETs with parylene gate dielectric are *zero threshold* devices at room temperature. The zero threshold operation suggests that the density of the charge traps at the rubrene/parylene interface is low ($< 10^9$ cm^{-2}) at room temperature [59]. However, the situation changes at low temperatures: the threshold voltage, measured in the 4-probe configuration, increases with cooling (see Fig. 23). This might signal depopulation of the surface traps that are filled at room temperature. Note that only the 4-probe measurements are essential to study the behavior of V_{th}; in the 2-probe measurements, an increase of the non-linear contact resistance with cooling might imitate the threshold shift.

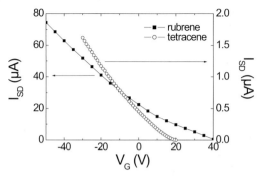

Figure 18 Gate-sweeps demonstrating a built-in channel for electrostatically bonded crystals, both for a tetracene (open circles) and for a rubrene (filled squares) single-crystal FET. The W/L ratio is the same for the two devices, $W/L = 0.14$.

For the OFETs fabricated by the electrostatic bonding of organic crystals, a relatively large (10 V or more) depletion gate voltage is often required to completely pinch off the channel (this V_G is positive for the p-type conductivity). This behavior is illustrated in Fig. 18 for rubrene and for tetracene crystals bonded to the RIE pre-cleaned SiO_2 surface [24], similar behavior has been observed for pentacene [25]. The positive threshold has been also observed for the rubrene crystals bonded electrostatically to the surface of PDMS rubber stamps [48]. These observations suggest that the same microscopic mechanism responsible for electrostatic bonding might be responsible for inducing the built-in channel on the organic surface.

14.3.3
Sub-threshold slope

The sharpness of the field-effect onset is characterized by the sub-threshold slope, $S \equiv dV_G/d(\log I_{SD})$. Since this quantity depends on the capacitance of the insulating layer C_i, it is also convenient to introduce the normalized slope, $S_i \equiv S \cdot C_i$, which permits to compare more directly the properties of different devices [23]. For single crystal FETs, even in the devices with relatively low mobility (the tetracene single-crystal FETs with $\mu = 0.05$ cm^2/Vs [24]), the observed normalized sub-threshold slope $S_i = 28$ V·nF/decade·cm^2 was comparable with that for the best pentacene TFTs ($S_i = 15 - 80$ V·nF/decade·cm^2 [60–62]). The high-mobility single-crystal rubrene OFETs with $\mu \approx 5 - 8$ cm^2/Vs exhibit a sub-threshold slope as small as $S = 0.85$ V/decade, which corresponds to $S_i = 1.7$ V·nF/decade·cm^2 [23]. This value is an order of magnitude better than what has been achieved in the best organic TFTs; it also compares favorably with α-Si:H FETs, for which $S_i \approx 10$ V·nF/decade·cm^2 has been reported [63].

It is commonly believed that the sub-threshold slope is mainly determined by the quality of insulator/semiconductor interface [7]. This is definitely the case for Si MOSFETs, where the resistance of source and drain contacts is low and does not

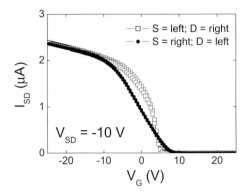

Figure 19 V_G-sweeps of a rubrene FET fabricated by electrostatic bonding. The two different curves are obtained by interchanging the source and the drain, while the source-drain voltage is the same (−10 V) in both cases. The influence of the contacts is visible in the sub-threshold region of the V_G-sweeps, which shows a clearly asymmetric behavior, in spite of the long channel length (1200 μm). At higher values of V_G, the electrical characteristics are independent of the source/drain configuration. The channel width of this device is approximately 200 μm.

depend on the gate voltage. In contrast, the contact resistance in the Schottky-type OFETs is high, it depends non-linearly on V_G – as the result, the sub-threshold slope might reflect the quality of contacts rather than the insulator/semiconductor interface. The effect of the contacts is illustrated in Fig. 19 for a rubrene device fabricated by electrostatic bonding. In this device, interchanging the source and drain contacts results in different sub-threshold V_G characteristics. At higher values of V_G, when the conducting channel is formed, the electrical characteristics are symmetric, i.e. they are not sensitive to the source/drain configuration. Note that in this device the contacts dominate the behavior in the sub-threshold region even though the length of the channel is considerable (1.2 mm; the channel width is approximately 0.2 mm).

14.3.4
Double-gated rubrene FETs

The conventional method for the fabrication of low-resistance contacts in Si MOSFETs is based on ion implantation of dopants beneath the contact area. Unfortunately, a similar technique has not been developed for OFETs yet. In this situation, the contact resistance can be reduced by using the so-called double gate, a trick that has been successfully applied for the study of high-mobility Si MOSFETs at low temperatures [64]. Schematic design of such device, fabricated at Rutgers, is shown in Fig. 20.

Two separately biased gate electrodes are deposited on the surface of a rubrene crystal with pre-formed source and drain contacts: the main gate and the complementary gate. These electrodes are isolated from the crystal and each other by a layer of parylene (~ 1 μm thick). The complimentary gate electrode (closest to the surface) consists of two stripes, connected together, that overlap with the source and drain

SIDE VIEW

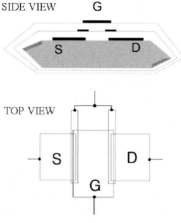

TOP VIEW

Figure 20 Schematic design of the double-gated FET. Additional to the conventional source, drain and gate-electrode, also a complimentary gate electrode is deposited, which is separated from the other electrodes by parylene layers. The complementary gate overlaps with the source and drain contacts, and it serves to control the contact resistance.

contacts. The resistance of Schottky barriers and the charge density in the channel of this device can be controlled separately by applying different voltages to the complimentary gate, V_{Gc}, and to the main gate electrode, V_G.

The trans-conductance characteristics of the double-gated rubrene FET, $I_{SD}(V_G)$, are shown in the Fig. 21 for several values of V_{Gc} and a fixed $V_{SD} = 5$ V. Large negative voltage V_{Gc} greatly reduces the contact resistance, and the regime of low carrier densities becomes easily accessible. However, the price for this is the non-linearity of trans-conductance characteristics at $V_G > V_{Gc}$: a portion of the semiconductor surface beneath the complimentary gate electrode is screened from the field of the main gate electrode, and its resistance does not depend on V_G.

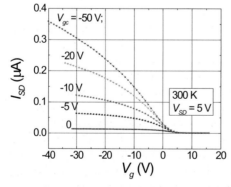

Figure 21 The trans-conductance characteristics of the double-gated rubrene FET. A set of $I_{SD}(V_G)$ curves is shown for a fixed $V_{SD} = 5$ V and several negative complimentary gate voltages V_{Gc}.

14.3.5
Mobility

The mobility of carriers at the surface of organic crystals can be estimated from the linear portion of the trans-conductance characteristics, where the conductivity of the channel, $\sigma = en\mu$, varies linearly with the density of mobile field-induced charges, n. The "intrinsic" mobility, not limited by the contact resistance, can be estimated from the 4-probe measurements as [7]

$$\mu = \left(\frac{L}{WC_iV}\right)\left(\frac{dI_{SD}}{dV_G}\right) \tag{4}$$

where V is the potential difference between the voltage probes at a distance L from each other, W is the channel width, and C_i is the specific capacitance between the gate electrode and the conduction channel. For the 2-probe measurements, L in Eq. 4 corresponds to the total length of the conduction channel, and V to the source-drain voltage V_{SD}. The latter measurements usually provide a lower estimate for μ, which approaches the intrinsic μ value with increasing V_G and V_{SD} (see Fig. 14). Eq. 4 is based on the assumption that all charge carriers induced by the transverse electric field above the threshold are mobile, and their density is given by:

$$n = \frac{C_i(V_G - V_{th})}{e} \tag{5}$$

This assumption has not been fully justified yet. For comparison, in a different type of FETs with comparable values of μ, amorphous silicon (a-Si:H) FETs, this is not the case: above the threshold, most of the induced charge in these devices goes into the "tail" (localized) states with only a small fraction going into the conduction band (see, e.g., [65]). The latter model of multiple thermal trapping and release of carriers involving shallow traps is not appropriate for OFETs, where charge transport cannot be described in terms of band transport owing to polaronic effects [1, 66]. Some justification of estimate (5) is provided by observations of V_G-independent mobility and the mobility increase with cooling, in a sharp contrast with the behavior of the a-Si:H FETs. Note that, contrary to the conventional inorganic FETs, the density of polaronic charge carriers in OFETs cannot be estimated from the Hall-type experiments, at least at high temperatures where hopping processes govern the charge transport (for discussion of the Hall effect in the polaronic hopping regime, see Ref. [67]).

The room temperature mobility of the field-induced carriers varies over a wide range for different organic single crystals. The following values of μ have been reported: tetracene – 0.4 cm^2/Vs [24], pentacene – 0.5 cm^2/Vs [25, 68], rubrene – 10 cm^2/Vs (up to 15 cm^2/Vs in recent unpublished measurements [48]), TCNQ – 1 cm^2/Vs [69]). For most of these materials, comparable values of μ have been obtained for FETs fabricated by both techniques of electrostatic bonding and direct fabrication on the crystal surface – another indication of the fact that, in many cases, the measurements with single-crystal OFETs probe the electronic properties of the crystals, at least at room temperature, and are not affected by artifacts due to the device fabrication.

The μ values for single-crystal devices are comparable or greater than the corresponding values of μ reported for the best thin-film devices (see, e.g., [24, 70]). To our knowledge the only exception is pentacene, for which the highest measured TFT mobility is 3 cm^2/Vs. Recently however, a room temperature mobility estimated by in-plane SCLC measurements as high as ~ 30 cm^2/Vs has been reported [71]. Work on the fabrication of FETs based on these pentacene single-crystals is in progress [72].

Two inter-related factors play an important role in the mobility improvement in single-crystal OFETs with respect to the organic TFTs. Firstly, the single crystal surfaces are free from the inter-grain boundaries that might limit significantly the mobility in the thin-film devices [73]. Secondly, the experiments with organic single crystals demonstrated for the first time that the mobility might be strongly anisotropic (see Sec. 14.3.6). Thus, in the experiments with single crystals, there is a possibility to choose the direction of the maximum mobility. In the case of elongated, needle-like crystals (rubrene, pentacene, etc.), the direction of maximum mobility coincides with the direction of the fastest crystal growth with the strongest inter-molecule interactions. At the same time, the mobility of carriers in OTFTs with the grains oriented randomly with respect to the current is an "angle-averaged" quantity: the grains oriented along the axis of the minimum mobility will have a much higher resistance.

The mobility in the single-crystal OFETs depends much less on the carrier density and the source-drain voltage than that in the organic TFTs. In the latter case, a pronounced increase of μ with V_G is observed due to the presence of structural defects [17]; because of the strong $\mu(V_G)$ dependence in the TFTs, a large $V_G \geq 100$ V (for a typical 0.2 μm thick SiO$_2$ gate insulator) is often required to realize higher mobilities, comparable to that of α-Si:H FETs (~ 0.5 cm^2/Vs). The typical dependence of the "2-probe" μ on the gate voltage for rubrene OFETs is shown in Fig. 22. The maximum of $\mu(V_G)$ near the zero V_G and an apparent increase of the mobility with V_{SD} are the artifacts of the 2-probe measurements; these artifacts are caused by a strong V_G-dependence of the resistances of the source and drain contacts [74]. At sufficiently large negative gate voltage ($V_G \leq -20$ V), μ becomes almost V_G-independent (the variations $\Delta\mu/\mu$ do not exceed 15%). The dependence of μ on V_{SD}, measured for the rubrene OFETs in the 2- and 4-probe configurations are compared in the inset to Fig. 14. The 4-probe data reflect the "intrinsic" charge carrier mobility, which is only weakly dependent on V_{SD}. The 2-probe data converge with the 4-probe data at high V_{SD} owing to a lower contact resistance. Similar independence of the mobility of V_G has been observed for tetracene single crystal FETs [24].

A non-monotonous temperature dependence of μ has been observed on the devices with highest (for a given material) mobilities: with cooling from room temperature, the mobility initially increases and then drops sharply below ~ 100 K. The temperature dependencies of the mobility for the rubrene (4-probe measurements) and tetracene (2-probe measurements) FETs are shown in Fig. 23 and 24. Similar trend has been observed for the high-mobility pentacene OFETs [21, 25]. There is no correlation between the absolute value of $\mu(300$ K) and its temperature dependence: similar dependencies $\mu(T)$ have been observed for the tetracene OFETs with $\mu \simeq 0.1$ cm²/Vs and rubrene OFETs with $\mu > 10$ cm²/Vs. The low-temperature drop of μ can be fitted by an exponential dependence $\mu = \mu_0 \exp(-T/T_0)$ with the activation energy $T_0 \approx 50 - 150$ meV. Observation of this drop in the 4-probe measurements indicates that the exponential decrease of the mobility at low temperatures is not an artifact of the 2-probe geometry and rapidly increasing contact resistance.

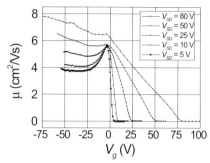

Figure 22 The dependencies $\mu(V_G)$ for the single-crystal rubrene OFETs, calculated from 2-probe measurements. The peak of μ near the zero V_G are the artifact of the 2-probe measurements; it is related to a rapid decrease of the total resistance of the source and drain contacts with increasing charge density. Note that the source is positive with respect to the grounded drain, so that for negative gate voltage the transistor is not in the saturation regime.

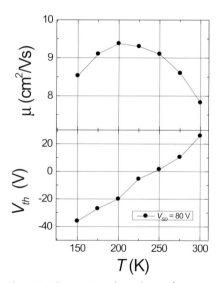

Figure 23 Temperature dependence of a rubrene FET with a room temperature mobility of 7.5 cm²/Vs. These measurements are performed in 4-probe configuration, at large positive source-drain voltage, $V_{SD} = 80$ V. Top panel: The mobility shows a non-monotonic behavior as a function of temperature, with an optimum value around ~ 200 K. Bottom panel: With lowering temperature the threshold voltage becomes smaller and eventually even changes sign.

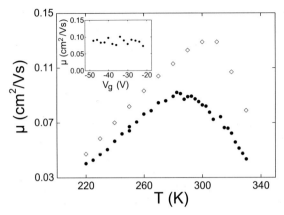

Figure 24 Temperature dependence of μ for two different tetracene FETs, measured at large negative gate voltage. The inset illustrates that at large negative gate voltage, −20 to −50 V, where the highest mobility is observed, μ is essentially independent of V_G.

There are some indirect indications that the room-temperature mobility in the devices with a low density of defects (high-purity crystals and high quality of organic surface) approaches its intrinsic value. The increase of μ with cooling, which is usually considered as a signature of the intrinsic transport, correlates with observation of the mobility anisotropy (see Sec. 14.3.6). The mobility drop, observed with further cooling, resembles the data obtained in the TOF experiments for not-so-pure crystals – this drop is likely caused by trapping of carriers by shallow traps, which can be active above the field-effect threshold due to thermal excitations in the system. An increase of the threshold voltage with cooling, clearly seen in Fig. 23, indicates that the trap concentration is relatively large even in the best crystals that have been used so far for the OFET fabrication. At present, the quantitative description of the polaronic transport, in general, and interaction of polarons with shallow traps in the FET experiments, in particular, is lacking. More systematic four-probe measurements of μ in different molecular crystals, along with the theoretical efforts on description of polaronic transport in systems with disorder, are needed to understand the origin of the observed temperature dependence of the mobility.

14.3.6
Mobility anisotropy on the surface of organic crystals.

Because of a low symmetry of organic crystals, one expects a strong anisotropy of their transport properties. Indeed, a strong anisotropy of the polaronic mobility with respect to the crystallographic orientation has been demonstrated by the TOF experiments [6, 42, 43]. Observation of the mobility anisotropy can be considered as a prerequisite for observation of the intrinsic (not limited by disorder) transport in organic semiconductors. The measurements on thin-film organic transistors never revealed such anisotropy. By eliminating grain boundaries and other types of defects, fabrication of the single-crystal OFETs allow for the first time to address the correlation between the molecular packing and the anisotropy of the charge transport on the surface of organic crystals.

To probe the mobility anisotropy on the *a-b* surface of the rubrene crystals, the researchers in Ref. [48] exploited an advantage of the PDMS stamp technique, which allows re-establishing the contact between the stamp and some organic crystals without damaging the surface. In this experiment, the crystal was rotated in a stepwise fashion, after each rotation the 2-probe stamp was re-applied, and the mobility was measured in both linear and saturation regime to exclude the effect of contacts. Figure 25b shows the data for two 360° rotations, to demonstrate reproducibility of the results. Black and red symbols correspond to the μ values extracted from the linear and saturation regimes of the FET *I–V* characteristics, respectively. The agreement between these values confirms that the contact effects can be neglected. Com-

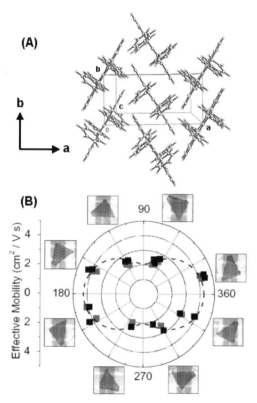

Figure 25 (a) Molecular packing in the *a-b* plane of the rubrene crystal. The direction of maximum mobility corresponds to the *b*-axis. (b) The angular dependence of the mobility for a rubrene crystal, measured by a 2-probe rubber-stamp device at room temperature. Black and red dots correspond to the μ values extracted from the linear and saturation regime of the FET *I–V* curves. The experiment was performed twice to ensure the reproducibility of this result.

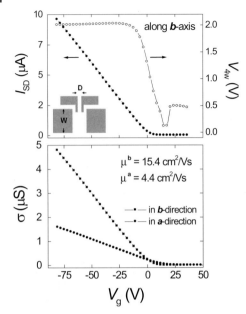

Figure 26 Top panel: Four-probe measurements of the charge transport along the *b*-axis of rubrene crystal: the source drain current I_{SD} (closed circles) and the voltage difference between the voltage contacts V_{4w} (open circles) are shown as a function of the gate voltage V_G. The inset illustrates the contact geometry. Lower panel: The channel conductivity σ as a function of V_G extracted from the 4-probe measurements along two crystallographic axis.

bination of the orientation-dependent field-effect transport measurements with the Laue x-ray analysis of the crystals shows that the direction of the highest mobility coincides with the *b*-axis (see Fig. 25a).

The channel conductivity, $\sigma(V_G) \equiv I_{SD}/V_{4w}$, extracted from the 4-probe measurements along the directions of maximum and minimum mobility (the *b* and *a* crystallographic axes, respectively) are shown in Fig. 26. The mobility $\mu = (D/WC_i)(d\sigma/dV_G)$ along the *b*-axis is almost 4 times greater than the mobility along the a axis.

14.3.7
Preliminary results for the OFETs with high-*k* dielectrics.

The characteristics of OFETs with high-*k* gate dielectrics (see Sec. 14.2.3.1) are in many respects similar to those for devices with the SiO_2 gate dielectric. The *I–V* curves measured for a tetracene FET electrostatically bonded onto a Ta_2O_5 insulating layer are shown in Fig. 27. Although the leakage gate current is normally higher than that for SiO_2, it is still significantly smaller than the source-drain current. The maximum charge density that has been reached so far is $\sim 5 \times 10^{13}$ cm^{-2}.

Despite apparent similarity, there is a significant difference between FETs fabricated on SiO_2 and on Ta_2O_5: the characteristics of high-*k* OFETs degrade substantially upon successive measurements. Specifically, the source-drain current mea-

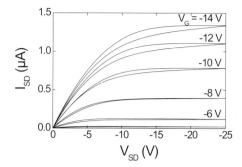

Figure 27 Two-probe characteristics of a single-crystal tetracene FET with Ta_2O_5 as a dielectric. The figure shows the source-drain current I_{SD} versus source-drain voltage V_{SD} measured at different values of V_G. The source-drain distance is 25 μm, the width of the conduction channel is 225 μm, and the Ta_2O_5 thickness is 0.35 μm. The mobility of the device is 0.03 cm^2/Vs.

sured at a fixed source-drain voltage is systematically decreased when subsequent scans of the gate voltage are performed. In general, the higher the applied gate voltage, the stronger the degradation. At the same time, the threshold voltage systematically increases to higher (negative) values.

For the case of Ta_2O_5, this degradation eventually results in a complete suppression of the field effect. This is shown in Fig. 28 where the source, the drain and the leakage current are plotted as a function of the gate voltage, for a fixed source/drain voltage ($V_{SD} = -25$ V). Instead of increasing linearly with V_G as in conventional FETs, the source/drain current exhibits a maximum (at $V_G \simeq -22$ V in this device) and then decreases to zero upon further increasing V_G. This degradation is completely irreversible: no measurable source/drain current was detected when V_G was decreased to 0 and swept again to high voltage.

The experiments show that current leaking from the gate is the cause of the FET degradation. This current, typically much smaller than the source-drain current (see Fig. 28), has a different effect: the electrons leaking through the gate insulator, accelerated in a strong electric field due to the voltage applied to the gate electrode, damage the organic crystals. The precise microscopic mechanism responsible for the damage is still unclear, but it is likely that the injection of high-energy electrons introduces surface defects (locally disrupting the crystal due to the accumulation of negative charge) and result in formation of traps (e.g., "broken" individual molecules). This would account for the observed decrease in mobility and shift in threshold voltage.

Optimization of the growth of high-k dielectrics in a near future is possible, since it is known how to reduce the leakage current by several orders of magnitude (by sputtering the high-k materials onto a heated substrate [51]). This is known to bring it to the level observed in SiO_2 ($\leq 10^{-10}$ A/cm^2 for a SiO_2-thickness of 200 nm), at which the irreversible damage introduced in the crystal is minimal. This improvement will allow stable FET operation at a charge density $\sim 1 \times 10^{14}$ cm^{-2}.

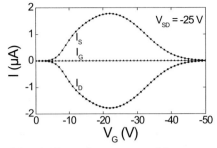

Figure 28 Two-probe gate-sweep of the same tetracene device as in figure 27. The source current I_S and the drain current I_D have the same magnitude and opposite sign, as expected. The leakage current to the gate, I_G, is typically much smaller than the source-drain current. Ramping the gate voltage up to high value results in a non-monotonous depen- dence of the source-drain current. This is due to degradation of the device caused by the small, but finite, current that is leaking through the gate insulator. The degradation is irreversi- ble: when the same scan is repeated again, there is no field-effect and the current through the device is zero.

14.4
Conclusion

The development of single-crystal OFETs offers a great opportunity to understand better the charge transport in organic semiconductors and to establish the physical limits on the performance of organic field-effect transistors. Below we briefly sum- marize the recent progress in this field, and attempt to formulate the research direc- tions that, from our viewpoint, need to be addressed in the near future.

One of the central issues in the field-effect experiments with organic semiconduc- tors is realization of the *intrinsic* (not limited by disorder) charge transport on the organic surface. In disorder-free organic semiconductors, several regimes of polar- onic transport are expected (see, e.g., [1, 75]), between the limiting cases of the coherent motion in extended states (band-like polaronic transport) at low tempera- tures and the incoherent hopping at high temperatures. All these regimes are intrin- sic in the aforementioned sense and all of them require better understanding: the incoherent polaronic hopping is pertinent, for instance, to the devices functioning at room temperature, whereas the realization of band-like polaronic transport is the Holy Grail of basic research.

We have witnessed significant progress towards realization of this goal. In this respect, an important characteristic of the first generation of single-crystal OFETs, whose significance should not be underestimated, is the device reproducibility and consistency of experimental observations – e.g., similar mobilities have been ob- served for the single-crystal OFETs based on the same organic compound but fabri- cated by different teams using different fabrication techniques. Although it might be taken for granted that this should be the case, such reproducibility has never been achieved in thin-film organic field effect transistors: the organic TFTs are notoriously known for a large spread of parameters, even if they were prepared under nominally identical conditions [17, 19]. This irreproducibility is still a problem despite the fact that OTFTs have been the subject of intense research for a much more extended period of time than single crystal OFETs.

The reproducibility of the OFET parameters is a necessary condition for investigation of the intrinsic electronic properties of organic semiconductors; the reproducibility, however, does not guarantee that the electrical behavior of the first-generation single crystal OFETs reflects the *intrinsic* electronic properties of organic molecular materials. Indeed, all OFETs investigated by different research groups have been fabricated on the basis of crystals grown under similar conditions, and, presumably, with a similar concentration of defects. Thus, the question remains: how close are we to the realization of the intrinsic transport in organic single crystal FETs?

For different temperature ranges, the answer to this central question will be different. There are many indications that at room temperature, the best single-crystal OFETs demonstrate the intrinsic behavior, and the room-temperature mobility of these devices approaches its intrinsic value. For example, observation of the anisotropy of the mobility of field-induced charges in the *a-b* plane [48] reveals for the first time (in the FET experiments) correlation between the charge transport and the underlying lattice structure. Another important observation is a moderate increase of the mobility with cooling (over a limited T range), similar to the case of "bulk" transport in the TOF experiments. Does this mean that these devices disorder-free? Of course, it does not. Since this characteristic time of the charge release by shallow traps decreases exponentially with the temperature, the shallow traps might be "invisible" at room temperature. However, with cooling, they start dominating the charge transport in single-crystal OFETs, as all the data obtained so far suggest: the mobility drops exponentially, typically below 100 K. In this respect, the low-temperature $\mu(T)$ dependence observed in the best single-crystal OFETs resemble the results of TOF experiments with not-so-pure "bulk" crystals. The observed exponential drop of $\mu(T)$ with cooling suggests that the density of shallow traps near the surface of organic crystals in single-crystal OFETs is still relatively large.

Though the low-temperature transport in single-crystal OFETs is, most likely, disorder-dominated, it is worth considering alternative explanations of a qualitative difference between the results of OFET and TOF experiments. It is usually tacitly assumed that the "surface" transport in OMCs is similar to the "bulk" transport, which has been comprehensively studied in the 70s and 80s (see, e.g., [2]). However, the transport in organic FETs differs from the "bulk" charge transport in the TOF experiments at least in two important aspects. Firstly, one cannot exclude *a priori* the existence of some surface modes that might affect the charge transport at the organic/insulator interface. Secondly, even at a low gate voltage, the density of charge carriers in a two-dimensional conducting channel of OFETs exceeds by far the density of charges in the TOF experiments. The interactions between polaronic excitations and their effect on the charge transport in organic semiconductors is still an open issue [76]. Thus, the experiments with single-crystal OFETs pose new problems, which have not been addressed by the theory yet.

Quantitative description of polaronic formation is a challenging many-body problem. Band-type calculations for OMC provide guidance in the search for new high-mobility materials, but do not have predictive power since they ignore polaronic effects [1]. For better characterization of the polaronic states at the organic surface, it is crucial to go beyond the *dc* experiments and to measure directly the characteristic energies involved in polaron formation. This information can be provided by the infra-red spectroscopy of the conduction channel in the single-crystal OFETs, a very promising direction of the future research.

Though the current efforts are mostly focused on realization of the intrinsic polaronic transport, equally important is to study interaction of polarons with disorder. From the history of physics of inorganic semiconductors, we know how important is this research for better understanding of charge transport and device characteristics. In organic conductors, small polarons interact strongly with disorder, and this interaction is qualitatively different from the interaction of electron-like excitations with disorder in inorganic semiconductors. This research, which encompasses such important issues as the spectroscopy of the states in the band gap, mechanisms of polaron trapping by shallow and deep traps, polaron scattering by defects in a strongly-anisotropic medium will be critically important for both basic science and applications. One of the aspects of this general problem is the effect of stress on polaronic motion. Indeed, in the field-effect experiments, it is difficult to avoid the build-up of mechanical stress, caused by the difference in the thermal expansion coefficients for the materials that form an OFET. The stress in organic crystals might affect strongly the overlap between electronic orbitals, the density of defects, and, as the result, the temperature dependence of the mobility in the FET experiments. More experiments with different FET structures are needed to clarify this issue.

Realization of the intrinsic transport, especially at low temperatures, requires better purification of starting material and improvement of the crystal growth techniques. There is ample room for improvement in this direction: the first experiments on fabrication of single-crystal OFET did not use a powerful zone-refinement technique, which has been proven to be very useful in the TOF experiments [4, 5]. In fact, many of the molecules used in FET experiments, including rubrene, can be zone refined.

To understand better the intrinsic transport in organic semiconductors and the relation between macroscopic electronic properties and molecular packing, the experiments with OFETs based on a broader class of organic crystals are required. Note that the experiments with any type of organic crystals, regardless of the magnitude of the *intrinsic* mobility, are important – in fact, only by exploring materials with different crystal structures and, hence, a wide range of μ, this problem can be adequately addressed. However, realization of the coherent band-like transport and improvement of the room-temperature mobility in OFETs will be helped by the development of new organic materials with a stronger overlap between electronic orbitals of adjacent molecules. In this regard, the materials with a non-planar molecular structure similar to that of rubrene (conjugated chains with non-planar side groups) seem to be very promising (see, e.g., [77]). Recent experiments with rubrene, which enabled a 10-fold increase of the OFET mobility, illustrates how important is the effect of molecular packing on the charge transport in organic crystals. To realize the enormous potential of organic chemistry in "tailoring" new organic compounds for organic electronics, combined efforts of chemists and physicists are required.

The use of a broader variety of organic compounds for the OFET fabrication may also offer the possibility to explore new electronic phenomena. An interesting example is provided by the metal-phthalocyanines (MPc's), a large class of materials with isostructural molecules that differ by only one atom – the metal element. This metal atom determines the electronic properties of the molecules, including the spin S in the molecular ground state [78, 79]. For instance, among the MPc's containing $3d$ transition metal elements, MnPc has S= 3/2 [80], FePc has S= 1 [81], CoPc and CuPc have S= 1/2 [79]. In FETs based on these materials, therefore, it may be possible to induce and control electrostatically the magnetic properties of organic materials, just by tuning the density of mobile charge carrier that mediate interactions between the local spins. Successful realization of 'magnetic' OFETs would open the way for organic spintronics.

14.5
Acknowledgments

We are grateful to M. Jochemsen, N. Karl, V. Kiryukhin, T. M. Klapwijk, Ch. Kloc, J. Niemax, J. Pflaum, V. Pudalov, J. Rogers, and A. F. Stassen for useful discussions and help with experiments. The work at Rutgers University was supported in part by the NSF grant DMR-0077825 and the DOD MURI grant DAAD19-99-1-0215. Work at Delft University was supported by the "Stichting FOM" (Fundamenteel Onderzoek der Materie) and by NWO via the Vernieuwingsimpuls 2000 program.

References

[1] E. A. Silinsh, and V. Čápek, *Organic Molecular Crystals: Interaction, Localization, and Transport Phenomena* (AIP Press, Woodbury, 1994)

[2] M. Pope, and C. E. Swenberg, *Electronic Processes in Organic Crystals and Polymers*, 2nd ed. (Oxford University Press, New York, London, 1999)

[3] R. Farchioni, and G. Grosso, *Organic electronic Materials* (Springer-Verlag, Berlin, 2001)

[4] N. Karl, In: *Crystals, Growth, Properties and Applications*, Vol. 4, ed. by H. C. Freyhardt (Springer-Verlag, Berlin 1980) pp. 1–100

[5] W. Warta, and N. Karl, Phys. Rev. B **32**, 1172 (1985)

[6] N. Karl, K.-H. Kraft, J. Marktanner, M. Münch, F. Schatz, R. Stehle, and H.-M. Uhde, J. Vac. Sci. Technol. A **17**, 2318 (1999)

[7] S. M. Sze, *Physics of Semiconductor Devices* (Wiley, New York, 1981)

[8] A. T. Fiory, A. F. Hebard, R. H. Eick, P. M. Mankiewich, R. E. Howard, and M. L. O'Malley, Phys. Rev. Lett. **65**, 3441 (1990); J. Mannhart, J. Bednorz, K. A. Muller, and D. G. Schlom, Z. Phys. B **83**, 307 (1991); J. Mannhart, Supercon. Sci. Technol. **9**, 49 (1996); N. Chandrasekhar, O. T. Valls, and A. M. Goldman, Phys. Rev. B **54**, 10218 (1996); P. Konsin, and B. Sorkin, Phys. Rev. B **58**, 5795 (1998); C. H. Ahn, S. Gariglio,

P. Paruch, T. Tybell, L. Antognazza, and J.-M. Triscone, Science **284**, 1152 (1999)

[9] S. Mathews, R. Ramesh, T. Vankatesan, and J. Benedetto, Science **276**, 238 (1997); H. Ohno, D. Chiba, F. Matsukura, T. Omiyra, E. Abe, T. Dietl, Y. Ohno, and K. Ohtani, Nature **408**, 944 (2000); H. Tanaka, J. Zhang, and T. Kawai, Phys. Rev. Lett. **88**, 027204 (2002)

[10] D. M. Newns, J. A. Misewich, C. C. Tsuei, A. Gupta, B. A. Scott, and A. Schrott, Appl. Phys. Lett. **73**, 780 (1998)

[11] H.-T. Kim, B. G. Chae, D. H. Youn, S. L. Maeng, and K. Y. Kang, cond-mat/0308042

[12] H. Ohno, D. Chiba, F. Matsukura, T. Omiyra, E. Abe, T. Dietl, Y. Ohno, and K. Ohtani, Nature **408**, 944 (2000)

[13] G. Horowitz, Adv. Mater. **10**, 365 (1998)

[14] H. E. Katz, and Z. Bao, J. Phys. Chem. B **104**, 671 (2000)

[15] Z. Bao, A. Dodabalapur, and A. J. Lovinger, Appl. Phys. Lett. **69**, 4108 (1996); H. Sirringhaus, P. J. Brown, R. H. Friend, M. M. Nielsen, K. Bechgaard, B. M. W. Langeveld-Voss, A. J. H. Spiering, R. A. J. Janssen, E. W. Meijer, P. Herwig, D. M. de Leeuw, Nature **401**, 685 (1999)

[16] G. H. Gelinck, T. C. T. Geuns, and D. M. de Leeuw, Appl. Phys. Lett. **77**, 1487 (2000)

[17] C. D. Dimitrakopoulos, and P. R. L. Malenfant, Adv. Mater. **14**, 99 (2002)

[18] G. Horowitz, Adv. Funct. Mater. **13**, 53 (2003)

[19] I. H. Campbell, and D. L. Smith, Solid State Phys. **55**, 1 (2001)

[20] M.C.J.M. Vissenberg, and M. Matters, Phys. Rev. B **57**, 12964 (1998)

[21] S. F. Nelson, Y. -Y. Lin, D. J. Gundlach, and T. N. Jackson, Appl. Phys. Lett. **72**, 1854 (1998)

[22] V. Podzorov, V. M. Pudalov, and M. E. Gershenson, Appl. Phys. Lett. **82**, 1739 (2003)

[23] V. Podzorov, S. E. Sysoev, E. Loginova, V. M. Pudalov, and M. E. Gershenson, Appl. Phys. Lett. **83**, 3504 (2003)

[24] R. W. I. de Boer, T. M. Klapwijk, and A. F. Morpurgo, Appl. Phys. Lett. **83**, 4345 (2003)

[25] J. Takeya, C. Goldmann, S. Haas, K. P. Pernstich, B. Ketterer, and B. Batlogg, J. Appl. Phys. **94**, 5800 (2003)

[26] V. V. Butko, X. Chi, D. V. Lang, and A. P. Ramirez, Appl. Phys. Lett. **83**, 4773 (2003)

[27] V. V. Butko, X. Chi, and A. P. Ramirez, Solid State Commun **128**, 431 (2003)

[28] G. Horowitz, F. Garnier, A. Yassar, R. Hajlaoui, and F. Kouki, Adv. Mater. **8**, 52 (1996)

[29] M. Ichikawa, H. Yanagi, Y. Shimizu, S. Hotta, N. Suganuma, T. Koyama, and Y. Taniguchi, Adv. Mater. **14**, 1272 (2002)

[30] A. F. Hebard, M. J. Rosseinsky, R. C. Haddon, D. W. Murphy, S. H. Glarum, T. T. M. Palstra, A. P. Ramirez, and A. R. Kortan, Nature **350**, 600 (1991); O. Gunnarsson, E. Koch, and R. M. Martin, Phys. Rev. B **54**, 11026 (1996)

[31] Ch. Kloc, P. G. Simpkins, T. Siegrist, and R. A. Laudise, J. Cryst. Growth **182**, 416 (1997)

[32] R. A. Laudise, Ch. Kloc, P. G. Simpkins, and T. Siegrist, J. Crystal Growth **187**, 449 (1998)

[33] M. Mas-Torrent, M. Durkut, P. Hadley, X. Ribas, and C. Rovira, J. Am. Chem. Soc. **126**, 984 (2004)

[34] R. Dabestani, M. Nelson, and M. E. Sigman **64**, 80 (1996)

[35] R. A. Laudise, *The Growth of Single Crystals* (Englewood Cliffs, N. J., Prentice-Hall, 1970)

[36] R. W. I. de Boer, M. Jochemsen, T. M. Klapwijk, A. F. Morpurgo, J. Niemax, A. K. Tripathi, and J. Pflaum, J. Appl. Phys. **95**, 1196 (2004)

[37] G. Horowitz, B. Bachet, A. Yassar, P. Lang, F. Demanze, J. -L. Fave, and F. Garnier, Chem. Mater. **7**, 1337 (1995); T. Siegrist, R. M. Fleming, R. C. Haddon, R. A. Laudise, A. J. Lovinger, H. E. Katz, P. M. Bridenbaugh, and D. D. Davis, J. Mater. Res. **10**, 2170 (1995)

[38] U. Sondermann, A. Kutoglu, and H. Bässler, J. Phys. Chem. **89**, 1735 (1985)

[39] J. Vrijmoeth, R. W. Stok, R. Veldman, W. A. Schoonveld, and T. M. Klapwijk, J. Appl. Phys. **83**, 3816 (1998)

[40] D. C. Hoesterey, and G. M. Letson, J. Phys. Chem. Solids **24**, 1609 (1963)

[41] J. Pflaum, private communication

[42] N. Karl, *Organic Semiconductors*, In: Land-olt-Boernstein (New Series), Group III, Vol. 17 *Semiconductors*, ed. by O. Madelung, M. Schulz, H. Weiss, (Springer-Verlag Berlin, Heidelberg, New York, Tokyo 1985) Subvolume 17i, pp. 106-218. Updated edition: Vol. 41E (2000)

[43] N. Karl, J. Martanner, Mol. Cryst. Liq. Cryst. **355**, 149 (2001)

[44] M. A. Lampert, and P. Mark, *Current Injection in Solids* (Academic Press Inc., New York, 1970)

[45] K. C. Kao, and W. Hwang, *Electrical Transport in Solids* (Pergamon Press, Oxford, 1981)

[46] J. Zaumseil, T. Someya, Z. N. Bao, Y. L. Loo, R. Cirelli, and J. A. Rogers, Appl. Phys. Lett. **82**, 793 (2003)

[47] O. Jurchescu, and T. T. M. Palstra, private communication

[48] V. C. Sundar, J. Zaumseil, V. Podzorov, E. Menard, R. L. Willett, T. Someya, M. E.Gershenson, and J. A. Rogers, Science **303**, 1644 (2004)

[49] R. B. van Dover, L. D. Schneemeyer, and R. M. Fleming, Nature **392**, 162 (1998)

[50] W. Ren, S. Trolier-McKinstry, C. A. Randall, and T. R. Shrout, J. Appl. Phys. **89**, 767 (2001)

[51] R. M. Fleming, D. V. Lang, C. D. W. Jones, M. L. Steigerwald, D. W. Murphy, G. B. Alers, Y. H. Wong, R. B. van Dover, J. R. Kwo, and A. M. Sergent, J. Appl. Phys. **88**, 850 (2000)

[52] C. Shen, and A. Kahn, J. Appl. Phys. **90**, 4549 (2001)

[53] J. Liu, T. Guo, Y. Shi, and Y. Yang, J. Appl. Phys. **89**, 3668 (2001)

[54] D. B. A. Rep, A. F. Morpurgo, and T. M. Klapwijk, Organic Electronics **4**, 201 (2003)

[55] W. Gao, and A. Kahn, Appl. Phys. Lett. **82**, 4815 (2003)

[56] R. W. I. de Boer, A. F. Stassen, T. M. Klapwijk, and A. F. Morpurgo, unpublished results

[57] H. Klauk, M. Halik, U. Zschieschang, G. Schmid, W. Radlik, and W. Weber, J. Appl. Phys. **92**, 5259 (2002)

[58] E. J. Meijer, D. M. de Leeuw, S. Setayesh, E. van Veenendaal, B. -H. Huisman, P. W. M. Blom, J. C. Hummelen, U. Scherf, and T. M. Klapwijk, Nature Materials **2**, 678 (2003)

[59] The total number of the "bulk" charge traps per the unit area of the conducting channel can be estimated as $N_t \times d$, where $N_t \simeq 10^{15}$ cm^{-3} is the bulk concentration of traps, and d is the effective channel thickness (≤ 10 nm). The corresponding threshold voltage, associated with the bulk traps in the studied devices, is therefore very small (< 0.1 V).

[60] C. D. Dimitrakopoulos, S. Purushothaman, J. Kymissis, A. Callegari, and J. M. Shaw, Science **283**, 822 (1999)

[61] Y. -Y. Lin, D. J. Gundlach, S. F. Nelson, and T. N. Jackson, IEEE Trans. Electron Devices **44**, 1325 (1997); H. Klauk, D. J. Gundlach, J. A. Nichols, and T. N. Jackson, IEEE Trans. Electron Devices **46**, 1258 (1999)

[62] C. D. Dimitrakopoulos, I. Kymissis, S. Purushothaman, D. A. Neumayer, P. R. Duncombe, and R. B. Laibowitz, Adv. Mater. **11**, 1372 (1999)

[63] J. Kanicki, F. R. Libsch, J. Griffith, and R. Polastre, J. Appl. Phys. **69**, 2339 (1991)

[64] R. Heemskerk, and T. M. Klapwijk, Phys. Rev. B **58**, R1754 (1998)

[65] M. Shur, M. Hack, and J. G. Shaw, J. Appl. Phys. **66**, 3371 (1989).

[66] M. W. Wu, and E. M. Conwell, Chem. Phys. Lett. **266**, 363 (1997).

[67] L. Friedman, Phil. Mag. **38**, 467 (1978)

[68] A. F. Stassen, R. W. I. de Boer, T. M. Klapwijk, and A. F. Morpurgo, work in progress

[69] V. Podzorov, and M. E. Gershenson, work in progress

[70] D.J. Gundlach, J.A. Nichols, L. Zhou, and T.N. Jackson, Appl. Phys. Lett. **80**, 2925 (2002)

[71] O. D. Jurchescu, J. Baas, and T. T. M. Palstra, Appl. Phys. Lett. **84**, 3061 (2004)

[72] O. D. Jurchescu, A. F. Stassen *et al.*, work in progress

[73] G. Horowitz, Synth. Met. **138**, 101 (2003)

[74] A. B. Chwang, and C. D. Frisbie, J. Phys. Chem. B **104**, 12202 (2000); P. V. Necliudov, M. S. Shur, D. J. Gundlach, and T. N. Jackson, Solid State Electron. **47**, 259 (2003); J. Zaumseil, K. W. Baldwin, and J. A. Rogers, J. Appl. Phys. **93**, 6117 (2003)

[75] S. Fratini, and S. Ciuchi, Phys. Rev. Lett. **91**, 256403 (2003)

[76] M. Capone, and S. Ciuchi, Phys. Rev. Lett. **91**, 186405 (2003)

[77] J. E. Anthony, D. L. Eaton, and S. R. Parkin, Org. Lett. **4**, 15 (2002)

[78] D. W. Clack, and J.R. Yandle, Inorg. Chem. **11**, 1738 (1972)

[79] M.S. Liao, and S. Scheiner, J. Chem Phys. **114**, 9780 (2001)

[80] K. Awaga and Y. Maruyama, Phys. Rev. B **44**, 2589 (1991)

[81] M. Evangelisti, J. Bartolome, L. J. de Jongh, and G. Filoti, Phys. Rev. B **66**, 144410 (2002)

15

Charge Carrier Photogeneration and Transport in Polymer-fullerene Bulk-heterojunction Solar Cells

I. Riedel, M. Pientka, and V. Dyakonov

15.1
Introduction

In recent years, the development of thin film plastic solar cells, using polymer-fullerene bulk heterojunctions as absorber, has made significant progress [1]. Efficiencies between 2.5% and 3.5% for laboratory cells under AM1.5 illumination conditions have been reported [2,3,4]. Most recently, polymer solar cells with efficiency above 5% were announced [5]. The typical structure of these devices consists of a composite of two materials with donor and acceptor properties, respectively, sandwiched between two electrodes with different work functions. The mostly investigated combination to date is the conjugated polymer poly [2-methoxy-5-(3′,7′-dimethyl-octyloxy-)1,4-phenylene-vinylene] (OC_1C_{10}-PPV) and the soluble fullerene derivative [6,6]-phenyl C_{61}-butyric acid methyl ester (PCBM). One advantage of this type of devices is the easy processability of absorber layer as well as of optional interface layer(s). They are solution processed by using spin casting or doctor-blade techniques. Significant breakthrough has been achieved by realizing that the morphology of the composites plays an important role for the device performance. In the case of OC_1C_{10}-PPV: PCBM solar cells, an improved morphology has been achieved by using chlorobenzene as a solvent [2]. However, a direct transfer of this recipe to other systems is not straightforward, therefore, a morphology optimization, in which the choice of solvent is only one parameter, should be performed for any combination of materials.

The charge carrier photogeneration yield in these composites is expected to be as high as 100% in contrast to a very low photogeneration yield in pure conjugated polymers. This is due to ultrafast photoinduced electron transfer between donor and acceptor [6]. The charge transfer occurs on a time scale of 40–50 fs [7], and the charge-separated state is metastable [8]. External quantum efficiency above 70% and an internal quantum efficiency of 96% in the main absorption peak have been demonstrated experimentally [9]. The high quantum yield of charge carrier generation is a precondition for an efficient energy conversion. Likewise, the electrical transport properties of the absorber material are important. We discuss the influence of electronic properties of two different material combinations on the performance of photovoltaic devices based on them.

Physics of Organic Semiconductors. Edited by W. Brütting
Copyright © 2005 WILEY-VCH Verlag GmbH & Co. KGaA, Weinheim
ISBN 3-527-40550-X

15.2
Experimental

Photoinduced absorption (PIA) measurements were performed on thin films spin cast onto sapphire substrates, which were mounted on the cold finger of a helium cryostat (20–293 K). During the measurements they were kept under dynamic vacuum to avoid photooxidation. The excitation source was a mechanically chopped Ar^+ laser at wavelength of 514 nm. A halogen lamp provided the probing white light illumination. Both light sources were focused on the same position of the sample. The transmitted light was collected by a large diameter concave mirror and focused on the entrance slit of a 300 mm monochromator. The detection was provided by a Si-diode (550 nm–1100 nm), or by a liquid nitrogen cooled InSb-detector (1100 nm–5550 nm). Therefore, a broad energy range 0.23–2.25 eV (with the sapphire cryostat windows and the sample substrate) was accessible. The signals were recorded with a standard phase sensitive lock-in technique synchronized with the chopping frequency of the laser. Photoinduced changes of the transmission, $\Delta T/T$, were monitored as function of probe energy. LESR (Light Induced Electron Spin Resonance) measurements were carried out with a commercial X-band Bruker E500 spectrometer equipped with an optically accessible microwave cavity and an Oxford helium flow cryostat. The principle of a LESR experiment is based on microwave induced transitions between Zeeman sublevels of an electronic state. The experimental details are described in [10]. The solutions were poured into quartz tubes, dried and sealed under vacuum. In PLDMR (Photoluminescence Detected Magnetic Resonance), the photoluminescence of the sample is monitored while sweeping the magnetic field. In all cases, an enhancement of the PL under resonant conditions is observed. This technique is a practical application for identifying spin carrying species like triplet excitons and polaron pairs [11–14].

Figure 1(a) schematically illustrates the configuration of a bulk-heterojunction solar cell. Blends of OC_1C_{10}-PPV, as an electron donor, and PCBM, as the electron accepting moiety, were blended (1:4, weight) in chlorobenzene. A 100 nm active layer of OC_1C_{10}-PPV: PCBM was spin cast from solution under inert atmosphere in a glove box. As a back electrode we used aluminum, which was thermally evaporated in a high vacuum at pressures lower than 10^{-6} mbar. As a hole extracting electrode, an 80 nm layer of poly-[ethylene dioxy-thiophene]:(poly-(styrene sulphonic acid) (PEDOT: PSS) (BAYTRON P, Bayer AG, Germany) was spin coated onto the substrate. Another set of devices consisted of regioregular poly(3-hexylthiophene-2,5-diyl) (P3HT) (Rieke Met. Inc.) and PCBM in the absorber blend. The active layer was made by casting a 1:2 solution of P3HT and PCBM onto dried PEDOT: PSS. The P3HT-based devices were annealed in a manner described in [4], however, without additional electric field applied during thermal treatment. Monochromatic incident photon conversion efficiency (IPCE) spectra were measured using a home made lock-in-based setup equipped with a calibrated Si/Ge photodiode, as reference detector. The light from the halogen and Xenon lamps was spectrally dispersed using a monochromator with a set of gratings to cover the range of wavelength from 250 to 1000 nm. The current-voltage profiling was automatically carried out with a

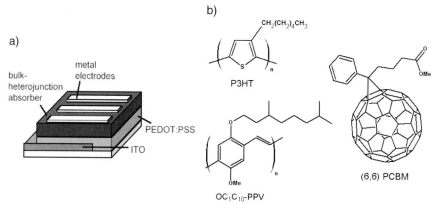

Figure 1 (a) Scheme of the polymer-fullerene bulk heterojunction solar cell. (b) Chemical formulae of donor and acceptor materials used as light absorber.

source-monitor unit (Avantest TR-6143) in a variable temperature cryostat equipped with a liquid nitrogen supply and a temperature controller (Lakeshore 330). The devices were illuminated through a sapphire window by white light from a Xenon arc lamp whose intensity is controlled by a set of neutral density filters with transmission coefficients from $5 \cdot 10^{-4}$ to 1. The maximum intensity was calibrated to $100\,\mathrm{mW/cm^2}$ inside the cryostat.

15.3
Photoinduced charge transfer in polymer: fullerene composites

Figure 2(a) shows photoinduced absorption spectra of OC_1C_{10}-PPV and OC_1C_{10}-PPV: PCBM films. Several transitions can be identified: The PIA spectrum of the pure polymer is dominated by a strong contribution at 1.35 eV (High-Energy peak, HE), whereas a very weak transition at 0.42 eV (Low-Energy peak, LE) can only be supposed. The "negative" signal around 2 eV corresponds to ground state depletion due to the intense excitation of the sample. Usually, the HE absorption is attributed to a transition from the T_1 level of the triplet exciton to a higher lying quasi-continuum T_n (see Fig. 2(b)), whereas the LE transition in the mid infrared is identified as a polaronic absorption [11]. Thus, the formation of free charge carriers (mainly due to interchain interactions) in OC_1C_{10}-PPV is an ineffective process as compared to the photoluminescence and triplet formation. The second polaronic transition is expected at 1.34 eV (see Fig. 2(a), inset). This absorption is superimposed with the strong triplet signal and therefore cannot be distinguished. Since the PIA technique is not sensitive to the spin state, PLDMR experiments might be helpful: Fig. 3(a) shows the PLDMR spectrum for OC_1C_{10}-PPV. It consists of a narrow line at $g \approx 2$ and a broad powder spectrum corresponding to transitions with $\Delta m = \pm 1$ from the triplet exciton. At half-field ($g \approx 4$), the forbidden transition with $\Delta m = \pm 2$ is seen, too. This result proves the existence of triplet excitons in the polymer due to their remarkable fingerprint in the wing-like structure [12,13,14]. Nevertheless, the exact origin of the signal and its connection to a PL enhancement (why should triplet excitons contribute to the PL?) is still a matter of controversy [15,16].

The situation changes dramatically when adding a strong electron acceptor like PCBM to the polymer. Fig. 2(a) shows the PIA spectrum of OC_1C_{10}-PPV: PCBM composite (80wt.% PCBM), exhibiting two strong transitions at 1.35 eV (HE) and 0.42 eV (LE). In contrast to pure OC_1C_{10}-PPV, both are of nearly the same intensity. Fig. 3(a) shows the PLDMR spectrum of the same composite: The broad triplet powder spectrum is quenched and only a weak narrow line remains [11]. Additionally, the photoluminescence of OC_1C_{10}-PPV is almost completely quenched (not shown). These results are interpreted in terms of a photoinduced charge transfer between polymer and fullerene. Accordingly, singlet and triplet excitons are quenched due to a transfer of the photoexcited electron from the polymer to the fullerene molecule. This process is known to occur within 45 fs after excitation [7], and results in a high quantum yield of charge carrier states or polarons on the polymer backbone. The PIA spectrum demonstrates two transitions (HE and LE) with nearly equal intensity, in contrast to the strongly asymmetric spectral shape measured in the pristine material. Nevertheless, we were unable to detect the optical transition originating from the fullerene anion, which makes the identification of a charge-transferred state unclear. However, LESR is a helpful technique to provide more insight into the separated states, since it directly probes the spin state of the charged species and makes the assignment on the g-factor [6,10]. Accordingly, it is well suited for proving the existence of photogenerated spins on both, polymer chain and fullerene molecule. Fig. 3(b) shows the LESR spectra for a OC_1C_{10}-PPV: PCBM composite (80 wt.-% PCBM) being the difference between the "dark" and the "light on" scan, therefore, showing exclusively the contribution of photogenerated spins. Two indications can be clearly distinguished: one in higher magnetic field (g<2) and a second one at lower field (g>2). In [17] it was shown, that the lower g-value belongs to a fullerene radical state with orthorhombic symmetry, whilst the signal with g>2 is attributed to the polymer cation with an axial symmetry. It is worth noting that in the pure materials, polymer and fullerene, no significant light induced signals were detectable.

Recapitulating these experiments the following scenario can be drawn: The pristine OC_1C_{10}-PPV shows strong photoluminescence, the formation of triplet excitons and a negligible yield of charge carrier states, which may be due to the interchain charge transfer. Doping with fullerenes results in a charge separated state between the polymer and the fullerene. A strong enhancement of the charge carrier formation is observed. This makes the composites well suitable for photovoltaic energy conversion.

Figure 4(a) shows photoinduced absorption spectra for pure P3HT and P3HT:PCBM composite. In contrast to OC_1C_{10}-PPV few more transitions are observed, which demand a detailed discussion: Photoinduced absorptions are present at 1.84 eV, 1.24 eV and 0.3 eV. It was shown that they all belong to charged states with spin ½ [18]. The transition at 1.06 eV is attributed to a trapped singlet exciton [19]. This result is surprising since no triplet excitons are present in the pure material [19]. PLDMR spectra, in principle, support this proposition: Fig. 5(a) shows an

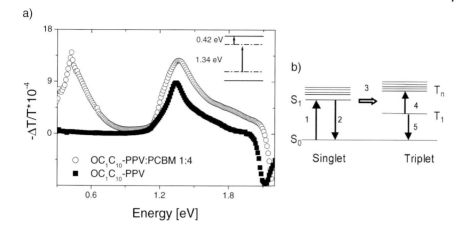

Figure 2 (a) Photoinduced absorption in OC₁C₁₀-PPV and OC₁C₁₀-PPV:PCBM. The inset illustrates the photoinduced transitions of polarons within the HOMO-LUMO gap of the polymer. (b) Possible optical transitions in the triplet manifold of a conjugated polymer: 1- excitation, 2- PL, 3- ISC, 4- T_1-T_n transitions, 5- monomolecular triplet decay.

intense signal at $g \approx 2$ and only a weak triplet powder spectrum showing two distinct transitions (probably connected with a different symmetry of the triplet exciton in P3HT (second pair of shoulders) as compared to OC₁C₁₀-PPV). Doping of the material with fullerenes results in a similar behaviour of the photoinduced transitions as observed for the pure material (Fig. 4(a)): Again, three transitions at 1.84 eV, 1.24 eV and 0.3 eV can be detected. The absorption at 1.06 eV is quenched being consistent with the photoinduced charge transfer (quenching of excitons). No shifts in the polaronic transitions are observed, which confirms the favoured formation of polarons already in the pure material. The polaron formation in pure P3HT seems to be

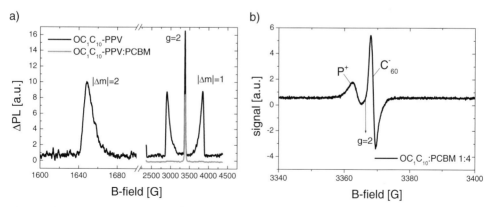

Figure 3 (a) PLDMR in OC₁C₁₀-PPV and OC₁C₁₀-PPV: PCBM
(b) LESR spectrum of OC₁C₁₀-PPV: PCBM composite.

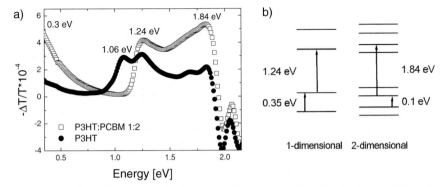

Figure 4 (a) Photoinduced absorption in P3HT and P3HT:PCBM: (b) Schematic representation of the energetic levels and photoinduced transitions in a chain or aggregate dominated structure as proposed in [19]: The aggregate formation leads to a splitting of the HOMO/LUMO levels, which makes the previously (1-dimensional case) forbidden polaronic transition allowed (1.84 eV transition in 2-dimensional case).

a decisive feature of this conjugated polymer. In [19] it is argued that regioregular P3HT films tend to form aggregates which result in delocalized primary photoexcitations. Hence singlet excitons may dissociate more easily into polaron pairs, which then separate into free charge carriers.

As we have seen there are fundamental differences between the photoinduced absorption transitions in OC_1C_{10}-PPV and P3HT. Nevertheless, doping the polymers with PCBM results in a charge separated state in both composites. Fig. 5(b) shows LESR spectra of P3HT: PCBM (67 wt.-% PCBM) composites. The spectral shape is identical to that obtained for the OC_1C_{10}-PPV: PCBM mixture (see above). The two distinct transitions belong to the polymer and the fullerene radicals, respectively.

Photophysical investigations may be summarized as follows: Two polymers, OC_1C_{10}-PPV and P3HT, were studied by means of photoinduced absorption spectroscopy and magnetic resonance techniques (PLDMR, LESR). OC_1C_{10}-PPV demon-

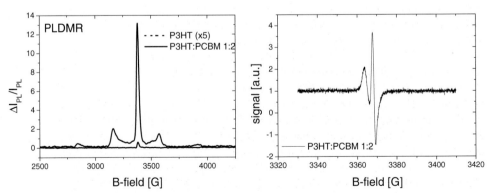

Figure 5 (a) PLDMR spectra of pure P3HT (stronger signal) and P3HT:PCBM. (b) LESR spectra of P3HT:PCBM composites.

strates strong triplet formation and only a weak charge carrier formation, whereas the yield of charge separated states is increased dramatically in OC_1C_{10}-PPV: PCBM composite. On the other hand, the main contribution in the photoinduced absorption spectra of P3HT arises from charged states; the yield of triplet excitons is negligible. The origin of this unusual behaviour may be due to a strong interchain interaction due to formation of ordered aggregates. By blending P3HT with PCBM - similar to the OC_1C_{10}-PPV: PCBM composite – additional charge carriers are generated effectively. These effects may be utilized for the development of polymer solar cells as we show in the next section.

15.4
Photovoltaic characteristics and charge carrier photogeneration efficiency of polymer-fullerene bulk heterojunction solar cells

Figure 6(a) shows the current-voltage characteristics for bulk-heterojunction solar cells based on different combinations of donor and acceptor. For OC_1C_{10}-PPV: PCBM samples with 100 nm active layer thickness photocurrents (J_{SC}) of 5–6 mA/cm^2 are measured under white light illumination of 100 mW/cm^2. The open-circuit voltage (V_{OC}) for solar cells with this material combination is typically above 0.8 V [3,20]. The corresponding values for P3HT: PCBM samples with the same active layer thickness are $J_{SC} \approx 8$–9 mA/cm^2 and $V_{OC} \approx 0.55$ V. The experimentally measured photocurrents for both systems fit well to the maximum values estimated in optical simulations [21].

Figure 6(b) shows the current-voltage (J-V) characteristics in a semi-logarithmic representation. The dark J-V characteristics can be separated into three regions: (i) under reverse bias and under forward bias up to turn-on voltage the characteristics behave almost symmetrically, however the origin of this "leakage current" is not well understood in bulk-heterojunction diodes; (ii) at a voltage corresponding to the condition, when the interface barriers can be overcome, the current starts to grow exponentially with applied forward bias; (iii) at voltages exceeding the flat band condition, the current saturates and is space charge limited, and follows the $J \propto V^m$, $m \geq 2$, law. In contrast to the OC_1C_{10}-PPV: PCBM diode, the current turn-on (exponential range) in non-illuminated P3HT:PCBM device occurs at much lower voltages (0.25 V versus 0.6 V for OC_1C_{10}-PPV). Similar values and, hence, differences were obtained in polymer only devices (not shown). This is surprising, as the electrode materials used in both devices are the same. The HOMO's for P3HT and OC_1C_{10}-PPV, determined electrochemically, are close to each other (5.1–5.2 eV), however, no direct measurements on solid films could be found in the literature.

Further, the LUMO levels difference of polymers can not account for the turn-on voltage shift, as the dark current through the polymer phase should be dominated by holes, not electrons. The PCBM is present in both types of devices and is known to be an electron transporting material. The pure PCBM device shows the turn-on of exponential range at 0.8 V [22], i.e., at higher voltage as measured for P3HT: PCBM and OC_1C_{10}-PPV:PCBM in this work. Therefore, to explain the difference in the turn-on voltages in both devices, the formation of different barriers at the PEDOT:

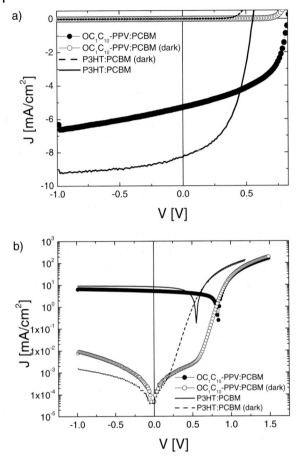

Figure 6 (a) Current-voltage characteristics for ITO/PEDOT: PSS/OC$_1$C$_{10}$-PPV: PCBM/Al and annealed ITO/PEDOT: PSS/P3HT: PCBM/Al in the fourth quadrant. (b) Semi-logarithmic representation of the dark and illuminated (100 mW/cm^2, white light) current-voltage characteristics.

PSS/polymer interface, being lower for P3HT, should be anticipated. However, we admit that the formation of interface dipole layer at the external interfaces may be responsible for the effect discussed as well. Moreover, the ground state charge transfer between the polymer and fullerene, proposed in [23], would lead to a variation of the HOMO-LUMO gap. In such a case, the formation of spin carrying states is expected. No dark electron spin resonance signals were detected by us in the polymer-fullerene blends studied, of course, within the sensitivity of the method.

The onset of the space charge limited current (corresponding to built-in voltage [22]) in ITO/PEDOT: PSS/P3HT: PCBM/Al device is 0.5 V, i.e. 0.3 V below the value obtained for ITO/PEDOT: PSS/OC$_1$C$_{10}$-PPV: PCBM/Al device. Further, the V$_{OC}$, being an estimate for V$_{bi}$, in P3HT: PCBM based solar cells is approximately 0.3 V smaller as compared to PPV based devices (Fig. 6). This seems to be a general feature of P3HT based solar cells, as was confirmed in blends of P3HT and OC_1C_{10}-PPV with different fullerene acceptor DPM-12 [24], and is not quite clear at the moment, however, the different HOMOs for P3HT and OC_1C_{10}-PPV, resulting in different effective band gaps, might be an explanation.

The spectral response of the devices with the two absorber materials was studied by measuring the incident-photon-conversion efficiency (IPCE). Fig. 7 shows the IPCE spectra for both devices. Two main absorption signatures were found in the spectra for OC_1C_{10}-PPV: PCBM (open squares) and P3HT: PCBM (solid line) (both as-cast): one in the range around 340 nm originating from fullerene absorption with subsequent hole transfer to the donor, the other at about 520 nm arising from the polymer absorption (electron transfer).

The IPCE spectra for PPV and P3HT (as cast) based devices have similar shape with peak values of 37% and 15% (at 340 nm), respectively. Upon annealing, the ITO/PEDOT: PSS/P3HT: PCBM/Al device, the IPCE spectral shape changes and becomes higher reaching 65% (at 540 nm). Further, the low energy edge is red shifted by 100 nm. Additionally, the signatures at 400 nm and 620 nm become more pronounced. Annealing the sample did not improve the device performance of the device with OC_1C_{10}-PPV.

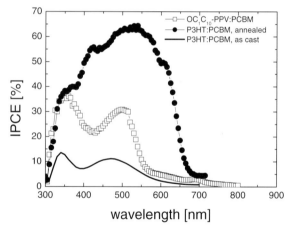

Figure 7 Incident-photon-conversion efficiencies for bulk-heterojunction solar cells based on OC_1C_{10}-PPV: PCBM and P3HT: PCBM. For P3HT: PCBM, the IPCE is shown for the device as cast (solid line) and after annealing the device (filled symbols).

The ratio of the photocurrents of both devices (Fig. 6(a)) corresponds to the ratio of the respective integrated IPCE spectra (Fig. 7) and is not solely due to the smaller band gap in the annealed P3HT. In order to identify the mechanism responsible for the increase of photocurrent in P3HT: PCBM device, we studied the electrical transport properties by means of illumination intensity and temperature dependent current-voltage characteristics for both device structures.

15.5
Analysis of the electrical transport properties

15.5.1
Illumination intensity dependence

The charge carriers in the solar cells under investigation are efficiently photogenerated via ultrafast electron transfer between donor and acceptor counterparts of the composite. The fate of the photogenerated electrons and holes is crucial for the device efficiency, therefore, the influence of light intensity P_{Light} on the J-V characteristics is important.

Figure 8(a) shows the fill factor FF as function of illumination intensity for ITO/PEDOT: PSS/OC_1C_{10}: PCBM/Al. Above T=140 K, the FF depends weakly on the illumination intensity, i.e., it decreases by 6–8%, as compared to 30% at T=125 K. The highest FF of 53% was measured at T=320 K. The fill factor depends on series (R_S) and parallel resistance (R_P). In contrast to the series resistance, the R_P reduces dramatically with light intensity, as shown in Fig. 8(b). The R_P, which is in the order of 100 kΩ·cm^2 at lowest light intensity (0.05 mW/cm^2), decreases by nearly three orders of magnitude when P_{Light} is increased to 100 mW/cm^2. Qualitatively similar behavior is observed for P3HT-blends. The intensity dependence of the FF is in qualitative agreement with the decrease of R_P at higher illumination levels.

The power conversion efficiency of polymer-fullerene bulk heterojunction solar cells also depends on the incident light intensity, and decreases at higher illumination levels due to the degradation of the fill factor with P_{Light} [20,24,25,26].

Figure 9(a) displays the short-circuit current density of ITO/PEDOT: PSS/OC_1C_{10}-PPV: PCBM/Al as function of light intensity, P_{Light}, in a double-logarithmic scale. The J_{SC} follows the power law dependence

$$J_{SC} = P_{Light}^{\alpha} \tag{1}$$

with scaling exponents α=0.94 and 0.86 at T=320 K and T=120 K, respectively. Similar results were obtained for ITO/PEDOT: PSS/P3HT: PCBM/Al with α=0.94 (T=320 K) and α=0.73 (T=130 K). The nearly linear dependence of the J_{SC} on P_{Light} indicates that charge carrier losses in the absorber bulk are dominated by monomolecular recombination via defects. Bimolecular recombination of electrons and holes, reflected by a square-root current-light-intensity-ratio (α=0.5), was not observed even at highest illumination intensities. This is in agreement with the picture of photogeneration and transport of charge carriers in a donor-acceptor bulk heterojunction: the ultrafast photogeneration of charge carriers is followed by the charge delocalization within spatially separated domains: electrons within percolated fullerene clusters and holes within the polymer network with subsequent transport towards electrodes. As both electrodes may be considered as ohmic (the barrier heights are less than 0.5 eV) for the respective charge carriers, no space charge limited conditions under short-circuit conditions exist even at highest generation rates.

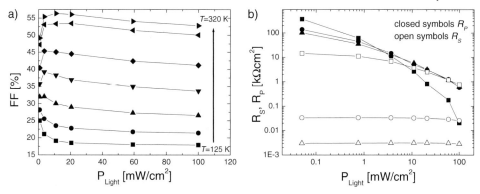

Figure 8 (a) Fill factor of ITO/PEDOT: PSS/OC$_1$C$_{10}$-PPV:
PCBM/Al as function of illumination intensity for temperatures
from 125 K – 320 K. (b) Calculated series (R_S, open symbols)
and parallel resistance (R_P, closed symbols) as function of light
intensity. T= 125 K (squares), 220 K (circles), 320 (triangles).

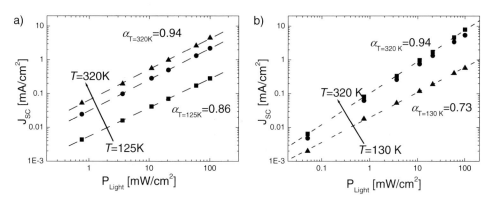

Figure 9 Scaling behavior of the short-circuit current density
with light intensity for (a) ITO/PEDOT: PSS/OC$_1$C$_{10}$-PPV:
PCBM/Al (T=125 K – 320 K), (b) ITO/PEDOT: PSS/P3HT:
PCBM/Al devices (T=130 K – 320 K). The scaling exponents α
were obtained from fitting the experimental data to Eq. (1).

15.5.2

Temperature dependence

Figure 10 shows the short-circuit current density J_{SC} of ITO/PEDOT: PSS/OC$_1$C$_{10}$-
PPV: PCBM/Al as a function of temperature at different illumination intensities.
The J_{SC} of the studied samples strongly increases with temperature. The temperature dependence can be due to either the temperature dependent mobility [27, 28],
or due to steady state charge carrier concentration. In the absence of mono- and bimolecular recombination, the J_{SC} is expected to be weakly temperature dependent,

mainly due to the thermal generation of charge carriers. The thermally activated charge carrier mobility will not influence the steady state photocurrent, since the total amount of photogenerated charges remains unaffected. In a real situation, the J_{SC} can be reduced due to recombination. The capture of charge carriers by traps with subsequent recombination or emission may lead to a variation of the current with temperature. In this case, an increase of temperature will promote the current through the device. This is a thermally activated process and is denoted as monomolecular recombination. The experimental data in Fig. 10 can be fitted by the expression

$$J_{SC}(T, P_{Light}) = J_0(P_{Light}) \exp\left(-\frac{\Delta}{kT}\right), \tag{2}$$

where $J_0(P_{Light})$ is a pre-exponential factor which contains the intrinsic charge carrier density, the charge carrier mobility, and the electric field; Δ is the energetic depth of the trap state with respect to the corresponding band.

From the fits we obtained the activation energy $\Delta=44.7 \pm 1.5$ meV for all the light intensities applied, and the $P_{Light}^{0.94}$ dependence of the pre-exponential factor on light intensity, i.e., similar to that for J_{SC} (Fig. 9(a) [20, 24]. The activation energy Δ is somewhat smaller at lower intensities being in the range of standard deviation.

The thermally activated mobility might be responsible for the temperature dependence of the short circuit current, e.g., due to poor transport and extraction of photogenerated electrons and holes forming space charge and, hence, recombining. In this case, the recombination of bimolecular type with a square-root current-intensity ratio will be dominant. A small deviation of scaling exponent (0.94) from unity at room temperature indicates that the bimolecular recombination is a negligible loss mechanism at elevated temperatures, being more important at low temperatures.

Figure 11 displays the short-circuit current density versus temperature of ITO/PEDOT: PSS/P3HT: PCBM/Al devices with different absorber thickness between

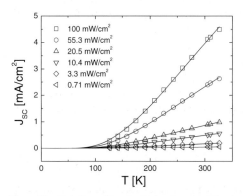

Figure 10 Variation of the short-circuit current density with temperature for an ITO/PEDOT: PSS/OC$_1$C$_{10}$-PPV: PCBM/Al solar cell. The $J_{SC}(T)$- curves were recorded at different light intensities as indicated by the legend. The solid lines are fits with Eq. (2) with an activation energy of $\Delta=44.7\pm1.5$ meV.

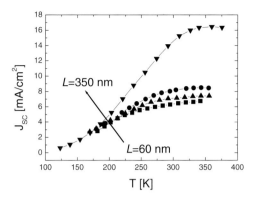

Figure 11 Variation of short-circuit current density with temperature for ITO/PEDOT: PSS/P3HT: PCBM/Al solar cells under illumination with P_{Light}=100 mW/cm² for different active layer thicknesses.

60 nm and 350 nm (illumination intensity: 100 mW/cm²). At low temperatures a thermally activated increase of the J_{SC} is observed for all samples studied and is similar to the OC_1C_{10}-PPV case. This behavior extends over a broader range for thicker samples, whereas the current saturates at particular temperatures and becomes almost constant for highest temperatures. The transition region shifts to high temperatures for increased active layer thickness. The device with L=350 nm demonstrated a broad transport limited regime, with an activation energy of Δ=62 meV, whereas the saturation occurs at T>320 K.

The values of the J_{SC} measured in the saturation region vary almost linearly with the incident light intensity (not shown) reflecting that bimolecular recombination is a negligible loss mechanism in this temperature range. The temperature independent behavior of J_{SC} indicates, that charge carriers traverse the active layer without recombining during their lifetime, i.e., the charge transport occurs without significant losses in the active layer. This interpretation is in agreement with high internal quantum efficiency of 96% in the main absorption of the donor material for samples with 70 nm active layer thickness [9, 29].

The charge carrier lifetime τ, the mobility μ and the internal field $E=V_{bi}/L$ (V_{bi} is determined by the work function difference of the electrode materials) define the mean drift length d_m, i.e., the path length the charge carriers cover prior to recombination:

$$d_m = \mu \cdot \tau \cdot \frac{V_{bi}}{L}. \tag{3}$$

To fulfil the condition of the loss free charge transport, d_m must exceed the active layer thickness L:

$$\mu \cdot \tau \cdot V_{bi} > L^2. \tag{4}$$

The results shown in Fig. 11 demonstrate that the thickness of the P3HT: PCBM layer can be significantly increased well above the 100 nm, widely used in polymer solar cells, without limitation of the short-circuit current due to, e.g., recombination losses. At the same time, the characteristics shown in Fig. 11 indicate that the optimal thickness for the P3HT: PCBM absorber lies below 350 nm, as the performance of this device is already transport limited at room temperature. In contrast, the OC_1C_{10}-PPV based device with 100 nm absorber seems to be transport limited up to 310 K, and the onset of saturation above T=326 K can only be anticipated (see Fig. 10).

As previously noted, the onset of the exponential range in dark J-V as well as the setting of the open circuit conditions in the illuminated devices occur at lower voltages in P3HT: PCBM devices. According to Eqs. (3) and (4), the mobility-lifetime product for P3HT: PCBM device can be roughly estimated to be 50 time larger that that for the OC_1C_{10}-PPV:PCBM device. Assuming the same built-in potential difference (0.7 V) in both types of devices is responsible for the electric field across the absorber and the drift lengths of 300 nm for P3HT and 50 nm for OC_1C_{10}-PPV based devices, the $\mu\tau$ products ratio will be as high as 36. (Note, the 70–100 nm OC_1C_{10}-PPV based cells tested by us were already transport limited.) If we consider the V_{OC} as an estimate for the V_{bi}, the ratio should be multiplied by the factor of 1.5 (0.8 V/0.55 V).

Hence, solar cells based on P3HT: PCBM have sufficient electrical transport properties allowing a higher absorber thickness than in case of OC_1C_{10}-PPV:PCBM.

15.5.3
Bulk-heterojunction solar cells with increased absorber thickness

Fig. 12 shows the current-voltage characteristics of an ITO/PEDOT: PSS/P3HT: PCBM/Al solar cell with 350 nm active layer thickness, displayed in the third and fourth quadrants (Fig. 12(a)) and in semi-logarithmic scale (Fig. 12(b)).

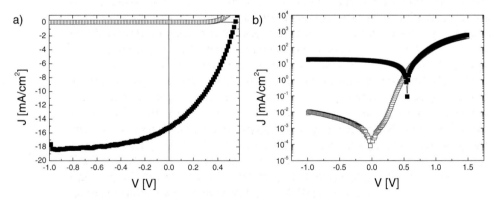

Figure 12 Current density – voltage characteristics for an annealed ITO/PEDOT: PSS/P3HT:PCBM/ Al solar cell with 350 nm active layer thickness under illumination with P_{Light}=100 mW/cm^2 (full symbols) and in the dark (open symbols). (a) J-V profiles in the photovoltaic regime, (b) semi-logarithmic representation in the full voltage range.

A large increase of the absorber thickness results in a large photocurrent of J_{SC}=15.2 mA/cm^2 due to higher absorption. This value is by nearly 7 mA/cm^2 higher than the value obtained for devices with a 100 nm thick active layer. The open-circuit voltage measured is 0.54 V. The fill factor of the *L*=350 nm device is reduced to 37%, as compared to 50%–60% for 100 nm samples. We calculated a resulting power conversion efficiency of 3.1% (P_{Light}=100 mW/cm^2, *T*=300 K). We note that this device operates in the transport-limited regime, as it follows from Fig. 7. The saturation (= loss free) region starts above room temperature, i.e., at *T* > 330 K, whereas the characteristics shown in Fig. 8 were measured at *T*=300 K.

Such value of short-circuit current is, to our knowledge, the highest reported value for organic solar cells so far. However, the low fill factor is definitely the main drawback of this device. R_S increases from 1 Ω·cm^2 to 3.1 Ω·cm^2 for 100 nm and 350 nm devices, respectively, which is still too low to reduce the fill factor so dramatically (see Fig. 8(a) and 8(b)). Alternatively, the *J-V* curve can be superimposed by an additional field-current through the illumination-dependent parallel resistance (see Fig. 8(b)). In the latter case, the total current in the reverse direction would decrease continuously with negative voltage. However, the photocurrent in our devices saturates at ~18 mA/cm^2. Similar is the behavior of a three times thinner (L≈100 nm) P3HT:PCBM device, as Fig. 6(a) shows. The 350 nm device operates in the transport limited regime, as the built-in field and, hence, the resulting mean drift length are smaller than the Eqs. (3) and (4) demand. It means, an additional negative bias supporting the built-in field is required in a thick cell to prevent charge carriers from recombination. This is different in thin cells. Further optimization of the devices thickness needs to be performed.

15.5.4
Dependence of IPCE on reverse bias

Figure 13 shows the results of the dc-bias dependent measurements of the IPCE in the range 300 nm – 550 nm for an ITO/PEDOT: PSS/P3HT: PCBM/Al, *L* = 80 nm, device. The curves correspond to different applied bias voltages, as indicated by the legend. The device under test did not change the performance after the measurement cycle, as the final scan at V_{Bias}=0 V (Fig. 13, closed triangles) clearly demonstrates. The IPCE increases with the reverse bias voltage. Consequently, the short-circuit current, being the integral of the IPCE curve, increases under reverse bias, as well.

The overall gain of approximately 1% observed is indeed small. As this particular device operates in a nearly loss free region (*α*=0.94), only a small field effect compensating the deviation from the unity might be expected. However, the calculated scaling exponents in Eq. (1) are identical at *V*= 0 V and *V*= –1 V, therefore, the field does not suppress the recombination of photogenerated charges. Similar measurements for thick devices are in progress. In the latter case, however, the contribution of both mechanisms, field-induced e-h separation and increased mean drift length, can not be separated.

The mechanism of the electric field on the IPCE remains, however, unclear. The fact that it is more pronounced in the high energy part of the spectrum might indi-

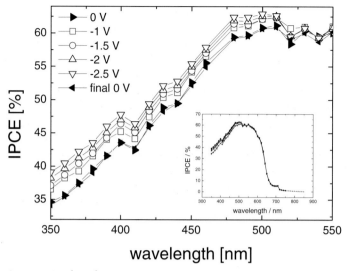

Figure 13 Incident photon conversion efficiency of an ITO/PEDOT:PSS/P3HT: PCBM/Al solar cell. Closed symbols represent measurements under short-circuit conditions (V=0 V). Open symbols are measurements with additionally applied negative bias V_{Bias}. The inset shows the IPCE spectra in full wavelength range.

cate a completely different charge carrier generation mechanisms, via, for example, photoexcitation of electrodes or a field supported charge carrier generation in the thin PEDOT:PSS interfacial layer.

15.6
Conclusions

The photoexcited states in OC_1C_{10}-PPV and P3HT and their blends with PCBM were investigated by means of photoinduced absorption spectroscopy and magnetic resonance techniques (PLDMR, LESR). Upon photoexcitation OC_1C_{10}-PPV demonstrates a strong formation of triplets and only weak charge carrier generation. However, the yield of charged states is strongly increased in the blend with PCBM. In contrast to pure OC_1C_{10}-PPV, the main contribution to the photoinduced absorption spectra of pure P3HT arises from polaronic states. By doping P3HT with PCBM further charge carriers are generated with high efficiency. Photovoltaic bulk-heterojunction devices based on OC_1C_{10}-PPV: PCBM and P3HT: PCBM demonstrated reasonable performance. However, a notable difference in the values of short-circuit currents was observed. This difference is attributed to the electronic transport properties of these two absorber materials studied.

A strong temperature dependence of short-circuit current of the ITO/PEDOT: PSS/OC_1C_{10}-PPV: PCBM/Al solar cells indicates that devices are limited by transport properties of the absorber. A thermally activated transport of photogenerated charge carriers, influenced by recombination with shallow traps, describes the variation of photocurrent at different illumination levels. In such transport limited solar cells, the active layer thickness must be kept low at the expense of the photogeneration of charge carriers. In contrast, devices with P3HT: PCBM active layer display a temperature independent photocurrent above certain temperature, indicating that the transport and extraction of charge carriers are not significantly attenuated by losses. It means, the mean drift length of photogenerated charge carriers exceeds the active layer thickness under short-circuit conditions and at given temperature. Therefore, the active layer thickness of photovoltaic devices based on P3HT: PCBM can be increased almost up to 350 nm without influencing the device performance. The resulting short-circuit current density of 15 mA/cm^2 is to our knowledge the highest reported value for polymer solar cells so far. By increasing the fill factor and the open circuit voltage, the full potential of solar cells based on P3HT-PCBM absorbers (power efficiencies in the range 5–10%) can be exploited.

Acknowledgements

The authors thank E. von Hauff, D. Chirvase, Z. Chiguvare, V. Mertens and J. Parisi (U. of Oldenburg, Germany), C. J. Brabec, P. Schilinsky, Ch. Waldauf (Siemens AG, Germany), J. C. Hummelen (U. of Groningen, The Netherlands), D. Vanderzande, L. Lutsen (L.U.C, Diepenbeek, Belgium) for stimulating ideas, material supply, experimental support and fruitful discussions. Financial support was granted by the Bundesministerium für Bildung und Forschung (BmB+F, support codes: 01SF0019 and 01SF0026), and the European Commission (EC, project: HPRN-CT-2000–00127).

References

[1] C. J. Brabec, V. Dyakonov, J. Parisi and N. S. Sariciftci (Eds.), Organic Photovoltaics: Concepts and Realization (Springer Verlag, Heidelberg, 2003).

[2] S. E. Shaheen, C. J. Brabec, F. Padinger, T. Fromherz, J. C. Hummelen, N. S. Sariciftci, Appl. Phys. Lett. **78**, 841 (2001).

[3] C. J. Brabec, J.C. Hummelen, N. S. Sariciftci, Adv. Func. Mat. **11**, 15 (2001).

[4] F. Padinger, R. Rittberger, N. S. Sariciftci, Adv. Func. Mat. **13**, 85 (2003).

[5] press release 07/01/04, Siemens AG, Germany.

[6] N. S. Sariciftci, L. Smilowitz, A. J. Heeger, F. Wudl, Science **258**, 1474 (1992).

[7] C. J. Brabec, G. Zerza, G. Cerullo, S. De Silvestri, S. Luzatti, J. C. Hummelen, N. S. Sariciftci, Chem. Phys. Lett. **340**, 232 (2001).

[8] I. Montanari, A. F. Nogueira, J. Nelson, J. R. Durrant, C. Winder, M. A. Loi, N. S. Sariciftci, C. J. Brabec, Appl. Phys. Lett. **81**, 3001 (2002).

[9] P. Schilinsky, C. Waldauf, C.J. Brabec, Appl. Phys. Lett. **81**, 3885 (2002).

[10] V. Dyakonov, G. Zoriniants, M. Scharber, C. J. Brabec, R. A. J. Janssen, J. C. Hummelen, N. S. Sariciftci, Phys. Rev. B **59**, 8019 (1999).

[11] X. Wei, Z.V. Vardeny, N. S. Sariciftci, A. J. Heeger, Phys. Rev. B **53**, 2187 (1996).

[12] V. Dyakonov, G. Rösler, M. Schwoerer, E. L. Frankevich, Phys. Rev. B **56**, 3852 (1997).

[13] E. J. W. List, U. Scherf, K. Müllen, W. Graupner, C. H. Kim, J. Shinar, Phys. Rev. B **66**, 235203 (2002).

[14] M. Wohlgenannt, C. Yang, Z. V.Vardeny, Phys. Rev. B **66**, 241201 (2002).

[15] E. L. Frankevich, Chem. Phys. **297**, 315 (2004).

[16] M. Wohlgenannt, phys. stat. sol. (a) **6**, 1188 (2004).

[17] J. De Ceuster, E. Goovaerts, A. Bouwen, J. C. Hummelen, and V. Dyakonov, Phys. Rev. B **64**, 195206 (2001).

[18] R. Österbacka, C. P. An, X. M. Jiang, Z. V. Vardeny, Science **287**, 839 (2000).

[19] X. M. Jiang, R. Österbacka, O. Korovyanko, C. P. An, B. Horovitz, R. A. J. Janssen, Z. V. Vardeny, Adv. Func. Mat. **12**, 587 (2002).

[20] I. Riedel, V. Dyakonov, J. Parisi, L. Lutsen, D. Vanderzande, J. C. Hummelen, Adv. Funct. Mat. **14**, 38 (2004).

[21] H. Hoppe, N. Arnold, D. Meissner, N. S. Sariciftci, Thin Solid Films, **451–452**, 589 (2004).

[22] V. D. Mihailetchi, P. W. M. Blom, J. C. Hummelen, M.T. Rispens, J. Appl. Phys. **94**, 6849 (2003)

[23] V. I. Archipov, P. Heremans, H. Bässler, Appl. Phys. Lett. **82**, 4605 (2003).

[24] I. Riedel, N. Martin , F. Giacalone , J. L. Segura , D. Chirvase , J. Parisi , V. Dyakonov, Thin Solid Films **451–452**, 43 (2004).

[25] V. Dyakonov, Appl. Phys. A, **79** (1), 21 (2004).

[26] C. Waldauf, P. Schilinsky, J. Hauch, C. J. Brabec, J. Appl. Phys. **95** (5), 2816 (2004).

[27] H. Bässler, phys. stat. sol. B **175**, 15 (1993).

[28] V. D. Mihailetchi, J. K. J. van Duren, P. W. M. Blom, J. C. Hummelen, R. A. J. Janssen, J. M. Kroon, M. T. Rispens, W. J. H. Verhees, M. M. Wienk, Adv. Funct. Mater. **13**, 43 (2003).

[29] P. Schilinsky, C. Waldauf, I. Riedel, V. Dyakonov, C.J. Brabec, Adv. Mat., unpublished (2004).

16

Modification of PEDOT:PSS as Hole Injection Layer in Polymer LEDs

M. M. de Kok, M. Buechel, S. I. E. Vulto, P. van de Weijer, E. A. Meulenkamp, S. H. P. M. de Winter, A. J. G. Mank, H. J. M. Vorstenbosch, C. H. L. Weijtens, and V. van Elsbergen

16.1
Introduction

Organic light-emitting diodes (OLED) form the basis of emissive displays that are eminently suited for high-quality, high-information content video applications. The inherent advantages include a wide viewing angle, fast response time, broad colour range and good form factor [1]. Two types of OLEDs are being industrialised. Small-molecule OLEDs (smOLED) use evaporated small molecules in the active light-emitting stack [2]. Polymer OLEDs (PLED) are based on semiconducting conjugated polymers deposited from solution [3]. PLEDs form the subject of this paper.

In a PLED, the light-emitting polymer is sandwiched between two electrodes. From the cathode electrons are injected, from the anode holes. The charge carriers recombine in the active layer to form an exciton, which can decay to the ground state by emission of a photon with an energy determined by the polymer composition. A typical PLED is a multi-layer device. The first layer is a shunt of indium tin oxide (ITO). A second layer of poly(3,4-ethylenedioxythiophene):poly(styrenesulphonic acid) (PEDOT:PSS) [4], a hole-conducting polymer that acts as the anode, is deposited from an aqueous dispersion. Then, the light-emitting polymer (LEP) is applied from an organic solution to ensure integrity of the underlying PEDOT:PSS, which is insoluble in an organic solvent. LEPs commonly used for PLED devices are poly-(*para*-phenylene vinylene) (PPV) [5], poly-fluorenes [6] and poly-spirobifluorenes [7]. The cathode is a low work-function metal (Ba, Ca), capped by aluminium to lower the resistance and to protect the electron-injecting metal [8].

In the early years of PLED development, the LEP was deposited directly on ITO, and lifetime was short. The lifetime was significantly increased after an UV-ozone or oxygen-plasma treatment of the ITO to modulate its work function and remove (organic) contaminants [9], but remained too short for practical applications. Insertion of a *p*-type hole-conducting polymer led to markedly improved device stability [9,10]. Moreover, a large number of catastrophic short-circuits occurred in devices using ITO as anode. This was also remedied by insertion of the hole-conducting polymer, as it functions as a planarisation layer. Thirdly, the hole-conducting polymer provided an anode with a work function independent of the substrate handling procedure, which is important from an industrial point of view.

Physics of Organic Semiconductors. Edited by W. Brütting
Copyright © 2005 WILEY-VCH Verlag GmbH & Co. KGaA, Weinheim
ISBN 3-527-40550-X

The most commonly used polymeric hole conductor is PEDOT:PSS, sold by H.C. Starck as Baytron® P [11]. The focus of this contribution is modification of the ionic content of PEDOT:PSS and the effect thereof on its properties and PLED device performance. To outline the background and relevance of this topic, we briefly review some of the key characteristics of PEDOT:PSS.

PEDOT belongs to the class of semiconducting polythiophenes, and was developed during the search for stable conducting polymers [4]. Introduction of the dioxy-ethylene substitution circumvents undesired (α,β) and (β,β) coupling of the thiophene ring, and leads to air-stable material in its doped, oxidised form due to its electron-rich character (low oxidation potential) [12]. PEDOT has a relatively high conductivity in its oxidised form, which can be ascribed to a planar structure allowing effective delocalisation of π electrons. It also has good optical transparency in the visible region, but is insoluble. Combination with a poly-electrolyte (PSS) resolved this problem. PSS acts as charge-compensating counter-ion to stabilise the p-doped conducting polymer, and forms a processable water-borne dispersion of negatively charged swollen colloidal particles consisting of PEDOT and excess PSS [13]. PEDOT and PSS chains are linked tightly by ionic interaction and form an ionic polymer complex, which cannot be separated by electrophoresis [13]. PSS is present during the oxidative polymerisation of monomer EDOT in water to yield doped PEDOT:PSS in a one-step synthesis. The success of this approach is amply demonstrated by commercial use in antistatic coatings, capacitors and polymer LEDs, and by possible applications in electrochromics [12, 14, 15].

The electrical characteristics of PEDOT:PSS can be varied widely. The original product developed for antistatic applications has a conductivity of about 10 S/cm. This can be improved up to several hundred S/cm by addition of poly-hydroxy compounds such as sorbitol and glycerol [16] or polar solvents like NMP (N-methyl pyrrolidone) [17, 18]. The enhancement is due to subtle changes in the morphology of dried PEDOT:PSS particles, and especially in how particles interpenetrate. An increase in inter-chain interaction has been deduced from a change in optical anisotropy upon reorientation of the polymer chains by addition of sorbitol [19]. XPS data pointed to a concomitant smaller extent of PSS segregation (see below) [18].

Lower conductivity can be obtained by adjustment of the particle size distribution and by increase of the PSS content [11, 20]. PEDOT:PSS applied in PLEDs is generally of this type. Commercial materials supplied by H.C. Starck have a conductivity of about 10^{-3} S/cm (1:6 PEDOT:PSS by weight) or 10^{-5} S/cm (1:20 PEDOT:PSS by weight) [11]. It is important to recognise that the conductivity of PEDOT:PSS used in PLEDs is, therefore, not determined by the intrinsic hole mobility, but rather by the contact resistance of holes moving from one isolated PEDOT segment to the next, through a matrix of non-conducting PSS [21]. The PEDOT chain length has been estimated at 5-10 repeat units [22]. The actual conductivity thus depends strongly on the 'macroscopic' morphology of dried particles in the film, which depends on the annealing conditions [23] and on the distribution of PEDOT segments in the matrix, which is supposedly determined by the polymerisation conditions.

X-ray Photoelectron Spectroscopy (XPS) and Ultraviolet Photoemission Spectroscopy (UPS) have been used, particularly by Salaneck and co-workers [21,23-25], to study the surface of PEDOT:PSS films. Four species were identified using XPS: (i) PSS-H, the acidic form of PSS. (ii) PSS-Na, the corresponding Na-salt. The origin of Na$^+$ ions was not commented upon, but is presumably related to the synthesis: Na$_2$S$_2$O$_8$ may be used as oxidising agent [12], similar to K$_2$S$_2$O$_8$ [26]. (iii) PEDOT^0, or neutral thiophene units. (iv) PEDOT$^+$. Actually, the PEDOT signal was rather complex due to the delocalised nature of the positive charge, giving rise to a range of sulphur binding energies. Importantly, XPS and UPS gave a consistent picture of the composition of the surface layer. Significant phase segregation of PSS occurs. It was estimated that the outer 4 nm are strongly enriched in PSS.

The schematic picture of the morphology of PEDOT:PSS and its chemistry, shown in Figure 1, summarises much of the above. The particle size in solution is 20–70 nm [20] and each particle contains many PEDOT chains. The thiophene doping level has been estimated at 0.33 [22], in line with spectroscopic investigations and comparable to other doped (electropolymerised) thiophenes. The Na$^+$ content of commercial product can vary between zero (*i.e.* below the detection limit) and a few hundred ppm in solution [27]. Taking into account the average chain length of 5–10 units, a wide distribution of electronic properties of PEDOT is present: the shortest chains harbour one, or even zero positive charges, while the longest chains carry 3–4 holes. It should also be noted that there is a large excess of protonated PSS-H, which is a key difference with electropolymerised PEDOT:PSS and, in particular, PEDOT:tosylate (or another small anion), where the number of anions equals the number of oxidised thiophene units [21]. Although there is ample evidence for strong optical anisotropy in thiophene-rich PEDOT:PSS [19], no such results have been found for PEDOT:PSS used in PLEDs [28], and the PEDOT chain orientation is drawn accordingly.

Figure 1 does not include possible interaction with the light-emitting polymer (LEP). The interface properties of PEDOT:PSS/LEP are of utmost importance for the performance of a PLED, since charge injection depends strongly on a possible energy barrier. Obviously, such a buried interface cannot be studied at the same level of detail as the PEDOT:PSS/vacuum (or air) interface, and only indirect evidence about PEDOT:PSS/LEP interaction is available. It is, therefore, unfortunate that there seems to be no study of the (surface) properties of PEDOT:PSS after contact to toluene, or another organic LEP solvent, and that many studies of the (surface) properties have used different processing conditions (annealing time and temperature, possible exposure to ambient) than those used in industrial PLED manufacture.

Two types of interaction between PEDOT:PSS and organic materials have been observed in OLED devices. One is purely electronic: when the work function (equal to the Fermi level) of PEDOT:PSS is larger than the ionisation energy of the LEP, oxidation of the LEP takes place, creating a *p*-type doped interface region that facilitates ohmic injection [29]. The second type of interaction is due to the acidic nature of PEDOT:PSS. Formation of TDAPB cations (1,3,5-tris-(N,N-bis(4,5-methoxy-phenyl)aminophenyl)benzene) has been observed with electromodulated absorbance for TDAPB in contact to PEDOT:PSS [30]. Ohmic injection into MEH-PPV (2-methoxy-5-2′-ethyl-hexyloxy-PPV) from PSS-H, *viz.* in the absence of PEDOT, was ascribed to

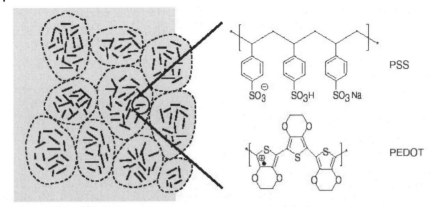

Figure 1 Schematic representation of the morphology of PEDOT:PSS. The left part shows PEDOT:PSS particles, surrounded by a thin PSS-rich surface layer. The PEDOT chains are pictured as short bars. The surface roughness is determined by the particle size. The zoom-in on the right gives the molecular structure of the species present, where PSS-anion, PSS-Na, and PSS-H are indicated.

acid doping of the LEP [31]. Such interactions are specific to particular materials, and the LEP properties can change significantly: one study showed a marked difference between formation of interfacial states for PPV and poly-spirobifluorene [32].

Summarizing, we can conclude that PEDOT:PSS is a rather complex material, from a chemical, morphological and an electronic point of view. PLED-grade material with low conductivity has received relatively little attention. In the course of our work on improving and commercialising PLEDs, we noticed an apparent effect of Na^+ content on device performance, which triggered more extensive investigations. Commercial PEDOT:PSS generally has a low Na^+ content as it is purified for electronic application. It is, however, possible to exchange the protons by addition of NaOH. Na^+-rich PEDOT:PSS is examined with various techniques that are sensitive to bulk and surface properties, such as XPS/UPS, Raman spectroscopy, Kelvin Probe measurements and UV-vis absorbance spectrometry. The changes observed are reflected in device performance for various classes of LEP materials. We find markedly different properties for medium to high Na^+ content and pH that have not been studied before, and that point to the important role played by the PSS poly-anion in stabilising charge carriers in PEDOT:PSS, and thereby in determining its electronic properties.

16.2
Experimental

PEDOT:PSS solution was modified using 1.0 M NaOH (p.a. from Merck), to obtain a particular Na^+ content or pH value. The pH is determined with a Metrohm 713 pH meter, with a combined LL pH glass electrode (Metrohm 6.0233.100). H.C. Starck

Baytron® P AI4083 and CH8000 are used [11], with a solid content of 2.8 wt%, and a PEDOT:PSS weight ratio of 1:6 and 1:20, respectively. Films were prepared by spin-coating in air. The typical thickness was 200 nm. The samples were heated in air for 6 minutes at 200 °C to remove water from the PEDOT:PSS layer, unless mentioned otherwise.

Optical absorption was recorded on a Perkin Elmer Lambda 9 spectrophotometer from 300 to 2500 nm. A scan speed of 120 nm min^{-1} and a slit width of 1 nm were used. Both aqueous dispersions and spin-coated films on glass were analysed. Raman spectra were collected using a Jobin Yvon Labram spectrometer, equipped with two lasers (20 mW 633 nm HeNe laser and 50 mW 785 nm diode laser). Spectra were collected through a 50 × (N.A.=0.8) LWD objective from Olympus with a 600 l/mm or 1800 l/mm grating for collection of overview and detailed spectra, respectively.

XPS and UPS measurements were performed in a multi-chamber ultrahigh vacuum (UHV) system with a base pressure in the 10^{-8} Pa range. The analysis chamber was equipped with an X-ray source with a Mg/Al double anode for XPS and a differentially-pumped, windowless discharge lamp for UPS. Energy distribution curves of photo-emitted electrons were measured with a concentric hemispherical analyser. UPS spectra were recorded using He-I radiation. The ionization energy I was calculated from the energetic difference between the onset of secondary electron emission and the high-energy cut-off at the highest occupied molecular orbital. The work function ϕ was derived from the onset of secondary electron emission in combination with the known position of the Fermi energy. Samples were prepared on an Au-coated Si substrate. After spinning, the samples were heated in a glove-box filled with argon. A portable fast-entry air lock system was used to transport the samples under vacuum to the surface science set-up, *i.e.* the samples were not exposed to air after the heating step. Kelvin probe (Besocke delta phi GmbH, Jülich, Germany) experiments were done under ambient conditions (50% humidity, 22 °C). The reference electrode was vibrating with a frequency of approximately 200Hz.

The standard PLED device structure was used, consisting of 150 nm ITO on glass, 200 nm PEDOT:PSS, 80 nm LEP, 5 nm barium and 100 nm aluminium. So-called hole-only devices comprise 100 nm gold instead of BaAl to avoid electron injection. LEP is spin-coated from toluene in air. Materials studied are a yellow-green emitting PPV [5, 33, 34] and blue poly-spirobifluorene [7, 34] from Covion Organic Semiconductors GmbH, and LUMATION™[1] Green 1300 Series poly-fluorene from The Dow Chemical Company [6,35]. Devices are protected from water and oxygen using a metal or glass lid with getter. Current-voltage-light (IVL) curves were measured using a Keithley 2420 SourceMeter and a Si photodiode. Calibration was performed with a Minolta LS100 luminance meter. Lifetime measurements were performed in a climate chamber at room temperature or at 80 °C. The devices were driven at a constant current of 6.3 mA/cm^2 and both luminance and voltage were measured continuously. Lifetime was defined as 50% loss of luminance.

[1] Trademark of The Dow Chemical Company

16.3
Results

16.3.1
Chemistry and bulk properties

The pH of the Na^+-free dispersions is about 1.45 and 1.55 for 1:20 and 1:6 PEDOT: PSS, respectively. Figure 2 shows results for a titration experiment with NaOH. Initially, PEDOT:PSS behaves as a buffer, owing to the presence of the acid, PSS-H. The equivalent point at pH $\approx$ 7 is within a few percent of the calculated value based on the solid content, taking into account the amount of Na^+ already present and incomplete protonation of PSS in the original dispersion, as PSS^- anions act as counter-ion to oxidised thiophene units. Slow pH relaxation effects associated with poor accessibility of the poly-acid are not observed. Figure 2 denotes not only the molar ratio of Na^+ to total PSS, but also the Na^+ concentration in ppm. Further useful numbers are the monomer MW for PEDOT (140) and PSS-H (184), Na^+ content in a thin film (100 ppm Na^+ in the solution is equivalent to 3570 ppm Na^+ in the solid) and Na^+ mole percentage (for 1:6 PEDOT:PSS, 100 ppm Na^+ in the solution is equivalent to 2.67 mole%, counting Na^+ and PSS-H and PEDOT monomers).

The optical absorbance of several PEDOT:PSS dispersions is given in Figure 3. The as-received material shows low absorbance in the visible region. A peak is found around 1.5 eV or 800 nm. The absorbance continues to rise for longer wavelength. The spectrum is unchanged by addition of ions at constant pH, see Figure 3A. Addition of 3,000 ppm Na^+ in the form of NaCl and neutralisation followed by re-acidification to the original pH value give the same absorbance spectrum as the reference material. Addition of HCl to the starting solution to lower the pH does also not result in changes in the absorbance spectrum.

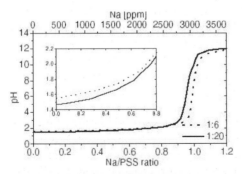

Figure 2 Titration curve of PEDOT:PSS (1:6 and 1:20) with NaOH. The bottom *x*-axis shows the ratio of the Na^+ to total PSS, including Na^+ already present in the dispersion. The top *x*-axis displays the Na^+ content in ppm (for 1:6 PEDOT:PSS). The inset is a zoom-in to the pH-Na^+ relationship at lower pH.

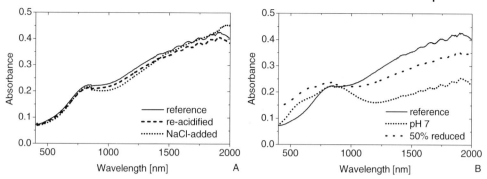

Figure 3 Optical absorbance of various PEDOT:PSS solutions. The absorbance has been normalised to the concentration of the starting solution with Lambert-Beer's law. The absorbance continues to rise until 2,500 nm (not shown). The oscillations for wavelengths longer than 1,400 nm are due to interference fringes caused by the optical path length (50-μm cuvette). Reference is the as-received material (Na⁺-free). Fig. 3A shows data at pH 1.4. Re-acidification from pH 7.0 was done with HCl; NaCl indicates the presence of 3,000 ppm Na⁺ added in the form of NaCl to the starting solution. Fig. 3B shows data at pH 7 obtained by NaOH addition, and for chemically reduced PEDOT:PSS, using 50% of the calculated equivalent amount of hydrazine.

However, a marked change is seen upon increasing the pH. Figure 3B shows data for pH 7: (i) the absorbance in the near infrared (NIR) is decreased; (ii) the absorbance around 800 nm is unaffected; (iii) the absorbance in the visible is increased. These changes do not occur simultaneously. The NIR absorbance decrease is already noticeable at pH 3, while the increase in the visible becomes apparent only at about pH 7. Increase of the pH to values above 10, needing little additional Na⁺, leads to small further changes, in line with the trends observed from pH 1.6 to pH 7.

Modification of 1:6 and 1:20 PEDOT:PSS yields similar results. Moreover, spectral changes in thin films paralleled the solution data. Thus, the changes observed are relevant to PLED devices. Whether they are governed by the pH increase alone or by the combination of pH and cation concentration cannot be determined at present, since cations are invariably formed by addition of base. In the following, when it is mentioned "pH" (or analogous statements), this is implied to mean pH or combined pH/cation concentration effect.

It is worth taking a closer look at the spectral changes. Absorption of undoped and doped, well-conducting PEDOT has been studied intensely. The NIR band is associated with the presence of bi-polarons [36]. Spectro-electrochemical data show that it is present only beyond some oxidation potential, or doping level [36–38]. The origin of the band around 800 nm is less clear. A recent study assigns it to interband transitions [38], in line with older work [39]. However, it has also been attributed to or associated with polarons, defects, and a distribution of polythiophene chain lengths. Neutral or de-doped PEDOT shows a π-π* transition centred around 600 nm. Careful comparison of literature data to the spectra measured here reveals that reference PEDOT:PSS is not fully doped and can be oxidised to higher doping level: spectro-electrochemical experiments show that, for high oxidation voltage close to the stability limit, the peak around 800 nm and the tail in the visible region are less pronounced than in the reference spectrum shown here. Gross and co-workers have used this property to modulate the PEDOT:PSS work function [29].

Surprisingly, the changes induced by the pH increase are reminiscent of those upon partial de-doping. Several workers have reported quite similar spectra as those found for pH 7 [12, 15, 36, 40]. In their case, these were obtained at a more negative potential than at the potential yielding a spectrum similar to the reference curve (pH 1.45). The potential difference involved is approximately a few tenths of a volt. Therefore, we also looked at PEDOT:PSS solutions treated with a small amount of hydrazine (Figure 3B), a chemical reducing agent. Indeed, the absorbance in the NIR decreases, that around 800 nm stays more or less constant, and absorbance in the visible region increases, resembling the changes observed for higher pH. For fully reduced PEDOT:PSS we find a further increased absorbance in the visible region, in line with earlier results [12, 15].

It is not mere coincidence that 'de-doping' induced by pH and by electrochemical reduction gives rise to similar changes in absorption features. The same is seen in Raman spectroscopy. Spectra were obtained for complete PLED devices, and for PEDOT:PSS films on glass, giving identical results. Only contributions from PEDOT, and not from PSS, are visible owing to the resonance Raman effect, *i.e.* the occurrence of a markedly enhanced Raman scattering cross section when absorption to an electronically excited state is possible at (or close to) the excitation wavelength, for phonon modes coupled to that transition [41]. Accordingly, several peaks in the spectra for PEDOT:PSS at higher pH, with increased absorption in the visible region, had considerably higher intensity when excited at 633 nm, while acidic and neutralised PEDOT:PSS had similar intensity when excited at 785 nm.

Figure 4A shows spectra from 1150 cm^{-1} to 1650 cm^{-1}, where the strongest changes are observed. For comparison, spectra measured at positive and negative potential in a two-electrode solid-state electrochromic device are included in Figure 4B. The standard material shows a complex series of peaks between 1350 and 1550 cm^{-1}, with a characteristic overlapping set of bands around 1450 cm^{-1}. Fully oxidised PEDOT:PSS (measured at 1.2 V in the solid-state electrochromic device) shows similar features. However, PEDOT:PSS at the highest pH shows an entirely different shape in this region, characterised by three distinct sharp peaks at 1365, 1421 and 1516 cm^{-1}. This pattern is very similar to the signal measured at –1.2 V in the solid-state electrochromic device, *viz.* in the reduced, de-doped state. The similarity between spectra at high pH and at a lower electrochemical potential also holds for bands outside the spectral region shown. For example, a sharp peak at 979 cm^{-1} and a broader, smaller peak at 1823 cm^{-1} are only observed for these samples. Spectra at intermediate pH can be described by a linear combination of the two extremes.

Lefrant and co-workers have performed extensive Raman studies of PEDOT at various doping levels [40], and their work is taken as a framework for more detailed discussion. The peak positions found here are sufficiently close to their values for PEDOT and poly-thiophenes to allow peak assignment [42], as studies on chemically

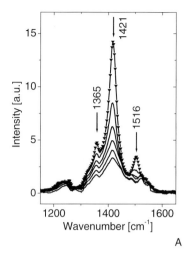
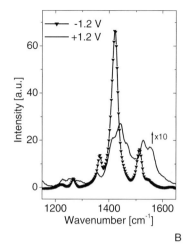

A

B

Figure 4 Raman spectra (633 nm excitation) of 1:20 PEDOT:PSS films as a function of pH (A, left), and as a function of oxidation level (B, right). In Fig. A the pH increases from bottom to top as 1.4 (reference), 2.0, 3.0, 5.0, 6.85 and 9.9. In Fig. B the line curve was obtained at +1.2 V (intensity multiplied by 10 for the sake of clarity), and the symbol curve at −1.2 V. The device build-up was glass/ITO/PEDOT:PSS/electrolyte/PEDOT:PSS/ITO/glass. Confocal microscopy was used to get data from one PEDOT:PSS layer only.

polymerised PEDOT:PSS are, unfortunately, not available to the best of our knowledge. Up to about 1200 cm^{-1} peaks correspond to deformation and bending modes, mainly associated with the dioxyethylene ring (not shown). Between 1200 and 1400 cm^{-1}, peaks have been assigned to C-C stretch vibrations, and between 1400 and 1600 cm^{-1} to C=C stretch vibrations, making them particularly sensitive to the presence and degree of (de-)localisation of holes. The results for the electrochromic device are in good agreement with data by Garreau, Lefrant et al. [40, 42], and by other workers [36, 37]. De-doped electropolymerised PEDOT showed the same set of peaks as found here. The peak at 1365 cm^{-1} is assigned to the C_β-C_β stretching deformation; that at 1421 cm^{-1} to the symmetric C=C in-plane stretch vibration, and that at 1516 cm^{-1} to the asymmetric C=C in-plane stretch vibration. Oxidised material showed a broadened band around 1430 cm^{-1}, shifted to higher wavenumber compared to the de-doped state, in line with present data. The peak at 1516 cm^{-1} was absent in the oxidised samples, and a new peak at 1546 cm^{-1} appeared. Generally, peak positions shifted to higher wavenumber with increased doping level. This effect was linked to the degree of backbone deformation during the oxidation steps forming polarons and bi-polarons, and associated transitions between quinoid- and benzoid-dominated structures.

We conclude that literature data are consistent with our spectra, and confirm the resemblance between electrochemically de-doped PEDOT:PSS and PEDOT:PSS at neutral pH. Kim and co-workers also reported similar spectra for doped and de-doped PEDOT:PSS in their study on black spots [43]. It should be noted that, because of the resonance Raman effect, optical absorbance and Raman spectroscopy actually look at the same part of the PEDOT:PSS system, *i.e.* the conjugated thiophene backbone. As electrical conductivity is also related to the conjugation, its pH dependence was also studied.

ITO/PEDOT:PSS/Au devices were used. Linear current-voltage characteristics are found at low voltage, from which the conductance was calculated. It decreased by a factor of about 3 when going to pH 7 for 1:6 and 1:20 PEDOT:PSS. This decrease is markedly smaller than that reported by others. Aleshin *et al.* [44] report a difference of two orders of magnitude for PEDOT:PSS. For so-called self-doped PEDOT, bearing a sulphonic acid side chain on the ethylene glycol ring, a similar decrease was found upon addition of Bu_4NOH [45]. Electrochemically de-doped PEDOT also shows a $>10^3$-fold lower conductance than its doped counterpart [37, 46]. These results all pertain to highly conductive (>1 S/cm) PEDOT. Conductivity in PLED-grade PEDOT:PSS is limited by hopping between PEDOT chains through the PSS matrix, as outlined in the introduction section, and not by the intrinsic mobility of holes along the PEDOT backbone. Hence, a significantly smaller effect of pH on conductivity can be expected for PLED-type low-conductive PEDOT:PSS.

16.3.2
Surface properties

Figure 5 shows three typical examples of S(2p) and O(1s) core-level XPS spectra. From the bottom to the top the Na^+ content increases from pH 1.4 to pH 1.6, and finally to pH 6.8. All spectra have been deconvoluted following the interpretation of Greczynski *et al.* [23,24].

The S(2p) spectra exhibit two peaks at 169 eV and 165 eV. The former consists of the spin-split doublet $S(2p_{1/2})$ and $S(2p_{3/2})$ of sulfur in PSS-H (labeled '3') and in PSS-Na (labeled '2'). The spin splitting is 1.2 eV and the shift from PSS-H to PSS-Na is 0.5 eV, in good agreement with the 0.4 eV observed by Greczynski *et al.* The PSS-H signal is entirely replaced by the PSS-Na signal at pH 9, as expected. The 165 eV peak consists of the spin-split doublet of sulfur atoms in PEDOT (labeled '1'). The asymmetric tail used by Greczynski *et al.* for fitting this peak has been neglected.

The measured atomic ratio of PEDOT-sulfur to PSS-sulfur for standard 1:6 PEDOT:PSS (curve *a*) is 1:7, corresponding to a weight ratio of 1:9. Thus, PEDOT/PSS is enriched in PSS at the surface, as has also been observed by Greczynski [23, 24]. The degree of enrichment is somewhat smaller here. This may be related to the lower PEDOT:PSS bulk ratio (1:1.25) used in [23–25]. Curves *b* and *c* in Figure 5 were obtained for 1:20 PEDOT:PSS. Hence, the signal from PEDOT-sulfur was very weak in these samples. The extent of surface segregation could not be determined accurately, but was in line with a similar degree of segregation as found with 1:6 PEDOT:PSS.

Figure 5B shows the O(1s) spectra. The spectrum for the lowest Na^+ content (curve *a*) shows contributions from PEDOT (labeled '1') and PSS-H (labeled '3'). PSS-H exhibits two types of oxygen bonding sites: oxygen in the S=O group, and hydroxyl oxygen at higher binding energy. The contribution of oxygen from PSS-Na

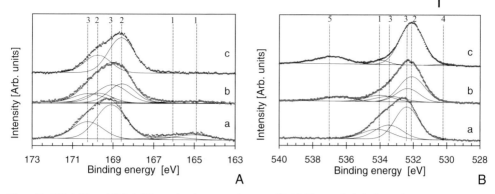

Figure 5 S(2p) (A) and O(1s) (B) core level spectra of three PEDOT-PSS films with different amounts of Na$^+$. From the bottom to the top the pH increases from 1.3 (*a*) to 1.6 (*b*), and finally to pH 6.8 (*c*). The first film has a nominal PEDOT:PSS weight ratio of 1:6, the other two of 1:20. The symbols indicate the experimental results while the solid lines represent the deconvoluted contributions. The numbers indicate: 1: O or S from PEDOT, 2: O or S from PSS-Na, 3: O or S from PSS-H, 4: O from NaOH, and 5: Na(KLL)-Auger line.

(labeled '2') increases with Na$^+$ content. An additional minor peak (labeled '4') is observed at high pH (*c*), which is ascribed to oxygen in NaOH. This confirms that the transition of the acidic form to the Na-salt has been completed and some excess NaOH remains present. Finally, the Na(KLL) Auger line is observed in the spectrum (labeled '5'). Inspection of the sulfur and oxygen spectra shows that significant segregation of PSS-H *vs* PSS-Na does not occur.

UPS was used to determine the ionization energy and work function. The UPS spectra of PEDOT-PSS films (not shown) are dominated by PSS features, except in the region close to the Fermi level [24]. The signal intensity between 2 and 0 eV binding energy is extremely low and originates from PEDOT as pointed out earlier by Greczynski *et al.* [23, 25]. Figure 6 shows the variation of ionization energy and work function with the pH. While the ionization energy is hardly affected, the work function strongly decreases with increasing pH, *i.e.* the UPS spectra shift to higher binding energy.

The difference between the ionization energy and the work function equals the gap between HOMO level and Fermi energy. For higher pH, the HOMO is clearly separated from the Fermi level. The material exhibits a semiconducting state. The situation is different at low pH. The Fermi level approaches the HOMO level around 5.0 eV and eventually coincides: the density of states extends up to zero binding energy and a Fermi edge is observed. Under these conditions the HOMO is better named valence band to stress the delocalised nature of the charge carriers. These PEDOT-PSS layers with very low pH exhibit metallic character and the ionization energy and the work function are identical. The transition from the metallic to the semiconducting state takes place at approximately pH 1.5 (corresponding to about 2 atom% Na$^+$ at the surface according to XPS). A more detailed discussion of the UPS and XPS results is given in a separate paper [47].

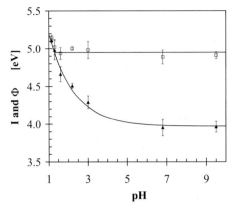

Figure 6 Ionization energy I (□) and work function ϕ (▲) as a function of pH. The error bars indicate the total variation in the ionization energy and work function. The solid lines are guides to the eye.

It is interesting to compare the UPS work function at low Na^+ content with literature values. Obviously, the Na^+ content has to be taken into account. For example, the value quoted by Salaneck and co-workers in [23] of 4.7 eV for PEDOT:PSS 1:1.25 samples is lower than what is found here, but was measured for material with one Na atom per two S atoms at the surface. In another paper, the work function for de-doped neutral PEDOT:PSS was given at 4.0 eV, and that of tosylate-doped conducting PEDOT at 4.4±0.2 eV [48]. Others have reported a value between 4.8 and 5.15 eV for 1:6 PEDOT:PSS [49], and ascribed the variation to changes in the surface composition owing to slight differences in preparation conditions and possible contamination. Recently, consensus about a value of 5.1 eV seems to have been reached [50], in good agreement with the present value for Na^+-poor, low pH PEDOT:PSS.

The electrochemical equilibrium potential of commercial PEDOT:PSS, which equals the Fermi level for a conductor, is approximately 0.3 V *vs* Ag/AgCl [29, 51]. This is equivalent to about 5.0 eV *vs* vacuum. The close agreement with the UPS values points to a small dipole contribution to the work function. Accordingly, PEDOT:PSS with various PSS excess (1:1.25, 1:6, 1:20) shows the same work function.

Finally, the UPS results were corroborated by Kelvin probe experiments. These were done in air under ambient conditions to determine the applicability of the UPS findings to PEDOT:PSS in standard PLED device processing. The decrease measured at pH 7 amounted to about 0.3 eV. The difference with the UPS data is ascribed to the effects of moisture, oxygen and organic contaminants, changing in particular the surface properties of the PEDOT:PSS layer.

16.3.3
Device performance

Hole-only devices (Au cathode) and double-carrier devices (Ba/Al cathode) were used. Hole-only devices are particularly useful to study hole injection from various PEDOT:PSS materials. A yellow-green emitting PPV, a blue poly-spirobifluorene polymer and a green poly-fluorene polymer were chosen as representative examples of LEPs. Data at low and high voltage were measured in *dc* and pulsed mode, respectively, using different set-ups, to avoid heating of the sample at high current density. This can introduce a discontinuity in the curves at some intermediate voltage, and explains the inclusion of two curves for the same device (see graphs). Graphs combining hole-only and double-carrier results are plotted *vs* bias, *i.e.* they have been corrected for the built-in voltage to allow more direct comparison.

IV-characteristics for both types of device with PPV are shown in Figure 7A. The hole-only current decreases markedly for higher pH: at pH 6 (not shown), the measured current is another two orders of magnitude lower than at pH 2, and dominated by leakage. With standard PEDOT:PSS and in the absence of PEDOT:PSS, a space-charge limited current (SCLC) is obtained [52]. As the bulk mobility of holes is, therefore, not affected by the PEDOT:PSS layer, the lower current in hole-only devices for the higher pH PEDOT:PSS formulations must be due to the presence of a hole injection barrier.

The double-carrier current also decreases for higher pH, in particular at lower voltage, but to a remarkably smaller extent than the hole-only current. The ratio of the current in a double carrier device to that in a hole-only device thus increases markedly when going from standard PEDOT:PSS to higher pH. Figure 7B shows a marked decrease in efficiency for pH 6 PEDOT:PSS. This points to poor matching of electron and hole partial currents. Thus, it is proposed that the samples with pH 6 PEDOT:PSS are actually electron-dominated devices. This is in line with the more than two orders of magnitude lower current at low bias with pH6 PEDOT:PSS, since a similar PPV derivative is known to possess much lower electron mobility than hole mobility [52]. Moreover, when using hydrogen plasma-treated ITO as hole injection layer [32], one also obtains an electron-dominated device. This showed similar current density as devices with pH6 PEDOT:PSS. We note that electron-aided injection of holes into PPV, which has been described in the literature [53], apparently does not take place at low bias in this case. In that study a wide variation in hole injection was observed in hole-only devices with various anodes, while such variation could be much less pronounced in double-carrier devices.

Lifetime tests were performed at 80 °C for double-carrier devices. Intermediate pH had only a small effect on lifetime. For neutral pH, lifetime showed more scatter and was generally somewhat lower than for reference PEDOT:PSS. It can be concluded that the performance of double-carrier devices based on this PPV is insensitive to the pH of PEDOT:PSS in the acidic region up to a pH of about 2, and that device efficiency at neutral pH (around 7) is strongly affected.

Results for the blue poly-spirofluorene are collected in Figure 8. Differences are seen already for a small increase in pH, *e.g.* from 1.4 to about 1.5 (1,500 ppm Na^+

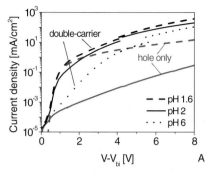
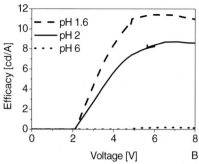

Figure 7 IVL-characteristics for devices based on a yellow PPV LEP using three types of 1:20 PEDOT:PSS with different pH. Fig. A gives current density *vs* bias for hole-only devices (dark grey curves) and double-carrier devices (black curves). The bias is given by the applied voltage minus the built-in voltage (about 0.4 V and 1.4 V for hole-only and double-carrier devices, respectively). Fig. B shows efficacy *vs* voltage for the double-carrier devices.

for 1:20 PEDOT:PSS, see Figure 2). Thus, here we adopt Na$^+$ content as a more discriminatory measure to describe PEDOT:PSS formulation. Standard PEDOT:PSS and PEDOT:PSS at intermediate Na$^+$ concentration are compared. The IV-characteristics in Figure 8A show that the current in hole-only devices is lowered by one to two orders of magnitude for the higher pH. Analogous to the case of PPV, this points to a lower hole injection rate. In double-carrier devices the current and efficiency are hardly affected by the choice of PEDOT:PSS, similar to the observation with PPV. This may be an indication that the current in a double-carrier device is dominated by electrons. Alternatively, electrons trapped at the PEDOT:PSS/LEP

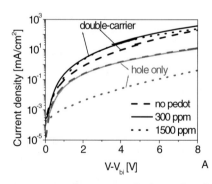
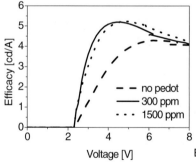

Figure 8 IVL-characteristics for devices based on a blue poly-spirobifluorene LEP using two types of 1:20 PEDOT:PSS. For comparison results for a device with no PEDOT:PSS are included. Fig. A gives current density *vs* bias for hole-only devices (dark grey curves) and double-carrier devices (black curves). The bias is given by the applied voltage minus the built-in voltage (about 0.7 V and 2.2 V for hole-only and double-carrier devices, respectively). Fig. B shows efficacy *vs* voltage. The double-carrier device with no PEDOT:PSS had a slightly larger LEP thickness, causing the lower current compared to the devices with 300 and 1,500 ppm Na$^+$.

interface may nullify any difference in hole injection, in line with electron-induced hole injection that was observed with blue poly-fluorenes [54]. The absence of any effect on efficiency (Figure 8B) favours the latter explanation, as an electron-domi-nated device would be expected to show poor efficiency, as seen with PPV using PEDOT:PSS at pH 6 in Figure 7B.

The devices with higher pH PEDOT:PSS showed a noteworthy reduction in life-time, measured at 20 °C. The voltage increase to maintain the constant stress cur-rent was also twice as high. This is an indication that hole injection becomes a prob-lem during the lifetime of these devices. This is confirmed by photoluminescence measurements. For devices with standard PEDOT:PSS a significant reduction in the photoluminescence (PL) intensity was observed after a lifetime stress experiment. Together with the spectral changes observed, this is an indication for bulk polymer deg-radation [55]. For the higher pH PEDOT:PSS, the PL intensity decrease was less and only minor spectral changes were observed, showing that the lifetime is limited by another mechanism, *viz.* hole injection. It can be concluded that double-carrier device characteristics and performance are more sensitive to pH and Na^+ content for blue poly-spirobifluorene based devices than for the PPV based devices discussed above.

The effect of the PEDOT:PSS composition was also tested for LUMATION™ Green 1300 Series poly-fluorene LEP. Similar to the blue poly-spirobifluorene based devices, the hole-only current, see Figure 9A, decays rapidly with increasing Na^+ concentration. For 1,000 ppm Na^+ the current has decreased by two orders of magni-tude, again pointing to the presence of a barrier to hole injection: standard PEDOT:PSS and a device without PEDOT:PSS (injection of holes by ITO), showed space charge limited behaviour. The lower current is not or hardly observed in dou-ble carrier devices. Similar to the devices discussed above, this can be explained by an electron-dominated current or by electron trap states at the PEDOT:PSS/LEP interface facilitating efficient hole injection.

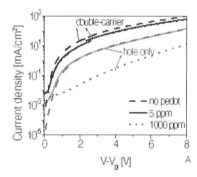 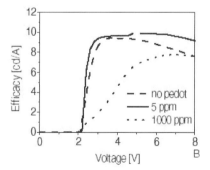

Figure 9 IVL-characteristics for devices based on LUMATION™ Green 1300 Series poly-fluor-ene LEP using two types of 1:6 PEDOT:PSS with different pH. For comparison results for a device with no PEDOT:PSS are included. Fig. A gives current density *vs* bias for hole-only devices (dark grey curves) and double-carrier devices (black curves). The bias is given by the applied voltage minus the built-in voltage (about 0.6 V and 2.0 V for hole-only and double-carrier devices, respectively). Fig. B shows efficacy *vs* voltage for the double-carrier devices.

The efficacy-voltage curve depends on the PEDOT:PSS composition (Figure 9B). At low Na^+ concentration the efficacy rises rapidly above 2.2V. The increase of the efficacy is much slower at high Na^+ concentration. The difference disappears at high bias. The combined results suggest that limited hole injection occurs at low bias and that the hole content in an operating device determines the efficiency, in line with the idea that electrons are probably the dominant charge carrier.

Lifetime measurements were performed at 80 °C. The best lifetime of a few hundred hours was observed at the lowest Na^+ concentration. Lifetime is slightly shorter at a few hundred ppm Na^+. Around 1,000 ppm Na^+ lifetime is reduced by a factor of three. By analogy to the results for the blue LEP, we ascribe this to development of a hole injection problem during electrical stress.

16.4
Discussion

16.4.1
Effect of NaOH on PEDOT:PSS properties

The influence of Na^+ content and pH were studied with four methods. Two of these, optical absorbance and Raman spectroscopy, are sensitive to the bulk properties of PEDOT:PSS. The other two, XPS/UPS and manufacture of PLED devices comprising PEDOT:PSS as hole injection layer, probe the surface and interface properties. The methods used have looked at PEDOT:PSS dispersions (optical absorbance) and films. The films were annealed in air and under argon (XPS/UPS). They were studied in contact with air (optical absorbance), nitrogen (Raman spectroscopy), vacuum (UPS/XPS) and various LEPs.

Despite these small variations in device structure or processing, both optical absorbance and Raman spectroscopy found results that are similar to those seen with electrochemical dedoping. UPS showed a decrease of the work function. XPS results did not show evidence of substantial Na^+-effects on surface segregation or of substantial segregation of Na^+ itself. Therefore, the observed changes in surface and interface characteristics can be considered a consequence of changes in the bulk properties. The UPS decrease in work function was corroborated by device data showing the existence of a barrier to hole injection.

In all experiments described in the previous section 1:6 and 1:20 PEDOT:PSS were investigated and very similar results were obtained. A single exception was the device work, where it was sometimes observed that the choice of PEDOT:PSS ratio had an impact on device performance and on the importance of the effect of NaOH addition. The influence of other cations than Na^+ was also investigated using LiOH and CsOH. With all methods similar results were found as with NaOH although, for example, the extent of the changes could be somewhat different at the same cation content or pH.

It can, therefore, be concluded that the various methods point to a single trend: the electronic and related optical properties change markedly when a significant amount of base is added to PEDOT:PSS. The combined observations actually fit in a picture of a transition, upon addition of NaOH, from a highly doped metallic state to a lower doping level semiconducting state. This transition is generic to PEDOT:PSS as it does, to a good approximation, not depend on the choice of cation, on the PEDOT:PSS ratio and on the actual device structure.

A first question is whether pH alone, or the combination of pH and cation is responsible for the changes. PSS-Na is probably dissociated in solution. Optical absorbance found similar changes for the dispersion and for a film. This indicates that Na^+ ions may not have to be present to induce the changes. However, PSS-Na ion pairing may still occur, and Na^+ can have an indirect effect on *e.g.* the packing of the PSS chains in an aqueous environment due to the change in ionic strength, making it difficult to draw definite conclusions. It is also important to realise that the discrimination between pH and Na^+ is no longer relevant in films where the key parameter is the PSS-Na/PSS-H ratio. Thus, we continue to refer to the key parameter as Na^+ content or pH, which is implied to mean pH or combined pH/Na^+ effect.

The central question in this discussion is thus how NaOH addition induces the transition from a doped, *i.e.* metallic state to a semiconducting state, which is akin to dedoping of PEDOT:PSS. PEDOT cannot be undoped by removal of protons, as is the case for poly-aniline (PANI). Figure 10 highlights the difference between both conducting polymers. For PEDOT dedoping an electrochemical reducing agent has to be present. However, NaOH and a higher pH do not constitute such a reducing agent. The electrochemical equilibrium potential of PEDOT:PSS is about 0.3 V *vs* Ag/AgCl [29,51]. The Nernst potential of water oxidation at pH 7 is 0.61 V *vs* Ag/AgCl, and higher still at lower pH [56]. Therefore, electrochemical reduction of PEDOT:PSS in the experiments discussed here is impossible according to thermodynamics, and true dedoping can not be the origin of the changes observed at higher pH.

Several papers have mentioned changes in PEDOT properties when the pH of the PEDOT:anion combination was increased. In [31], formation of a hole injection layer in a PLED after exposure of PEDOT:PSS to an amine base was reported. No attention was paid to the mechanism thereof, although it was referred to as dedoping. In [44] PEDOT:PSS 1:1.5 was modified with NaOH. This led to a two and three orders of magnitude lower conductivity at pH 4 and 12, respectively. The temperature dependence of the conductivity had also changed, and was no longer typical for an organic metal. An increased distance between conducting clusters was put forward as explanation. In self-doped PEDOT with a sulphonic acid attached to a side chain [45], the conductivity decreased from 0.5 S/cm to 10^{-3} S/cm after deprotonation by Bu_4NOH in acetonitrile. This was attributed to the bulkiness of the Bu_4N^+ cation. Finally, in [21] electropolymerisation of PEDOT was carried out in the presence of PSS-H and PSS-Na. The Na-salt led to an 80× lower conductance, whereas the conductance of PEDOT:tosylate-H and PEDOT:tosylate-Na was the same. The origin of the difference was not discussed.

Bulkiness of cations or separation of conducting domains upon pH increase is not a probable explanation in our case. First, 1:6 and 1:20 PEDOT:PSS already con-

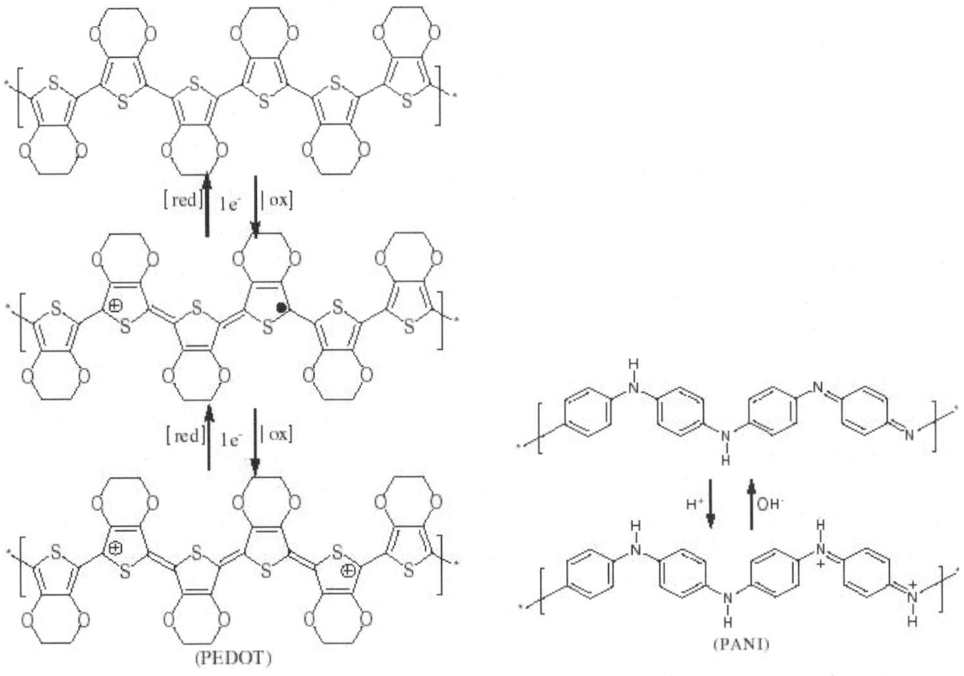

Figure 10 Comparison of doping processes in polyaniline (PANI) and PEDOT.

sist of individual conducting segments as outlined in the Introduction section and Figure 1. Second, such explanations are of a *macro*scopic nature, and fail to explain the observed changes in *e.g.* work function and Raman spectra, which are sensitive to the *micro*scopic properties of the individual PEDOT chains. Below we propose a different mechanism for the pH-induced changes, which is based on two aspects that form a recurrent theme in the literature on PEDOT: the presence of (bi-)polarons and the role of the surroundings, in particular the charge-compensating anion.

It is widely accepted that three oxidation states exist for PEDOT: a neutral, undoped state; a singly charged polaron state, and a doubly charged bipolaron state. Key to our discussion below is the notion of three states. We describe them in terms of the polaron model, but note that our conclusions are not altered when other models are applied to understand charge carriers and their transport in conducting polymers. The neutral state is probably of aromatic (benzenic) nature [57], while the charged states have significant quinoid character as evidenced by, for example, Raman spectroscopy [40]. These are also depicted in Figure 10. The conversion between the states can be clearly seen in cyclovoltammetry [58, 59] and by the existence of isosbestic points in spectroelectrochemistry [12, 15, 36, 40]. Accordingly, electron paramagnetic resonance (EPR) spectroscopy has shown the presence of a maximum concentration of unpaired spins (polarons) at intermediate doping level [14, 60]. For the present discussion it is of particular relevance that the relative stability of polarons and bipolarons depends on the chain length [59] and appears to be dependent on the environment, *i.e.* the solvent [40]. Furthermore, EPR has shown that polarons are present in PEDOT even at high doping level [60].

The relative abundance of polarons and bipolarons depends on the doping level. In fact, polarons and bipolarons have characteristic optical and Raman signals, bipolarons being generally associated with the fully doped state, and polarons with a partially dedoped state. The optical and Raman spectra at high pH resemble those of polarons. Thus, it is argued that the pH-induced changes are not due to true dedoping by electrochemical reduction but, instead, are caused by a redistribution of the positive charges over polaron and bipolaron states. The change in relative population of polarons and bipolarons reflects a change in the morphology of the PEDOT chains effected by the presence of Na^+ ions.

Several reports have expressed the important role of the surroundings, and in particular of the anions on the electronic properties of PEDOT. For example, the variable distance between the negative charge on the anion and the positive charge was used to explain results for a series of electropolymerised alkylated PEDOT [61]. The same argument was applied in [24] to PEDOT:PSS. The anion also appears to have a direct impact on the doping level [36], with higher levels being found for ClO_4^- than for PSS^- in electropolymerised PEDOT. Finally, the doping level decreases from 0.35 with PSS-H to 0.25 when PSS-Na is used instead in electropolymerisation of PEDOT films that were grown at the same electrochemical potential [21].

From these reports it is clear that the nature and distribution of countercharges determines the electronic properties of PEDOT to a considerable degree, in accordance with the strong effect of pH found here. The exact mechanism by which Na^+ ion changes the PEDOT surroundings remains, however, unclear. One hypothesis is that it acts on the conformation of the poly-electrolyte PSS which, in turn, determines the stabilisation of holes on PEDOT. For example, replacing PSS-H by PSS-Na increases the electron density in PSS since H^+ is more electronegative than Na^+. The ionic character is, therefore, higher for Na-O bonds than for H-O bonds. Na^+ ions can form multidentate bonds to sulphonate groups [21, 23], possibly inducing a weak ionic cross-linking between neighbouring PSS chains, similar to what has been seen with *e.g.* Mg^{2+} ions [12], and destroying a hydrogen bond-mediated cross-linking. Some of these mechanisms should act differently on a simple counterion such as tosylate and it would be interesting to investigate pH effects in more detail for such systems. Small differences in bound water content (solid state solvation) may also exist and affect the electronics properties. An alternative hypothesis is that small cations such as Na^+ are associated with the ether oxygens in the dioxyethylene substituent [36]. This would bring Na^+ ions in close proximity of the oxidised PEDOT moieties, with significant changes in PEDOT chain morphology. The planarity of the conjugated thiophene system is of particular interest in this respect.

Application of other methods than those used here will be needed to elucidate the role of Na^+ ion in more detail. For example, a technique such as EXAFS may be helpful since it can probe the local chemical environment. Calculation of the electronic properties of a PEDOT chain embedded in PSS-Na and PSS-H would also be very interesting.

16.4.2
Effect of NaOH addition on PLED device performance

In order to understand the influence of the PEDOT:PSS properties on hole injection, some information on the relevant energy levels is required. The PEDOT:PSS work function and ionisation potential were obtained by UPS. The position of the HOMO of the LEPs was derived from cyclovoltammetry [62, 63]. The value for the PPV and for the other green- and blue-emitting LEPs is 5.2-5.3 eV. The barrier for hole injection is given by the difference between the PEDOT:PSS Fermi energy and the LEP HOMO level. Obviously, no significant barrier exists with standard Na^+-free PEDOT:PSS with a work function of 5.1 eV. However, a large barrier is present for high pH PEDOT:PSS. This is in good agreement with the results shown in Figures 7–9 for the hole-only devices, where limited hole injection is, indeed, observed.

While the position of the HOMO is very close for all three LEPs studied, differences in the effect of Na^+ on hole injection and lifetime were observed with the various LEPs. This is related to neglect of possible additional important contributions to the interface energetics. For example, interfacial dipoles, chemical reactions between LEP and PEDOT:PSS, and creation of (filled) traps are not taken into account. A more detailed interpretation is thus not possible at present without undue speculation.

Another interesting result was the lowered lifetime with high pH PEDOT:PSS for some LEPs. This was ascribed to development of a hole injection problem during electrical stress. Devices with a high Na^+ concentration are thought to be more susceptible to this because hole injection is already relatively difficult for a fresh device, and may only be possible due to electrons trapped at the interface [53, 54]. Any change in the interface properties would, therefore, easily result in a barrier to hole injection. This hypothesis is in accordance with a recently proposed model for degradation, where electrons injected in the LEP reach the PEDOT:PSS/LEP interface, causing damage [64]. A paper that describes oxidation of amine units [65] during electrical stress, starting from the PEDOT:PSS interface, provides another example of a degradation mechanism that may be especially detrimental to high pH PEDOT.

Concluding, we have looked at a limited set of polymers, yet found significant impact of reasonably low Na levels (a few 100 ppm). From this we infer that subtle changes in interface characteristics can already be at work for <100 ppm Na. Other polymers may be even more sensitive to the Na level. Thus, it is interesting for all work in the area of PLEDs that uses PEDOT:PSS to test the sensitivity of results to variations in the Na level.

16.5
Conclusions

The Na^+ concentration and pH of aqueous PEDOT:PSS dispersions have a marked effect on the electronic and optical properties of spin-coated films. The properties change with increasing pH in a similar way as during electrochemical dedoping: the intensity of the IR absorbance, which is typical of highly doped PEDOT, decreases. The Raman peaks that probe the conjugation in PEDOT, shift, sharpen and increase in intensity. The work function measured by UPS decreases while the ionisation potential is unaffected. We argue that the mechanism of the pH-effect is, however, different from electrochemical dedoping since no reducing agent is added. Instead, it is proposed to originate from a change in the relative stability of polarons and bipolarons on the doped thiophene that is induced by the replacement of PSS-H with PSS-Na. The changes in the electronic properties of PEDOT:PSS point to the determining role of the counterion in the stabilisation of oxidised thiophene units.

As a consequence of the lower work function, high pH PEDOT:PSS introduces a barrier for hole injection with several classes of light-emitting polymers. Polymer LEDs comprising Na$^+$-rich, proton poor PEDOT:PSS can show lower lifetime and efficiency than the corresponding Na$^+$-free, proton rich devices. For light emitting polymers which suffer from the addition of sodium to the hole injecting PEDOT:PSS, the decreased lifetime hints at hole injection as limiting factor in the degradation of these PLEDs.

Acknowledgements

We would like to thank all project members in Philips Research Eindhoven (The Netherlands), Philips Research Aachen (Germany) and Philips Mobile Display Systems in Heerlen (The Netherlands). We also warmly thank our collaborators at Covion Organic Semiconductors GmbH, The Dow Chemical Company, and H.C. Starck for discussion and for supplying materials. This work was funded by the European Commission under GDR1-2000-25820 (POWERPLAY).

References

[1] E. I. Haskal, M. Büchel, P. C. Duineveld, A. Sempel and P. Van de Weijer, MRS Bull. **27**, 864 (2002)

[2] C. W. Tang and S. A. Van Slyke, Appl. Phys. Lett. **51**, 913 (1987)

[3] J. H. Burroughes, D. D. C. Bradley, A. R. Brown, R. N. Marks, K. Mackay, R. H. Friend, P.L. Burn, and A.B. Holmes, Nature **347**, 539 (1990).

[4] G. Heywang, and F. Jonas, Adv. Mater. **4**, 116 (1992).

[5] H. Becker, H. Spreitzer, W. Kreuder, E. Kluge, H. Schenk, I. Parker, and Y. Cao, Adv. Mater. **12**, 42 (2000).

[6] M. T. Bernius, M. Inbasekaran, J. O'Brien, and W. Wu, Adv. Mater. **12**, 1737 (2000).

[7] H. Becker, A. Büsing, A. Falcou, S. Heun, A. Parham, P. Stoessel, H. Spreitzer, K. Treacher, and H. Vestweber, Proceedings of the International Society for Optical Engineering (SPIE) (USA), Meeting San Diego 2001, **4464**, 49 (2002).

[8] D. Braun, and A. Heeger, Appl. Phys. Lett. **58**, 1982 (1991).

[9] A. Berntsen, Y. Croonen, C. T. H. F. Liedenbaum, H. Schoo, R.-J. Visser, J. Vleggaar and P. van de Weijer, Opt. Mater. **9**, 12 (1998).

[10] S. Karg, J. C. Scott, J. R. Salem, and M. Angelopoulos, Synth. Met. **80**, 111 (1996); S. A. Carter, M. Angelopoulos, S. Karg, P. J. Brock, and J. C. Scott, Appl. Phys. Lett. **70**, 2067 (1997); J. C. Scott,

S. A. Carter, S. Karg, and M. Angelopoulos, Synth. Met. **85**, 1197 (1997); Y. Cao, G. Yu, C. Zhang, R. Menon, and A. J. Heeger, Synth. Met. **87** 171 (1997); J. C. Carter, I. Grizzi, S. K. Heeks, D. J. Lacey, S. G. Latham, P. G. May, O. Ruiz de los Paños, K. Pichler, C. R. Towns, and H. F. Wittmann, Appl. Phys. Lett. **71** 34 (1997); I. D. Parker, Y. Cao, and C. Y. Yang, J. Appl. Phys. **85** 2441 (1999).

[11] Commercial information from H. C. Starck can be found at www.hcstarck.com

[12] L. Groenendaal, F. Jonas, D. Freitag, H. Pielartzik, and J. R. Reynolds, Adv. Mater. **12**, 481 (2000).

[13] S. Ghosh, and O. Inganäs, Synth. Met. **101**, 413 (1999).

[14] L. Groenendaal, G. Zotti, P.-H. Aubert, S. M. Waybright, and J. R. Reynolds, Adv. Mater. **15**, 855 (2003).

[15] H. W. Heuer, R. Wehrmann, and S. Kirchmeyer, Adv. Funct. Mater. **12**, 89 (2002).

[16] F. Jonas, EP 686662 A2; F. J. Touwslager, N. P. Willard, and D. M. de Leeuw, Synth. Met. **135-136**, 53 (2003).

[17] J. Y. Kim, J. H. Jung, D. E. Lee, and J. Joo, Synth. Met. **126**, 311 (2002); F. Louwet, L. Groenendaal, J. Dhaen, J. Manca, J. Van Luppen, E. Verdonck, and L. Leenders, Synth. Met. **135–136**, 115 (2003).

[18] G. Zotti, S. Zecchin, G. Schlavon, F. Louwet, L. Groenendaal, X. Crispin, W. Osikowicz, W. Salaneck, and M. Fahlman, Macromol. **36**, 3337 (2003).

[19] L. A. A. Petterson, S. Ghosh, and O. Inganäs, Org. Electronics **3**, 143 (2002).

[20] A. Elschner, F. Jonas, and S. Kirchmeyer, K. Wussow, in: Proceedings of the Asia Display IDW 2001, 1427–1429 (2001)

[21] G. Zotti, S. Zecchin, G. Schiavon, F. Louwet, L. Groenendaal, X. Crispin, W. Osikowicz, W. Salaneck, and M. Fahlman, Macromolecules **36**, 3337 (2003); X. Crispin, S. Marciniak, W. Osikowicz, G. Zotti, A. W. Denier van der Gon, F. Louwet, M. Fahlman, L. Groenendaal, F. de Schryver, and W. R. Salaneck, J. Pol. Science Part B: Polymer Physics **41**, 2561 (2003).

[22] ref. 22 in J. Lu, N. J. Pinto, and A. G. MacDiarmid, J. Appl. Phys. **92**, 6033 (2002); M. Dietrich, J. Heinze, G. Heywang, and F. Jonas, J. Electroanal. Chem. **369**, 87 (1994).

[23] G. Greczynski, Th. Kugler, M. Keil, W. Osikowicz, M. Fahlman, and W. R. Salaneck, J. Electron. Spectrosc. Relat. Phenom. **121**, 1 (2001)

[24] G. Greczynski, Th. Kugler, and W. R. Salaneck, Thin Solid Films, **354**, 129 (1999)

[25] G. Greczynski, Th. Kugler, and W. R. Salaneck, J. Appl. Phys. **88**, 7187 (2000).

[26] F. Jonas, W. Krafft, and B. Muys, Macromol. Symp. **100**, 169 (1995)

[27] M. M. de Kok and S. H. P. M. de Winter, unpublished results

[28] A. Ricksen, H. Bechtel, and E. A. Meulenkamp, unpublished results.

[29] M. Gross, D. C. Müller, H.-G. Nothofer, U. Scherf, D. Neher, C. Bräuchle, and K. Meerholz, Nature **405**, 661 (2000); H. Frohne, D.C. Müller, and K. Meerholz, ChemPhysChem. **8**, 707 (2002).

[30] K. Book, H. Bässler, A. Elschner, and S. Kirchmeyer, Org. Electronics **4**, 227 (2003).

[31] J. M. Bharathan, and Y. Yang, J. Appl. Phys. **84**, 3207 (1998).

[32] A. van Dijken, A. Perro, E. A. Meulenkamp, and K. Brunner, Org. Electronics **4**, 131 (2003).

[33] H. J. Bolink, M. Büchel, B. Jacobs, M. M. de Kok, M. Ligter, E. A. Meulenkamp, S. Vulto, and P. van de Weijer, Proceedings of The International Society for Optical Engineering (SPIE) (USA), Meeting Seattle August 2002, **4800**, 1 (2003).

[34] Commercial information from Covion Organic Semiconductors can be found at www.covion.com.

[35] Commercial information from The Dow Chemical Company can be found at www.lumation.com.

[36] J. C. Gustafsson, B. Liedberg, and O. Inganäs, Solid State Ionics **69**, 145 (1994).

[37] M. Lapkowski, and A. Prón, Synth. Met. **110**, 79 (2000).

[38] S. C. J. Meskers, J. K. J. van Duren, and R. A. J. Janssen, Adv. Funct. Mat. **13**, 805 (2003).

[39] Y. Chang, K. Lee, R. Kiebooms, A. Aleshin, and A. J. Heeger, Synth. Met. **105**, 203 (1999).

[40] S. Garreau, G. Louarn, J. P. Buisson, G. Froyer, and S. Lefrant, Macromolecules **32**, 6807 (1999); S. Garreau, J. L. Duvail, and G. Louarn, Synth. Met. **125**, 325 (2002).

[41] B. Schrader (Ed.), Infrared and Raman Analysis, VCH, Weinheim 1995, Ch. 4.8, pp. 372–410.

[42] A. Pron, G. Louarn, M. Lapkowski, M. Zagorska, J. Glowczyk-Zubek, and S. Lefrant, Macromolecules **28**, 4644 (1994); S. Garreau, G. Louarn, G. Froyer, M. Lapkowski, and O. Chauvet, Electrochim. Acta **46**, 1207 (2001)

[43] J.-S. Kim, P. K. H. Ho, C. E. Murphy, N. Baynes, and R. H. Friend, Adv. Mater. **14**, 206 (2002).

[44] A. N. Aleshin, S. R. Williams, and A. J. Heeger, Synth. Met. **94**, 173 (1998).

[45] G. Zotti, S. Zecchin, G. Schiavon, and L. Groenendaal, Macromol. Chem. Phys. **203**, 1958 (2002).

[46] T. Johansson, L. A. A. Petterson, and O. Inganäs, Synth. Met. **129**, 269 (2002); M. C. Morvant and J. R. Reynolds, Synth. Met. **92**, 57 (1998)

[47] C. H. L. Weijtens, V. van Elsbergen, M. M. de Kok, S. P. H. M. de Winter, Org. Electr. to be published 2005.

[48] K. Z. Xing, M. Fahlman, X. W. Chen, O. Inganäs, and W. R. Salaneck, Synth. Met. **89**, 161 (1997).

[49] N. Koch, A. Kahn, J. Ghijsen, J.-J. Pireaux, J. Schwartz, R. L. Johnson, and A. Elschner, Appl. Phys. Lett. **82**, 70 (2003).

[50] A. J. Mäkinen, I. G. Hill, R. Shashidlar, N. Nikolov, and Z. H. Kafafi, Appl. Phys. Lett. **70**, 557 (2001); T. M. Brown, J.-S. Kim, R. H. Friend, F. Cacialli, R. Daik, and W. J. Feast, Appl. Phys. Lett. **75**, 1679 (1999).

[51] F. Zhang, A. Petr, and L. Dunsch, Appl. Phys. Lett. **82**, 4587 (2003).

[52] P. W. M. Blom and M. J. M. de Jong, IEEE J. Sel. Top. Quantum Electron. **4**, 1077 (1998) .

[53] T. van Woudenbergh, P. W. M. Blom, and J. N. Huiberts, Appl. Phys. Lett. **82**, 985 (2003) .

[54] D. Poplavskyy, J. Nelson, and D. D. C. Bradley, Appl. Phys. Lett. **83**, 707 (2003).

[55] S. I. E. Vulto, M. Büchel, P. C. Duineveld, F. Dijksman, M. Hack, M. Kilitziraki, M. M. de Kok, E. A. Meulenkamp, J.-E. Rubingh, P. van de Weijer, and S. H. P. M. de Winter, Proceedings of The International Society for Optical Engineering (SPIE) (USA), Meeting San Diego August 2003, in the press (2004).

[56] A. J. Bard, R. Parsons, and J. Jordan (Eds.), Standard Potentials in Aqueous Solution, Marcel Dekker, Inc., New York 1985

[57] A. Dkhissi, F. Louwet, L. Groenendaal, D. Beljonne, R. Lazzaroni, and J. L. Brédas, Chem. Phys. Lett. **359**, 466 (2002).

[58] X. Chen, and O. Inganäs, J. Phys. Chem. **100**, 15202 (1996).

[59] J. J. Apperloo, L. Groenendaal, H. Verheyen, M. Jayakannan, R. A. J. Janssen, A. Dkhissi, D. Beljonne, R. Lazzaroni, and J.-L. Brédas, Chem. Eur. J. **8**, 2384 (2002).

[60] A. Zykwinska, W. Domagala, A. Czardybon, B. Pilawa, and M. Lapkowski, Chem. Phys. **292**, 31 (2003).

[61] L. Groenendaal, G. Zotti, and F. Jonas, Synth. Met. **118**, 105 (2001).

[62] D. M. Welsh, unpublished results; M. Bernius, M. Inbasekaran, E. Woo, W. Wu, and L. Wujkowski, J. Mat. Sci.: Mat. in Electronics, **11**, 111 (2000).

[63] H. Spreitzer and S. Heun, unpublished results.

[64] J. Burroughes, Society for Information Display (SID), 23rd International Display Research Conference (IDRC), Septmeber 2003, Phoenix, Arizona, USA, Tutorial on "Progress in Light Emitting Polymer Technology", pp. T-1/27-1/97.

[65] J.-S. Kim, P. Ho, C. Murphy, I. Grizzi, A. Seeley, M. Leadbeater, J. Burroughes, and R. Friend, in Proceedings of the 11th International Workshop on Inorganic and Organic Electroluminescence & International Conference on the Science and Technology of Emissive Display and Lighting, September 2002, Ghent, Belgium, Academic Press, pp. 551–553.

17

Insights into OLED Functioning Through Coordinated Experimental Measurements and Numerical Model Simulations

D. Berner, H. Houili, W. Leo, and L. Zuppiroli

17.1
Introduction

Flat panel displays play a key role in the design of many if not most modern electronic products today. Certainly the most "visible" component of any product, the display must meet an ever increasing range of requirements. As the communication interface between man and machine, the display is called upon to communicate large quantities of information. Resolution, color, brightness, angle of view, response time are thus critical performance parameters. Today's electronic products are also becoming more and more portable, so that light weight and thinness are now also essential. Further, portability also implies many different environments which means increased reliability and resistance to temperature, humidity, etc. If the current trend continues, flexibility, i.e. "bendable" displays, will also become a major requirement in the future. And finally of course, there is low production cost, an all important overall factor imposed by increasing global competition.

The liquid crystal display (LCD) is currently the dominant technology. Although limited in terms of angle of view and brightness, the LCD continues to make impressive progress and is now replacing the cathode ray tube in more traditional mass market products such as desktop computer monitors and television. However newer display technologies are emerging and are also making rapid progress, in particular, the OLED (*O*rganic *L*ight-*E*mitting *D*evice) based on organic electroluminescent materials. Touting superior performance in terms of brightness, response time, viewing angle and potentially cheaper production cost, the OLED has been the subject of steadily increasing attention in both academic and industrial circles over the last 15 years. In this period, OLED technology has proven its feasibility and more recently has made commercial débuts in car radios, mobile telephones and digital cameras. Nevertheless, there is still a long way to go before the OLED can challenge the LCD. Indeed, many critical commercial and industrial issues such as cost, reliability and reproducibility are still open and far from resolution. As well, a detailed understanding of many fundamental processes is still missing. For example, the injection characteristics at the electrode interfaces or the voltage drop over the different organic layers. Knowledge of aging processes and light generation, especially at high temperatures and for the different colors is also far from being complete.

Physics of Organic Semiconductors. Edited by W. Brütting
Copyright © 2005 WILEY-VCH Verlag GmbH & Co. KGaA, Weinheim
ISBN 3-527-40550-X

Some lifetime limiting factors such as a low glass transition temperature, anode surface quality, oxygen and water contamination are already known [1-3], for example, but the underlying mechanisms are only partially understood. Dye doping is also known as a way to increase lifetime [4,5] or to tune color [6]. Indeed, concentration and current dependent color changes have been experimentally observed for a partially doped emission layer and associated with solvatation effects or inefficient energy transfer from host to dopant [7]. However, general considerations of the effects of partial doping on the transport characteristics including the microscopic light generation mechanisms have not been addressed.

Except for very special cases, there are no analytic solutions to the equations describing the functioning of bipolar organic devices. For this reason, recourse is being made to simulation models of various types and complexity in the hope of gaining further insight into the basic mechanisms of device operation. For example, numerical simulation models based on master equations have been developed by several different groups. These models are generally one-dimensional and include among other things; injection mechanisms at the electrodes and field dependent mobility for both single layer [8–10] and bilayer devices [11–14]. Theoretical studies and Monte Carlo simulations of the organic/organic interfaces, e.g. injection and recombination, have also been considered by Bässler and co-workers [15, 16]. An overall review of the most commonly used transport models is presented by Walker et al. [17].

In this article, we will concentrate on our own "MOLED" program, a comprehensive modular structured device simulation model which was developed, in fact, to facilitate the process of OLED device optimization. This process is generally long and tedious if the usual systematic experimental approach is used but can be significantly shortened if a realistic simulation model is available. The source code for MOLED is currently available in a public software library [18–20].

"MOLED" is a complex model which considers many of the physical processes and mechanisms involved in charge transport through organic materials. Although the theoretical expressions describing these processes are used, many of the formulae depend on free input parameters which are not necessarily well-known beforehand. Crucial to the validity of the simulation, therefore, are the values used for these parameters. Our resolution to this problem was to design a series of experiments of increasing complexity, each focused on a particular aspect of OLED functioning and to simulate these experiments with MOLED, adjusting the appropriate parameters in the model to reproduce the results. At each experiment, a limited number of parameters could be determined and at the same time some interesting general information concerning OLED device processes could be gleaned. With each step also, experience was gained with handling the parameters and a feeling for their influence on the model obtained. In this way, the model was gradually developed so that in the end we could treat complex structures such as multi-layer devices with partially doped layers.

In the following sections, we will present the general procedure used for device fabrication in our experiments (Section 17.2) and describe the "MOLED" simulation model, its underlying architecture and the simulation parameters needed (Section

17.3). In Section 17.4, some applications and case studies are presented. We will discuss, in particular, results for the electric field distribution inside the device and the effect of selective charge trapping introduced by dye doping. We will outline how doping can be introduced into the device model, the possible mechanisms introduced and their effects, and finalize with a general picture of recombination zone evolution as a function of different dye parameters.

17.2
Device Fabrication

All our multi-layer light-emitting devices were fabricated by thermal evaporation in high vacuum ($< 5 \cdot 10^{-7}$ mbar) on indium tin oxide (ITO, AFC) glass substrates with a sheet resistance of <20 Ohm/square. The ITO substrates were first cleaned in ethanol, acetone and soap ultrasonic bathes.

The parent device architecture used as a starting point for our model simulations was composed as follows: ITO anode covered by a 10 nm thick copper phthalocyanine (CuPc) layer for hole injection improvement followed by a variable thickness hole-transporting layer of N,N'-diphenyl -N,N'-bis(1-naphthyl)-1,1'biphenyl-4,4''dia-mine (α-NPD) and a variable thickness electron transport layer of tris (8-hydroxyquinolinato) aluminium (Alq_3). All organics were purified by gradient sublimation before use and were thermally evaporated at a rate of about 1.0 Å/s. The top cathode was identical for all devices and was deposited in another chamber without breaking the vacuum. This consisted of a 0.8 nm thick LiF layer followed by a 75 nm layer of aluminium. The deposition rates were 0.2 Å/s and 10 Å/s for LiF and aluminium, respectively.

For the injection study, the experimental results for the metal/Alq3/metal structures from Ref. [21] were used. Additionally different device architectures were fabricated: ITO (indium-tin oxide) covered by a 200 nm thick Alq_3 layer finished with an Al electrode. LiF layers with thicknesses of 0.6, 1.2, and 2 nm were deposited right after Alq_3 in three of four devices, leaving the last as a reference device. Two types of hole-only devices were also fabricated: one with a 200 nm thick CuPc layer between the bottom ITO electrode and a top Au electrode and the other with an ITO substrate covered by 10 nm of CuPc followed by a 100 nm layer of α-NPD topped off with a Au electrode. A reference sample without the CuPc layer was also fabricated simultaneously [22].

For the internal electric field study, the experimental devices were fabricated on the same substrate, in the matrix geometry shown in Fig. 1. In this way, all devices were prepared simultaneously under the same conditions. Such an approach has already been used in a number of studies [23-29] aimed at the optimization of the OLED structure. Our matrix consisted of 4 x 4 segments, each segment having a surface area of 9 mm^2. A sliding shutter was used to achieve a systematic variation of the evaporated layer thickness over different rows and columns. Thus, among the rows the α-NPD layer thickness was varied (assuming values of 30, 40, 50 and 60 nm), while amongst the columns the thickness of the Alq_3 layer was varied

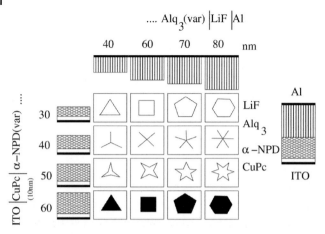

Figure 1 The architecture of the device is shown on the right side. Sixteen diodes are prepared simultaneously on the same glass substrate in the matrix geometry, as shown. The various α-NPD and Alq$_3$ layer thicknesses are schematically illustrated at the borders. Each diode is labelled by a separate symbol. In the following figures these symbols are used to denote the data related to the respective segments.

(40, 60, 70 and 80 nm). The electrodes were structured in the usual manner for the passive matrix operation to provide independently addressable segments: the ITO electrodes were arranged in a stripe configuration in one direction, prior to evaporation, while the Al electrode was evaporated through a mask with stripe openings in the perpendicular direction.

For the doping study the α-NPD layer thickness of the parent device was fixed at 40 nm and the Alq$_3$ layer thickness at 60 nm. The first 20 nm of the Alq$_3$ layer were used as a host with 1 % wt of a DCM derivative, i.e., either 4-(dicyanomethylene)-2-methyl-6-{2-[(4-diphenylamino)phenyl]ethyl}-4H-pyran (DCM-TPA) or 4-(dicyano-methylene)-2-methyl-6-[2-(julolidin-9-yl)phenyl]ethenyl]-4H-pyran (DCM-II). Additional doping concentrations of 0.3 %, 3 % and 8 % were used to study concentration effects. In another device, an additional 10 nm thick bathocuproine (BCP) hole blocking layer was evaporated after the 20 nm thick doped Alq$_3$ layer. The cathode was made in the same way as for the field study. All electroactive materials used were purchased commercially except for the doping molecule DCM-TPA, which was synthesized for this study [30]. The molecular structures of the two pyran derivatives used in this work are given in Fig. 2. The active area in these cases was 12 mm^2.

The completed devices were conveyed to a nitrogen glove box without being exposed to the ambient atmosphere. The current-voltage and emission characteristics were measured inside the glove-box using a Keithley 236 source-measure unit and the light emission was simultaneously recorded using a Grasby Optronics 247 calibrated radiometric photodiode. The absolute luminance was measured with a Minolta LS-110 luminance meter and the external quantum efficiency was obtained by using an integrating sphere. To measure the emission spectra, an Ocean Optics grating spectrometer was used. The spectra were corrected for CCD response.

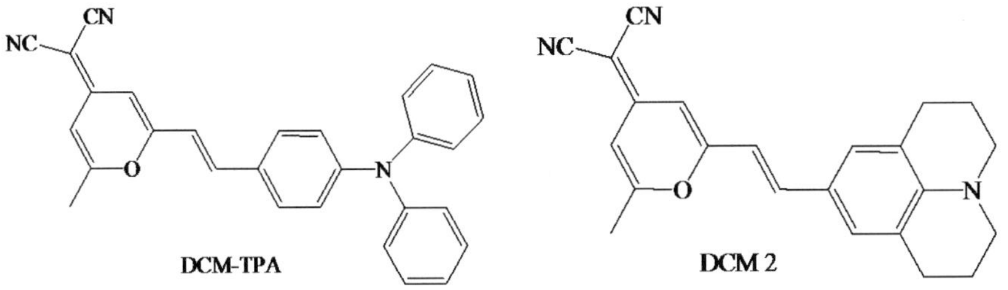

Figure 2 The structure of the dye molecule 4-(dicyanomethy-lene)-2-methyl-6-{2-[(4-diphenylamino)phenyl]ethyl}-4H-pyran (DCM-TPA) and 4-(dicyanomethylene)-2-methyl-6-[2-(julolidin-9-yl)phenyl]ethenyl]-4H-pyran (DCM 2) are presented.

For the electric field study, the resistivity contribution of the ITO lines on the I(U) curves was subtracted. This subtraction is important for the matrix geometry and for high performance (low impedance) diodes at higher current densities.

17.3
The Model

MOLED assumes a typical OLED architecture consisting of an organic multi-layered structure sandwiched between two planar, parallel electrodes. Since the distance between the two electrodes is small compared to the other two dimensions and the system is uniform in directions parallel to the electrodes, an effective one-dimensional simplification of the device is applied with the model axis perpendicular to the electrodes. This one-dimensional representation, of course, is advantageous for exploring various multilayer structures.

To describe the recombination of electrons and holes, the Langevin model is used – suitably generalized to cases where strong carrier density gradients are expected. However, three-dimensional effects are very pronounced close to the electrodes due to the image force potential which is very different for localized and smeared charges. MOLED thus solves the dynamics of electrons and holes in a one-dimensional geometry but includes three-dimensional effects close to the electrodes.

To simulate transport through multi-layer devices, MOLED defines discrete nodes along the model axis. In a broad sense, these nodes represent the average of molecular sites in the plane perpendicular to the axis. For example, the charge attributed to a particular node would represent the average charge in that particular molecular monolayer. Within the model only charge carrier hops between nearest neighbors are allowed. The hopping mechanism involves the microbalance formula, with an attempt-to-jump frequency derived from the imposed mobility law. The carrier dynamics at the electrodes is governed by the tunneling to molecular sites, as well as by the connection to the external battery which imposes the overall voltage drop between the electrodes. The reaction of the system is followed through a transient, until a steady state is established.

17.3.1
Coulomb effects

The charge distributions are represented by the number of carriers per site, i.e. n_m for the electrons and p_m for the holes, with $m = 1,..., N$. The surface electron densities at the electrodes are represented by n_0/A_1 and n_{N+1}/A_1, where A_1 stands for the surface that one molecule occupies in the plane perpendicular to the MOLED axis. The positions of the sites are denoted by x_m, $m = 1,..., N$. Additionally, $x_0 = 0$ and $x_{N+1} = L$ stand for the positions of the left (usually anode) and the right (usually cathode) electrodes, respectively. In an operating device the electron energy levels are influenced by the charge distribution within the organic layers (space charge effects) and by the electrodes (total charge at electrodes and image force potential). The discrete (3D) nature of the interaction of carriers localized at molecules is particularly pronounced in the molecular monolayers closest to the electrodes. For instance, it has been shown in Refs. [31, 32] that carriers may be trapped in the generalized image force potential and thus increase the Schottky barrier by several tenths of eV. These effects are small for the lesser charged monolayers far from the electrodes. The molecular energy levels are thus shifted according to

$$\tilde{E}_m = E_{0m} + E_{IF}(x_m) + \frac{q^2}{\varepsilon_0 \varepsilon_r A_1} \sum_{i=0...m, i \neq 1,N} (x_i - x_m)(n_i - p_i) - \frac{q^2}{\varepsilon_0 \varepsilon_r A_1} x_m (n_1 - p_1)$$

$$- \frac{q^2}{2\varepsilon_0 \varepsilon_r A_1} \left[\sqrt{(x_m - x_1)^2 + r^2} - \sqrt{(x_m + x_1)^2 + r^2} \right] (n_1 - p_1)$$

$$- \frac{q^2}{2\varepsilon_0 \varepsilon_r A_1} \left[\sqrt{(x_m - x_N)^2 + r^2} - \sqrt{(x_m - x_{RI})^2 + r^2} \right] (n_N - p_N) \tag{1}$$

$$+ \frac{q^2}{2\varepsilon_0 \varepsilon_r A_1} \left[\sqrt{x_N^2 + r^2} - \sqrt{x_{RI}^2 + r^2} \right] (n_N - p_N)$$

where E_{0m} is the bare LUMO(HOMO) energy level and $x_{RI} \equiv x_{N+1} + (x_{N+1} - x_N)$. The third and fourth terms stand for the Coulomb shift of the energy level caused by the homogeneously charged sheets at positions x_i, with charge density $q(n_i-p_i)/A_1$. The remaining terms represent the effect of discreteness within monolayers at $m=1$ and $m=N$. Here the discreteness in these monolayers is considered as a series of holes of radius r (see Ref. [19] for typical values of this parameter) within the uniformly charged sheets. Finally, $E_{IF}(x_m)$ is the image force potential given by

$$E_{IF}(x_m) = -\frac{q^2}{16\pi\varepsilon_0 \varepsilon_r x_m} - \frac{q^2}{16\pi\varepsilon_0 \varepsilon_r (L - x_m)}. \tag{2}$$

17.3.2
Equations of motion and injection

The equations of motion for the electron and hole distributions at the site m consist of three contributions. These are respectively: the injection of carriers from electrodes to nodes in the bulk, the hopping of carriers among organic molecules, and electron–hole recombination,

$$\frac{dn_m}{dt} = \left(\frac{dn_m}{dt}\right)_{inj} + \left(\frac{dn_m}{dt}\right)_{hop} + \left(\frac{dn_m}{dt}\right)_{rec},$$

$$\frac{dp_m}{dt} = \left(\frac{dp_m}{dt}\right)_{inj} + \left(\frac{dp_m}{dt}\right)_{hop} + \left(\frac{dp_m}{dt}\right)_{rec}. \tag{3}$$

The injection term arising from tunneling between the electrode and the mth layer is given by (for the left electrode in this example):

$$\left(\frac{dn_m}{dt}\right)_{inj} = g_L \left|J_{0,m}^{tun}\right|^2 \left\{ f\left(\tilde{E}_m - \mu_L\right) - n_m \right\} \tag{4}$$

where g_L is the density of states at the metal electrode and f is the Fermi–Dirac function. The tunneling integral $J_{0,m}^{tun}$ is given, in the WKB approximation, as

$$\left|J_{0,m}^{tun}\right|^2 = \gamma_L \cdot p^{(m-1)} \exp\left(-2I_{1,m}\tilde{E}_m\right) \tag{5}$$

where the factor γ_L corresponds to the tunneling integral between the metal and the first organic layer. The factor $p^{(m-1)}$ comes from intermolecular tunneling. The exponent

$$I_{r,s}(E) = \int_{x_r}^{x_s} dx \sqrt{\frac{2m_e}{\hbar^2} \left[\tilde{E}(x) - E\right] \theta\left[\tilde{E}(x) - E\right]} \tag{6}$$

reflects the overall slow variation of the position of the LUMO energy in space.

The equation of motion for the charge at an electrode includes the tunneling to molecular sites as well as the inflow of charge from the external circuit (i.e. battery),

$$\left(\frac{dn_0}{dt}\right) = -\sum_{m=1}^{N} g_L^{(n)} \gamma_L^{(n)} \cdot p_n^{(m-1)} \exp\left[-2I_{1,m}^{(n)}\tilde{E}_m^{(n)}\right] \left[f\left(\tilde{E}_m^{(n)} - \tilde{E}_0\right) - n_m\right]$$

$$+ \sum_{m=1}^{N} g_L^{(p)} \gamma_L^{(p)} \cdot p_p^{(m-1)} \exp\left[-2I_{1,m}^{(p)}\tilde{E}_m^{(p)}\right] \left[f\left(\tilde{E}_m^{(p)} - \tilde{E}_0\right) - p_m\right] + \left(\frac{dn_0}{dt}\right)_{bat}. \tag{7}$$

A similar equation applies to the right electrode. The last term in the equation is related to the current in the external circuit per unit OLED surface, J_{bat}, through the relation

$$J_{bat} A_1 = -\left(\frac{dn_0}{dt}\right)_{bat} = \left(\frac{dn_{N+1}}{dt}\right)_{bat}. \tag{8}$$

The term $(dn_0/dt)_{bat}$ is determined in such a way that at all times the difference between chemical potentials of the electrodes $\left(\tilde{E}_0, \tilde{E}_{N+1}\right)$ satisfies $qU = \tilde{E}_{N+1} - \tilde{E}_0$.

In the model, the injection related parameters are adjusted to the results from material specific experiments (section 17.4.1). Therefore, some charge injection effects due to the disordered nature of the material are implicitly taken into account.

Investigations of disorder related charge injection have been performed in detail by Bässler and coworkers [33, 34].

17.3.3
Transport in the organic material

The time evolution of the carrier density at site m in the bulk is determined by the hopping processes towards neighboring sites,

$$
\left(\frac{dn_m}{dt}\right)_{hop} = \Omega_m^{(n)}\left\{-n_m\exp\left[\left(\tilde{E}_m - \tilde{E}_{m-1}\right)/2kT\right] + n_{m-1}\exp\left[\left(\tilde{E}_{m-1} - \tilde{E}_m\right)/2kT\right]\right\}
$$
$$
- \Omega_{m+1}^{(n)}\left\{-n_{m+1}\exp\left[\left(\tilde{E}_m - \tilde{E}_{m+1}\right)/2kT\right] + n_m\exp\left[\left(\tilde{E}_{m+1} - \tilde{E}_m\right)/2kT\right]\right\}
$$

(9)

with a similar formula applying to holes. The hopping frequency $\Omega_m^{(n,p)} = \Omega_0(F, T)$ may vary in space because it may depend on the local electric field or may vary as the composition of the device changes. The microbalance formula used here usually represents a fundamental starting point for field-dependent mobility studies (see, for instance, Ref. [35]). However, the opposite standpoint is taken here. In MOLED device simulations, an effective hopping frequency $\Omega_0^{(n,p)}(F, T)$ is calculated from the mobility, preferably known from experiments, through the relation

$$
\mu_{n,p}(F, T) = \frac{a^2 q}{kT}\Omega_0^{(n,p)}(F, T)\frac{\sinh(Fqa/2kT)}{Fqa/2kT}.
$$

(10)

Here a stands for the inter-node distance, q is the carrier charge and F denotes the local electric field. Note that the change in mobility due to field and temperature may always be absorbed in $\Omega_0^{(n,p)}(F, T)$. Thus the introduction of any mobility law is very simple and requires coding only a few lines. In fact, four mobility formulae are already included in the code:

- symmetric detailed balance (Microbalance I) given by

$$
\mu(F, T) = \mu_0\frac{\sinh(Fqa/2kT)}{Fqa/2kT}
$$

(11)

- polaron mobility defined by μ_0 and the polaron binding energy W_0,

$$
\mu(F, T) = \mu_0\exp\left(-\frac{(qaF)^2}{8W_0 kT}\right)\frac{\sinh(Fqa/2kT)}{Fqa/2kT}
$$

(12)

- asymmetric detailed balance (Microbalance II) given by

$$
\mu(F, T) = \mu_0\frac{1-\exp(-Fqa/kT)}{Fqa/kT}
$$

(13)

- the Poole–Frenkel mobility law defined by the two phenomenological parameters, μ_0 and F_0,

$$\mu(F, T) = \mu_0 \exp\left(\sqrt{\frac{F}{F_0}}\right), \tag{14}$$

Formulae (11) and (12) verify the condition of symmetric detailed balance and are generally applied to the small polaron hopping case. Formula (12) reduces to (11) when the polaron binding energy W_0 is larger than the potential drop due to the field i.e: $W_0 \gg qaF$ and leads under this condition to a mobility that increases with increasing field. The opposite is found for Eq. (13), where the mobility decreases with increasing field. Formula (13) is adequate for weak charge-lattice coupling and satisfies the asymmetric detailed balance condition. However, when energetic disorder is present, which is generally the case in organic semiconductors, all three formulas (11–13) lead to the same qualitative behavior that

is described by a Poole-Frenkel law (Eq. 14). The latter is characterized by an increase of the mobility with field and is adjusted through the fit parameters μ_0 and F_0 [36] (see also section 17.4.2).

For most of the materials used in OLEDs, the experimental results (e.g. Time of flight) suggest a Poole-Frenkel type of mobility behavior. However some materials like CuPc shows less energetic disorder and the use of formula (11) or (13) is recommended instead. In addition, one can introduce alternative forms of mobility into MOLED such as the large polaron mobility formula relevant for polymers [37].

For further investigations in the height density and disorder strength regime the Boltzmann statistics in Eq. (9) should be replaced by Fermic-Dirac statistics. In fact, it has recently been shown that even for the relatively small charge density, as is the case in OLED's, the degenerate state of the carriers is attainable even at moderate energetic disorder strength [38, 39].

It is worth noting here that tunneling injection, combined with inter-site hopping (Fig. 3) accounts for two injection mechanisms, which are usually regarded as separate in literature. In fact, MOLED allows one to limit the tunneling distance from the electrodes into the organic material [19]. If only tunneling to the first monolayer is allowed then the carrier must climb the energy barrier, formed by the image force and the applied field. This corresponds to the thermionic or Richardson-Schottky injection mechanism. If the tunneling is not restricted to the first monolayer then additionally the Fowler-Nordheim tunneling injection is taken into account. However, MOLED does not include far-site tunneling at the organic/organic interface.

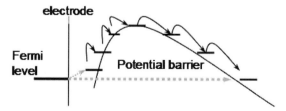

Figure 3   Tunnelling mechanisms at electrodes.

17.3.4
Recombination

A very simple way to picture the recombination of holes and electrons is to assume that it takes place whenever the attraction between an electron and a hole overcomes the electric forces produced by the external electric field and other charges. Then the change of electron density, $\rho_n(x_m) = n_m/aA_1$, due to recombination, may be written as

$$\frac{d\rho_n(x)}{dt} = -\left[J_n(x)\rho_p(x) + J_p(x)\rho_n(x)\right] s. \tag{15}$$

Here ρ_p stands for the hole density with J_n and J_p denoting the electron and hole particle current densities, respectively. The cross section for the capture (recombination) process, $s = \pi r_s^2$ depends on the local electric field,

$$\frac{1}{4\pi\varepsilon_0\,\varepsilon_r}\frac{q}{r_s^2} = \eta F(x). \tag{16}$$

The geometrical factor η is expected to be of the order of unity. In the case where the drift component of the current, $J_{n,p}^{drift} = \rho_{n,p}\mu_{n,p}(F)F$, is dominant over the diffusion component, the equation is nothing else than the Langevin recombination formula. However the presence of interface barriers in multilayer OLED's causes a build up of charge density which is in general not far from recombination regions and, as a consequence, the diffusion component of the current no longer remains negligible.

17.3.5
Numerical procedure

The charge configuration must satisfy the device neutrality condition $\sum_{i=0}^{N+1}(n_i + p_i) = 0$ as the simulation proceeds. It should be pointed out that, apart from this constraint, the steady-state charge configuration will not depend on the initial charge distribution. The calculation goes until the steady state is achieved for a given voltage. Usually, the code is used to calculate the steady states for a number of voltage points to get I(U) curves.

MOLED uses a split-time advance, divided for every time step into the advance of the hopping term, the charge injection term and the recombination of electrons and holes. Using the charge distribution at a particular time, MOLED evaluates the energy levels at every molecular site due to Coulomb effects (Eq. (1)). The charge supplied from the battery required to impose $qU = \tilde{E}_{N+1} - \tilde{E}_0$ is calculated. The time step is implemented using the implicit procedure for hopping and tunneling separately. The differential equation for the electron–hole recombination is solved exactly for the time step, at each node separately. The process of time-step advance is repeated until a stationary state is reached at a desired accuracy.

For simplicity, the dielectric constant was set to 3 for the whole device. In fact, dielectric constants of the organic materials used in OLED fabrication do not vary very much and are in the range of 2.5~3.5 [40]. This may lead to slight variations of the absolute values of the transport parameters in the injection and hole transporting layers as pointed out by Song et al. [41], but this will not be considered in our discussion.

To simulate a multi-layer OLED, the most important input parameters required are the HOMO and LUMO energy levels. We have therefore measured the HOMO and LUMO levels for the dopants, Alq₃ and α-NPD using a cyclovoltametric method. Our results are in good agreement with those found in the literature [42, 43]. The values for CuPc, a material difficult to dissolve in standard solvents, were taken from Refs. [44–46]. The most important features of the energy level arrangement in our devices were: *(i)* the barrier for electrons at the Alq₃/α-NPD interface which was on the order of 0.6 eV and much higher than the barrier for holes which cross the same interface in the opposite direction, and *(ii)* the barrier for holes at the CuPc/α-NPD interface which was higher [47] than that at the α-NPD/Alq₃ interface.

The data on electron and hole mobility are equally important and were taken from Refs. [47–52]. The main features of these data are: *(i)* the electron mobility of Alq₃ at 1 MV/cm is approximately $2*10^{-5}$ cm²/Vs, *(ii)* the mobility of holes in Alq₃ is approximately two orders of magnitude lower than the electron mobility in the same material, and *(iii)* the mobility of holes, in both CuPc and α-NPD, is at least ten times higher then the electron mobility in Alq₃. The mobilities, as obtained from TOF experiments, are strongly dependent on the electric field. In our device model this dependence is introduced through the field dependent hopping frequency [18] (see also section 17.3.3). In the present work the hopping frequencies at the CuPc/α-NPD and the α-NPD/Alq₃ interfaces were estimated as the geometric means of the hopping frequencies of the layers involved. For a summary, see Fig. 4.

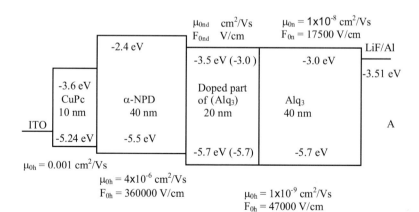

Figure 4 Energy levels and mobility data used in our numerical device simulations. The variabled is used as placeholders for the energy levels and mobility characteristics of the dopants, the values of which can be located inside the gap, equal to or outside that of the matrix. In our case The HOMO and LUMO levels are for DCM-II: –5.38; –3.39 and for DCM-TPA: –5.78; –3.48.

The effect of LiF layer on the Al cathode was taken into account through a shift in the electron injection energy barrier from Al to Alq$_3$ of the order of 0.5 eV. The tunneling integral between Al and Alq$_3$ was also reduced [53–55]. Details of the modeling of this improvement in charge injection will be presented below [53–55].

17.4
Case Studies

As we have seen, the model considers many processes and mechanisms and a rather elaborate set of parameters is required to describe them. We have found that a full parameterization of the model is best done progressively in several planned steps, each of increasing complexity. To begin, different metallic electrodes for monolayer devices were studied to get a feeling for the related parameter range and their effects, e.g., image force effects, residual energy barrier as well as dipole related injection improvements [52–55]. Then, step by step, the complexity of the devices was increased. Each time the results from an experiment were simulated, a certain number of parameters could be determined, but experience was also gained and the reliability of the model increased. Of course, such an evolutionary procedure also requires direct cooperation between the theoreticians and the experimentalists.

After considering the bipolar single layer device, a logical continuation was to study the field drop inside an organic bilayer device. Here it was useful to combine the theoretical simulations with a 2-dimensional combinatorial method which allowed the varying of two layer thicknesses at the same time [56, 57]. The next step of increasing complexity was to take into account partial doping of the emission layer. The doped part of the emission layer was simulated by the introduction of an additional layer into the parent diode architecture and a careful adjustment of the parameters.

17.4.1
Electrode effects

The first step in our approach was to simulate single-layer charge only devices. The first experiments were with Ca/Alq$_3$/Ca devices which enabled us to determine the first few model parameters, for example, the tunneling integral and the position of the first node. In a second step, bilayer devices with either an additional CuPc hole injection layer or with LiF on the cathode side were fabricated to determine further parameters. The numerical simulation of these devices revealed that LiF and CuPc act differently but confirmed the dramatic influence they have on the charge injection interface and on general device behavior. Indeed, the main effect of both materials can be reproduced by a reduction of their respective injection barriers. In the case of LiF, the reduction in the injection barrier, induced by only a few monolayers, is essentially related to the creation of a static charge double layer across the fluoride resulting in a surface dipole. This in turn leads to an energy level shift of up to 0.5 eV. In contrast, the barrier reduction induced by a thin CuPc film is dynamic, i.e. related to the injection itself [52–55]. This results from a large hole accumulation at the CuPc/α-NPD interface as well as a reduced ITO/CuPc energy barrier. As the external field increases, the interface charge increases and lowers the energy barrier between CuPc and α-NPD. Hole transport is therefore made easier. However, the simulation shows that the general device behavior is mainly controlled by the characteristics of the CuPc/α-NPD interface. The simulation also shows that image force effects are important in both situations. In the case of LiF, the thin insulating layer moves the first organic layer away from the metal surface reducing the image force zone [31, 32]. This drastically decreases the return current to the electrode, but without greatly affecting the direct current through the insulating film.

17.4.2
Time of flight simulations

In MOLED, the carrier mobility enters as an input parameter while the charge and current distribution come out as the result of the calculations. The electric field dependence of the mobility is of great importance to the simulation however, and is generally obtained through time of flight experiments (TOF) [58]. In these experiments, a sample of material with thickness L is sandwiched between two electrodes under voltage and the flight time for an electron or hole is determined by measuring the time it takes for a photo-induced charge carrier sheet to move across the sample. This is seen as a kink in the photo-generated current. The mobility is then obtained through the relation $\mu(F) = L^2/Vt_{tr}$ where L is the sample thickness, V the applied voltage and t_{tr} is the arrival or transit time.

As the mobility law is dictated by the microscopic hopping rates and mostly by the energetic and spatial disorder in the organic material, the simulation of TOF experiments can help to understand the results. In addition, when experimental data are not available or incomplete, the TOF simulation can give valuable information on mobility behavior if reasonable assumptions are made about the energetic and spatial distributions of the molecular sites in the organic material. Of special importance are the energetic correlations which have been shown to give the correct mobility behavior in the field range of interest [36, 59].

The MOLED code turns out to be an excellent tool for studying issues related to the TOF signal. In fact, MOLED is designed to model both the steady state in OLED's as well as the transients that may occur, e.g., as the external voltage changes. The statistics over many carriers is included automatically since a cloud of particles is studied instead of an averaging over many separate particles as in the Monte Carlo approach. More important, the distribution of both electrons and holes produced by the initial light pulse in experiments, is easily implemented in the simulation as the initial charge distribution. Further, we are dealing with a full device model, therefore, the TOF signal in the simulation acts exactly as in the experiment – as the electron current from the battery to one of the electrodes. Many of the experimental parameters can also be input into these simulations, e.g. the injection barriers at the electrodes.

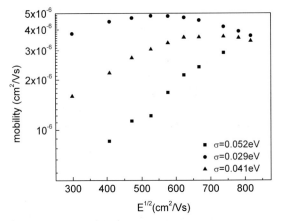

Figure 5 Field dependence of mobility at room temperature in samples with short-range correlated energetic disorder (circles: $\sigma=0.029$, triangles: $\sigma=0.041$, squares: 0.052eV).

Different types of disorder relevant to organic materials can be studied by generating the appropriate LUMO and HOMO energy configurations. For instance, dipolar correlated disorder [36] is generated by randomly oriented dipoles in a 3D cell of $50 \times 50 \times 50$ lattice sites and taking the site energies along the longitudinal direction as the 1D energy configuration. The energy distribution is Gaussian and the correlations are of long range decaying as $1/r$. Short-range correlations can also be generated in a 3D lattice. The energy sites in the sample are given by random sampling of a Gaussian distribution with variance σ and mean value equal to the LUMO and

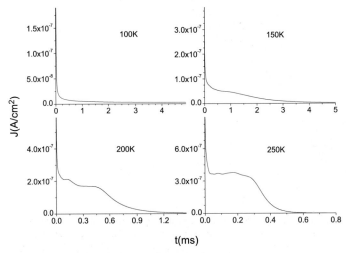

Figure 6 Current transients for different temperatures in the case of short-range correlated disorder, $E=2.77 \times 10^5$ V/cm, $\sigma=0.041$eV.

HOMO levels. Short-range correlations between the site energies are introduced by taking the energy of the ith site to be the average over the energies of the first neighbors including the self-energy [59] i.e. $\varepsilon_i = (\varepsilon_{i-1} + \varepsilon_i + \varepsilon_{i+1})/3$. Fig. 5 shows the result of the MOLED simulations for the previous energy configuration and the microbalance II formula. The resulting mobility law follows a Pool-Frenkel behavior for various disorder strengths. Above the field value given by $qEa \approx \sigma$, the mobility begins decreasing. Examples of TOF transients are displayed in Fig. 6 for various temperatures. Clearly, important aspects of TOF experiments and mobility laws can be understood by MOLED.

17.4.3
Internal electric field

In this section, we discuss a family of OLED devices in which two specific device parameters were systematically varied. As presented in the experimental part, the variation leads to 16 I(U) curves, which are presented in Fig. 7. Using the experimental data as input into our model, a detailed picture of the current, charge, field, and recombination distribution in the device is obtained. The subsequent analysis readily highlights the processes that dominate the functioning of the particular device architecture.

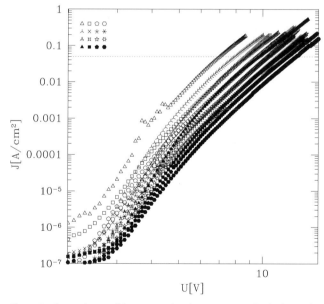

Figure 7 Dependence of the current density on voltage, for all sixteen segments, is shown in a logarithmic plot. The symbols correspond to the segment labels of Fig. 1. From top to bottom increase of α-NPD layer thickness and from left to right increase of Alq$_3$ layer thickness. In the logarithmic plot one can clearly see some grouping. The data for sixteen segments separate into eight groups in the range between 0.004 A/cm^2 and 0.04 A/cm^2. The groups correspond to devices of the same total thickness.

17.4.3.1 **Experimental analysis**

The following procedure was used to analyze the experimental results. For a given value of the current density, the voltage values were extracted from all sixteen I(U) curves. These sixteen voltage values were plotted against the corresponding thicknesses of the Alq_3 layer as well as the α-NPD layer. The resulting graphs are shown in Fig. 8 for $J=0.05A/cm^2$

The first graph in Fig. 8 shows that the voltage drop across the device increases linearly with the thickness of the Alq_3 layer. The line connects the points that correspond to devices with equal α-NPD layer thicknesses. These lines turn out to be straight, with slopes independent of the thickness of the α-NPD layer. We associate the value of the slope with the average electric field in the Alq_3 layer. A similar linear behavior is observed in the second graph, with the roles of Alq_3 and α-NPD interchanged. The four slopes in the second graph are related to the value of the electric field in the α-NPD layer. The same procedure was applied for many values of the current.

The data presented in Fig. 7 were analyzed over four orders of magnitude in current density, from 10^{-5} A/cm^2 to 0.1 A/cm^2. The lower limit for the analysis was fixed by the onset of noise in the experimental data at very low current. It turns out that the linear behavior, demonstrated for $J=0.05$ A/cm^2 in Fig. 8, persists over the entire current range. Therefore, the values for the average electric fields, $F_{\alpha-NPD}(J)$ and $F_{Alq3}(J)$ are virtually independent of the particular segment combination used for the calculation. The results for the average fields, as obtained from the experiment, are shown in Fig. 9. Similar results were obtained by Martin et al [60] even though a CuPc hole injection layer was not used. An electroabsorption method to calculate the voltage drop in each layer was used.

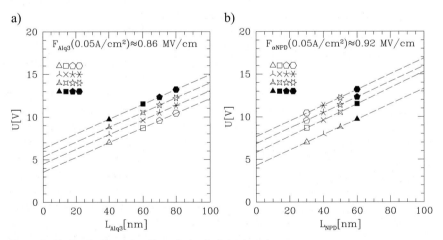

Figure 8 The total voltage drop through the diode is plotted against the thickness of the (a) Alq_3 layer (b) α-NPD layer for the current density of 0.05 A/cm^2. The dashed lines are guides to the eye. Their slopes are related to the average electric fields in (a) Alq_3 and (b) α-NPD layers.

17.4.3.2 Discussion of Simulations

From the qualitative features of Fig. 9, it is clearly seen that the average electric field in α-NPD is larger than or equal to the average field in Alq$_3$ over the entire current range. In the range above 0.001 A/cm^2 both fields are close in value. Given the obvious linearity in Fig. 8, the space charge effects in both Alq$_3$ and α-NPD layers seem to be negligible. Therefore, one tends to interpret the data in Fig. 9 as a sign of a rather well balanced electron and hole injection at their respective electrodes. It should be noted that, even for devices based on the same or similar bilayer structures, this type of balance has not always been observed by other authors. This is attributed to the different electrodes used for these studies. For example, in Ref. [61], it was concluded from the electro-absorption experiment that the field in Alq$_3$ is much higher than in α-NPD. However, in that work a Mg:Ag cathode was used instead of Al/LiF and no CuPc was used on the ITO side. In Ref. [62] it was argued that CuPc may be used to reduce the hole injection from O$_2$-plasma treated ITO, thus balancing the hole injection against the electron injection at the Al/LiF/Alq$_3$ junction. Using differently prepared ITO, we have observed [55] the opposite. However, this difference is not surprising, considering the different starting points and the effect that O$_2$-plasma treatment has on the workfunction of ITO [63].

A more detailed look at the present device architecture should go beyond the injection at the electrodes. First, it is well established that the mobility of holes in Alq$_3$ is very low [47+48]. In general, low mobility leads to increased space charge accumulation. Therefore, in order to avoid such space charge accumulation in Alq$_3$, the hole current in Alq$_3$ should be very small. It is also very difficult for electrons to cross the entire device. This is due to the electron blocking energy barrier at the Alq$_3$/α-NPD interface [42, 43]. This barrier is much higher than the barrier for holes

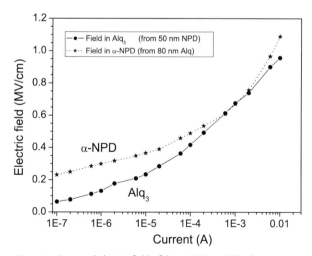

Figure 9 Averaged electric field of the α-NPD and Alq$_3$ layers inside our device, calculated from the experimental results. For the other thickness values qualitatively similar results are obtained.

at the same interface. The requirement of negligible space charge effects therefore translates into a requirement on recombination: the hole current should convert entirely into the electron current within a narrow region near the α-NPD/Alq$_3$ interface. This requirement is particularly difficult to fulfill at very low currents because the accumulation of carriers is low in the device while the recombination probability goes roughly as the product of the electron density and the hole density. But it turns out that the space charge effects are already well developed in the Alq$_3$ layer at low current densities above 0.004 A/cm^2. In spite of this, the linearity observed in Fig. 8 is reproduced.

On the anode side, the workfunction for ITO is taken to be around 5 eV, while a finer adjustment was left for the simulation. We also allowed for some adjustment of the CuPc HOMO level within experimental uncertainty. This was assumed to be of the order of ± 0.1 eV. These adjustments allowed us to get a good quantitative agreement between the calculated and experimental curves. The final set of parameters used for this simulation is shown schematically in Fig. 4. Once chosen, this set of parameters was kept fixed for all simulations related to sixteen segments of the OLED matrix.

Motivated by the experiment, we first considered the average electric fields over the α-NPD and Alq$_3$ layers. This average is calculated as the ratio of the voltage drop across the layer and the respective layer thickness. The calculation of $F_{\alpha\text{-NPD}}$ and F_{Alq3} was performed over the same current range as in the experiment (Fig. 9). The experimental results of Fig. 7 as well as the averaged field dependencies of Fig. 9 are well reproduced within the model simulations.

17.4.3.3 Charge and current distribution inside the device

The simulated carrier density distributions in the steady state for low and high voltage are shown in Fig. 10. As may be expected, most carriers accumulate in the region close to the α-NPD/Alq$_3$ interface. This accumulation is based on the high barrier for electrons at the interface and on the low mobility of holes in the Alq$_3$ bulk. The space charge effect caused by holes accumulating in Alq$_3$ is very pronounced, contrary to the simple expectations in section 17.4.3.1. Nevertheless, the electron density dominates the hole density at the α-NPD/Alq$_3$ interface. The space charge related to holes in the bulk of Alq$_3$ is more pronounced at low voltage (Fig. 10a) than at high voltage (Fig. 10b).

At low voltage a significant current of unrecombined holes (*"leakage current"*) reaches the cathode. As the voltage increases, the total concentration of electrons increases on the Alq$_3$ side of the α-NPD /Alq$_3$ interface. This provides an increasing cross-section for the recombination of holes. Therefore the recombination intensity distribution becomes more localized on the Alq$_3$ side of the α-NPD /Alq$_3$ interface. Accordingly, the leakage of holes diminishes and the space charge effect decreases in the Alq$_3$ layer.

The situation in the α-NPD layer is simpler. The space charge effect in α-NPD is negligible since the hole mobility is high. Also, only very few electrons enter the α-NPD layer due to the high electron barrier at the Alq$_3$/α-NPD interface. More precisely, what turns out to be important is that this barrier is much higher than the

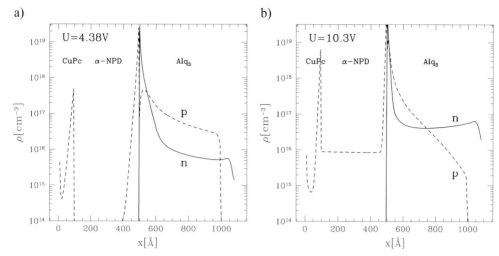

Figure 10 Carrier density distribution inside the 10 nm/40 nm/60 nm device for (a) 4.38V and (b) 10.3V. Most of the charge accumulates in the region around the α-NPD/Alq$_3$ interface. Notice the logarithmic scale. Electrons are blocked at that interface and barely enter the α-NPD layer. In contrast, the holes readily enter the Alq$_3$ layer at low voltage. They even dominate the charge density over the larger part of the Alq$_3$ layer. This situation gradually disappears as voltage increases. At higher voltages, the region over which holes extend into the Alq$_3$ shrinks towards the α-NPD/Alq$_3$ interface. Most of the carriers recombine at that interface. Holes also accumulate on the CuPc side of the CuPc/α-NPD interface.

barrier for holes at the same interface. For comparable bare barriers, the Coulomb potential of the accumulated charges would effectively reduce the barrier, causing both carriers to cross the interface.

We should point out that the sensitivity of the modeled I(U) curves to the α-NPD/Alq$_3$ interface parameters is rather low. This is true in spite of the fact that this interface accommodates most of the carriers in the device. The changes in the hopping rates, the variation in the effective distance among monolayers and the barrier heights are readily compensated by the change of hole and electron densities on both sides of the interface, as long as the barrier for electrons is kept significantly higher than the barrier for holes. The changes have very little effect on the total interface charge, the current through the interface, as well as the overall recombination rate.

The behavior observed for the CuPc/α-NPD interface is quite different however. The holes accumulate at the CuPc side of this interface, the energy barrier determining the amount of the accumulated charge. However, not being compensated by the electron charge, the accumulated holes significantly influence the electric field in the α-NPD layer. As a result, the CuPc/α-NPD interface parameters (i.e., the energy barrier and the bare hopping frequency) strongly affect the I(U) curve of the whole device [55].

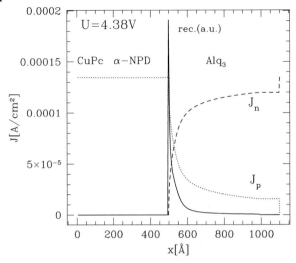

Figure 11 The current density distribution inside the 10 nm/40 nm/60 nm device in the steady state at 4.38V. The total current, $J_{tot}=J_n+J_p$, is constant throughout the sample. The conversion of the hole current into the electron one goes through recombination. The recombination intensity is shown in the full line (arbitrary units). At low voltage the conversion is incomplete and a fraction of holes reach the cathode without being recombined. Much of the recombination is confined to the first few Alq$_3$ monolayers at the α-NPD/Alq$_3$ interface. For higher voltages the leakage current goes to zero.

17.4.3.4 Electric field density distribution

The electric field distribution is shown in Fig. 12. The space charge effects in the Alq$_3$ layer are quite visible here. In addition to the space charge effects due to holes, one also notices the effect of the space charge of electrons at higher voltage. The strong electric field variation in the Alq$_3$ layer is related to both, hole and electron density distribution. When the thickness of the Alq$_3$ layer is changed, the field profile at a given current changes through the entire Alq$_3$ layer. Therefore, a thinner device should not be considered as a subsystem of a thicker one. The opposite would be true for a unipolar device with strong space charge effects.

17.4.4
Dye doping

17.4.4.1 Experimental results:

Dye doping of OLED is done for three reasons: 1) color tuning, 2) to increase lifetime and 3) to increase efficiency. Color tuning is achieved by selecting dye materials with a lower energy gap than the matrix thereby permitting energy transfer from the host to the dopant. But consequences arising from the location of the dye energy levels relative to those of the matrix are often not addressed. In general, the energy levels of the dopants can be located above or below (LUMO, HOMO) those of the

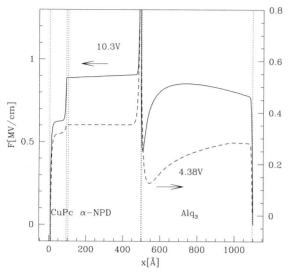

Figure 12 The electric field distribution inside the device, as obtained in the numerical simulation. The pronounced increase of the electric field at the CuPc/α-NPD interface is the consequence of the accumulation of holes at the CuPc side. The big change in the electric field at the α-NPD/Alq$_3$ interface comes from the charge accumulation (mostly electrons) in that region. In contrast, the holes are responsible for the space charge effect inside the Alq$_3$ layer, particularly pronounced at low voltage.

Alq$_3$ matrix and the variation of the energy level parameters in space can lead to localizations effects.

To investigate how partial doping of the emission layer in the parent device affects the transport and recombination characteristics, we used for the experimental part two different pyran containing donor-acceptor laser dyes: one with electron trapping capabilities (DCM-TPA, trap depth ~0.5 eV) and the other with both electron and hole trapping capabilities (DCM-II, trap depths ~0.5 eV each). This allowed us to check charge trapping effects induced by the guest molecules in the electro-active host material. Fig. 13 presents the energy level parameters of the two dopants used in our study relative to the Alq$_3$ matrix. In spite of the fact that the present discussion relies only on a specific device architecture, we believe that most of the effects can be regarded in a general framework and adapted to other systems.

As discussed in Ref. [64], the devices with DCM-TPA doping, show a high quantum efficiency which depends only weakly on the doping concentration over a large range of current densities. The quantum efficiency varies between 3.7 % for the lowest and 3.24 % for the highest doping concentrations. With increasing doping ratio, a red shift of the main emission peak is observed. In Fig. 14 the emission spectra as a function of doping ratio from 0.3 mol% to 8 mol% is presented for equal current densities. This effect was explained by Nuesch et al. [64] and attributed to polarization effects. Here we will focus on the appearance of the shoulder at about 520 nm.

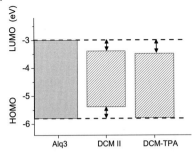

Figure 13 Energy levels of the dopants used in this study. All measurements were made with glassy carbon and PT electrodes against a Ag/AgCl reference electrode (0.205 eV vers NHE). As solvent MeCl$_2$ (0.1M TBAP) was used.

This shoulder can be attributed to a green Alq$_3$ emission component, as has already been remarked by Bulovic et al. [65] for a similar device structure. At the lowest doping concentration this shoulder is either washed out by the dopant emission or is too small to be noticeable. For increasing doping concentration, however, it becomes increasingly visible relative to the main peak. In Fig. 15 the current dependence of the emission spectra is shown for the 8 mol% doped device where a remarkable color shift effect is noticed. At low current densities, a red-orange emission is observed, which then shifts to green-orange at higher current densities. From the spectra, we see that this color shift is related to the size of the Alq$_3$ side-peak relative to

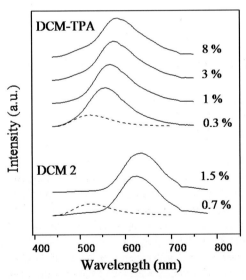

Figure 14 Normalized electroluminescence emission spectra as a function of doping concentration for DCM-TPA and DCM-II. For each spectrum, the doping concentration is indicated in the graph. For each concentration, the spectrum was measured at a current density of 80 mA/cm^2. For comparison, the pure Alq$_3$ electroluminescence emission spectrum is also indicated in the graph (dashed line).

a)

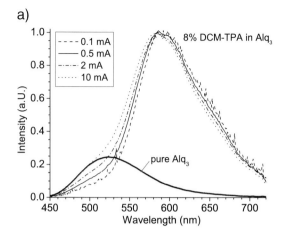

b)

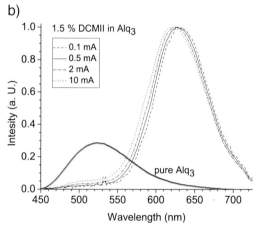

Figure 15 Normalized emission spectra as a function of the applied current. Left for the 8 mol% DCM-TPA doped device and right for the 1.5 mol% DCM-II device. The solid state emission spectra of Alq$_3$ is fitted to the high energy shoulder.

the peak of the doping molecule, the ratio of which increases with increasing current. This indicates that the mechanism responsible for this Alq$_3$ shoulder is voltage (current) dependent. Similar current dependencies, although less pronounced or less visible, are measured for the 3 mol% and 1 mol% doped devices. For DCM-II doped samples this current dependence is negligible (Fig 15).

As we will show later, this current dependence of the side shoulder is related to a redistribution of the recombination zone. This means that the recombination zone moves at least partially out of the doped area. To check this hypothesis experimentally, we introduced an additional hole blocking layer into one sample, i.e., 10 nm of 2,9-dimethyl-4,7-diphenylphenatroline (BCP) evaporated right after the doped part. The value of its highest occupied molecular orbital, −6.7 eV (Alq$_3$ −5.7 eV), leads to

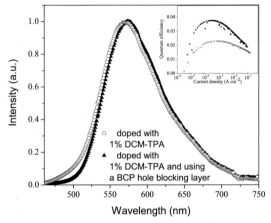

Figure 16 Normalized electroluminescence spectra of devices with and without additional bathocuproine hole blocking layer. The emission layer of the samples is 1/3 doped in thickness with 1 mol% DCM-TPA. In the inset the corresponding quantum efficiencies as a function of the current density are shown.

an energy barrier for the holes of about 1.0 eV which is considered high enough to completely block them in front of this interface. This then leads to an electron hole confinement zone inside the doped part, if we neglect tunneling effects. This means that the recombination zone in these samples could not move outside the doped part. As the spectra in Fig. 16 shows, the green shoulder disappears for those devices with the hole blocking layer. As well, the quantum efficiency jumps by nearly a factor two (inset).The result confirms our hypothesis of a partial recombination zone shift outside the doped part and that the saturation of the host-to-guest energy transfer can be excluded as the dominant effect [66–69].

17.4.4.2 Analysis Through Simulation

To explain the presented experimental results based on a partially doped emission layer, we once more used our simulation model "MOLED". This time the doped part was introduced as an additional layer with suitable properties. To get a better understanding of the mechanisms which can be attributed to dye doping, we will consider five different cases. These are discussed in the general context of mobility and energy level related effects so as to be able to compare with experiment:

Case 1: Undoped device as reference.

Case 2: The doped sublayer has reduced electron mobility with respect to the undoped part. A reduction factor of 10 is used. This 10-fold reduction of the mobility in the doped region corresponds to our example of a DCM-TPA doped Alq_3 layer [66].

Case 3: The doped sublayer has reduced hole mobility with respect to the undoped case. We use a reduction factor of only 5, as the parent hole mobility is already very low. Dopant energy levels outside of the matrix are

not considered because they should not influence the transport characteristic at the doping concentrations generally used.

Case 4: The doped sublayer has reduced electron and hole mobility with respect
to the undoped case. This reduction of the mobility in the doped region
corresponds to the example of the DCM-II doped Alq_3 layer. A reduction
factor of 10 is used for both.

Case 5: The HOMO level of the doped sublayer is increased with respect to the
undoped device. We take 0.2 eV for this shift. The case of a lowered
LUMO level inside the doped part has no particular effect on the device
performance. Therefore we do not consider this case separately.

17.4.4.3 Mobility variation inside the doped part

In principle, doping is expected to diminish the mobility and to increase its electric
field dependence for electrons, holes or both, depending on the energy levels of the
dopant relative to the matrix. For sure, the introduction of the dopant into the matrix
can also lead to an energy level shift. Nevertheless trapping effects remain dominant
so that a fraction of the charges moving inside the Alq_3 host material are trapped by
the dopant molecules thereby reducing the effective macroscopic mobility. Such a
reduction in mobility is introduced in our model by a decrease of the Poole-Frenkel
(PF) parameters μ_0 and F_0 of Eq. 14 inside the doped regions.

For completeness, we outline an alternative and more detailed way of modeling
doping. In this method, the mobility is directly parameterized through the trap levels and the trapping/detrapping rate equations. This type of modeling has been
done by Staudigel et al. [12], for example, but it often requires specifying some not
very well-known parameters such as the trapping energy ΔE (trap depth), the trap-
ping/detrapping attempt frequency as well as the trap density. If these parameters
are very well known, then this model can be useful. However, if this is not the case it is
best not to use this approach since the resulting effective mobility can be rather sensitive
to changes in these parameters even within experimental error. At this point, we should
mention that we have carefully examined our model against those device simulations in
the literature in which trapping/detrapping is modeled. The results for the charge and
current distribution generally agree over the available voltage ranges.

Often, the field dependence of the mobility in organic materials is accessible
directly by experiment, e.g. time of flight measurements. As for our doped layers,
however, we did not have the precise values of the carrier mobility from separate
experiment. On the other hand, we were confident of the values for other parameters entering into the model, including those for the mobilities of the undoped parts
of the device, as it was presented in the previous chapters. In the cases modeled, it
was clear that trapping reduced the mobility of carriers in the doped layer. Reduction
factors of 0.1 (electrons), 0.2 and 0.1 (holes) were used to account for this, leading to
a reasonable agreement with experiment regarding the I(U) curves, and a qualitatively and quantitatively good explanation of the observed spectral effect [66–68].

First we consider the case of reduced electron mobility inside the doped part
(case 2). Here, it is less important to know the real origin of the decreased mobility
because the effect relates only to the mobility drop for electrons entering the doped

part. The new voltage dependence is now described by the decreased PF parameters μ_0 and F_0. More precisely, the PF parameters were set as $\mu_0 = 1*10^{-9}$ cm^2/Vs and for the field dependence, $F_0 = 60000$ V/cm. To consider interface related traps with a strong trap concentration variation, the doped layer can be further divided into sub-layers each with its own respective mobility, but this goes too far for the moment.

Figure 17 presents the voltage dependent charge distribution around the doped/undoped interface. At low voltages (< 2.7 V) we see only a large charge accumulation at the α-NPD /doped-Alq$_3$ interface just as in the case of the undoped reference device (x-position 500 Å). In this voltage range, the transport behavior is contact limited and space charge effects do not yet play an important role, so the charge densities are only related to energy barriers. With increasing voltage, a space charge barrier begins to form around the doped/undoped interface inside the Alq$_3$ layer (x-position 700 Å). This is due to the electron mobility drop in the doped region which slows down the evacuation of the electrons from the doped side of the interface. An accumulation of charge thus results, hindering further electrons from entering. At even higher voltages (> ~5 V), the development of two zones with high electron density is clearly seen. The first stays inside the doped area near the α-NPD /doped-Alq$_3$ interface as before, while a new one is now located around the doped/undoped interface of the Alq$_3$ layer. At these high voltages, the majority of electrons are blocked at the doped/undoped interface as described above while those electrons which do manage to pass through get blocked at the α-NPD /doped-Alq$_3$ interface. Nevertheless, the total electron density at the latter one continues to increase with increasing voltage due to the positive field dependence of the electron mobility.

Let us now look at the hole density distribution in Fig. 18. We see, as expected, an increased accumulation at the α-NPD/Alq$_3$ interface due to the energy barrier and the low mobility inside the Alq$_3$ layer. However, an increased hole density at the

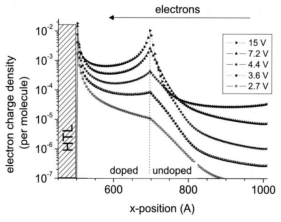

Figure 17 Local electron density distribution around the doped/undoped interface as a function of the bias.

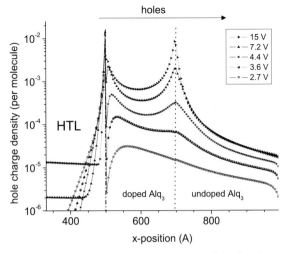

Figure 18   Local hole density distribution around the doped/undoped interface

doped/undoped interface also develops, especially at higher voltages. This is very surprising and cannot be explained in a hole-only picture since the increase is related to a change in *electron* mobility.

The origin of this unexpected behavior may be better seen in Fig. 19, which shows the local electric field within the device at different applied voltages. The variations in the field strength reflect the charge distribution in the device. As can be seen, there is a marked decrease in the electric field on the undoped side of the doped/undoped interface. This is due to the screening effect from the *electrons* accumulated about the interface. The accumulation is caused by the reduced electron mobility in the doped region. We might remark here that a small decrease of only 0.5 orders of magnitude in the electron mobility of the doped area is already enough to significantly increase the charge densities at the doped/undoped interface for voltages above 6 V.

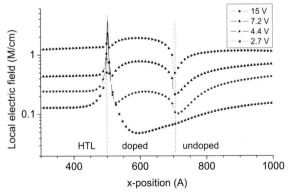

Figure 19   Simulated local electric field distribution over the entire device is shown for different voltages.

If we now consider what happens if a hole mobility decrease inside the doped region is made, this turns out to be much easier. In this case, the second charge density peak remains negligible compared to the main peak at the α-NPD /doped-Alq$_3$ interface.

17.4.4.4 Recombination rate distribution

The charge density distribution discussed in the previous section is essential for understanding the variations in the local recombination rate (Eqs: (15) and (16)). Here the recombination rate is dimensionless and represents the radiative relaxation of excited molecules in the singlet state.

From those equations, we see that the recombination rate is proportional to the hole and electron densities. Thus for the DCM-TPA case and at low voltages, where the hole and electron density peaks around the doped/undoped interface are still undeveloped, recombination is predominantly at the α-NPD /doped-Alq$_3$ interface where the charge densities are the highest. The light seen is thus dopant color emission. With increasing voltage and as the second charge density peak develops, the recombination rate becomes significant in this zone as well. The recombination area is thus split into two essential zones: the left-hand side emitting predominantly from the dopant, while the right-hand side emits from the undoped Alq$_3$ host. Furthermore, as Figure 20a shows, the light emission from the undoped Alq$_3$ becomes more significant relative to the light from the doped part, as the voltage increases. Another aspect is that with increasing voltage, the recombination peaks become tall and narrow with widths as small as a few monolayers, for voltages higher than 8–10 V.

In the previous discussion we have only been concerned with the effects of mobility variation. In the most real cases, however, other effects must also be considered. For example, we have not taken into account that the molecular photoluminescence of the host molecules is generally less efficient than that of the dopant molecule. The contribution of a recombination channel outside the doped area can therefore be accompanied by a drastic decrease in the quantum efficiency. Another effect is the possibility of an energy transfer of the Förster type [69] from the first undoped Alq$_3$ monolayers to the dopant molecules. It is obvious, that such a mechanism can suppress color change completely if the Förster radius exceeds the recombination zone width and can maintain a high quantum efficiency. In the same way, exciton diffusion from the undoped region into the doped region followed by an effective energy transfer to the dopant can also maintain high quantum efficiency and prevent color change.

If the hole mobility is also reduced, as in our case study 5 with DCM-II doping, the additional hole space charge compensates the effect and the second recombination peak at the doped/undoped interface is reduced, as seen in Fig 20b).

17.4.4.5 Energy level shift inside the doped region

The effective LUMO and HOMO energy levels, at least of the dopant, may be redshifted due to polarization effects. However, the frontier orbital energy levels measured in solution are a precious indication of the relative energy level positions of

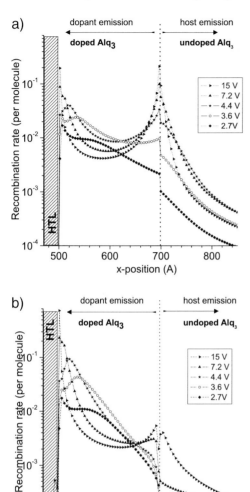

Figure 20 Calculated local recombination rate distribution inside the doped an undoped Alq₃ layer as a function of applied bias. The simulation was carried out for reduced electron mobility of one order of magnitude (left) and for reduced electron and holes mobilities, one order of magnitude each (right).

the materials used in our devices. The fact that electronic orbital levels shift of the dye depend on the polarity of the surrounding molecule is most clearly evidenced by solvatochromic effects observed for numerous families [71]. Such dyes were therefore taken to define solvent polarity scales [72]. As a first approximation, the shift of the absorbance maximum can be understood by comparing the solvation of the ground and excited state dipole moment of the dye. If the ground state dipole moment $\vec{\mu}_g$ is more important than the excited state dipole moment $\vec{\mu}_e$, the ground state will be better stabilized in polar solvents than the excited state and a hypsochromic (blue) shift is expected with increasing polarity. On the other hand, if $\vec{\mu}_e$ is more important that $\vec{\mu}_g$, a bathochromic (red) shift of the absorption spectrum is expected with increasing solvent polarity. This simple model is meaningful only in the absence of specific solvent-solute interactions. In this case, polarity induced shifts of the molecular frontier orbitals can be calculated by semi-empirical methods treating the solvent as a continuum [73].

In the solid state, similar considerations prevail as in solution. Indeed, solvation effects are observed for donor-acceptor guests when the polarity of the host matrix is changed [74]. This effect may be more important in systems of weakly correlated dipoles. For example, Baldo et al. [75] have calculated the spectral shift of the guest molecules in a domain of correlated dipoles with different dipolar moments.

In our example of DCM-TPA doped into Alq_3 the spectral bathochromic shift reaches up to 0.3 eV, as is shown in Fig. 14. A qualitatively similar effect is seen for DCM-II. Specific chemical interaction between the guest and host is unexpected and in such a system the averaged polarity of the host depends on the concentration of guest molecules. Typically, the dipole moment of the dye is higher than that of Alq_3, leading to a polarity increase with doping concentration. Since we observe a bathochromic shift with increasing concentration, we deduce that the excited state dipole moment is stronger that the ground state dipole moment of the doping molecule [76]. Therefore, we can assume that the LUMO of our doping molecules will decrease more rapidly than the HOMO.

The effective LUMO and HOMO shift of the dopant further into the gap of the host material is important mainly when considering the emission spectra. For low doping concentration, as is generally the case for laser dyes, the charge transport occurs through the host molecules, whose energy levels are less affected as discussed. For high dye concentration, as hopping between trap sites starts to dominate the transport mechanism inside the doped part, these energy level shifts have an important impact factor on the calculation and should be adequately considered, as we will do in simulation case 5.

Therefore, we consider mainly the energy levels of the transport related parameters of the matrix material. A variation of the molecular energy levels of the matrix in the doped region results in a change of the energy barriers at the surrounding organic/organic interfaces. It is obvious, that only an increase in an energy barrier leads to additional charge accumulation, so that a decreasing barrier can be neglected in this analysis. As well, a decreasing LUMO level which leads to an effective barrier height increase for electrons at the α-NPD/Alq_3 interface is also not relevant because the electrons are already considered as completely blocked at this interface.

The other possibility considers an increase in the HOMO level which is in general accompanied by a decrease of the hole energy barrier at the HTL/Alq_3 interface. Such a variation has only a small impact on the macroscopic transport characteristic as shown in Ref. [56] and Fig. 23. But at the other side, at the doped/undoped inter-

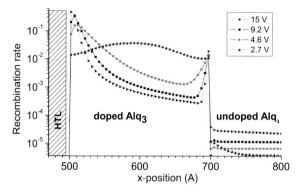

Figure 21 The recombination rate for a reduced HOMO level of about 0.2 eV inside the doped layer. It is clear, that in this case an additional recombination channel appears inside the doped layer at the doped/undoped interface, but the contribution to light emission remains negligible. No recombination takes place outside the doped part, so that a color shift is not expected.

face a new barrier appears leading to hole accumulation already at low voltages. Such an increase of hole density results in an increase of local recombination rate, as illustrated in Fig. 21. This calculation was done for a HOMO level increase of 0.2 eV. The additional recombination rate peak stays small and appears only inside the doped layer, so no color tuning is expected based on this mechanism.

17.4.4.6 General recombination picture
In Fig 22, the possible recombination channels are summarized for the low and high voltage cases. In the low voltage case, space charge effects are still not developed so that the only possibility of having charge accumulation leading to light generation is related to energy barriers. The classical recombination zone RZ1 is the same as for the undoped devices. The RZ2 appears only if an energy barrier preventing holes from leaving the doped area is present, leading to local charge accumulation. However, this barrier should be related to the doping itself.

In the high voltage regime (*) the space charge barriers become dominant. Both, HOMO and LUMO level shifts are less relevant. RZ1* still behaves like RZ1, but with a reduced recombination rate. For a reduced electron mobility inside the doped part of the Alq_3 layer, a space charge barrier for electrons develops around the doped/undoped interface, as the electrons inside the doped part are not evacuated quickly enough. The resulting high rate recombination has the emission color of the dopant for the part RZ2* inside the doped layer and that of the host outside the doped part RZ3*. A color change can be observed under this condition. For hole and electron trapping RZ3* is less developed.

17.4.4.7 Transport characteristics
In this subsection, we will show how the different doping parameter variations can affect the current/voltage characteristics and the leakage current.

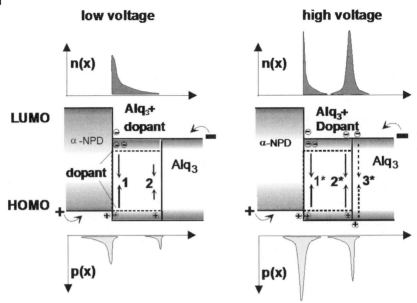

Figure 22 Schemes for possible recombination zones in partially dye doped OLED's at low and high voltages. p(x) and n(x) stand for the charge densities and the arrows mark the resulting recombination which leads to light generation. Dotted arrows indicate color emission from the matrix, while line arrows indicate dopant emission.

In Fig. 23a the relative hole leakage current as a function of the bias is presented for four cases. It is clear, that for voltages below 5 V and without an additional energy barrier between the doped and undoped part of the matrix a large amount of holes can escape to the cathode side without any contribution to light generation. In this low voltage range, we have low charge densities and the space charge barriers are not present. The huge leakage current for low voltages can be significantly reduced by using dopant molecules which lead to a hole energy barrier at the doped/undoped interface. A small energy barrier of 0.2 eV can already reduce the leakage current by nearly one order of magnitude. In contrast, a reduction of the charge mobility inside the doped part does not significantly affect the amount of leakage current. This is related to the fact that a mobility decrease leads to space charge barriers only at higher voltages where an efficient recombination already takes place and the leakage current is anyway negligible. Note also that charge balance between holes and electrons is a major criterion for high quantum efficiency, so that as the leakage current approaches zero, the quantum efficiency will reach its maximum value.

This strong variation in the leakage current forces us to look at how the current voltage characteristics behave for these cases. In Fig. 23b the I(U) curves of the four cases are plotted. Despite huge differences in the leakage current, the I(U) curves show no significant difference. This shows clearly, that the effects introduced by doping are not the dominant transport limiting factors in the voltage range considered, even when the doping related space charge barrier becomes significant.

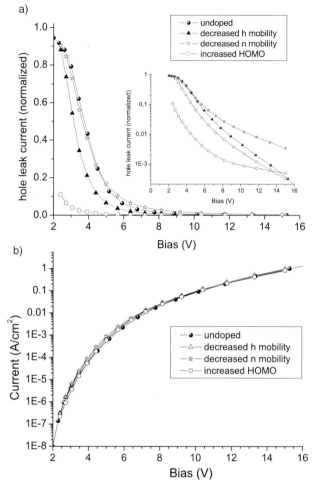

Figure 23 Left: Leakage current for four cases. For low voltages a small energy barrier at the doped/undoped interface is enough to reduce the leakage current by nearly one order of magnitude. Right: I(U) curves of cases 1-4. No relevant changes are seen.

17.5
Summary

Our device simulation model "MOLED" has proven to be a very useful analytic tool for investigating, electrode effects, voltage drop and charge distribution in multi-layer OLED's and dye doping related effects in a standard device architecture. The results show that device modeling helps to glean new insights into the microscopic charge transport and recombination mechanisms within doped OLED's.

Energy level adjustment and image force effects dominate charge injection into molecular organic layers. At low voltages, hole leakage current is a major factor affecting device efficiency. Electrode architecture, organic/organic interface barriers and charge mobilities are the main parameters when considering the voltage drop inside the organic layers. When CuPc is used, the voltage drop in the HTL layer exceeds that of the ETL. Mobility variations and the related space charge accumulation at higher voltages are very important mechanisms when considering the transport and recombination characteristics due to dye doping. In particular, an electron mobility decrease due to dye doping or the formation of an energy barrier preventing holes from leaving the doped region leads to a split of the recombination zone. Under some special conditions, the new recombination zone can be composed of two emission colors resulting in a color change of the emission spectra. The second peak becomes more pronounced with increasing voltage. Such current density dependent color change is observed for DCM-TPA which shows only electron trapping capabilities. This experimental result can be modeled by making a simple trapping related 10-fold reduction of the zero field mobility inside the doped part. This indicates the importance of trapping related space charge effects. If holes are trapped too, as in DCM-II, the locally increased hole density compensates the screening effect of the trapped electrons and the current density dependent color change becomes negligible. This splitting of the recombination zone may be used perhaps to create white light-emitting diodes.

Acknowledgements

This work was supported in part by the SCOPES project number 7KRPJ065619.01 and the project number 20-67929.02 of the Swiss National Science Foundation.

References

[1] J. S. Kim, R. H. Friend, F. Cacialli, Appl. Phys. Lett., **74**, 3084 (1999).

[2] Y. Sato, T. Ogata, J. Kido, Proc. SPIE **4105**, 134 (2000).

[3] M. Schaer, F. Nuesch, D. Berner, W. Leo, L. Zuppiroli, Adv. Funct. Mater. **11**, 116 (2001).

[4] G. Sakamoto, C. Adachi, T. Koyama, Y. Taniguchi, C. D. Merritt, H. Murata, Z. H. Kafafi, Appl. Phys. Lett., **75**, 766 (1999).

[5] A. A. Shoustikov, Y. J. You, M. E. Thompson, IEEE Journal of Selected Topics in Quantum Electronics **4**, 3 (1998).

[6] C. W. Tang, S. A. Vanslyke, C. H. Chen, J. Appl. Phys. **65**, 3610, (1989).

[7] F. Nuesch, L. Zuppiroli, D. Berner, C. Ma, X. Wang, Y. Cao, B. Zhang, Res. on Chem. Intermediat. **30**, 4–5, 495 (2004).

[8] P. S. Davids, I. H. Campbell, and D. L. Smith, J. Appl. Phys. **82**, 6319 (1997).

[9] G. G. Malliaras and J. C. Scott, J. Appl. Phys. **85**, 7426 (1999).

[10] G. G. Malliaras, J. C. Scott, J. Appl. Phys. **83**, 5399 (1998).

[11] T. Ogawa, D. C. Cho, K. Kneko, T. Mori, T. Mizutani, IEICE Trans. **E85-C**(6) 1239 (2002).

[12] J. Staudigel, M. Stössel, F. Steuber, and J. Simmerer, J. Appl. Phys. **86**, 3895 (1999).

[13] B. K. Crone, P. S. Davids, I. H. Campbell, and D. L. Smith, J. Appl. Phys. **87**, 1974 (2000).

[14] B. Ruhstaller, S. A. Carter, S. Barth, H. Riel, W. Riess, J. C. Scott, J. Appl. Phys., **89**, 4575 (2001).

[15] V. I. Arkhipov, E. V. Emelianova, H. Bässler, J. Appl. Phys. **90**, 2352 (2001).

[16] V. R. Nikitenko, O. V. Salata, H. Bässler, J. Appl. Phys. **92**, 2359 (2002).

[17] A. B. Walker, A. Kambili, S. J. Martin, J. Phys.: Condens. Matter **14**, 9825 (2002).

[18] E. Tutiš, M. N. Bussac, B. Masenelli, M. Carrard, L. Zuppiroli, J. Appl. Phys. **89**, 430 (2001).

[19] H. Houili, E. Tutis, H. Lutjens, M. N. Bussac, and L. Zuppiroli, Comp. Phys. Commun. **156**, 108 (2003).

[20] Computer Physics Communications Program Library, http://cpc.cs.qub.ac.uk/summaries/ADSG

[21] I. H. Campbell and D. L. Smith, Appl. Phys. Lett. **74**, 561 (1999).

[22] B. Masenelli, D. Berner, M. N. Bussac, F. Nuesch, L. Zuppiroli, Appl. Phys. Lett. **79**, 4438 (2001).

[23] C. Schmitz, M. Thelakkat, and H. W. Schmidt, Adv. Mater. **11**, 821 (1999).

[24] C. Schmitz, P. Posch, M. Thelakkat, and H. W. Schmidt, Phys. Chem. Chem. Phys. **1**, 1777 (1999).

[25] C. Schmitz, P. Pösch, M. Thelakkat, and H.-W. Schmidt, Macromolecular Symposia **154**, 209 (2000).

[26] C. Schmitz, H. W. Schmidt, and M. Thelakkat, Chemistry of Materials **12**, 3012 (2000).

[27] L. Zou, V. Savvate'ev, and J. Booher, C.-H. Kim and J. Shinar, Appl.Phys. Lett. **79**, 2282 (2001).
[28] S.M. Sze, *Physics of semiconductor devices* (Wiley, 1967), see e.g. page 496.

[29] T. A. Beierlein, H.-P. Ott, H. Hofmann, H. Riel, B. Ruhstaller, B. Crone, and S. Karg, W. Riess, *Proc. SPIE 4464*, **178** (2001) (SPIE Conference on Organic Light Emitting Materials and Devices V, 2001, San Diego, California, USA).

[30] C. Ma, B. Zhang, Z. Liang, P. Xie, P. Xie, X. Wang, B. Zhang, Y. Cao, X. Jiang, Z. Zhang, J. Mater. Chem. **12**, 1671 (2002).

[31] M. N. Bussac, D. Michoud, L. Zuppiroli, Phys. Rev. Lett. **81**, 1678(1998).

[32] E. Tutiš, M. N. Bussac, L. Zuppiroli, Appl. Phys. Lett. **75**, 3880 (1999).

[33] V. I. Arkhipov, E. V. Emelianova, Y. H. Tak and H. Bässler, J. Appl. Phys. **84**, 848, (1998).

[34] S. Barth, U. Wolf, H. Bässler, P. Müller, H. Riel, H. Vestweber, P.F. Seidler and W. Riess, Phys. Rev. B. **60**, p.8791, (1999).

[35] D. H. Dunlap, P. E. Parris, V. M. Kenkre, Phys. Rev. Lett. **77**, 542(1996).

[36] S. V. Novikov, D. H. Dunlap, V. M. Kenkre, P. E. Parris, A. V. Vannikov, Phys. Rev. Lett. **81**, 4472(1998).

[37] M. N. Bussac, L. Zuppiroli, Phys. Rev. B **54**, 4674(1996).

[38] Y. Roichman, N. Tessler, Appl. Phys. Lett. **80**, 1948, (2002).

[39] Y. Roichman, Y. Preezant, N. Tessler, Phys. Stat. Sol. A **201**, 1246, (2004).

[40] W. Xie, Y. Zhao, J. Hou and S. Liu, Jpn. J. Appl. Phys. **42**, 1466 (2003).

[41] S. B. Song, J. H. Lee, J. S. Oh, J. Appl. Phys., **93**, 9404 (2003).

[42] A. Rajagopal, C. I. Wu, and A. Kahn, J. Appl. Phys. **83**, 2649 (1998).

[43] W. Brütting, S. Berleb, and A. G. Muckl, Organic Electronics **2**, 1 (2001).

[44] J. Simon, J. J. Andre, *Molecular Semiconductors* (Springer-Verlag, Berlin, Heidelberg, 1985).

[45] R. O. Loutfy and Y. C. Cheng, J. Chem. Phys. **73**, 2902 (1980).

[46] D. W. Clack, N. S. Hush, and I. S. Woolsey, Inorg. Chem. Acta **19**, 129 (1976).

[47] S. Naka, H. Okada, H. Onnagawa, Y. Yamaguchi, T. Tsutsui, Synth. Metals **111–112**, 331 (2000).

[48] G. Kepler, P. M. Beeson, S. J. Jacobs, R. A. Anderson, M. B. Sinclair, V. S. Valencia, and P. A. Cahill, Appl. Phys.Lett. **66**, 3618 (1995).

[49] A. Hosokawa, H. Tokailin, H. Higashi, and T. Kusumoto, Appl. Phys. Lett. **60**, 1220 (1992).

[50] P. Devaux, P. Quedec, Phys. Lett. **28** A, 537 (1969).

[51] W. Mycielski, B. Ziolkowska, and A. Lipinski, Thin Solid Films **91**, 335 (1982).

[52] R. D. Gould, Thin Solid Films **125**, 63 (1985).

[53] B. Masenelli, E. Tutiš, M. N. Bussac, L. Zuppiroli, Synth. Metals **122**, 141 (2001).

[54] B. Masenelli, E. Tutiš, M. N. Bussac, L. Zuppiroli, Synth. Metals **121**, 1513 (2001).

[55] B. Masenelli, D. Berner, M. N. Bussac, F. Nuesch, L. Zuppiroli, Appl. Phys. Lett. **79**, 4438 (2001).

[56] D. Berner, E. Tutiš, L. Zuppiroli, Proceedings on "International Conference on the Science and technology of Emissive Displays and Lighting 2002", pages 503–506.

[57] Tutiš, D. Berner, L. Zuppiroli, J. Appl. Phys., **93**, 4594 (2003).

[58] W. E. Spear, J. Non-Cryst. Solids **1**, 197(1969).

[59] S. V. Rakhmanova, E. M. Conwell, Appl. Phys. Lett. **76**, 3822 (2000).

[60] S. J. Martin, G. L. B. Verschoor, M. A. Webster, A. B. Walker, Organic Electronics **3**, 129–141 (2002).

[61] T. Yamada, F. Rohlfing, D. Zou, and T. Tsutsui, Synth. Metals **111–112**, 281 (2000).

[62] H. Vestweber and W. Riess, Synth. Metals **91**, 181(1997).

[63] F. Nuesch, M. Carrara, M. Schaer, D. B. Romero, and L. Zuppiroli, Chem. Phys. Lett. **347**, 311 (2001).

[64] F. Nuesch, D. Berner, L. Zuppiroli, Res. Chem. Intermediat. 30 (4–5): 495–507 (2004).

[65] V. Bulovic, R. Deshpande, M. E. Thompson and S. R. Forrest, Chem. Phys. Lett. **308**, 317 (1999).

[66] D. Berner, F. Nuesch, E. Tutiš, C. Ma, X. Wang, B. Zhang, L. Zuppiroli, J. Appl. Phys. **95**, 3794 (2004).

[67] D. Berner, F. Nuesch, E. Tutiš, C. Ma, X. Wang, B. Zhang, L. Zuppiroli, Proc. of SPIE Vol **5464**, 72 (2004).

[68] D. Berner, E. Tutiš, F. Nuesch, L. Zuppiroli, Proceedings on "International Conference on the Science and technology of Emissive Displays and Lighting 2004", pages 69–72.

[69] T. Förster, Discussion Faraday Soc. **27**, 7 (1959).

[70] E. Tutiš, D. Berner, L. Zuppiroli, J. Appl. Phys., **93**, 4594 (2003).

[71] S. Hunig, G. Bernhard, W. Liptay and W. Brenning, Annalen Der Chemie – Justus Liebig **690**, 9 (1965).

[72] E. Buncel and S. Rajagopal, Accounts Chem. Res. **23**, 226 (1990).

[73] G. J. Tawa, R. L. Martin and L. R. Pratt, Int. J. Quantum Chem. **64**, 143 (1997).

[74] V. Bulovic, R. Deshpande, M. E. Thompson and S. R. Forrest, Chem. Phys. Lett. **308**, 317 (1999).

[75] M. A. Baldo, Z. G. Soos, and S. R. Forrest, Chem. Phys. Lett. **347**, 297 (2001).

[76] B. W. Domagalska, K. A. Wilk and S. Wysocki, Phys. Chem. Chem. Phys. **5**, 696 (2003).

18

Optimizing OLED Structures for a-Si Display Applications via Combinatorial Methods and Enhanced Outcoupling

W. Rieß, T. A. Beierlein, and H. Riel

18.1
Introduction

In recent years, organic light-emitting devices (OLEDs) have made tremendous progress. They are now emerging as the leading candidate among the many technologies under development for next-generation flat-panel displays. The progress was recently highlighted by the first demonstration of the world's largest (20-inch) full-color active-matrix OLED display (AMOLED) driven by amorphous silicon (a-Si) thin-film transistors (TFT) [1]. Such an achievement was only possible through systematic device optimization, which involved tackling critical performance and lifetime issues and benefited from significant improvements in performance of OLEDs in recent years, including improved charge-carrier balance by tuning the injection barrier [2–4], and from the use of novel emitting materials, e.g. phosphorescent emitters [5, 6].

Device optimization of OLEDs based on small molecules largely exploits the possibility of tailoring the electrical and electro-optical characteristics by sequential evaporation of multiple layers from various materials. However, because the performance of these multilayer devices depends on the complex interplay between different materials, individual layer thicknesses, dopant concentrations and electrode configurations, a huge parameter space has to be covered to understand and optimize these devices. As conventional one-by-one OLED preparation is very time-consuming, we use the combinatorial approach [7, 8] to obtain systematic and reliable datasets. Also, because the internal quantum efficiency of optimized OLEDs is gradually approaching the theoretical limits [9], further improvements of the external quantum efficiency can only be expected by increasing the outcoupling efficiency [10, 11]. Owing to optical interference effects, both the external quantum efficiency and the spectral characteristic depend significantly on the OLED architecture, in particular on the layer thicknesses [12, 13, 8]. In top-emitting devices, in which electroluminescence (EL) is outcoupled through a semitransparent thin metal cathode, interference effects can be very strong. Consequently, the strength of the optical interference effects in such OLEDs depends critically on the reflectivity of the metal cathode. Thus, the control of the cathode reflectivity is a further degree of freedom to tailor the emission characteristics. The thickness variation of the cathode metal is one pos-

Physics of Organic Semiconductors. Edited by W. Brütting
Copyright © 2005 WILEY-VCH Verlag GmbH & Co. KGaA, Weinheim
ISBN 3-527-40550-X

sibility to change the reflectivity. However, this approach has the disadvantage of increasing absorption losses in the case of thicker metal layers and of insufficient conductivity in the case of thinner metal cathodes. A preferred concept, known from the optics of metal coatings, employs a thin dielectric layer on top of a thin metal film to modulate the transmittance of the cathode [14, 15]. In this paper we will first show the possibilities provided by combinatorial device fabrication for optimizing electrical characteristics and emission spectra. In the second part we will focus on enhancing the outcoupling and tailoring the EL emission of top-emitting multilayer OLEDs. Finally, we demonstrate the advantages of device optimization via combinatorial methods and the capping concept by presenting red, green and blue (RGB) device performance characteristics, where we achieve an efficiency of NTSC white (National Television Standard Committee) from separate RGB pixels exceeding 22 cd/A under realistic display-driving conditions, thus enabling the world largest (20-inch) a-Si AMOLED display.

18.2
Experimental

Devices have been fabricated in two different evaporation systems. Combinatorial device fabrication is carried out in a custom-made evaporation vacuum system under ultra-high vacuum conditions ($p < 1 \times 10^{-8}$ mbar) [8, 16]. Combinatorial substrates are 80×80 mm^2 glass plates (Schott AF45). The anode pads are arranged in a matrix of 10×10 independent devices with an active area of 2×2 mm^2 each. The substrate is attached to a rotating sample holder, which has a shutter directly in front of the substrate. This shutter can be moved stepwise during evaporation, allowing staircase-like film deposition. Moreover, it can be rotated in 90° steps with respect to the substrate, enabling the fabrication of the three types of device configurations described in Fig. 1. The basic element of the combinatorial technique is the evaporation of a simple staircase of organic or electrode material (Fig. 1(a)). Such a step-like film can also be combined with one or more uniform layers. A two-layer

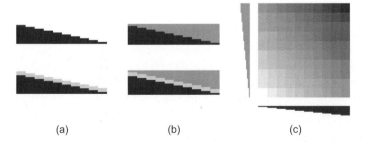

(a)   (b)   (c)

Figure 1 Examples of possible device configurations: (a) Simple staircase and staircase combined with a uniform layer for thickness-dependent device series. (b) Opposing staircases with/without uniform layer for position-dependent device series. (c) Combinatorial matrix, consisting of two perpendicular staircase structures.

structure with two opposing staircases can be deposited if the shutter is rotated by 180° after evaporation of the first layer (Fig. 1(b)), which, for example, results in a two-layer device with constant total device thickness, but with varying positions of the interface between these layers. At the interface a uniform layer, e.g. a doped region, can be introduced. If the shutter is rotated by 90° in a two-layer staircase structure, all possible thickness combinations can be realized (Fig. 1(c)). It is clear that such a matrix allows systematic and comparable data to be obtained for device optimization and further analysis. Each of the 100 test structures can be individually addressed via an automated *X-Y* stage.

Depositions for the capping experiments were carried out in a high-vacuum system at a chamber base pressure of less than 8×10^{-7} mbar by thermal evaporation from resistively heated boats. Devices with an active area of 2×3 mm^2 are fabricated on 50×50 mm^2 glass (Schott AF45) substrates. After device fabrication, all devices are transferred directly to argon glove boxes with $O_2 < 1$ ppm and $H_2O < 1$ ppm. Current–voltage (*I–V*) and *EL–V* characteristics are measured with a Hewlett Packard parameter analyzer (HP 4145) and Si photodiodes (Hamamatsu S1337, S2281). Luminance calibration of the photodiode and spectral characterization are performed with PR704 and PR705 spectroradiometers (PhotoResearch, Inc.). The schematics of the device structures investigated are shown in Fig. 2. High-work-function anode metals such as Ni, Pt, Pd, Ir, and Mo, are e-beam-evaporated through shadow masks. The first structure is the most widely investigated three-layer device consisting of copper phthalocyanine (CuPc) as buffer layer, N,N'-di(naphthalene-1-yl)-N,N'-diphenyl-benzidine (NPB) as hole-transporting layer and tris(8-hydroxyquinolinato)aluminum (Alq$_3$) as electron-transporting and emission layer. The phosphorescent structure also uses NPB and Alq$_3$, and in addition, 4,4'-N,N'-dicarbazole-1,1'-biphenyl

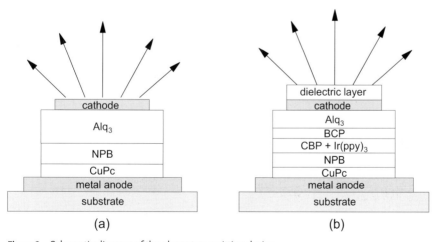

(a)   (b)

Figure 2 Schematic diagram of the planar top-emitting device structure studied. (a) Conventional CuPc/NPB/Alq$_3$ tri-layer structure. (b) Phosphorescent five-layer structure. A dielectric capping layer is applied on top of the semitransparent metal cathode to tune the emission characteristics.

(CBP) doped with 6% tris(2-phenylpyridine)iridium (Ir(ppy)$_3$) as emission layer, and 2,9-dimethyl-4,7-diphenyl 1,10-phenanthrolin (BCP) as hole-blocking layer (see Fig. 2). The organic-layer thicknesses were optimized with the combinatorial method to be 20 nm CuPc, 40 nm NPB, 20 nm CBP:Ir(ppy)$_3$, 10 nm BCP, and 40 nm Alq$_3$ [16]. Typical evaporation rates are on the order of 0.05 and 1.0 Å/s for dopants and host materials, respectively. Semitransparent layers of either Ca (20 nm) or Ca (12 nm)/Mg (12 nm) were deposited as cathodes. For the capping experiments, the multilayer structures were covered with ZnSe layers of various thicknesses. The angular dependence of the spectral characteristics and the intensity distribution of the OLED were measured with the spectroradiometer under constant current conditions at various emission angles up to $\pm 80°$.

18.3
Results and Discussion

18.3.1
Combinatorial Device Optimization

The CuPc/NPB/Alq$_3$ OLED [17] is one of the most widely investigated structures [18–23]. It is therefore an ideal system to demonstrate the power of combinatorial device fabrication and to elucidate the injection, transport and recombination processes. A combinatorial CuPc/NPB/Alq$_3$ matrix is fabricated as described in Section 18.2; in this case, the CuPc thickness was fixed to 15 nm, and the NPB and Alq$_3$ layer thicknesses were varied from 10 to 100 nm, in increments of 10 nm.

Figure 3(a) shows the efficiency (cd/A) of the NPB/Alq$_3$ matrix devices measured at 20 mA/cm^2. This two-dimensional gray-scale plot is based on 100 values, and the contour lines are interpolated between these values. The device with 60 nm Alq$_3$ and 40 nm NPB exhibits the highest efficiency of 2.9 cd/A at 8.5 V with a luminance of 1920 cd/m^2 through the semi-transparent Ca electrode. The shape of the efficiency maps changes only slightly with voltage, and at all voltages the device with 60 nm Alq$_3$ and 40 nm NPB is the most efficient one. At Alq$_3$ thicknesses of less than 30 nm, a clear drop in intensity is observed. This decrease can be explained by exciton-quenching effects due to dipole coupling to the metallic cathode [25, 26]. At thin NPB thicknesses, this effect is not as pronounced because the CuPc layer provides an additional separation of the emission zone from the anode metal.

In display applications, minimization of the operating voltage is always an important issue. For the 100 devices investigated, Fig. 3(b) shows the voltage that is necessary to obtain a current density of 20 mA/cm^2 as a function of individual layer thicknesses. A monotonic rise in voltage from 2.9 to 12.9 V for increasing total device thickness (35 to 215 nm) is observed. This indicates that for a given material system there will always be a trade-off between maximum efficiency (Fig. 3(a)) and lowest voltage operation (Fig. 3(b)). Figure 3(b) shows that at a current density of 20 mA/cm^2 the dependence of the voltage on the NPB and Alq$_3$ layer thicknesses is almost equivalent. Therefore, the voltage drops across the NPB and Alq$_3$ layers must

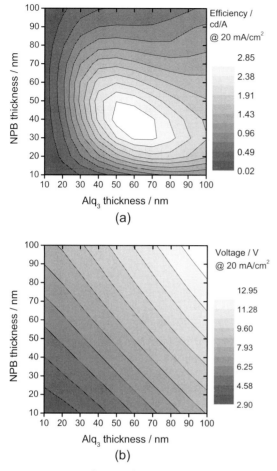

Figure 3 2D maps of 100 combinatorial CuPc/NPB/Alq$_3$ devices. The *x*- and *y*-axes represent the Alq$_3$ and NPB layer thicknesses, respectively. (a) Efficiency (cd/A) and (b) driving voltage at 20 mA/cm^2.

be similar at this current density. However, for different current densities, an electrical field redistribution between the NPB and Alq$_3$ layers occurs [16]. As shown in the following, insights into the limiting mechanisms can be gained by analyzing a certain subset of devices of the combinatorial matrix.

18.3.1.1
Layer-Thickness Variations: Charge-Carrier Balance

The following discussion concentrates on ten devices with a constant total organic-layer thickness of 125 nm. Figure 4 shows the *I–V* EL characteristics of devices with the following structure: CuPc 15 nm/NPB *x* nm/Alq$_3$ (110-*x*) nm. Because the

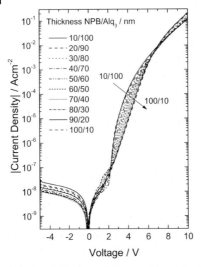

Figure 4 *I–V* curves of devices with an identical total thickness of 125 nm. Device structure: CuPc 15 nm/ NPB *x* nm/Alq₃ (110-*x*) nm.

anode, cathode and CuPc layer of all devices on this substrate are identical, these devices differ only in the position of the NPB/Alq$_3$ interface within the organic stack (see Fig. 1(b)). All the devices exhibit an identical *I–V* onset, which is characterized by a sharp kink in the *I–V* curves at 2.2 V. The second feature of this subset is a crossover of the *I–V* curves at a voltage of about 7 V. At low voltages (2–7 V), the curves for the devices with the thin NPB and thick Alq$_3$ exhibit the highest current flow, whereas at voltages higher than 7 V this behavior is reversed. A similar behavior can also be observed for device subsets with other constant total thicknesses. Because oxidized Ni is a high-work-function anode material (~5.5 eV), it is justified to assume that the injection of holes into CuPc is ohmic and that holes are the first carriers to enter the device. The kink in the *I–V* curves at 2.2 V therefore corresponds to the injection of electrons from the cathode into Alq$_3$. Moreover, as the devices differ only with regard to the position of the NPB/Alq$_3$ interface, the changes in current flow can be directly associated with different internal charge and field distributions. The steeper *I–V* characteristics at low voltages in Fig. 4 for the devices with a thin NPB layer point to limitation by internal injection barriers [24]. Once holes are injected via the CuPc/NPB interface, transport to the NPB/Alq$_3$ interface is not limiting because of the relatively high hole mobility $\mu_h = 10^{-3}$ cm^2/Vs in NPB. At higher voltages (> 7 V), especially the devices with a thicker Alq$_3$ layer will exhibit a reduced current flow, which suggests that electron transport in Alq$_3$ is limiting. The processes described above control the balance between hole and electron currents, which in turn determines the device efficiency. One way to compare the charge-carrier balance of different devices is to normalize the efficiency. Figure 5 shows the efficiency normalized to the maximum as a function of current density. All efficiency curves exhibit an increase with current density and have their maximum at current densities in the range from 10 to 100 mA/cm². Devices with 10 and 20 nm NPB thicknesses did not reach a maximum in efficiency up to the current density measured; therefore, their maximum values are extrapolated. With increasing NPB/Alq$_3$ thickness ratio, the position of the maximum in efficiency shows a monotonic shift from 100 to 10 mA/cm², i.e., devices with a thick NPB layer reach their maximum efficiency at lower current densities than those with a thin NPB layer. Generally speaking, the shape of the efficiency curve (Fig. 5) is indicative of the current balance γ, which is the ratio of the recombination current density j_r and the total current density j_{total} in the device. Unbalanced electron and hole currents reduce the efficiency in the device simply because injected charge carriers reach the opposite electrode without recombining and without generating light. To first order, it can be assumed that the maximum in the efficiency curve corresponds to the point of charge balance ($\gamma \approx 1$). The absolute values of the maximum efficiency for the ten individual devices, however, vary considerably between 0.05 and 2.7 cd/A. Optical interference effects are one reason for this because in these devices the position of the emission zone, which is close to the NPB/Alq$_3$ interface, is shifted over a large distance. Another reason for the reduced efficiency of thin NPB or Alq$_3$ layers is the above-mentioned exciton quenching near the electrodes. In contrast to the absolute efficiency value, the shape of the efficiency curve for an individual device should not be influenced by interference and quenching effects as long as the position of the emitting zone in Alq$_3$ does not shift with different driving conditions. Note that at high current densities (> 100 mA/cm²), high concentrations of ionic species, especially of Alq$_3$ cations, lead to luminescence quenching and hence to a reduction in efficiency [27]. Nevertheless, the normalized efficiencies allow a comparison of the current balance as it is shown in Fig. 5. The device with the thinnest NPB layer (NPB 10 nm/Alq$_3$ 100 nm) is the least balanced device, and the current balance increases steadily with increasing NPB layer thickness. This type of analysis

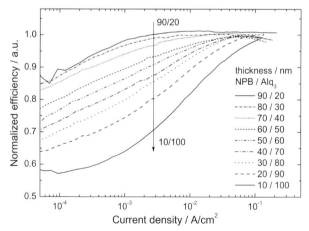

Figure 5 Normalized efficiency vs. current density of devices with constant organic-layer thickness: CuPc 15 nm/NPB x nm/Alq$_3$ (110-x) nm.

of the current balance in different device architectures is ideally suited to understand the limiting mechanisms and allows the selection of optimum materials with respect to injection, transport, and emission.

18.3.1.2
Layer-Thickness Variation: Optical Effects

Our top-emitting OLED consists of an emitting region embedded between organic layers as well as of a highly reflective and a semi-transparent metal mirror, and can therefore be regarded as a weak cavity. Accordingly, varying the individual layer thicknesses in such a device will not only affect the electrical but also the optical properties, especially the emission characteristics. In Fig. 6(a) the peak wavelengths

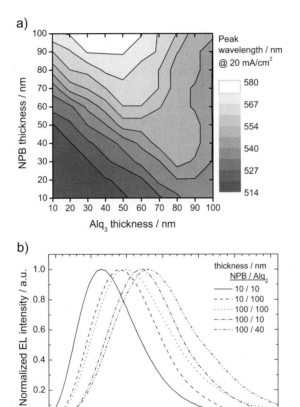

Figure 6 Spectral characteristics of CuPc/NPB/Alq$_3$ OLEDs. (a) 2D plot of peak wavelength. The *x*- and *y*-axes represent the Alq$_3$ and NPB layer thicknesses, respectively. (b) Normalized spectra of selected devices. All spectra were taken at 20 mA/cm^2.

of the 100 combinatorial devices as a function of NPB and Alq$_3$ thicknesses are shown as a gray-scale-coded contour plot. The emission spectra show a broad distribution of peak wavelengths ranging from 514 to 580 nm. Figure 6(b) shows a selection of five EL spectra out of the hundred. The spectra are chosen such as to cover the limiting cases, i.e., the thinnest and the thickest device, and a combination of thinnest NPB/thickest Alq$_3$ layer and vice versa. The plot also contains the device with the most red-shifted spectrum, which consists of 100 nm NPB and 40 nm Alq$_3$. As expected from cavity effects, the thinnest device (15 nm CuPc, 10 nm NPB, and 10 nm Alq$_3$) exhibits the shortest peak wavelength. Increasing the thickness of the Alq$_3$ layer from 10 to 100 nm (NPB: 10 nm) leads to a red shift of the spectrum by 26 nm from 514 to 540 nm. However, when the thickness of the NPB layer is increased from 10 to 100 nm (Alq$_3$: 10 nm) the peak redshifts by 58 nm (see Fig. 6(b)). By varying the thicknesses of the individual layers, the emission zone, which is located close to the NPB/Alq$_3$ interface, is shifted over a large distance within the cavity. As a consequence, the efficiency and the spectral shape will be determined by interference effects [12, 28]. The side-by-side comparison of Figs. 3(a), 3(b) and 6(a) illustrates that device optimization for color coordinates and efficiency via layer-thickness variations will require a trade-off of the desired properties. It would be therefore desirable to tune the spectral characteristics without changing the *I–V* characteristics and the charge-carrier balance, and, in addition, improve the light outcoupling. One concept for achieving this will be discussed in the following.

18.3.2
Optical Outcoupling

A preferred concept known from thin-film optics utilizes a thin dielectric layer on top of a thin metal film to modulate the transmittance of the cathode [14, 15]. The main goal of the second part of our paper is the quantitative understanding of optical effects of OLEDs, in particular the effect a dielectric capping layer on top of a semitransparent metal cathode has on the OLED performance. For this reason the EL emission characteristics of phosphorescent OLEDs were investigated as a function of ZnSe thickness. The wide-band-gap semiconductor ZnSe with a refractive index of $n = 2.6$ was deposited on top of the Ca/Mg cathode of an OLED consisting of 20 nm CuPc, 40 nm NPB, 20 nm CBP:Ir(ppy)$_3$, 10 nm BCP, and 40 nm Alq$_3$ (see Fig. 2(b)). The thickness of the dielectric capping layer was sequentially increased from 0 to 110 nm on the same OLED. After each ZnSe growth sequence, electrical and optical measurements were performed under inert conditions. Figure 7 shows the efficiency of the phosphorescent top-emitting OLED as a function of current density for various ZnSe-layer thicknesses. A significant influence of the ZnSe thickness on the outcoupled EL intensity can be observed. Already without capping layer can an efficiency of 38 cd/A at 10 μ A/cm^2 be achieved owing to the efficient phosphorescent device structure. The efficiency increased to a maximum value of 64 cd/A when a 60-nm-thick ZnSe layer is used. Note that the general functional behavior of the efficiency curves and also the current–voltage characteristics [29] are not influenced by the deposition of the dielectric layer. This indicates that the

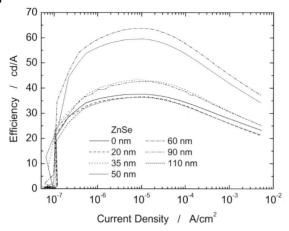

Figure 7 Efficiency (cd/A) versus current density of the OLED for various capping-layer thicknesses. OLED structure: Al/Ni/ 20 nm CuPc/40 nm NPB/20 nm CBP + Ir(ppy)$_3$/10 nm BCP/40 nm Alq$_3$/12 nm Ca/12 nm Mg/X nm ZnSe.

electrical properties of the OLED, including the charge-balance factor, are not affected by the capping layer. The EL efficiency enhancement of a factor of 1.7 is therefore purely due to the modified optical architecture. In other words, the number of excitons generated in the OLED is identical, whereas the number of photons detected externally in the solid angle considered is significantly affected by modifications of the optical structure.

In [29] it was demonstrated that the transmittance of the Ca/Mg/ZnSe layer sequence changes periodically with the ZnSe thickness. The transmittance of the uncovered cathode consisting of 12 nm Ca and 12 nm Mg is about 0.52. With increasing capping-layer thickness, the transmittance reaches a maximum of 0.78 at about 20 nm and a minimum of 0.32 at about 65 nm ZnSe. The experimentally found thickness of 60 nm ZnSe for maximum efficiency therefore does not coincide with the value of 20 nm for the highest transmittance of the cathode configuration, but matches well with the value at which minimum transmittance occurs. This discrepancy substantiates that the simple assumption that maximum outcoupling is obtained at maximum transmittance of one electrode, here the cathode, is not valid. The correlation between the two parameters is much more complex. To explain the experimental results qualitatively as well as quantitatively, interference effects present in structures of this type have to be taken into consideration. An optical model developed by Neyts [28] inherently takes into account all these effects and is thus ideally suited to calculate the emission characteristics of OLEDs.

This model can, for example, be used to calculate the spectral characteristics and intensity emitted in forward direction as a function of the capping-layer thickness [30]. In Fig. 8 the dependence of the EL intensity emitted in forward direction on the ZnSe thickness ise shown together with the corresponding simulated values. With increasing ZnSe thickness, the outcoupled EL intensity first decreases slightly, ex-

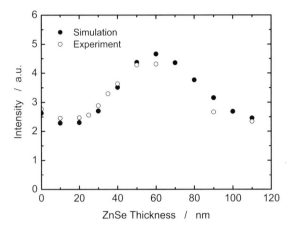

Figure 8 Simulated (solid circles) and measured (open circles) EL intensity (radiometric) emitted in forward direction as a function of ZnSe thickness. The simulated values are scaled with the same constant factor to best fit the measurements. The measurements were performed under constant current condition using 12.5 mA/cm².

hibits a minimum between 10 and 20 nm, and finally reaches a maximum at 60 nm. For thicker ZnSe layers, the intensity falls off again, and for 110 nm ZnSe a value even lower than the one without capping is obtained. The data simulated by the optical model [30] is in perfect agreement with the measurement presented here. This proves that interference effects within the device are responsible for the oscillatory behavior of the EL intensity emitted in forward direction. Furthermore, it substantiates that to optimize the outcoupled light it is by far not sufficient to consider only the transparency of the cathode [30].

The interplay between the different interference effects directly affects the outcoupled intensity and also influences the spectral characteristics significantly. Figure 9(a) depicts a set of normalized EL spectra for selected ZnSe thicknesses. The spectral weight of EL emission shifts strongly by varying the capping-layer thickness. Without ZnSe, the EL intensity shows a maximum at 512 nm and a full width at half maximum (FWHM) of 72 nm. A comparison with the photoluminescence (PL) spectrum of a thin-film reference sample of CBP doped with 6% Ir(ppy)$_3$ reveals that the emitted EL spectrum of the uncapped OLED is already modified by the weak microcavity structure in which Al/Ni and Ca/Mg are used as highly and partially reflective electrodes, respectively. The corresponding EL spectrum possesses a second peak at 536 nm whereas the PL only shows a weak shoulder (see Fig. 9(a)). The best agreement between the EL and PL spectra is found for 25 nm ZnSe, for which the cathode transmittance is very close to its maximum value [29]. With 50 nm ZnSe, a pure green emission with a peak at 508 nm and an extremely narrow FWHM of only 36 nm is observed, resulting in CIE 1931 color coordinates of $x = 0.19$ and $y = 0.67$. Upon further increasing the thickness of the dielectric layer, the spectrum becomes broader again, accompanied by a shift of the peak wavelength to 544 nm at 90 nm ZnSe thickness. At 110 nm ZnSe, the peak wavelength has

a)

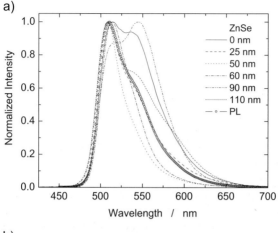

b)

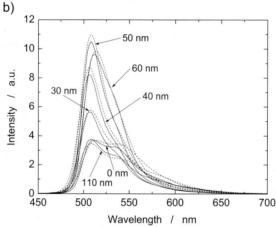

Figure 9 (a) Normalized EL spectra measured at constant current density for various ZnSe thicknesses of the OLED shown in Fig. 7. The PL spectrum of Ir(ppy)$_3$ doped with CBP is shown as reference. (b) Comparison of the simulated (solid lines) and experimentally measured (dashed lines) spectral characteristics for ZnSe thicknesses of 0, 30, 40, 50, 60, and 110 nm. The simulated spectra are all scaled with the same constant factor to best fit the experimental data.

moved back to 508 nm. In Fig. 9(b) the simulated spectral characteristics (solid lines) for six different ZnSe thicknesses and the corresponding measured EL intensities (dashed lines) are compared. The optical simulation excellently describes the dependence of the spectral characteristic on the ZnSe thickness. Remarkable is also that the calculated intensity ratios agree perfectly with the experiment. Note that all spectra are simulated with an identical set of optical input parameters. This result is strong evidence that microcavity effects induced by a change of the cathode reflectivity due to the capping layer are responsible for the spectral shift.

The spatial emission pattern strongly depends on the details of the optical architecture, in particular on the organic-layer thicknesses and the electrode reflectivity [31]. Figure 10 displays the EL intensity measured at 12.5 mA/cm² for selected ZnSe thicknesses as a function of the viewing angle. The emission pattern of the OLED without ZnSe is nearly Lambertian, and therefore has an almost constant intensity at all angles in this representation. With increasing ZnSe thickness, the externally detected EL intensity is enhanced in forward direction, and reduced for larger viewing angles. The most strongly directed emission pattern is found at a capping-layer thickness of 50 nm ZnSe, where an enhancement of the radiation by a factor of 1.6 is measured at 0° viewing angle. Under these conditions, also the integrated intensity (integration between −70° and +70°) is increased, i.e., it amounts to 110% of the value without capping. However, for 25 and 35 nm ZnSe, this integrated light output is reduced to 85% and 90%, respectively. By further increasing the ZnSe thickness, the spatial emission pattern returns to its original form, and at 90 nm almost corresponds to the one without capping. Even the integrated intensities are nearly identical.

This dataset demonstrates that the capping layer and its thickness significantly influence the optical properties of the top-emitting OLED. As the device structure used represents a weak microcavity, interference effects occur, and it is their interplay that defines the optical properties and determines the emission characteristics of the OLED. By varying the capping-layer thickness the reflectivity of the top mirror and thus the optical interference can be controlled. The excellent agreement between the experimental results and the data simulated by an optical model shows that the variation of the EL emission due to the capping layer can be entirely accounted for by a change in the optical interference effects. Consequently, these interference effects can be exploited to tune the spectral characteristics and the angular intensity distribution as well as to improve light outcoupling.

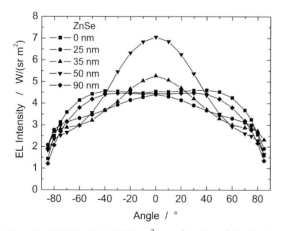

Figure 10 EL intensity in W/(sr m²) as a function of the viewing angle of the OLED described with 0, 25, 35, 50, and 90 nm ZnSe as capping layer, measured at 12.5 mA/cm².

18.3.3
20-inch a-Si AMOLED Display

The power of combinatorial device optimization and of our concept for improving outcoupling lies in the fact that they can be used independently and for any type of device architecture. Figure 11 shows the spectral characteristics of RGB devices with extremely high efficiency, excellent color coordinates and lifetime. The performance data of the RGB devices are listed in Table 1. It is clear that to achieve such performance data the proper material sets and optimized electrodes have to be chosen. In particular, the thickness of the individual active organic layers, the position of the emission zone and the thickness of the capping layer have to be fine-tuned with nanometer precision. The high-efficiency RGB sub-pixels result in NTSC white with more than 22 cd/A under realistic display-driving conditions. Such a high efficiency significantly reduces the current flow through the a-Si TFTs and therefore allows bright and long-term stable emission. Because OLEDs are current-driven devices, it has so far been believed that active-matrix displays could only be driven with polycrystalline or crystalline silicon transistors. Figure 12 shows the 20-inch prototype top-emitting full-color a-Si AMOLED display presented at the May 2003 SID Meeting in Baltimore, MD, by IBM Research, IDTech and CMO [1]. It has WXGA resolution (1280 × 768 pixels) and a power consumption of 25 Watt at a brightness of 300 cd/m^2 (its desktop display brightness can exceed 500 cd/m^2). The organic layers are fabricated on a substrate with an array of a-Si TFTs. An improved color reproduction, which is better than that of cathode-ray tubes and LCDs, and an extremely high efficiency have been achieved specifically by the use of custom-made organic materials in combination with the optical-device architecture described. In conclusion, we have demonstrated that a-Si TFTs can deliver sufficient driving current for bright

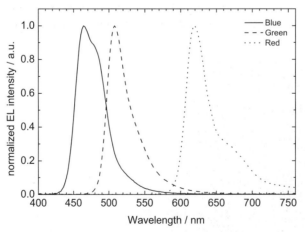

Figure 11 Spectra of the optimized RGB device structures. The efficiencies of the OLED structures are 22, 64, and 10 cd/A for R, G, and B, respectively. The CIE *xy* coordinates are listed in Table 1.

Table 1 Performance data of optimized OLED structures. Efficiency values achieved in test structures.

	Red	Green	Blue	White
NTSC	0.67, 0.33	0.21, 0.71	0.14, 0.08	0.33, 0.33
CIE x,y	0.68, 0.32	0.24, 0.68	0.13, 0.12	0.31, 0.32
cd/A (max.)	22	64	10	25.6
cd/A (@ 10 mA/cm^2)	16	55	10	22

OLED emission and are long-term stable under the operation conditions needed for high-end display applications. Hence, a-Si AMOLED can now be considered to represent a challenge to the large-sized active-matrix liquid-crystal display (AMLCD) business.

18.4
Conclusions

To summarize, combinatorial device fabrication is a powerful tool to provide reliable and systematic data of complex organic light-emitting structures, as are invaluable for a detailed understanding and modelling of the device as well as for its fast and efficient optimization. In addition, we have demonstrated that the concept of dielectric capping considerably improves OLED performance in top-emitting devices without modifying the electrical properties. Furthermore, we have shown that precise predictions of the emission characteristics of planar OLEDs can be made through numerical simulations. The demonstration of the 20-inch a-Si AMOLED panel of this quality is a breakthrough for the emerging AMOLED display technology. It will have tremendous impact on the entire display industry and the future display market

Figure12 World's largest full-color AMOLED display (20-inch), WXGA (1280 × 768) driven by amorphous-silicon TFTs.

because a-Si TFTs are the cheapest driving technology for medium- and large-sized active-matrix displays. It also represents a milestone in the industry as it establishes AMOLED displays as competitor of the liquid-crystal displays in all sizes and form factors.

Acknowledgments

We are grateful to K. Neyts for his simulations and thank S. Karg, C. Rost, M. Tschudy, M. Sousa, S. Alvarado, D. Gundlach, T. Brunschwiler, B. Ruhstaller, and P. Müller for useful discussions. We thank P. F. Seidler for his continuous support.

References

[1] T. Tsujimura *et al.*, *Journal of the SID* **XXXIV**, 6 (2003).

[2] C. W. Tang and S. A. Van Slyke, *Appl. Phys. Lett.* **51**, 913 (1987).

[3] J. Kido and Y. Iizumi, *Appl. Phys. Lett.* **73**, 2721 (1998).

[4] L. S. Hung, C. W. Tang, and M. G. Mason, *Appl. Phys. Lett.* **70**, 152 (1997).

[5] M. A. Baldo, S. Lamansky, P. E. Burrows, M. E. Thompson, and S. R. Forrest, *Appl. Phys. Lett.* **75**, 4 (1999).

[6] S. Lamansky, P. Djurovich, D. Murphy, F. Abdel-Razzaq, H.-E. Lee, C. Adachi, P. E. Burrows, S. R. Forrest, and M. E. Thompson, *J. Am. Chem. Soc.* **123**, 4304 (2001).

[7] C. Schmitz, M. Thelakkat, and H.-W. Schmidt, *Adv. Mater.* **11**, 821 (1999).

[8] T. A. Beierlein, H.-P. Ott, H. Hofmann, H. Riel, B. Ruhstaller, B. Crone, S. Karg, and W. Rieß, in *Proc. SPIE*, vol. 4464, pp. 178–186 (2002). Proc. SPIE Vol. 4464, p. 178–186, Organic Light-Emitting Materials and Devices V; Zakya H. Kafafi; Ed.

[9] C. Adachi, M. A. Baldo, M. E. Thompson, and S. R. Forrest, *J. Appl. Phys.* **90**, 5048 (2001).

[10] C. F. Madigan, M.-H. Lu, and J. C. Sturm, *Appl. Phys. Lett.* **76**, 1650 (2000).

[11] T. Tsutsui, M. Yahiro, H. Yokogawa, K. Kawano, and M. Yokoyama, *Adv. Mater.* **13**, 1149 (2001).

[12] S. K. So, W. K. Choi, L. M. Leung, and K. Neyts, *Appl. Phys. Lett.* **74**, 1939 (1999).

[13] Y. Fukuda, T. Watanabe, T. Wakimoto, S. Miyaguchi, and M. Tsuchida, *Synth. Met.* **111–112**, 1 (2000).

[14] A. Vasicek, *Optics of Thin Films* (North Holland, Amsterdam, 1960).

[15] L. S. Hung, C. W. Tang, M. G. Mason, P. Raychaudhuri, and J. Madathil, *Appl. Phys. Lett.* **78**, 544 (2001).

[16] T. A. Beierlein, Ph. D. Thesis, University of Bayreuth (Cuvillier Verlag, Göttingen, Germany, 2003).

[17] S. A. Van Slyke, C. H. Chen, and C. W. Tang, *Appl. Phys. Lett.* **69**, 2160 (1996).

[18] J. Shi and C. W. Tang, *Appl. Phys. Lett.* **70**, 1665 (1997).

[19] H. Riel, H. Vestweber, and W. Rieß, in *Polymer Photonic Devices*, ed. B. Kippelen and D.D. Bradley, *Proc. SPIE*, vol. 3281, pp. 240–247 (1998).

[20] M. Matsumura and Y. Miyamae, in *Organic Light-Emitting Materials and Devices III*, ed. Z.H. Kafafi, *Proc. SPIE*, vol. 3797, pp. 283–289 (1999).

[21] E. W. Forsythe, M. A. Abkowitz, Y. Gao, and C. W. Tang, *J. Vac. Sci. Technol. A* **18**, 1869 (2000).

[22] W. Brütting, H. Riel, T. Beierlein, and W. Rieß, *J. Appl. Phys.* **89**, 1704 (2001).

[23] W. Brütting, S. Berleb, and A. G. Mückl, *Org. Electron.* **2**, 1 (2001).

[24] H. Riel, T. A. Beierlein, S. Karg, and W. Rieß, in *Organic Light-Emitting Materials and Devices VI*, eds. Z.H. Kafafi and

H. Antoniadis, *Proc. SPIE* vol. 4800, pp. 148–155 (2003).

[25] H. Becker, S. E. Burns, and R. H. Friend, *Phys. Rev. B* **56**, 1893 (1997).

[26] A. L. Burin and M. A. Ratner, *J. Phys. Chem. A* **104**, 4704 (2000).

[27] Z. D. Popovic, H. Aziz, N.-X. Hu, and A. I. ad P. N. M. dos Anjos, *J. Appl. Phys.* **89**, 4673 (2001).

[28] K. A. Neyts, *J. Opt. Soc. Am. A* **15**, 962 (1998).

[29] H. Riel, S. Karg, T. Beierlein, B. Ruhstaller, and W. Rieß, *Appl. Phys. Lett.* **82**, 466 (2003).

[30] H. Riel, S. Karg, T. Beierlein, W. Rieß, and K. Neyts, *J. Appl. Phys.* **94**, 5290 (2003).

[31] T. Tsutsui and K. Yamamato, *Jpn. J. Appl. Phys.* **38**, 2799 (1999).

Index

a

absorption coefficient   135
acceptor   433, 439, 442
accumulation   345
accumulation contact   351
action spectra   209
activation energy   315, 444
active-matrix OLED (AMOLED) display   511
aggregate emission   155, 168 f
alkoxy-substituted poly-phenylenevinylene (PhPPV)   186 ff
Alq_3 95 ff, 121, 261, 265 f, 477
– blue luminescence   103 ff
– crystalline phases   97 ff
– doped with Rubrene   285
– energy diagram at interfaces   87
– high-temperature phase   104
– δ-phase   100
– symmetry groups   112
– trap distribution   279 ff, 282, 294 f
Alq_3 film, polarization   85
annealing   441

b

band bending   58, 69
– C_{60}/metal interface   78
– definition   72
– in Alq_3 film   84
– in nonequilibrium   75
– in thermal equilibrium   74
– TPD/metal interface   81
band transport   7
barrier for hole injection   470
bathochromic shift   504
BCP   478
bimolecular process   119
bimolecular recombination   442 ff, 445
binding energy   142, 483
bipolaron   276, 347, 459

blocking energy barrier   491
blue light emitting polymer   153 f
built-in field   447
built-in potential   69
bulk-heterojunction solar cells   433 f, 439 ff, 442

c

C_{60}, bulk Fermi level   80
– energy diagrams on metals   81
capacitance, depletion-layer   355
– insulator   345
– MIS   355
carbonyl   161, 164
β-carotene   137 ff
carotenoids   139 ff
carrier density distributions   492 ff
channel length   345, 371
charge carrier density   9, 310 ff
charge carrier generation   448
charge carrier injection   9
charge carrier mobility   307 ff, 444
charge carrier transport   6
charge photogeneration   142 ff
charge transfer center   192
charge transfer state   186
– singlet exciton   184
charge transport   305 ff, 319, 321, 445
– density dependence of   320
charge trapping   266
charge-carrier balance   515
charged encounter complex   260, 264
chelate complex   95, 125
chemical interaction   58
color tuning   494
combinatorial device fabrication   512
combinatorial method   486
conjugated polymers (CP)   1, 131 ff
conjugation-length   236, 242 ff, 251

Physics of Organic Semiconductors. Edited by W. Brütting
Copyright © 2005 WILEY-VCH Verlag GmbH & Co. KGaA, Weinheim
ISBN 3-527-40550-X

contact workfunction   330, 332
continuity equation   350
correlated dipoles   504
Coulomb effects   480
CBP   263, 265
CuPc   477
cut-off frequency   356, 372
cyclovoltametric method   485

d
DCM   203 ff, 478
defects   442 ff
delayed fluorescence   118, 126
delayed luminescence   126
delayed phosphorescence   265 f
Density of occupied states s. DOOS
Density of states (DOS)   319, 347
– Gaussian DOS   319, 322, 328
dephasing   137 ff
depletion length   352
device efficiency   442, 517
device simulation   442, 517
– ATLAS   351
– MOLED   476, 479
– ISETCAD   351
device structures   9
dielectric constants   484
differential scanning calorimetry (DSC)   100
differential transmission   135 ff
diffusion   320
dipolar layer   227
dipole   41, 43, 56, 58
disorder, correlated disorder   488
– dipolar   488
– energetic   487
– short-range correlations   488
– spatial   487
DMO-PPV, trap distribution   298 ff
donor   433, 439, 442
donor/acceptor interface, dipolar layer   230
– exciton dissociation   226 ff
DOOS   278
– α-NPD doped with 1-NaphDATA   283
– of Alq$_3$ 295
– of DMO-PPV   299
– of small molecules   281
dopant assisted charge photogeneration   215 ff, 220 ff
doped conjugated polymers   195 ff
doping   9, 348, 362, 436 ff
double-cable polymers   230
double-excitation   145 ff
– s. a. spectroscopy, pump-push-probe

drift-diffusion-model   350
dye doping   494 ff
dynamic CV curves   362

e
effective carrier mass   223, 228 ff
effective density of states   347
efficiency   494
Einstein relation   351
Einstein relation – generalized   324
– density dependence of   324, 326, 330, 337
electric field density distribution   494
electrical potential   350
electrical transport properties   446
electroabsorption   278
electrode, insulating layer   487
– monolayers   486
– surface dipole   486
electrode-sensitized photoinjection   209
electroluminescence   236 ff, 252, 486
– delayed   118
– efficiency   235, 237
– efficiency, maximum possible   238, 245
– spin statistics   238 ff
electron affinity   346
electron hole confinement   498
electron scavenger   204
π-electron system   2
electron trapping   495
electronic relaxation   139 ff
electron-phonon interaction   235 ff
electro-phosphorescence   118
electrostatic bonding   405
emission layer   495
emission pattern   523
energetic disorder   483
energetic position   118
energy gap   346
energy level alignment at organic/electrode interfaces   69
energy level shift   502
equations of motion   480
equilibrium   322 f, 331
equivalent circuit   356
escape energy   274
exchange interactions   264
excimer emission   155, 168 f
exciton   131, 183, 235, 237 ff, 245 ff
– bimolecular exciton annihilation   189
– binding energy   6, 185, 239
– breaking rate   189
– density-of-states   198 ff
– diffusion   265 f

– diffusion length 11
– dissociation 185, 189, 193 ff, 224, 226
– dissociation at a charge transfer center
  221
– dissociation potential 205
– dissociation probability 221
– energy relaxation 198 ff, 221
– excitonic DOS distribution 205
– field induced exciton quenching 214
– field-assisted dissociation 191
– formation 245
– formation cross-section 239
– formation rate 239
– fusion 119
– off-chain exciton dissociation 221
– optical absorption 241, 247 ff
– quenching 198, 259, 261, 266 f
– quenching, at charge transfer centers 204 ff
– singlet 235, 237 ff, 242, 245 ff
– singlet exciton 249 ff
– spin-statistics 245
– triplet 235, 237 ff, 245 ff, 250 ff
– ultrafast on-chain dissociation 215
exponential density of states 305
external quantum efficiency 433
extraction 449
extrinsic charge photogeneration 209

f
facial isomer 125
Fermi distribution 347
Fermi energy 347
Fermi level 323
Fermi level alignment 69, 73 f
field-effect doping 9
field-effect mobility 310 ff
field-effect transistor (FET) 305 ff, 334, 393
field-induced exciton dissociation 185 ff
field-induced PL quenching 187
fill factor 442, 447
flat band voltage 354
fluorescence 5, 118
fluorescence quenching 185 ff
– spectral dependence 188
– time resolved 188, 502
Förster energy transfer 260
Förster radius 198, 205
Förster type 502
Fourier-transform infrared spectroscopy
  (FT-IR) 112, 240 ff
Franck-Condon region 136 ff
Franck-Condon state 184, 187 ff, 191, 215,
  218

Frank-van-der-Merwe growth 18
Frenkel exciton 184
FT-IR 112, 240 ff

g
gain 135
gate insulator 345, 383
Gaussian density of states 8, 305, 348
Gaussian distribution 387
geminate pair 184 ff, 191 ff, 195 ff, 209,
  212 ff, 220, 226 ff, 230, 276 f
– interfacial 228 ff
geminate recombination 192
general recombination picture 505
geometric isomers 114
g-factor, g-value 436
glass transition temperature 476
ground state depletion 435

h
height difference correlation function 19
hexabenzocoronene (HBC) 224 ff
high-k dielectrics OFETs 408, 424
hole accumulation 487
hole mobility 306 ff
hole trapping 495
hole-conducting polymer 451 ff
HOMO 441
hopping frequency 482
hopping rate 325
hopping transport 7
hot exciton 184
hot exciton dissociation 191, 218 ff
hydroxyquinoline 95, 112, 125
hydroxyquinoline ligands 108
hyperthermal growth 34
hypsochromic shift 504
hysteresis 377

i
image charge screening 54
image force 332
image force potential, three-dimensional
  effects 479 ff
impedance 356
impurity 275
indium-tin oxide (ITO) 58, 451
infrared (IR) spectroscopy 112, 240 ff
– s. a. IR
injection 480
– density of states 481
– disorder 482
– Fermi-Dirac function 481

– Fowler-Nordheim 483
– Langevin recombination 484
– metal electrode 481
– Richardson-Schottky 483
– thermionic 483
– tunneling distance 483
– tunneling integral 481
– WKB approximation 481
intensity oscillations 32
interaction potential 20
interchain charge transfer 436 ff
interchain interaction 235
interdiffusion 17
interface 17, 41, 50
interface charges 354, 362, 365
interface dipole 70
interface morphology 229
interference effects 521
intermolecular interactions 117
internal conversion 136 ff
internal electric field 445, 489
internal quantum efficiency 433, 445
intersystem crossing (ISC) 115, 126
intrinsic charge carrier density 444
intrinsic photogeneration 210 ff
inverse subthreshold slope 354, 365, 368,
   372
inversion 379
ionisation energy 460
IPCE 441, 447
IR (infrared) spectroscopy 112, 240 ff
Ir(ppy)$_3$ 263, 265
islanding 37
isomer, facial 96 ff
– meridional 96 ff
– transformation 111
isomerism 96, 125
ITO 58, 451

k
Kelvin probe (KP) method 72, 76 f, 461
keto defect 153 f

l
layer-by-layer growth 32
leakage current 492 ff
lifetime 445, 494
– of OLED devices 465
– triplet states 121
light emitting diode 305 ff, 330
light induced electron spin resonance
   (LESR) 434, 448
liquid crystal display (LCD) 475

localisation 479
localizations effects 494
luminescence 4
luminescent efficiency 258
LUMO 439 ff

m
macroscopic transport 504
magnetic resonance 245
– electroluminescence detected 251 ff
– optically detected 244 ff
– photoinduced absorption detected 245 ff,
   250 ff
– photoluminescence detected 244 ff, 248 ff,
   434 ff, 448
matrix geometry 478
mean drift length 445 ff
MEH-PPV 263, 365
MeLPPP 143 ff, 209 ff, 218, 245
– charge generation efficiency 213
– field-induced fluorescence quenching 210
– photocurrent 207
– transient photocurrent 211, 212
– transient photogeneration 210
meridional isomer 125
migration 143, 145 ff
mini-exciton 118
miscut 28
mobility 7, 324, 346 f, 357, 393, 404, 419, 445
– anisotropy 363, 368, 423
– asymmetric detailed balance 482
– density dependence 326, 330, 337
– electric field dependence 327, 330
– extraction 335, 339
– gate-voltage-dependence 420
– lateral 363
– perpendicular 362, 365
– polaron mobility 482
– Poole-Frenkel mobility 483
– symmetric detailed balance 482
– temperature dependence 400, 405, 421,
   422
mobility-lifetime product 446
molecular crystals 1
molecular exciton 184
molecular orientation 46
molecular packing 109, 125
molecular structure, δ-Alq$_3$ 106
molecular symmetry 111
MOLED, OLED device simulation 476, 479
monomolecular recombination 442 ff
monotropic transition 101
Monte Carlo 487

morphology   103, 117, 433
MOS capacitor   360
Mott-Schottky (MS) model   69, 74

n
1-NaphDATA, TSC   279 ff
noncollinear optical parametric amplifier
   (NOPA)   132 ff
α-NPD   477
– I-V characteristics   289
– TOF   290
– TSC   279 ff, 296

o
OC$_1$C$_{10}$-PPV   306 ff, 433 ff, 439 ff
ODMR   245
ohmic contact   382
OLED   126, 235 ff, 451 ff
oligomer   236, 240, 242
– oligomer-length   242 ff
– oligophenyl   242
– oligophenylene-vinylene (OPV)   242
– oligothienylene-vinylene (OTV)   242
– oligothiophene (OT)   242
Onsager's theory   143, 184, 214 ff, 218, 228,
   230
– Brown's adaptation   220
– of geminate recombination   209
open-circuit voltage   439, 447
O$_2$-plasma   491
optical absorbance   456
optical outcoupling   519
optical properties   4
optical transition   235 ff
optically stimulated current (OSC)   276
π-orbital   3
σ-orbital   3
organic field-effect transistor (OFET)   10
organic insulator   359
organic light emitting diode (OLED)   10,
   235 ff, 245, 251 ff, 451 ff, 511
– delayed fluorescence   122
– delayed phosphorescence   122
organic photovoltaic cell (OPVC)   10
organic solar cells   447
orientational degrees of freedom   20

p
P3HT   434, 439 ff
P3OT   360, 378
PANI   468
parallel resistance   442
parity   140

parylene   410, 414
PCBM   433 ff, 439 ff
PDA   139
PDMS   406, 423
PEDOT:PSS   58, 61, 448, 451 ff
perylenediimide (PdI)   224 ff
PFO   245
phase transition   101, 125 f
phenyl-substituted PPV (PPPV)   186 ff
phosphorescence   5, 126
– Alq$_3$ 118 ff
photo cell   339
– quantum efficiency of   339
photoconductivity   208 ff
photocurrent   444, 447, 449
photocurrent action spectrum   209
photoemission   51, 54
photoemission spectroscopy   42, 43
photoexcitation   114 ff, 115, 448
photogeneration   9, 184, 433
photo-induced absorption   136, 158, 173,
   240 ff, 278, 448, 494
– s. a. PIA
photoinduced charge transfer   11, 435
photoinduced electron transfer   433
photoluminescence   125, 245, 248 ff
– Alq$_3$ 102 ff
photoluminescence decay time   197
photoluminescence quenching   194 ff, 201 ff,
   206
photomodulation   240 ff
photo-oxidative degradation   154, 165 f
photophysics   6
photovoltaic devices   449
PhPPV   196 ff, 215 ff, 218 ff, 223 ff
– charge carrier photogeneration quantum
   yield   216 ff
– photoluminescence intensity   208
PhPPV:PdI, charge photogeneration quantum
   yield   225
phthalocyanine   42, 50
PIA   136, 158, 173, 240 ff, 278, 448, 494
pi-conjugated oligomer   236
pi-conjugated polymer, degenerate ground
   state system   242
– non-degenerate ground state system   242
PL s. photoluminescence
plastic solar cells   433
PLDMR   245 ff, 448
– photoluminescence detected magnetic
   resonance   434 ff
PLED   153, 451 ff
Poisson equation   321, 350

polarization 495
polarization energies 7
polaron 140, 145 ff, 235 ff, 245, 248 ff, 276, 347, 459
– confinement parameter 242
– interchain-interaction 245
– energy level, localized 235 ff
– mobility 235
– optical absorption 235 ff, 240 ff, 244
– recombination 235, 245, 251
– recombination, spin-dependent 245, 247, 251
– Su-Schrieffer-Heeger model 235, 242
poly(3,4-ehtylenedioxythiophene):poly(styrene-sulphonic acid) (PEDOT:PSS) 451 ff
polyacetylene 140 ff
poly(3-alkylthiophene) (P3AT) 358
polyaniline 468
polycrystalline phases 125
polyenes 139 ff
polyfluorene (PFO) 148, 153 f, 245, 251, 451, 455, 463
– β-phase 154, 157 f
poly(fluorene-co-fluorenone) 166
polymer, conjugated 131 ff
polymer donor/acceptor blends 226
– photoconductivity 224 ff
polymer light emitting diode (PLED) 153, 451 ff
polymer solar cell 449
polymer/acceptor blend 185
polymer/acceptor interface 227
polymorphism 17, 100
poly-para-phenylene (MeLPPP) 186 ff
– ladder-type 245, 251
poly-phenylene-ethenylene (PPE) 244
poly-phenylene-vinylene (PPV) 207 ff, 244, 249 ff, 252, 305 ff, 358
– MeH-PPV 244
poly(phenylphenylene vinylene) (PPPV), fluorescence quenching 202 ff
poly-spirobifluorenes 451, 455, 463
polythiophene 452 ff
– polythiophene, regio-random 244
– polythiophene, regio-regular 244
poly(4-vinylphenol) (P4VP) 359
Poole-Frenkel mechanism 143, 499
potential energy surfaces 136
powder diffraction 106
powder spectrum 436
power conversion efficiency 442, 447
PPV 138, 209, 244, 451, 455, 463
PtOEP 265 f

π–π*-transitions 2
pulse radiolysis 259
pump-probe spectroscopy 189 ff

q
quantum efficiency 224
– PL 103
quantum yield 436
quasi-Fermi potential 350
quenching parameter 193 ff

r
Raman spectroscopy 458 ff
real-time X-ray diffraction 31
recombination 447, 449
– bimolecular 248, 250 ff
– direct radiative 350
– Langevin 350
– non-geminate 248
– Shockley-Read-Hall 350
recombination centre 273
recombination rate distribution 502
recombination zone 497
recrystallization 103
μ–σ relation 349
relaxation, electronic 139 ff
– vibrational 137 ff
Rietveld refinement 108
roughness 17
rubrene 398, 403, 406, 419, 423
– doped in Alq$_3$ 285

s
scaling 345, 383
scaling exponent 444
scaling theory 19
Schottky barrier 480
Schottky-type contact 351, 382
screening effect 501
self-assembled monolayers (SAMs) 30
self-trapping 276
semiconducting conjugated polymer 451 ff
series resistance 442
Shockley model 345, 353
short-channel effect 357, 383
short-channel transistor 370
short-circuit current 443 ff, 447
simulation model 379, 476
single-crystal 393
– characterization 399
– contact 404, 412, 417
– growth 396

– OFET characteristics 412
– OFET fabrication 395, 405
singlet exciton 186
singlet fission 139, 141
singlet state 5, 114
singlet-polaron quenching 249
singlet-singlet annihilation 230
solar cell 439 ff, 449
soliton 140 ff
solvation 503
solvatochromic effect 503
solvent polarity 503
space charge effect 480
space charge layer 69, 74
space-charge limited current (SCLC) 11,
  329 f, 402, 440
spectroscopy, coherent vibrational 136 ff
– field-assisted pump-probe 144 ff
– photocurrent cross-correlation 146
– pump-probe 132 ff, 147
– pump-push-probe 146 ff
– ultrafast 131 ff
spin mixing process 260 f
spin state 436
spin statistics 260 ff
states, charge-transfer 143
– dark 140 ff
– emissive 140 ff
– excited 140 ff, 145
– ground 139 ff
– singlet 140 ff
– triplet 140 ff
Stranski-Krastanov growth 19
structural defect 275
structural properties, Alq$_3$ 95 ff
sublimation 97
sub-threshold current 373
sub-threshold slope 416
surface and interface energies 18
surface electric field 352
surface potential 352
Su-Schrieffer-Heeger (SSH) model 142 ff,
  235
Suzuki-type coupling 153

t

tandem excitation 230
tetracene 399, 403, 419, 424
thermal degradation 154, 161 f
thermal expansion 31
thermal properties, Alq$_3$ 95 ff
thermalization 135 ff
thermally stimulated current s. TSC

thermally stimulated luminescence s. TSL
thin-film transistor 345, 363
threshold voltage 353, 355, 383, 415
time-of-flight (TOF) 400 f, 487
– on α-NPD 290
TNF 215 ff, 218 ff
TPD 264 f
– energy diagram of –metal 83
train sublimation 97, 105
transconductance 345, 356
transfer characteristics 309
transient optical absorption 188 ff
transient PL 115, 118
transit time 357
transmission, differential 135 ff
transport 433, 449
– carrier density 482
transport energy 274
transport gap 55
transport properties 433
transport states 274
transport-limited 447
trap 273 ff, 373, 388, 449
– acceptor-like 374
– bulk 374
– detection techniques 276 ff
– distribution 381
– donor-like 376
– effect on electrical and optical properties
  288 ff
– in Alq$_3$ 295
– in polymeric semiconductors 297 ff
– in small molecule semiconductors
  279 ff
– interface 374
– origin of 275 ff
– structural defects 293 ff
trapping energy 499
trinitrofluorene (TNF) 196 ff, 207
triphenylamine 154
triplet, lifetime 118
– temperature dependence 122
triplet energy 121, 126
triplet exciton 126, 435 ff
triplet state 5
– excited electronic 95 ff
– population 114
tris(8-hydroxyquinoline)aluminum s. Alq$_3$
TSC 277
– fractional 278
– on Alq$_3$ 279 ff, 282, 285, 294
– on DMO-PPV 298
– on 1-NaphDATA 279 ff

– on α-NPD   279 ff
TSL   277
– fractional   278
– on Alq$_3$ 285
T-T annihilation   119
turn-on voltage   439

u
ultraviolet photoemission spectroscopy
   (UPS)   453, 460
underechting   371
UPS   453, 460

v
vacuum level alignment   69
vacuum level shift   70
variable range hopping   348
vibrational analysis   111 ff
vibrational degrees of freedom   20
vibrational modes   121
vibrational properties, Alq$_3$ 125
vibrational relaxation   137 ff

vibronic progression   120
Vollmer-Weber growth   19

w
Wannier exciton   185 ff
wavepacket   136 ff
weak charge-lattice coupling   483
work function   43, 56, 351, 354, 381, 445,
   460 ff

x
X-ray absorption spectroscopy (XAS)   42, 45
X-ray photoelectron spectroscopy (XPS)   453,
   460
X-ray powder diffraction   98

y
Yamamoto-type coupling   153

z
zero-field splitting parameters   115, 118
zero-point oscillations   221 ff, 228